Recent Trends in Engineering, Science and Technology

Dr. Jyoti Sekhar Banerjee is currently serving as the Head of the Department in the Computer Science and Engineering (AI & ML) Department at the Bengal Institute of Technology, Kolkata, India. He is also the Professor-in-Charge, R & D and Consultancy Cell & Nodal Officer of the IPR Cell of BIT. Since 2024, he also works as a Remote Researcher in the Internet of THings & AppliCAtions Lab (ITHACA) at the Department of Electrical and Computer Engineering, University of Western Macedonia, Greece. He is the former remote research fellow of the Cognitive Computing and Brain Informatics Research Group (CCBI) at Nottingham Trent University (NTU), UK. Dr. Banerjee completed his post-doctoral fellowship at Nottingham Trent University, UK, in the Department of Computer Science. He also completed a post graduate diploma in IPR & TBM from MAKAUT, WB. He has teaching and research experience spanning 20 years and completed one IEI funded project. He is a member of the CSI, IEEE, ISTE, IEI, ISOC, IAENG and fellow of IETE. He is present Secretary-cum-Treasurer of the ISTE WB Section and Secretary of the IETE, Kolkata Centre. He is the immediate past Secretary of the Computer Society of India, Kolkata Chapter. Dr. Banerjee has also been elected as the Vice Chairman Cum Chairman Elect in the Computer Society of India, Kolkata Chapter for the year 2025-2027.

Dr. Siddhartha Bhattacharyya [FRSA, FIET (UK), FIEI, FIETE, FSCRS, LFOSI, SMIEEE, SMACM, SMAAIA, SMIETI, LMCSI, LMISTE] is currently a senior researcher at VSB - Technical University of Ostrava, Ostrava, Czech Republic. He is also a scientific advisor at Algebra University, Zagreb, Croatia. Prior to this, he was the Principal of Rajnagar Mahavidyalaya, Birbhum, India. Before this, he was a professor at CHRIST (Deemed to be University), Bangalore, India. He also served as the Principal of RCC Institute of Information Technology, Kolkata, India. He has served VSB Technical University of Ostrava, Czech Republic, as a Senior Research Scientist. He is the recipient of several coveted national and international awards. He received the Honorary Doctorate Award (D. Litt.) from The University of South America and the SEARCC International Digital Award ICT Educator of the Year in 2017. He was appointed as the ACM Distinguished Speaker for the tenure of 2018-2020. He has been appointed as the IEEE Computer Society Distinguished Visitor for 2021-2024. He is a co-author of six books and the co-editor of 106 books and has more than 400 research publications in international journals and conference proceedings to his credit.

Dr. Debashis De is a professor in the Department of Computer Science and Engineering at the Maulana Abul Kalam Azad University of Technology, West Bengal, India. He received an M.Tech. degree from the University of Calcutta, in 2002. and a Ph.D. from Jadavpur University in 2005. He is a senior member-IEEE, fellow IETE, and life member CSI. He was awarded the prestigious Boyscast Fellowship by the Department of Science and Technology, Government of India, to work at the Heriot-Watt University, Scotland, UK. He received the Endeavour Fellowship Award from 2008–2009 by DEST Australia to work at the University of Western Australia. He received the Young Scientist award both in 2005 in New Delhi and in 2011 in Istanbul, Turkey, from the International Union of Radio Science, Belgium. In 2016, he received the JC Bose research award from IETE, New Delhi.

Dr. Jinia Datta received her Bachelor of Technology degree in Electrical Engineering in 2006 from Birbhum Institute of Engineering and Technology, Birbhum, West Bengal. She received his Master of Technology degree in Electrical Engineering in the year of 2008 from Calcutta University. She was awarded PhD in the year 2015 from Calcutta University. At present, she is working as Principal of ABACUS Institute of Engineering and Management, Magra, West Bengal, India. Prior to this, she was engaged as Principal in charge Camellia Institute of Engineering and Technology, Budbud, West Bengal form (April 2018–August 2020) and as an Assistant Professor of Electrical Engineering Department (B.Tech) in Birbhum Institute of Engineering and Technology, Birbhum, West Bengal form (July 2010–April 2018). She was also working as an Assistant Professor of Electrical Engineering Department (B.Tech) in Bengal Institute of Technology and Management, Shantiniketan, West Bengal form (July 2007–June 2010). Her current areas of interest are control system based on modeling of controller, fuzzy control etc.

Dr. Panagiotis Sarigiannidis is the Director of the ITHACA lab, co-founder of the 1st spin-off of the University of Western Macedonia: MetaMind Innovations P.C., and Full Professor in the Department of Electrical and Computer Engineering in the University of Western Macedonia, Kozani, Greece. He received the B.Sc. and Ph.D. degrees in computer science from the Aristotle University of Thessaloniki, Thessaloniki, Greece, in 2001 and 2007, respectively. He has published over 360 papers in international journals, conferences and book chapters, including IEEE Communications Surveys and Tutorials, IEEE Transactions on Communications, IEEE Internet of Things, IEEE Transactions on Broadcasting, IEEE Systems Journal, IEEE Wireless Communications Magazine, IEEE Open Journal of the Communications Society, IEEE/OSA Journal of Lightwave Technology, IEEE Transactions on Industrial Informatics, IEEE Access and Computer Networks. He received six best paper awards and the IEEE SMC TCHS Research and Innovation Award 2023. He has been involved in several national, European and international projects, coordinating and technically leading numerous national and European projects including H2020, Horizon Europe, Erasmus+ and operational programs. His research interests include telecommunication networks, internet of things and network security. He is an IEEE member and participates in the Editorial Boards of various journals like IEEE Transactions on Communications, IET Networks, International Journal of Communication Systems and International Journal of Information Security.

Dr. Jan Platos is a professor at the Department of Computer Science and the Dean of the Faculty of Electrical Engineering and Computer Science, VSB-TUO, Czech Republic. He has co-authored more than 240 scientific articles published in proceedings and journals. His primary fields of interest are machine learning, artificial intelligence, industrial data processing, text processing, data compression, bioinspired algorithms, information retrieval, data mining, data structures, and data prediction.

Dr. Muhammad Mujtaba Asad is serving as an Assistant Professor and Lead Researcher of Educational Technologies and TVET Research at Department of Education, Sukkur IBA University, Pakistan. He is also serving as an honorary international faculty at the American University of Nigeria. Dr. Asad has national and international working experience of more than 13 years in the industrial and educational sectors in the domain of educational technologies for teacher education, digital equity and inclusion, virtual and augmented reality for experiential learning, work and product-based education, and technical and vocational education. Dr. Asad has published more than 80 research articles (WoS & Scopus Indexed) in reputed international journals and conferences worldwide. He has also published three books that focus on the observational skills of competent educators and Innovative Educational Technologies and pedagogies for Higher Education 4.0.

Recent Trends in Engineering, Science and Technology

Edited by

Jyoti Sekhar Banerjee

Siddhartha Bhattacharyya

Debashis De

Jinia Datta

Panagiotis Sarigiannidis

Jan Platos

Muhammad Mujtaba Asad

CRC Press
Taylor & Francis Group
Boca Raton London New York

CRC Press is an imprint of the
Taylor & Francis Group, an **informa** business

First edition published 2025
by CRC Press
4 Park Square, Milton Park, Abingdon, Oxon, OX14 4RN

and by CRC Press
2385 NW Executive Center Drive, Suite 320, Boca Raton FL 33431

British Library Cataloguing-in-Publication Data
A catalogue record for this book is available from the British Library

ISBN: 9781041121619 (hbk)
ISBN: 9781041121633 (pbk)
ISBN: 9781003663348 (ebk)

DOI: 10.1201/9781003663348

Typeset in Sabon LT Std
by HBK Digital

Conference Proceedings Series on Futuristic Intelligent and Smart Technologies (FIST)

Series Editors

Prof. (Dr.) Jyoti Sekhar Banerjee,
Department of Computer Science and Engineering (AI & ML),
Bengal Institute of Technology, Kolkata, India and
Remote Researcher, Internet of THings & AppliCAtions Lab (ITHACA),
Department of Electrical and Computer Engineering,
University of Western Macedonia, Greece

Prof. (Dr.) Siddhartha Bhattacharyya,
Senior Researcher, Faculty of Electrical Engineering and Computer Science,
VSB Technical University of Ostrava, Czech Republic and
Algebra Bernays University, Zagreb, Croatia

About the Series

This series serves as a beacon for scholars, researchers, and innovators navigating the dynamic landscape of technological evolution. This series stands at the forefront of interdisciplinary discourse, providing a platform for the exchange of ideas, discoveries, and insights that propel society towards a smarter, more intelligent future. Encompassing a diverse array of topics, this series delves into the realms of artificial intelligence, machine learning, robotics, Internet of Things (IoT), smart cities, and beyond. Each volume within the series is a testament to the relentless pursuit of innovation and the relentless quest for solutions to the complex challenges of the modern world.

Contributions to this series emanate from the minds of visionaries across the globe, representing a rich tapestry of perspectives and expertise. From pioneering research papers to visionary keynote addresses, each piece of content within the series reflects the cutting-edge advancements and transformative potential of futuristic intelligent and smart technologies.

Aim and Scope of this Series

The proceedings serve not only as a repository of knowledge but also as a catalyst for collaboration and networking. By fostering connections among researchers, practitioners, industry leaders, and policymakers, the series cultivates a vibrant ecosystem of innovation, where ideas are shared, synergies are discovered, and partnerships are forged.

The integration of traditional and modern intelligent and smart techniques continues to play an increasingly important role in various fields, shaping our city and society through data analysis, optimization, decision-making, and system evaluation and analysis. Artificial intelligence, Machine Learning, Big Data, and the Internet of Things, etc., are examples of futuristic, intelligent, and smart technologies that are enabling practically every area in the future.

As the pace of technological innovation accelerates, the Conference Proceedings Series on Futuristic Intelligent and Smart Technologies stands as a beacon guiding society towards a future imbued with intelligence, efficiency, and sustainability. Through its commitment to excellence, collaboration, and forward-thinking, the series remains an indispensable resource for those shaping the course of technological evolution.

Dedication

This volume of conference proceedings is dedicated to the relentless pursuit of knowledge, innovation, and collaboration that drives progress in various engineering fields.

May this collection serve as a beacon of insight, fostering further discoveries and meaningful discussions in the years to come.

Preface

AIEST is a leading conference focused on providing a platform to researchers, scholars, engineers, scientists and industrial professionals to gather knowledge and bridge the gap between academia and its industrial aspects, around the world. This conference will be an immersive experience primarily focusing on the latest advancements and researchers in various fields of engineering, including but not limited to Mechanical Engineering, Civil Engineering, Electrical Engineering, Electronics and Communications Engineering, Computer Science Engineering, Information Technology and other interdisciplinary areas. AIEST will cater to the transitional practices where industrial knowledge would be conveyed to academia regarding real-time scenarios and practical findings, thus fostering collaboration and the development of innovative solutions to counter contemporary challenges in engineering and technology.

The Abacus Institute of Engineering & Management, Hooghly, India (a joint venture of JIS Group and Techno India Group), organized the First International Conference on Advanced Innovations in Engineering, Science and Technology (AIEST-2024) at its campus from October 28th to 29th, 2024, though it was primarily scheduled during 25–26 October 2024, it was delayed due to cyclone.

The conference offered a platform for the discussion of recent advancements in a diverse array of subjects, such as different areas of core engineering as well as emerging areas of Computer science and engineering. The platform of AIEST will be an open forum for discourses related to the various regional and international concerns in reference with the recent advances in the field of engineering and IT. With striking research seeking innovations in modern sciences and technology, the conference will eventually lead to solving current societal problems.

The AIEST 2024 conference wishes to acknowledge the exceptional efforts of the Scientific, Advisory, and Technical Program Committee members, together with the Keynote speakers who spoke on 28–29 October 2024. We would also like to express our gratitude to the reviewers for their exceptional contributions in terms of their time, effort, and dedication. Additionally, we would like to extend our gratitude to the session chairs for their assistance in ensuring the smooth operation of the sessions.

We extend our gratitude to all attendees of the conference, including the presenters and participants. We really appreciate all those who contributed to and supported this conference. We would like to express our profound thanks and admiration for the diligent efforts and commitment shown by the members of the organizing committee, program committee, volunteers, and staff involved in AIEST 2024. We extend our gratitude to our Chief Patron, Patron, and the members of the different conference committees for their assistance.

Finally, we would like to express our gratitude to Mrs. Shatakshi Mishra, the Editorial Manager of CRC Press, for her kind support in publishing this book.

Kolkata	Prof. Dr. Jyoti Sekhar Banerjee
Ostrava	Prof. Dr. Siddhartha Bhattacharyya
Kolkata	Prof. Dr. Debashis De
Mogra	Prof. Dr. Jinia Datta
Kozani	Prof. Dr. Panagiotis Sarigiannidis
Ostrava	Prof. Dr. Jan Platos
Sukkur	Prof. Dr. Muhammad Mujtaba Asad

Advisory Committee

NAME	AFFILIATION
Prof. (Dr.) Ivan Zelinka	VSB Technical University of Ostrava, Czech Republic
Prof. (Dr.) Leo Mrsic	Algebra University College, Croatia
Prof. (Dr.) Jerzy Ryszard Szymanski	Kazimierz Pulaski University of Technology & Humanities, Radom, Poland
Prof. (Dr.) Indranil Manna	BIT Mesra, India
Prof. (Dr.) Chandan Chakraborty	IIT-KGP, India
Prof. (Dr.) Siddhartha Bhattacharya	VSB Technical University, Ostrava, Czech Republic Algebra University College, Croatia
Prof. Mete Yaganoglu	Ataturk University, Turkey
Prof. Nazim Hajiyev	UNEC Business School, Azerbaijan
Prof. Valliappan Raju	Perdana University, Malayesia
Prof. Santipada Gon Choudhuri	Director, DCL Group, India
Dr. Amit Dutta	IIT-KGP, India
Dr. Srimanti Roy Choudhury	Daffodil International University, Dhaka
Prof. Robin Augustin	Upsala University, Sweden
Dr. Pritam Das	Binghamton University, USA
Prof. (Dr.) Sudip Roy	IIEST Shibpur, India
Dr. Debasis Mitra	IIEST Shibpur, India
Prof. (Dr.) Nandini Mukherjee	Jadavpur University, India
Prof. (Dr.) Monojit Mitra	IIEST Shibpur, India
Dr. Susant Roy	Jadavpur University, India
Dr. Tarun Kr. Bera	Thapar Institute of Engineering and technology, India
Prof. (Dr.) Jitendra Nath Bera	University of Calcutta, India
Prof. (Dr.) Samarjit Sengupta	University of Calcutta, India
Dr. Prithwiraj Das	GCETT, Berhampore, India
Dr. Wasim Arif	NIT-Shilchar, India
Prof. (Dr.) Debes Das	Jadavpur University, India
Dr. Raju Basak	Chairperson, WBSC IEI, India
Prof. Alok Kumar Das	IIT(ISM) Dhanbad, India
Prof. Debabrata Dhupal	Veer Surendra Sai University of Technology, Burla, India
Prof. (Dr.) Biswanath Doloi	Jadavpur University, India
Dr. Taha Selim Ustun	Fukushima Renewable Energy Institute, Japan
Dr. Furkan Ahmad	Hamad Bin Khalifa University, Qatar
Mr. Abhijit Mandal	General Manager (GM), Primetals Technologies, India
Prof. (Dr.) Pavitra Sandilya	IIT-KGP, India
Dr. Abhinandan De	IIEST Shibpur, India
Prof. Arokiaswami Alphones	Nanyang Technological University, Singapore
Prof. Vecchi Giuseppe	Politecnico di Torino Corso Duca degli Abruzzi, Italy
Prof. Massimiliano Casaletti	Sorbonne Universities, France

Organizing Committee

Honorary Chair(s)	Prof. (Dr.) Jan Platos, VSB Technical University of Ostrava, Czech Republic
	Prof. (Dr.) Leo Mrsic, Algebra University College, Croatia
	Prof. (Dr.) Jerzy Ryszard Szymanski, Kazimierz Pulaski University of Technology & Humanities, Radom, Poland
	Prof. (Dr.) Sandeep Singh Sengar, Cardiff Metropolitan University, United Kingdom
Chief Patron	Sardar Taranjit Singh, Managing Director, JIS Group, India
Patron(s)	Mr. Tapan Kumar Ghosh, Executive Director, Techno India Group, India
	Sardar Simarpreet Singh, Director, JIS Group, India
Co-Patron(s)	Prof. (Dr.) Jinia Datta, Principal, AIEM, India
	Prof. Devmalya Banerjee, Registrar, AIEM, India
General Chair(s)	Prof. (Dr.) Siddhartha Bhattacharya, VSB Technical University, Ostrava, Czech Republic & Algebra University College, Zagreb, Croatia
	Prof. (Dr.) Jyoti Sekhar Banerjee, Bengal Institute of Technology, Kolkata, India
	Prof. (Dr.) Debashis De, Maulana Abul Kalam Azad University of Technology, West Bengal, India
Program Chair(s)	Prof. (Dr.) Panagiotis Sarigiannidis, University of Western Macedonia, Kozani, Greece
	Prof. (Dr.) Ahmed J. Obaid, University of Kufa, Iraq
TPC Chair	Prof. (Dr.) Jyoti Sekhar Banerjee, Bengal Institute of Technology, Kolkata, India
TPC Co-Chair	Prof. (Dr.) Goutam Kumar Das, AIEM, India
Industry Chair(s)	Mr. Snehasis Banerjee, TCS Research & Innovation, Kolkata, India
	Mr. Sumit Goswami, Senior Director of Engineering, Qualcomm, India
Convener(s)	Prof. (Dr.) Goutam Kumar Das, AIEM, India
	Prof. Bikash Banerjee, AIEM, India
Finance Chair	Prof. Shounak Bandyopadhyay, AIEM, India
	Prof. Sumit Banerjee, AIEM, India
Registration Chair	Prof. Avishek Gupta, AIEM, India
	Prof. Sankha Shubhra Goswami, AIEM, India
	Prof. Sohini Banerjee, AIEM, India
Publicity Chair(s)	Dr. Pritish Ghosh, AIEM, India
	Prof. Debraj Modok, AIEM, India
Publication Chair(s)	Dr. Soumendu Bhattacharya, AIEM, India
Publication Co-Chair(s)	Dr. Monalisa Halder, AIEM, India
	Prof. Sumangal Bhaumik, AIEM, India
	Prof. Titas Bhaumik, AIEM, India
Hospitality Chair(s)	Prof. Debobrata Mazumder, AIEM, India
	Prof. Suvendu Kar, AIEM, India

Technical Program Committee

Sl No	Name	Affiliation
1	Prof. (Dr.) Panagiotis Sarigiannidis	University of Western Macedonia, Kozani, Greece
2	Prof. (Dr.) Ahmed J. Obaid	University of Kufa, Iraq
3	Prof. (Dr.) Jyoti Sekhar Banerjee	Bengal Institute of Technology, Kolkata, India
4	Prof. (Dr.) Goutam Kumar Das	Abacus Institute of Engineering and Management, India
5	Dr. Krishna Sarkar	St. Thomas' College of Engineering & Technology, India
6	Dr. Jayanti Sarker Bhattacharjee	Techno Main Saltlake, India
7	Dr. Subhajit Das	SR University, Warangal, India
8	Dr. Rumpa Saha	Aliah university, India
9	Dr. Shamim Haider	Aliah university, India
10	Dr. Aveek Chattopadhyaya	Guru Nanak Institute of Technology, India
11	Dr. Priyanka Bera	GLA University, Mathura, India
12	Dr. Anwesha Sengupta	NIT-Rourkela, India
13	Dr. Atin Mukherjee	NIT-Rourkela, India
14	Dr. Barnali Kundu (Sarkar)	Guru Nanak Institute of Technology, India
15	Dr. Tanaya Datta Das	Techno India University, India
16	Dr Alok Srivastava	JIS College of Engineering, India
17	Dr. Tamal Roy	MCKV Engineering College, India
18	Dr. Surajit Kundu	NIT-Jamshedpur, India
19	Dr. Chaitali Koley	NIT-Mizoram, India
20	Dr. Munmun Bhattacharya	Jadavpur University, India
21	Dr. Rajat Jyoti Sarkar	Chandannagar Govt. College, India
22	Dr. Amitava Mandal	IIT(ISM) Dhanbad, India
23	Dr. Mrinal Sen	IIT(ISM) Dhanbad, India
24	Dr. Dhiren Kumar Behera	Indira Gandhi Institute of Technology, Sarang, India
25	Dr. Ujjal Sur	University of Delhi, India
26	Dr. Tista Banerjee	NIET Alwar, India
27	Dr. Sananda Pal	Jadavpur University, India
28	Mr. Subhadeep Koley	University of Surrey, UK
29	Prof. Pallav Dutta	Aliah University, India
30	Prof. Bidishna Bhattacharya	Techno Main Saltlake, India
31	Prof. Sudeep Samanta	MCKV Engineering College, India
32	Dr. Debarati Dey Roy	BPPIMT, India
33	Mr. A.K. Assad	Amazon FC, France
34	Dr.Bishaljit Paul	NiT, Agarpara, India
35	Mr. Debasis Sanki	JIS College of Engineering, India
36	Dr. Soumyabrata Saha	JIS College of Engineering, India
37	Dr. Sandip Ghosh	JIS College of Engineering, India
38	Dr.Biswarup Neogi	JIS College of Engineering, India

Sl No	Name	Affiliation
39	Mr. Indrajit Pandey	Techno International, New Town, India
40	Dr. Chiranjit Sain	Ghani Khan Choudhury Institute of Engineering & Technology, India
41	Dr. Pavitra Kumar Biswas	NIT Mizoram, India
42	Dr. Partha Sarathi Bishnu	BIT Mesra, India
43	Prof. Abhra Mukherjee	Techno India University, India
44	Prof. Ankur Ganguly	Royal Global University, India
45	Dr. Heranmoy Maity	Symbiosis Institute of Technology, India
46	Dr. Sumanta Karmakar	Assansol Engineering College, India
47	Dr. Nilanjan Mukhopadhyay	Global Institute of Management and Technology, India
48	Prof. (Dr.) Arijit Saha	Dumdum Motijheel Rabindra Mahavidyalaya, India
49	Mr. Pradipta Roy	Dr. B. C. Roy Academy of Professional Courses, India
50	Dr. Diganta Sengupta	Heritage Institute of Technology, India
51	Dr. Manas Chanda	Megnath Saha Institute of Technology, India
52	Dr. Priti Deb	Institute of Engineering and Management, India
53	Dr. Arpan Deyasi	RCC Institute of Information Technology, India
54	Prof. (Dr.) Chandan Kr. Bhattacharyya	Megnath Saha Institute of Technology, India
55	Dr. Arindam Biswas	Dr. B. C. Roy Engineering College, India
56	Prof. (Dr.) Subhranil Som	Bhairab Ganguly College, India
57	Prof. Prakash Narayan Mahanty	Kingston School of Management and Science, India
58	Dr. Sandeep Mukherjee	BIT Mesra, India
59	Dr. Joan Pons Llinares	Universitat Politechnica de Valencia, Spain
60	Prof. Joao Gama	University of Porto, Portugal
61	Dr.Zuhaina Zakaria	University Teknology MARA, Malyasia
62	Dr. Arindam Mondal	Dr. B. C. Roy Engineering College, India
63	Dr. Madhabi Ganguly	West Bengal State University, India
64	Prof. (Dr.) Partha Pratim Sarkar	Kalyani University, India
65	Dr. Pratik Mondal	SRM Institute of Science and Technology, Chennai, India
66	Dr. Jeet Ghosh	LNMIIT, Jaipur, India
67	Dr. Gopinath Samanta	LNMIIT, Jaipur, India
68	Dr. Ishita Chakraborty	Royal Global University, India
69	Dr. Anupam Das	Royal Global University, India
70	Dr. Soumik Podder	SR University, Warangal, India
71	Dr. Sabyasachi Mondal	NEHU, Shilong, India
72	Prof. Annabeth Aagaard	Aarhus University, Herning, Denmark
73	Prof. Koustav Routh	NIT-Agartala, India
74	Prof. Swapan Bhowmik	NIT-Agartala, India
75	Prof. Johannes M. Schleicher	Vienna University of Technology, Austria
76	Prof. Mega Shatila	Universitas Indonesia, Depok, Indonesia
77	Prof. Sebastien Gerard	Universit´e Paris-Saclay, France
78	Er. Rakesh Jain	IBM - Almaden Research Center, San Jose, CA, USA
79	Prof. Reinhard Herzogz	CSG, University of Zurich UZH, Zurich, Switzerland
80	Prof. Nicola Terrenghix	Fraunhofer-Institut IOSB, Karlsruhe, Germany

Sl No	Name	Affiliation
81	Prof. Maryam Bagheri	Universitadegli Studio di Torino, Turin, Italy
82	Prof. Wang Wei	Harbin University of Science and Technology, Harbin, China
83	Prof. Liu Liping	Harbin University of Science and Technology, Harbin, China
84	Prof. Frank Berkers	TNO, Delft, The Netherlands
85	Er. Toshihiko Yamakami	ACCESS, Kanda-Neribei-cho 3, Chiyoda-ku, Tokyo, Japan
86	Prof. V. Tikhvinskiy	Moscow Technical University of Communications & Informatics, Russia
87	Prof. Edgar Tello-Leal	Autonomous University of Tamaulipas, Victoria, Tamaulipas, Mexico
88	Prof. Akram Al-Hourani	RMIT University, Melbourne, Australia
89	Prof. Karina Gomez Chavez	RMIT University, Melbourne, Australia
90	Prof. M. Viju Prakash	College of Science Knowledge University Erbil, Kurdistan Region, Iraq
91	Prof. Banar Fareed Ibrahim	Cyprus International University, Cyprus
92	Prof. Muhammad Intizar Ali	National University of Ireland Galway, Ireland
93	Prof. Tamim Ahmed Khan	Bahria University, Islamabad, Pakistan
94	Prof. Luis Guijarro	ITACA, Universitat Polit`ecnica de Val`encia, Spain
95	Prof. Ambuj Kumar	Aarhus University, Herning, Denmark
96	Prof. Bilal Muhammad	Aarhus University, Herning, Denmark
97	Prof. Charilaos Akasiadis	Institute of Informatics and Telecommunications NCSR, Greece
98	Prof. Frank Eliassen	University of Oslo, Norway
99	Prof. Yu-Shan Lin	National Taitung University, Taitung, Taiwan
100	Prof. Md. A. Hossain	BRAC University, Bangladesh
101	Prof. Bruno Costa	PPGI Universidade Federal do Rio de Janeiro Rio de Janeiro, Brazil
102	Prof. Rasool Asal	Khalifa University, Abu Dhabi, UAE
103	Prof. Mark Allen Gray	University of Maryland Baltimore County
104	Er. Vesa Jordan	Mikkelin Puhelin Oy Ltd, Mikkeli, Finland
105	Dr. Vusala Nabi Jafarova	Azerbaijan State Oil and Industry University, Azerbaijan
106	Prof. Sevda Rzayeva	Azerbaijan State University of Economics, Azerbaijan
107	Dr. Rajesh Dey	Gopal Narayan Singh University,India
108	Prof. Nandan Banerjee	Sikkim Manipal University, Sikkim
109	Prof. Udit Kr. Chakraborty	Sikkim Manipal University, Sikkim

Contents

List of Figures

List of Tables

1 Demand-driven automatic soil cleaning for PR enhancement of a roof-top solar PV power plant

Saheli Sengupta[1,a], Indrajit Bose[1,2,b], Madhusree Bhattacharjee[1,c], Riya Paul[1,d] and Sugato Ghosh[1,e]

[1]Centre for Renewable and Sustainable Energy Studies, JIS Institute of Advanced Studies & Research (JISIASR), Kolkata, JIS University, West Bengal, India

[2]Agni Green Power Ltd., Kolkata, West Bengal, India

Abstract

Photovoltaic (PV) soiling reduces generation guarantee in arid and semiarid countries including India. Since ambient conditions and dust constituents are interrelated to its deposition, energy output based fixed interval cleaning schedule is not enough for optimum output of a PV plant. In this paper, a thorough study has been conducted to characterize dust particles and also examine its effect on efficiency of glass cover. Optical loss is observed for weekly cleaned module, no manual cleaning in a month and within six months' periods. When weekly cleaning maintains transmittance within 80%, and cumulative soiling causes large reduction, reflectance sustains a value within a limit. The change in the performance ratio of a 10kWp roof-top PV plant before and after cleaning over one month has been calculated. A laboratory developed dust detector-based cleaning arrangement has been used to conduct demand-driven cleaning by measuring three different parameters: optical loss index (OLI), insolation and PR of the plant. Three different cases are observed for triggering the cleaning mechanism as and when required which establishes successful implementation of sensor.

Keywords: Cleaning, optical loss index, performance ratio, PV soiling, transmittance loss

Introduction

In the recent past, the penetration of photovoltaic (PV) power has increased globally at a remarkable pace. The soiling is found to be a predominant threat to PV power generation in many parts of the globe. As the PV modules are exposed to field conditions, various environmental stressors affect their life and performance. Dust is the single most important parameter for hot and humid area like India which impairs the power output of PV module. Maxime Mussard and Amara [15] have made a review on the effect of dust on PV installation under different arid and semiarid conditions. Since dust accumulation process depends on various environmental factors, it is very site specific [16]. Deposited dust types in terms of size and chemical properties are analyzed by researchers [1, 13]. But the characteristics and content of aerosol particles varies with seasonal variation of climatic factors [3]. The main disadvantage of dust deposition is that it reduces the transmittance of module glass cover, allowing less amount of radiation Singh et al. [23] reaching the solar cell and, hence, the power output. As the property and amount of dust accumulation are affected by various environmental factors

Semaoui et al. [17], Gholami et al. [7], tilt angle, and the time of exposure, they also give an impact on optical property (transmittance, reflectance) of module glass. A comprehensive mathematical model has been developed considering composite climatic condition Sengupta et al. [18], Sengupta et al. [19] and validated for different locations in India [21]. Different methods for dust detection in field condition e.g. mass of soil accumulated; light transmission loss and response of a PV device are described [6]. Attempts are made to develop dust detectors Javed and Guo [10], Aïssa et al. [2], Micheli et al. [14], among them, some of which are also commercially available Gostein et al. [8] but do not consider the effect of AOI in dust deposition. Artificial neural network (ANN) has been used to estimate daily and cumulative soiling loss from weather data for cleaning of the module [11]. Images of soiled module captured by the surveillance camera are analysed to find the loss [24]. Different models are reviewed, and their advantages and disadvantages are discussed in [22, 4]. A novel cleaning method has been developed in Bose et al. [5] for non-contact dust detection. The effect of cumulative dust deposition and cleaning is observed in grid integrated PV-based

[a]saheli.s@jisiasr.org, [b]indrajitbose98@gmail.com, [c]madhusree.bhattacharjee13@gmail.com, [d]riya.me1993@gmail.com, [e]sugato@jisiasr.org

DOI: 10.1201/9781003663348-1

microgrid scheduling [20]. Therefore, an optimal cleaning frequency Makkar et al.[12], Hammad et al. [9] is required to maintain the generation at minimum usage of water. In this work, an experimentation has been carried out to examine the amount of transmittance loss due to dust deposition and then a laboratory developed smart dust detection technique is used to maintain an optimized cleaning of a PV plant. The following are the contributions made in this paper:

(i)   Dust is gathered from the test locations at different tilt angles and is characterized in the laboratory.
(ii) Change in transmittance of module glass at different tilt angles for long term field exposure without manual cleaning is studied.
(iii) Effect of weekly cleaning and cumulative deposition with no cleaning on transmittance and reflectance for one month period has been analyzed.
(iv) A laboratory-developed dust detector is used to measure and calculate a newly proposed indicator, "Optical Loss Index (OLI)". This parameter along with solar radiation and performance ratio (PR) of PV plant are utilized to actuate automatic cleaning mechanism as and when required.
(v)   Ultimately, improvement in performance ratio is demonstrated for 10kWp roof-top Solar Photovoltaic Power Plant (SPVPP) at IIEST, Shibpur.

Section 2 of this paper presents experimentation, whereas section 3 describes the results and discussions. An extensive conclusion is presented in section 4.

Experimentation

The experimentation has been carried out in two phases at IIEST, Shibpur as shown in Figure 1.1. In first phase of experimentation, three sets of low iron glass coupons are placed at three different angles e.g. 0°, 25°, 90° as shown in Figure 1.2 for different time period. One set is allowed for cumulative dust deposition without cleaning for six months. Between rest two one set is cleaned weekly. The last one is kept for monthly cumulative soiling. Change in transmittance due to soiling of both the sets is measured by UV/Vis spectrophotometer. Collected dust is characterized by optical microscope OPTIKA-600 and scanning electron microscope (SEM).

From the laboratory experiment, it is observed that, to maintain the generation, it is essential to clean the modules as and when required. To serve this purpose, a novel smart dust detector has been developed in the laboratory. This is based on the principle of reflected light measurement from dusty module surface [5].

At the field condition, due to deposited dust on the surface, transmittance reduces but the reflectance does not change drastically for a particular type of dust. The total reflectance is the summation of specular and diffuse reflectance. According to the refractive medium, the fraction of linear (specular) and diffuse components changes. The detector is made to capture mostly the linear component to take the cleaning decision.

There are two cells in the sensor. The first is the reference cell, known as the "R-cell," which measures incoming radiation in terms of current. Another is test cell, known as a "T-cell," which collects collimated

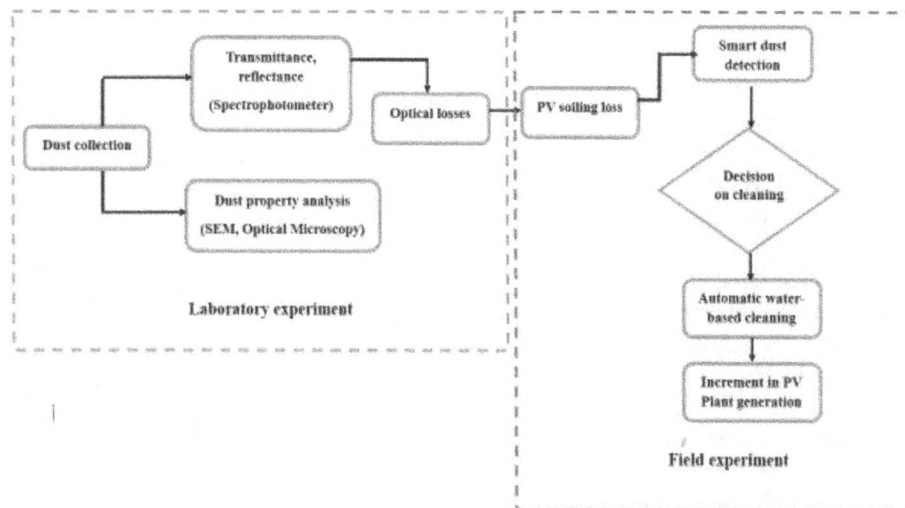

Figure 1.1 Block diagram of flow of experimentation
Source: Author

reflected light from the glass surface of the module to produce current. The following is the dust sensor's operational algorithm:

The incident solar energy is detected by the R-cell. If the radiation is less than 400W/m² during sun hour, the detector takes the decision as weather condition is not good and cleaning operation will not be performed.

If the radiation is high, then calculate optical loss index (OLI). It is the ratio of a fraction of reflected light from clean module cover to soiled one. I^{ij}_{Tdust} and I^{ij}_{R} are the test cell and reference cell current and at ith time and jth day.

$$OLI_{ij} = \frac{I^{ij}_{Tdust}}{I^{ij}_{R}} / \frac{I^{i}_{Tclean}}{I^{i}_{R}} \qquad (1)$$

$$OLI_{ij} \leq OLI_{thresh}$$

The PR of the plant is calculated at the same time as,

$$PR = \frac{actual\ generation}{\eta_{PV}PV_{area}I_{ins}} \qquad (2)$$

$$PR \leq PR_{thresh}$$

where, η_{PV} is the efficiency of the module, PV_{area} is the actual area of PV generator, and I_{ins} is the insolation.

When both *PR* and *OLI* are less than the threshold values and radiation is higher than 400W/m², cleaning initiation signal is generated.

Initially the dust detector is designed and simulated in AutoCAD and sketches up platform. Thereafter it is fabricated in 3D printer and assembled with its hardware accessories. The whole set-up is mounted on the modules of a 10 kWp roof-top SPVPP from 1/12/2023 to 31/12/2023 for field evaluation of performance of cleaning system as shown in Figures 1.3 and 1.4.

Results and Discussions

Laboratory experimentation
The microscopic image of collected dust at test location is presented in Figure 1.5. This image is used

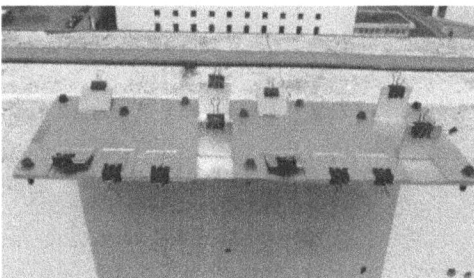

Figure 1.3 Smart dust detection system mounted on PV module of the PV plant
Source: Author

Figure 1.4 Cleaning arrangement for 10 kWp PV power plant
Source: Author

Figure 1.2 Glass coupons at different tilt angles for dust collection
Source: Author

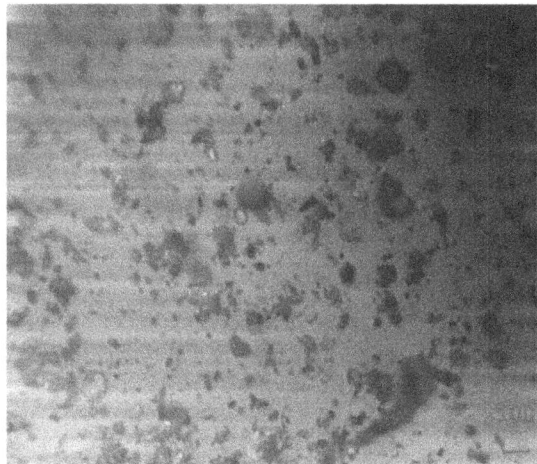

Figure 1.5 Microscopic image of deposited dust on the glass coupon
Source: Author

to analyze the particle size distribution in the place by ImageJ software and it is found that up to 10um particles are mostly deposited. Figure 1.6 shows the frequency count and probability distribution of the detected particle size. It is a normal distribution with mean 2.05 and standard deviation of 5.05. The individual shape and size of particles are shown in SEM image of Figure 1.7.

The transmittance loss due to cumulative soiling without manual cleaning over a six months period is presented in Figure 1.8. The reduced transmittance is 72.9%, 73.3% and 80.3% for 0°, 25° and 90°. It is observed that maximum transmittance reduction occurred at 0° tilt angle.

Figure 1.9 shows the solar radiation and PM concentration (2.5µm and 10µm) from 1/01/2020 to 25/01/2020. On 7/01/2020, 17/01/2020 and 24/01/2020 (dates of measurement) concentration of PM 2.5 is 99.7, 137.3, 121.3 µg/m³ whereas, that of PM 10 is 149.55, 193.39, 176.99 µg/m³. On the same dates, transmittance of glass coupons after weekly cleaning is measured to be 78%, 79% and 80%

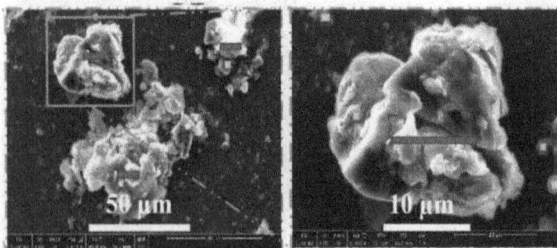

Figure 1.6 Distribution of deposited dust particles
Source: Author

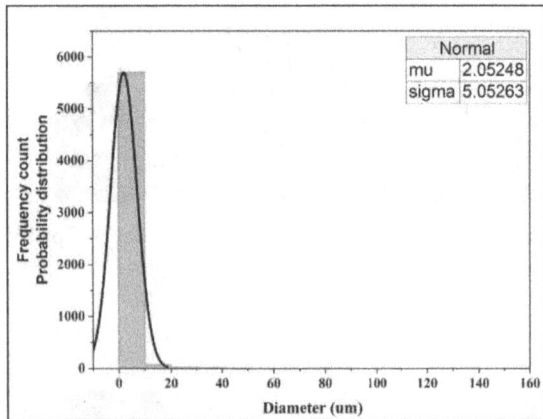

Figure 1.7 SEM image of collected dust particles
Source: Author

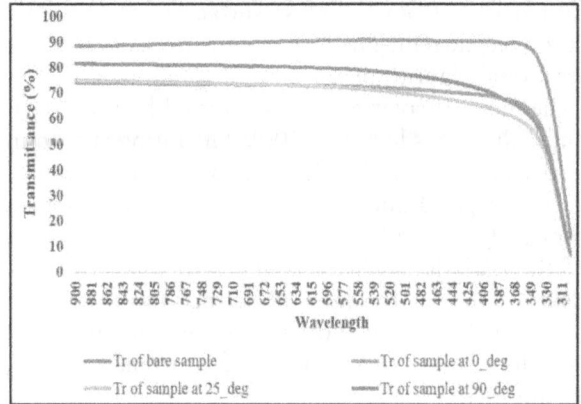

Figure 1.8 Transmittance of clean glass and soiled glass at different tilt angles without manual cleaning for over six months
Source: Author

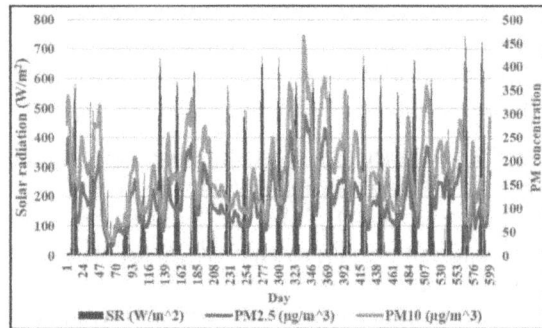

Figure 1.9 Weather parameters of the days of measurements 01/01/2020 to 25/01/2020
Source: Author

when reflectance is measured to be 8.71%, 9.60% and 9.63% respectively. If the glass coupons are not cleaned, cumulative soiling is demonstrated through transmittance measurements of 77.7%, 71.6%, and 64.8% as depicted in Figure 1.10. The change in reflectance for this case is obtained as 8.67%, 9.22%, 9.79%.

Field experimentation
The change in the PR of a PV plant due to dust deposition is shown in Figure 1.11. PR of the plant with dusted PV module and standard insolation is very low. The plant is cleaned on the 15th day, and the PR value is increased from 52% to 84%. The energy generation and change in the PR with smart detector-based cleaning arrangement for one month period is shown in Figure 1.12. From field experimentation, the minimum values of PR and OLI are selected as 80% and

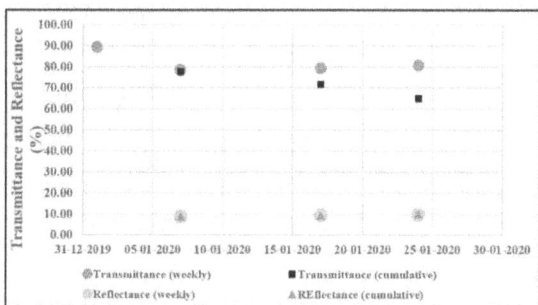

Figure 1.10 Weekly and cumulative transmittance and reflectance measurements from 01/01/2020 to 25/01/2020
Source: Author

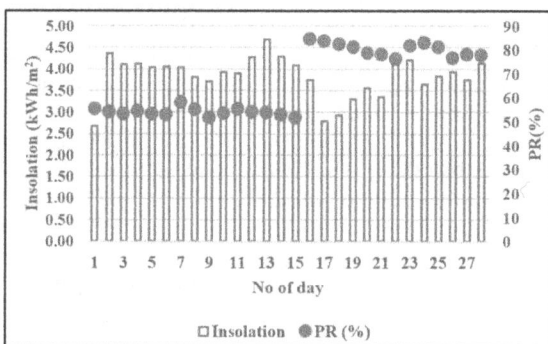

Figure 1.11 Change of PR from soiled to cleaned module
Source: Author

85% respectively in the place of measurement. Three cases are observed from this experimentation:

(i) Case 1
On 7/12/2023, solar insolation is very less 1.21kWh/m². Sensor detects it as bad weather condition, and the OLI is calculated as zero. Although the value of PR is 79.68%, the cleaning initiation will not start. The PR enhancement on the next day is due to natural cleaning of modules.

(ii) Case 2
Insolation on 16/12/2023 is 4.42kWh/m². OLI is found to be 84.43%. The value of PR is 78.85%, the cleaning initiation starts. As a result, the performance ratio becomes 85% on the next day.

(iii) Case 3
On 20/12/2023, average isolation becomes 3.96 kWh/m². Although the value of PR is low (<80%), OLI is 87%. This will prohibit the cleaning initiation. On the next day (21/12/2023), both PR and OLI values become lower than their predefined threshold values

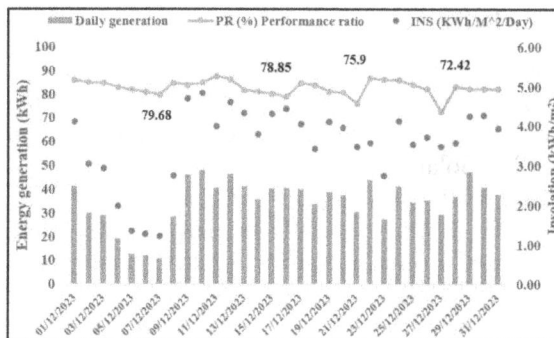

Figure 1.12 Algorithm for cleaning mechanism
Source: Author

with a good insolation condition. As three conditions are satisfied, cleaning operation is initiated and PR increases to 87% on 22/12/2023.

Since this mechanism considers insolation, PR and OLI simultaneously, it replaces the periodic cleaning frequency by an automatic and necessary basis, reducing water and energy usage for cleaning. Table 1.1 shows a sample calculation of optical loss index.

Conclusion

High performance ratio of a PV power plant is an indication of its good energy delivery. In this paper, PR is thus considered as an indicator with OLI and insolation to initiate automatic cleaning arrangement. To achieve this, a set of experiments is carried out in laboratory and field. Observation shows that, particles of sizes between 2.5-10µm are abundant at the place of experimentation and maximum transmission reduction occurs for horizontal lay up of module. Weekly cleaning causes a transmission loss of 12% whereas it comes down to 27% maximum for a cumulative non-cleaning period of three weeks at the place of experimentation. Field data also supports this situation by demonstrating an increment of plant PR by 61.5%. Thus, status of PR, OLI and radiation added together to initiate cleaning offers the best option.

Table 1.1 Sample OLI calculation

Glass coupon condition	Irradiation (W/m²)	R-cell current (mA)	T-cell current (mA)	OLI
Clean	581	18.5	1.69	
Dusty glass (naturally deposited dust amount: 2.3mg/cm²)	733	23.4	1.73	0.074 / 0.091 = 0.81

Source: Author

Acknowledgement

The first author Saheli Sengupta would like to acknowledge the fellowship support given by DST, GoI, during this research. The experimentation is carried out in DST-IIEST Solar PV Hub, IIEST, Shibpur. The zig is provided by IIT Kharagpur and glass coupons are supplied by University of Exeter, UK. Weather data are collected from Central Pollution Control Board (CPCB). The authors acknowledge with thanks these supports. The authors express their gratitude to the Director of JISIASR and the Managing Director of JIS Group for their unwavering cooperation.

References

[1] Abderrezek, M., & Fathi, M. (2017). Experimental study of the dust effect on photovoltaic panels' energy yield. *Solar Energy*, 142, 308–320.

[2] Aïssa, B., Scabbia, G., Figgis, B. W., Lopez, J. G., & Benito, V. B. (2022). PV-soiling field-assessment of mars™ optical sensor operating in the harsh desert environment of the state of Qatar. *Solar Energy*, 239, 139–146.

[3] Azarov, V., Sergina, N., Sidyakin, P., & Kovtunov, I. (2017). Seasonal variations in the content of dust particles Pm10 and Pm2.5 in the air of resort cities depending on intensity transport traffic and other conditions. In IOP Conference Series: Earth and Environmental Science, (Vol. 90, p. 012015).

[4] Bessa, J. G., Micheli, L., Montes-Romero, J., Almonacid, F., & Fernández, E. F. (2022). Estimation of photovoltaic soiling using environmental parameters: a comparative analysis of existing models. *Advanced Sustainable Systems*, 6(2100335), 1–12.

[5] Bose, I., Sengupta, S., Ghosh, S., Saha, H., & Sengupta, S. (2024). Development of smart dust detector for optimal generation of SPV power plant by cleaning initiation. *Solar Energy*, 276(112643), 1–10.

[6] Figgis, B., Ennaoui, A., Ahzi, S., & Remond, Y. (2016). Review of PV soiling measurement methods. In International Renewable and Sustainable Energy Conference, (IRSEC), (pp. 176–180).

[7] Gholami, A., Saboonchi, A., & Alemrajabi, A. A. (2017). Experimental study of factors affecting dust accumulation and their effects on the transmission coefficient of glass for solar applications. *Renewable Energy*, 112, 466–473.

[8] Gostein, M., Bourne, B., Farina, F., & Stueve, B. (2020). Field testing of mars™ soiling sensor. In 2020 47th IEEE Photovoltaic Specialists Conference (PVSC), (pp. 0524–0527). IEEE.

[9] Hammad, B., Al–Abed, M., Al–Ghandoor, A., Al–Sardeah, A., & Al–Bashir, A. (2018). Modeling and analysis of dust and temperature effects on photovoltaic systems' performance and optimal cleaning frequency: jordan case study. *Renewable and Sustainable Energy Reviews*, 82, 2218–2234.

[10] Javed, W., & Guo, B. (2022). Laboratory calibration of a light scattering soiling sensor. *Solar Energy*, 236, 569–575.

[11] Javed, W., Guo, B., & Figgis, B. (2017). Modeling of photovoltaic soiling loss as a function of environmental variables. *Solar Energy*, 157, 397–407.

[12] Makkar, A., Raheja, A., Chawla, R., & Gupta, S. (2019). IoT based framework: mathematical modelling and analysis of dust impact on solar panels. *3D Research*, 10(1), 3.

[13] McTainsh, G. H., Nickling, W. G., & Lynch, A. W. (1997). Dust deposition and particle size in mali, West Africa. *Catena*, 29, 307–322.

[14] Micheli, L., Morse, J., Fernandez, E. F., Almonacid, F., & Muller, M. (2018). Design and indoor validation of DUSST: a novel low-maintenance soiling station. In 35th European PV Solar Energy Conference & Exhibition (EU PVSEC), September 2018, pp.1–5

[15] Mussard, M., & Amara, M. (2018). Performance of solar photovoltaic modules under arid climatic conditions: a review. *Solar Energy*, 174, 409–421.

[16] Said, S. A. M., Hassan, G., Walwil, H. M., & Al-Aqeeli, N. (2018). The effect of environmental factors and dust accumulation on photovoltaic modules and dust-accumulation mitigation strategies. *Renewable and Sustainable Energy Reviews*, 82, 743–760.

[17] Semaoui, S., Arab, A. H., Boudjelthia, E. K., Bacha, S., & Zeraia, H. (2015). Dust effect on optical transmittance of photovoltaic module glazing in a desert region. *Energy Procedia*, 74, 1347–1357.

[18] Sengupta, S., Sengupta, S., & Saha, H. (2020). Comprehensive modeling of dust accumulation on PV modules through dry deposition processes. *IEEE Journal of Photovoltaics*, 10(4), 1148–1157.

[19] Sengupta, S., Ghosh, A., Mallick, T. K., Chanda, C. K., Saha, H., Bose, I., et al. (2021). Model based generation prediction of SPV power plant due to weather stressed soiling. *Energies*, 14, 5305.

[20] Sengupta, S., Chanda, C. K., & Saha, H. (2024). Cost effective operation scheduling of a solar PV-biogas-wind-grid integrated microgrid considering PV soiling loss. In 2024 IEEE 3rd International Conference on Control, Instrumentation, Energy & Communication (CIEC), (pp. 147–152). IEEE.

[21] Sengupta, S., Sengupta, S., Chanda, C. K., & Saha, H. (2021). Modeling the effect of relative humidity and precipitation on photovoltaic dust accumulation processes. *IEEE Journal of Photovoltaics*, 11(4), 1069–1077.

[22] Sharma, S., Raina, G., Yadav, S., & Sinha, S. (2023). A comparative evaluation of different PV soiling estimation models using experimental investigations. *Energy for Sustainable Development*, 73, 280–291.

[23] Singh, R. P., Dey, S., Tripathi, S. N., Tare, V., & Holben, B. (2004). Variability of aerosol parameters over Kanpur, Northern India. *Journal of Geophysical Research: Atmospheres*, 109 (D23206), pp. 1–14.

[24] Yang, M., Ji, J., & Guo, B. (2020). Soiling quantification using an image-based method: effects of imaging conditions. *IEEE Journal of Photovoltaics*, 10(6), 1780–1787.

2 Application of artificial neural networking for the estimation of illuminance and energy efficiency parameters of public road lighting systems

Sourin Bhattacharya[1,a], Saheb Samanta[2,b] and Parthasarathi Satvaya[2,c]

[1]Transport Department, Government of West Bengal, Kolkata, India

[2]School of Illumination Science, Engineering and Design, Jadavpur University, Kolkata, India

Abstract

Functional road lighting systems ensure public safety and provide visual guidance after dusk. In India, light-emitting diode (LED) luminaires are finding increased usage for public road lighting purposes as compared to the conventional discharge lamp counterparts. It is expedient for municipal authorities to assess the illuminance and energy efficiency parameters of excogitated road lighting projects before commissioning to check compliance with regional, provincial, or national energy codes and guidelines. This study performed photometric simulations of type II LED-based road lighting systems for 1,20,000 different design configurations and recorded outputs in terms of average illuminance, overall uniformity of illuminance, and installation energy efficiency. The generated data were utilized to develop, train, test, and validate artificial neural network (ANN)-based models with 10, 20, and 30 neurons in the hidden layer. For some standard designs conforming to the appurtenant Indian Standard for the illumination of public thoroughfares (IS:1944 of 1970), the performances of the ANN-based models were compared. The 30-neuron ANN-based model in general performed satisfactorily with a maximum error margin of 2.30% for the prediction of average illuminance, 20.17% for that of overall uniformity, and 2.38% for that of installation energy efficiency, and it can be utilized for road lighting system design.

Keywords: ANN, illuminance, LED, lighting design, road lighting

Introduction

Public road lighting systems, essentially regulated by local planning standards and state agency guidelines [11], are an indispensable part of urban planning and transportation engineering. Astutely planned public road illumination systems can ensure satisfactory human visual performance amidst foot traffic as well as motorized traffic [13], help in the estimation of contours of objects and color perception [5], and enable roadway and sidewalk users to perceive obstacles, observe the surroundings more effectively and choose their paths of locomotion with caution. Several studies correlated effective road and street illumination systems with a reduced probability of occurrence of acts in transgression of the law or crimes [6, 8, 9]. In addition, the likely impact of road lighting on road safety and the reduction of the possibility of accidents is well documented in literature and inefficient road lighting can exacerbate road accidents [14, 16, 18]. In India, high-pressure sodium (HPS) vapour and metal halide (MH) lamps are commonly deployed in road lighting systems although the usage of light-emitting

diodes (LEDs) has grown manifold in the past decade primarily owing to the necessity of assuring energy efficiency of installations and one recent study [15] highlighted LED-based smart street lighting systems can have a payback period of less than 5 years. With a view of the changing scenario of municipal engineering practices emphasizing upon proposing neoteric energy-efficient solutions for road illumination projects, this study aims to propose a simple and effective strategy to readily estimate photometric and energy efficiency parameters of proposed municipal road illumination projects by an artificial neural network (ANN)-based approach which is explicated in the subsequent sections.

The rest of this study is organized as follows: Section 2 discusses relevant literature germane to road lighting simulation and modelling, Section 3 deliberates upon the simulation method and the development of ANN-based models, Section 4 analyzes the performance of the proposed ANN-based models for some standard designs to commentate on their desirability and appositeness and Section 5 sums up this study with insights for future studies.

[a]sourinrcb@ieee.org, [b]saheb.samanta362@gmail.com, [c]parthas.satvaya@jadavpuruniversity.in

DOI: 10.1201/9781003663348-2

Literature Review

In recent literature, photometric simulation has been a recurrent theme in the academic discourse of road lighting studies. One photometric simulation-based study [4] explored the luminaire luminous flux output to road surface average illuminance characteristics of roadways illuminated with type II LEDs and type II HPS luminaires and came to a conclusion that the LEDs directed more of the output luminous flux onto the road surface as compared to the HPS luminaires. Another study [17] performed optimization of road lighting with the assistance of some state-of-the-art photometric simulation software and fuzzy logic and concluded that optimized solutions lead to 58.6% energy savings. Some recent studies [1, 2, 12] utilized multiple regression analysis of simulation data to promulgate models for rapid estimation of luminance, illuminance, and energy efficiency parameters for HPS and LED-based conceptual road lighting systems, expounded the effects of geometrical factors on lighting quality parameters and provided an in-depth assessment of different simulation methods. One study [7] employed an ANN-based approach for analyzing the overall uniformity of road lighting and optimizing the lighting quality and came to a conclusion that the Levenberg-Marquardt back-propagation algorithm was the best-suited one for the training of ANNs. Withal, another study [10] utilized deep learning to predict luminance values of different points of road lighting scenes through the mathematical correlation between luminance and pixel colour (R, G, B) of images taken of such scenes. Thus, with a view of pertinent literature germane to road lighting simulation and the application of machine learning methods for road illumination systems, this study performed photometric simulation of LED-based road lighting systems and adopted an ANN-based approach to propound models the methodology of which is described in the succeeding section (Section 3).

Photometric Simulation and Development of ANN Models

Considering single-sided siting of LED luminaires of different power ratings, photometric simulations were performed in Relux, a standard software package. Among the prerequisites were the assumptions that the pole spacing (span) was uniform, the luminaires were operative without dimming and the luminous intensity distribution patterns were consistent with that of type II luminaires. Simulations were performed for as many as 1,20,000 design configurations and Table 2.1 gives a compendium of input parameters and luminaire arrangement characteristics. Figure 2.1 provides a representational view of a typical road lighting scene with the labelling of pertinent design variables for enhanced visualization. Figure 2.2 demonstrates the luminous intensity distribution polar plots of the selected LED luminaires with luminous intensity expressed in cd/klm. The output parameters were: average illuminance (E_{av}, in lx), overall uniformity of illuminance (U_0), and installation energy efficiency (ε_x, in m²*lx/W). With the data generated by photometric simulation, ANN models were created, trained, tested, and validated in MATLAB. The ANN models

Figure 2.1 A representational road illumination system conceptualized in a photometric simulation environment with added graphics and labelling
Source: Author

were created with one hidden layer comprising sigmoid neurons and one output layer comprising linear neurons. Three ANN models were created in total and the hidden layers consisted of 10, 20, and 30 neurons respectively. In essence, the ANNs were two-layer feed-forward networks and were trained utilizing the Levenberg-Marquardt backpropagation algorithm which requires less time for training. 84000 (70%) samples were allocated for training and 18000 (15%) samples were allocated for network generalization or validation. In addition, 18000 (15%) samples were allocated for testing. The data division was random in nature. The ANN models were retrained once to achieve exiguous improvements in their performances, as gauged by the mean square error (MSE) and coefficient of correlation (R) values. Figure 2.3 shows a conceptualized representation of the ANN models and Figures 2.4 and 2.5 provide the performance plots in terms of mean square error (MSE) and coefficient of correlation (R).

Performance Analysis for Some Standard Installation Layouts

For some standard installation layouts with default luminaire tilt and overhang conforming to the primary requirements of the group A2 Indian Standard lighting class ($E_{av} \geq 15$ lx and $U_0 \geq 0.4$) as per the Indian Standard for the illumination of public thoroughfares (IS:1944 of 1970) [3], the performances of the developed ANN models were compared to each other and to the outputs of some standard photometric simulation software and the data is given in Table 2.2 (rounded off to two or lower decimal places). It is pertinent to mention that the developed ANN models were retrained before extracting the outputs.

For the prediction of average illuminance (E_{av}), the 10-neuron ANN model demonstrated a maximum error of 3.29%, the 20-neuron ANN model demonstrated a maximum error of 2.39% and the 30-neuron ANN model demonstrated a maximum error of 2.30%. For the prediction of overall uniformity (U_0), the 10-neuron ANN model demonstrated a maximum error of 23.68%, the 20-neuron ANN model demonstrated a maximum error of 23.77% and the 30-neuron ANN model demonstrated a maximum error of 20.17%. For the prediction of installation energy efficiency (ε_x), the 10-neuron ANN model demonstrated a maximum error of 2.71%, the 20-neuron ANN model demonstrated a maximum error of 2.77% and the 30-neuron ANN model demonstrated a maximum error of 2.38%. Therefore, it can be cogently stated that the performance of the 30-neuron ANN model in particular, trained with the application of the Levenberg-Marquardt backpropagation algorithm, could be viewed as satisfactory for the prediction of pertinent illuminance and energy efficiency parameters. It is concomitant to mention that the results were obtained for idealized scenarios assuming no atmospheric attenuation of light and no contribution from peripheral sources of lighting.

Table 2.1 A compendium of input parameters and luminaire layout features for the conduction of photometric simulation of road lighting

Input parameters/luminaire layout features/configurations	Values/description/particulars
Type of Luminaire	LED
Luminous Intensity Distribution Pattern	Type II
Power Ratings (W, in J/s or W)	88, 110, 149 and 189
Road Width (b, in m)	6.5, 7, 7.5 and 8
Overhang (y_l, in m)	0, 0.25, 0.5, 0.75 and 1
Mounting Height (h, in m)	6, 6,5, 7, … to 11.5
Pole Spacing/Span (S, in m)	15, 16, 17, … to 39
Tilt (δ, in °)	0, 3, 6, 9 and 12
Siting Type	Single-sided, single mounting
Maintenance Factor	0.8
Luminaire Aiming	Default Configuration

Source: Authors' Compilation

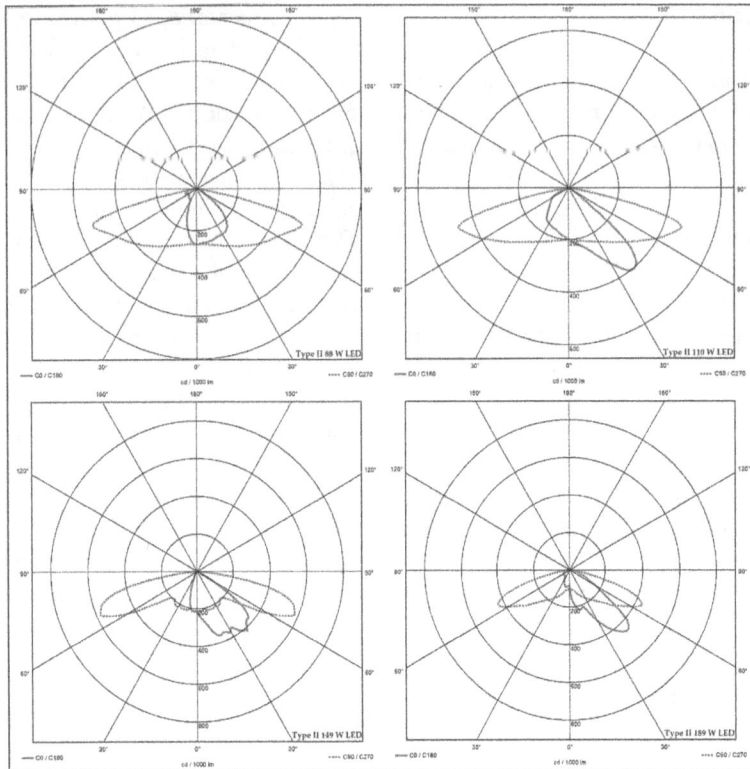

Figure 2.2 Polar plots of luminous intensity distribution patterns of the selected LED luminaires
Source: Author

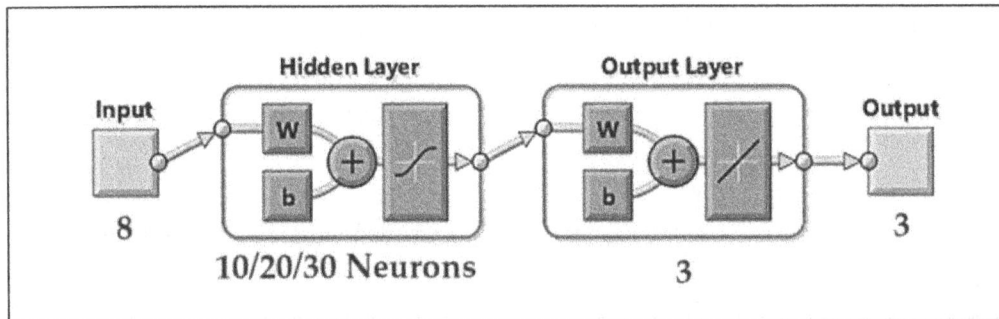

Figure 2.3 A conceptualized representation of the developed two-layer feed-forward ANN models
Source: Author

Conclusion

This study performed photometric simulation of single-sided light-emitting diodes (LEDs)-based road illumination systems of 88 – 189 W nominal power ratings and generated substantial photometric data which were used to create, train, validate, and test artificial neural network (ANN) models with 10, 20, and 30 neurons respectively in the hidden layer. Adjudicating their performances for certain standard installation layouts, the 30-neuron ANN model in particular performed satisfactorily with very small error margins of < 3% for the predictions of average illuminance and installation energy efficiency. However, this study did not conduct sensitivity analysis germane to ANN hyperparameters, did not incorporate data from field measurements, or did not consider environmental attenuation of light and these remain underexplored areas of research. Moreover, peripheral light sources or contributions of vehicular headlights were pretermitted from its purview and these represent the limitations of this study. As a future scope of work, it

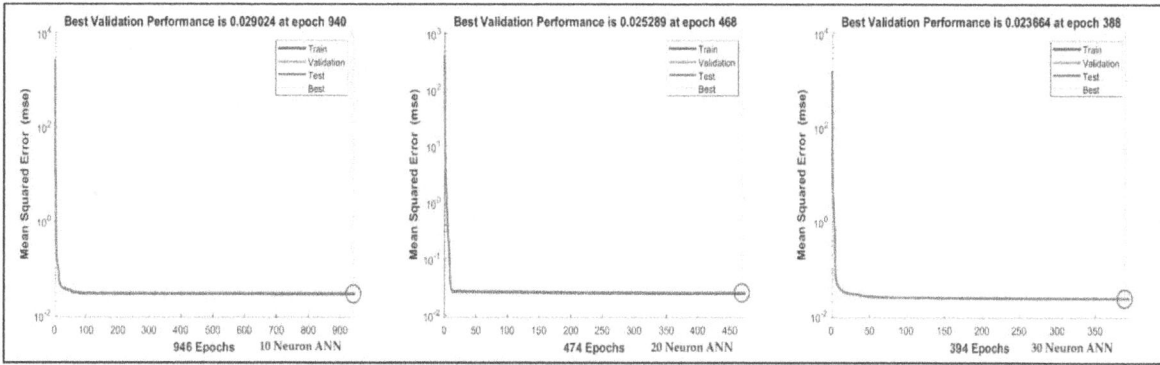

Figure 2.4 The performance plots of the created ANN models demonstrating the mean square error margins
Source: Author

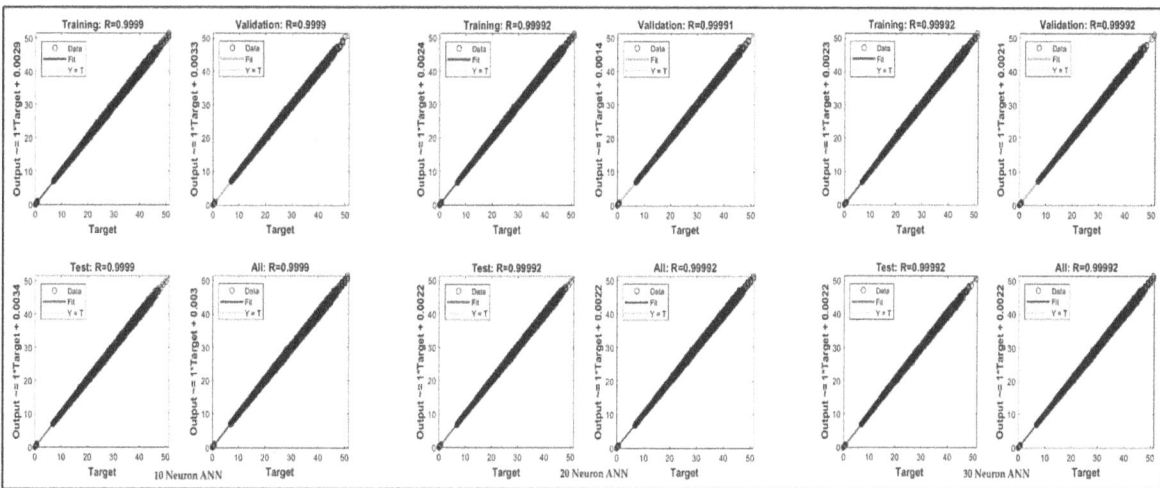

Figure 2.5 The performance plots of the created ANN models demonstrating the coefficient of correlation values
Source: Author

Table 2.2 A comparison of the performances of the ANN models with the outputs generated by the application of a standard photometric simulation software

Standard road lighting system design configurations				Outputs of a standard simulation software			Outputs of the 10 - neuron ANN model			Outputs of the 20 - neuron ANN model			Outputs of the 30 - neuron ANN model		
W	b	h	S	E_{av}	U_0	ε_x	E_{av}	U_0	ε_x	E_{av}	U_0	ε_x	E_{av}	U_0	ε_x
88	7.5	9	21	20	0.45	28.1	20.4	0.42	28.7	20.5	0.42	28.9	20.4	0.41	28.8
88	9	9	24	20	0.47	32.7	20.2	0.41	33	20.3	0.43	33.2	20.3	0.43	33.1
110	7.5	10.5	25	20	0.48	27.9	20	0.46	28	20	0.50	28	20	0.51	28
110	9	10	24	20	0.4	32.7	20.3	0.41	33.2	20.3	0.45	33.1	20.4	0.48	33.2
149	7.5	9.5	27	20	0.44	18.6	20.7	0.37	19.1	20.5	0.40	19	20.5	0.41	19
149	9	9.5	25	20	0.48	21.1	20.4	0.39	21.5	20.1	0.47	21.3	20.2	0.47	21.4
189	7.5	10.5	27	20	0.42	17.9	20	0.45	18	20.2	0.46	18.1	20.3	0.43	18.1
189	9	11	30	20	0.44	20	20	0.54	20.2	20.3	0.54	20.2	20.2	0.50	20.2

Source: Authors' Compilation

may be posited that ANNs may be developed with different numbers of neurons in the hidden layer and different training regimes involving algorithms such as Bayesian Regularization or Scaled Conjugate Gradient and impetus may be placed on reducing the margin of error of predicting overall uniformity which was about 20% in this study. This study is denotative of the utility and appositeness of ANNs in road lighting design and analysis and may serve as a stanchion for the conduction of future studies germane to road lighting simulation, modelling, and analysis.

References

[1] Ayaz, R., Roy, S., & Bhattacharya, S. (2024). An assessment of general road illumination system simulation methods and comparison of simulation outcomes with photometric measurements conducted on a public road with anthropogenic sources of peripheral illumination. *Journal of Optics*, 53(5), 4405–4422. https://doi.org/10.1007/s12596-023-01645-5.

[2] Bhattacharya, S., Majumder, S., & Roy, S. (2023). Modelling of the effects of luminaire installation geometries and other factors on road illumination system photometric parameters and energy efficiency. *World Journal of Engineering*. https://doi.org/10.1108/WJE-09-2022-0372

[3] Bhattacharya, S., Satvaya, P., & Roy, S. (2024). Predictive modelling of lighting quality parameters and energy efficiency of light emitting diode-based general road illumination systems with special emphasis on luminaire tilt and bracket length. *Sādhanā*, 49(1), 18. https://doi.org/10.1007/s12046-023-02382-y.

[4] Brons, J. A., Bullough, J. D., & Frering, D. C. (2021). Rational basis for light emitting diode street lighting retrofit luminaire selection. *Transportation Research Record: Journal of the Transportation Research Board*, 2675(9), 634–638. https://doi.org/10.1177/03611981211003890.

[5] Chakraborty, S., Dutta, P., Dey, S., & Anjum, S. N. (2022). A laboratory-based study on influence of peripheral source on on-axis object detection under different correlated color temperatures. *Optik*, 249, 168258. https://doi.org/10.1016/j.ijleo.2021.168258.

[6] Chalfin, A., Hansen, B., Lerner, J., & Parker, L. (2022). Reducing crime through environmental design: evidence from a randomized experiment of street lighting in New York City. *Journal of Quantitative Criminology*, 38(1), 127–157. https://doi.org/10.1007/s10940-020-09490-6.

[7] Corte-Valiente, A., Castillo-Sequera, J., Castillo-Martinez, A., Gómez-Pulido, J., & Gutierrez-Martinez, J.-M. (2017). An artificial neural network for analyzing overall uniformity in outdoor lighting systems. *Energies*, 10(2), 175. https://doi.org/10.3390/en10020175.

[8] Cozens, P. M., Neale, R. H., Whitaker, J., Hillier, D., & Graham, M. (2003). A critical review of street lighting, crime and fear of crime in the British City. *Crime Prevention and Community Safety*, 5(2), 7–24. https://doi.org/10.1057/palgrave.cpcs.8140143.

[9] Fotios, S. A., Robbins, C. J., & Farrall, S. (2021). The effect of lighting on crime counts. *Energies*, 14(14), 4099. https://doi.org/10.3390/en14144099.

[10] Kayakuş, M., & Çevik, K. K. (2020). Estimating luminance measurements in road lighting by deep learning method. In Artificial Intelligence and Applied Mathematics in Engineering Problems: Proceedings of the International Conference on Artificial Intelligence and Applied Mathematics in Engineering (ICAIAME 2019) (pp. 940-948). Springer International Publishing. https://doi.org/10.1007/978-3-030-36178-5_83.

[11] Murray, A. T., & Feng, X. (2016). Public street lighting service standard assessment and achievement. *Socio-Economic Planning Sciences*, 53, 14–22. https://doi.org/10.1016/j.seps.2015.12.001.

[12] Roy, S., Satvaya, P., Bhattacharya, S., Majumder, S., Majumder, S., & Sardar, I. H. (2022). An exposition of a road lighting model to facilitate simple estimation of road surface illuminance parameters for conventional system specifications and recommendations for retrofitting of luminaires. *Journal of Optics*, 51(2), 444–455. https://doi.org/10.1007/s12596-021-00792-x.

[13] Sefer, T., Ayaz, R., Ajder, A., & Nakir, I. (2023). Performance investigation of different headlights used in vehicles under foggy conditions. *Scientific Reports*, 13(1), 4698. https://doi.org/10.1038/s41598-023-31883-3.

[14] Tamakloe, R., Sam, E. F., Bencekri, M., Das, S., & Park, D. (2022). Mining groups of factors influencing bus/minibus crash severities on poor pavement condition roads considering different lighting status. *Traffic Injury Prevention*, 23(5), 308–314. https://doi.org/10.1080/15389588.2022.2066658.

[15] Viswanathan, S., Momand, S., Fruten, M., & Alcantar, A. (2021). A model for the assessment of energy-efficient smart street lighting—a case study. *Energy Efficiency*, 14(6), 52. https://doi.org/10.1007/s12053-021-09957-w.

[16] Wanvik, P. O. (2009). Effects of road lighting: An analysis based on Dutch accident statistics 1987–2006. *Accident Analysis and Prevention*, 41(1), 123–128. https://doi.org/10.1016/j.aap.2008.10.003.

[17] Wicaksono, H. I. S., Abdullah, A. G., & Hakim, D. L. (2021). Optimizing public street lighting and redesign of public road lighting based on DIALux and fuzzy logic. *IOP Conference Series: Materials Science and Engineering*, 1098(4), 042013. https://doi.org/10.1088/1757-899X/1098/4/042013.

[18] Yannis, G., Kondyli, A., & Mitzalis, N. (2013). Effect of lighting on frequency and severity of road accidents. *Proceedings of the Institution of Civil Engineers - Transport*, 166(5), 271–281. https://doi.org/10.1680/tran.11.00047.

3 LED based energy efficient lighting design in a metro railway system

Ankit Ghosh[a] and Parthasarathi Satvaya[b]

School of Illumination Science, Engineering and Design, Jadavpur University, West Bengal, India

Abstract

Metro Railway system is a part of rapid transportation system. Providing proper energy efficient lighting solution in a metro railway system not only improves visibility among the people but also gives a bright aesthetic looks to the system. Here an elevated metro railway station is simulated in DIALux 4.13 by using conventional lighting systems and energy efficient LED lighting systems. The results are based on average illuminance, uniformity and lighting power density. The results are based on the target value of DMRC standards.

Keywords: Illuminance, lighting power density, luminaires, metro railway, uniformity

Introduction

The metro railway system is an integral part of modern day. The rapid life of modern civilization requires fast movement in which Metro Railway- Underground as well as Elevated metro saves time as well as public money. Where rapidness is involved, lighting is required for visual guidance and for safe movement. The illuminance requirement in metro railway station is relatively higher than other areas as there are an ample amount of accumulation of people and their rapid movement. Not only people, but lighting also requires for the movement of train. Special emphasis is given on the vibration proof and proper impact protection on luminaires which are required in tunnel as well as station. In India, 19% of total generated electrical energy is spent on illumination related application [1]. In the USA, for illumination, energy consumption is 6.6% and domestic commercial lighting consumption [2] is 4.3% of total consumption [10]. Rapid growth in our economy also increases the consumption of electrical energy in lighting. So alternative energy efficient luminaires which have high luminous efficacy are required in these infrastructure areas. The illumination system should be aesthetic and proper as per standards. Passengers should be alerted by lighting. The lighting system should provide as per standard adequate lighting levels rated for hustle-free operation and other visibility criteria necessary to stimulate productivity, ease of rapid operation, alertness, safety, proper surveillance. The security must be alerted by the use of proper lighting levels. The tasks must be done adequately by the use of proper luminaires which create proper light levels on the floor as well as proper reflection on the wall ceiling and screen if any. It creates sufficient bright cheerfulness and alertness in light conditions.

This study shows how energy efficient LED luminaires create optimum lighting level for metro railways and also shows the comparison with conventional luminaires.

Metro Railway Application Area

The metro railway application area are broadly divided in two categories i.e. public areas (that are used by passengers- public parking areas and approach roads, entrance or exit areas and public assembly, platform, concourse area, restrooms and toilets [4]) and non-public area (that are restricted for passenger uses and only used by metro railway staffs for operation of metro railway services- railway yard, workshop and car shed, inspection bays, tunnels, offices, signaling room, station control room, store rooms, battery rooms [4].

General Requirement of Metro Railway Lighting

The energy-efficient lighting solution should be at all platform areas, tunnels, parking areas and drop-off points. The factors that are taken into while designing are energy efficient, glare, glare to the CTV cameras, Computer screen environment (Veiling reflectance) and easy maintenance. Also, there should be bright look throughout the design and no dark patches [4]. An emergency lighting system should be installed in all escape routes, tunnels. All stations and passageways should have emergency lighting powered from

[a]gankit044@gmail.com, [b]parthasatvaya@gmail.com

DOI: 10.1201/9781003663348-3

the essential power switchboard. The lighting should comply with DMRC or relevant metro railway standards [4].

Criteria for Selecting Lighting Equipment

Selection of lighting equipment will depend on factors mentioned below: 1. Ceiling type; 2. Mounting height and mounting possibility; 3. Application areas (for the selection of ingress protection rating and impact protection rating); 4. Criticality of restricting glare; 5. Importance of color rendering index (CRI) and 6. Use of lamps with proper correlated color temperature (CCT) [3].

Design Methodology

Here the illumination of a metro station is simulated by the use of conventional luminaires and LED luminaires and then the results are compares. The steps are:

For design simulating software DIALux 4.13 is used for following aspects: 1. Effective and professional light lay outing; 2. World's leading manufacturer's latest luminaire data are updated and available; 3.Free of charge software availability; 4. Energy calculation

facility and lighting control system facility; 5. Lux level calculation in the presence of integrated daylight and electrical light sources; 6. Availability of colorful rendering with color changing features in luminaires.

Design Specification

For this study, Station Platform level, concourse level are taken into account. First a simulation has done by using conventional luminaires and then another simulation is done by using LED luminaires by maintaining the proper illuminance level as per standard Figure 3.1–3.4.

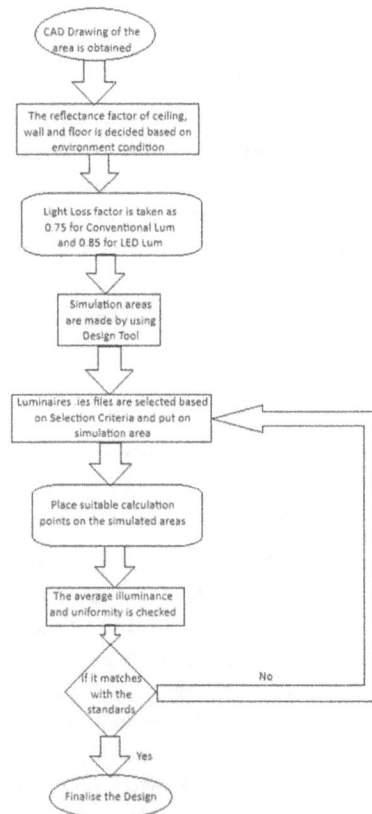

Figure 3.2 Platform level plan view
Source: Author

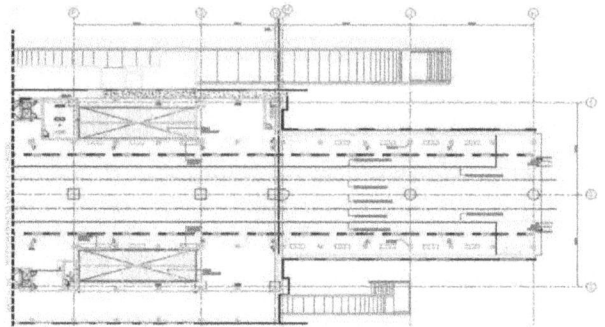

SECTION A-A

Figure 3.3 Platform level elevation
Source: Author

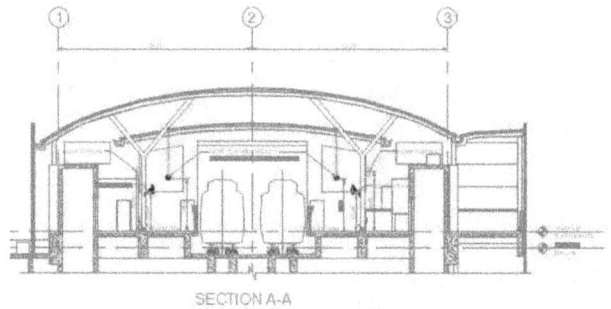

Figure 3.4 Plan view of concourse area
Source: Author

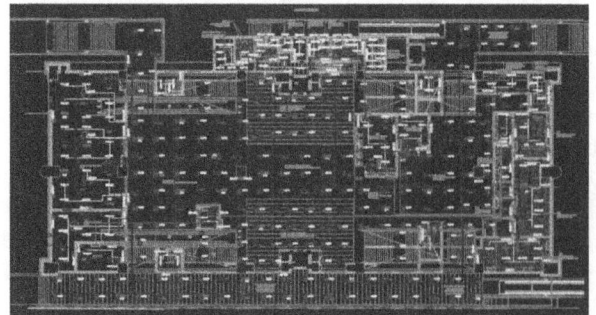

Figure 3.1 Lighting design flow diagram
Source: Author

Choice of Luminaires

The choice of luminaires for each area of a metro railway station are summarized in Table 3.1. In case of false ceiling areas, recessed mounted luminaires are used and where no false ceiling is present, surface mounted luminaires are used. Table 3.2 depicts the polar curve distribution of each luminaires used in simulation. All luminaires having correlated color temperature (CCT) value of 6500K and color rendering index (CRI) > 80 as all areas are indoor areas and visibility is better in that CCT and CRI value. CCT 6500K has some blue color component in spectrum which creates alertness which in turn creates more visibility in that areas.

Table 3.1 Choice of luminaires

Platform areas				
Platform area	Canopy luminaire 150W MH lamp	80	72W LED canopy IP66 luminaire	53
	2X36W FTL IP66 luminaire	149	2X20W LED tube IP66 luminaire	130
			36W IP66 LED integrated batten luminaire	60
Concourse Area				
Auxiliary substation	1X36W FTL IP66 luminaire	80	36W IP66 LED integrated batten luminaire	36
Lobby area	1X18W 2pin CFL surface mounted downlighter	22	12W LED surface mounted downlight	14
Unpaid area	4X14W T5 2X2 recessed mounted luminaire	31	36W LED 2X2 recessed mounted luminaire	26
Concourse paid area	4X14W T5 2X2 recessed mounted luminaire	22	36W LED 2X2 recessed mounted luminaire	17
Ticket office machine	4X14W T5 2X2 recessed mounted luminaire	2	36W LED 2X2 recessed Mounted Luminaire	2
Station entrances	4X14W T5 2X2 recessed mounted luminaire	11	36W LED 2X2 recessed mounted luminaire	11
Staff mess room	1X18W 2pin CFL surface mounted downlighter	17	12W LED surface mounted downlight	11
Station control room	4X14W T5 2X2 recessed mounted luminaire	6	36W LED 2X2 recessed mounted luminaire	6

Source: Author

Table 3.2 Polar curve distribution of luminaires

Canopy luminaire 150W MH lamp	72W LED canopy IP66 Luminaire	2X36W FTL IP66 Luminaire	1X36W FTL IP66 Luminaire
1x 18W 2pin CFL Surface Mounted Downlighter	12W LED SurfaceMounted Downlight	36W LED 2X2 Recessed mounted Luminaire	4X14W T5 2X2 Recessed Mounted Luminaire

Source: Author

Result Analysis

The simulation is run by using the .ies files of conventional luminaires and energy efficient LED luminaire by targeting the standard average illuminance and uniformity in each area [5]. An .ies file is a text file that describes the intensity of light source at a points on a spherical grid [6]. It provides more photorealistic lighting effects in rendered images than other types of lighting distribution [6]. The reflectance factor is taken as 70% for ceiling, 50% for walls and 20% for floors during simulation as metro stations generally have bright colors in ceiling and walls and the maintenance

schedule is very frequent. The light loss factor is taken as 0.70 for conventional luminaires and 0.80 for LED luminaires due to the high burning hours and low maintenance of LED luminaires. Figure 3.5 shows the lighting power density comparison, Figure 3.6 shows the average illuminance comparison and Figure 3.7 shows the uniformity comparison analysis between two simulations and target criteria.

LPD Evaluation of Metro Railway

Total lighting power density (LPD) by building area method: 3.93 W/m². According to Delhi Metro Rail Corporation Electrical and Design Wing (comparing values with energy conservation building code 2016) [5], LPD value should not exceed 7.5 W/m². Hence according to building area method by using LED luminaire, the LPD value is 52.4% less than standard LPD value. So, it can be stated that the design with LED luminaires are feasible and also it can be said as an energy efficient design.

Economic Comparison

By using the market value of LED luminaires and conventional luminaires, per year ownership cost is depicted in Figure 3.8 for simulations considering Rs. 7 per unit energy cost [9].

From Figure 3.8 it is observed that the intersection of two lines is approximately 6 months. So, by break even method calculation, payback period is approximately 6 months.

Conclusion

Electrical energy usage by using LED luminaires is found to be very less than using conventional

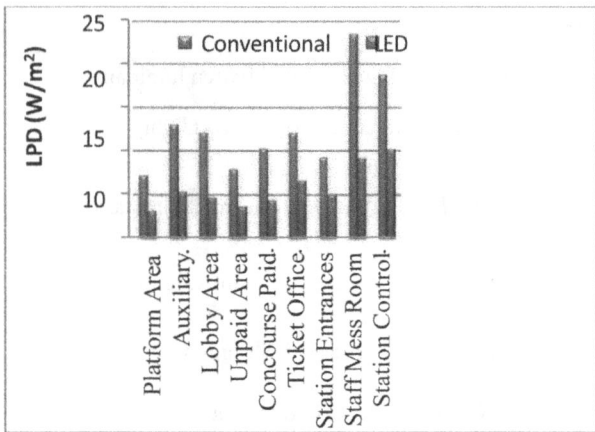

Figure 3.5 Lighting power density (LPD) comparison between two simulations
Source: Author

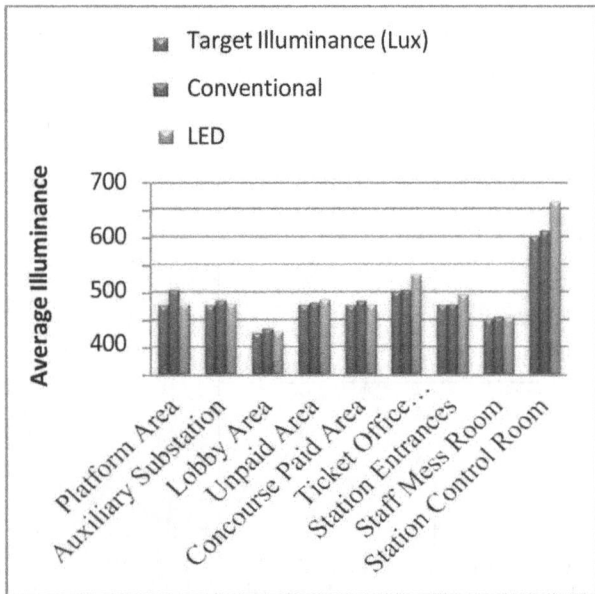

Figure 3.6 Average illuminance comparison between two simulations and DMRC standard target value
Source: Author

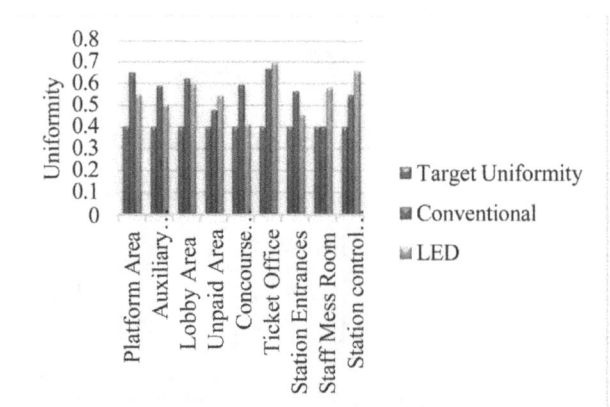

Figure 3.7 Uniformity comparison between two simulations and DMRC Standard Target value
Source: Author

Payback Period

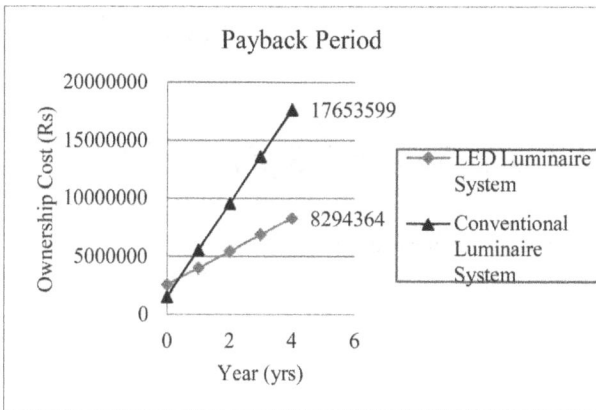

Figure 3.8 Per year ownership cost
Source: Author

luminaires. After conducting several simulations using different LED fixtures, it can be safely concluded that LED fixtures resulted in lower power consumption compared to other fixtures, keeping all other standards values intact. That being said, currently LED fixtures have higher prices compared to conventional fixtures, which may result in a higher payback period. But due to competitive market in luminaire B2B industry [7], the LED luminaire prices are decreasing rapidly from previous year and the discounted price is given to customer which in turn decreased payback period in near future.

Though the target illuminance and target uniformity is same between two simulations, the achieved values are not the same in between them which gives us the conclusion that retrofitting is only possible when the system lumen packages and the photometric distribution of the LED and conventional luminaires are nearly same. In case of LED, the photometric distributions are controlled by lens and diffusers and in case of conventional luminaires, the photometric distributions are controlled by reflectors and diffusers. The photometric distribution of conventional luminaires are more spread whereas the photometric distribution of LED luminaires are more controlled.

More energy efficient LEDs with better system luminous efficacy and less maintenance can be developed, which can further reduce power consumption. Here only the market is available lowest LED luminaires system luminous efficacy of 100 Lumen/Watt is used. In Indian market, now a day up to 160 Lumen/Watt system luminous efficacy luminaire has been available [8]. Platform area has maximum power consumption i.e. 47% of total power by using conventional luminaires and 48% of total power by using LED luminaires. The operational hours in platform areas is maximum, so the highest available system efficacy luminaires should be used for energy conservation.

Here as rapid transport, so the vibration force occurs during train movement. Being an electronic device, LED luminaires must have proper Impact Protection (IK Value). While using the tunnel, there are more air pressure and wind force present along with vibrations. Luminaires are used along with vibration dampers to sustain the vibration along with IK10 protection and having aerodynamic structure to reduce the air pressure through it.

References

[1]   Economic Survey of India, 2017.
[2]   https://www.energybot.com/energy-faq/how- much-electricity-is-used-for-lighting-in-the-united-   states.html#:~:text=Lighting%20accounts%20for%206.6%25%20of,lighting%20in%20the%20United%20States.
[3]   https://en.wikipedia.org/wiki/Rapid_transit.
[4]   National Lighting Code- 2010 Part 6 Section 7.
[5]   Delhi Metro Railway Electrical Standards and Design Wing.
[6]   https://help.autodesk.com/view/RVTLT/2022/ENU/?guid=GUID-4F1C8175-0F14-4C45-AA08-CE0304A15700.
[7]   Diao, Z. (2022). Research of the B2B consumers behavior of the lighting industry in the U.S. E-commerce market. Warren College, University of California San Diego, San Diego, 92093, USA *Corresponding author. Email: zdiao@ucsd.edu.
[8]   https://www.dial.de/en-GB/projects/efficiency-of-leds-the-highest-luminous-efficacy-of-a-white-l.
[9]   https://www.cesc.co.in/tariff.
[10] https://www.energybot.com/energy-faq/how-much-electricity-is-used-for-lighting-in-the-united-states.html#:~:text=Lighting%20accounts%20for%206.6%25%20of,lighting%20in%20the%20United%20States.

4 FFT based comparative performance study of electric vehicle charging system

Tapash Kr. Das[1], Ayan Banik[1,a], Bikash Gopal Tewary[1], Sudarshan Das[1] and Ayan Das Sarkar[2]

[1]Department of Electrical Engineering, Ghani Khan Choudhury Institute of Engineering and Technology, Malda, West Bengal, India

[2]Department of Electrical Engineering, Kalyani Government Engineering College, West Bengal, India

Abstract

The increasing use of electric vehicles (EVs) has the immense potential to revolutionize transportation in the coming future, leading to a zero carbon footprint and promoting sustainability. However, the success of this transition hinges on the development of a robust and capable infrastructure for charging. In this paper, the authors conduct a performance analysis of two types of charging systems for EVs: renewable (wind and solar) and grid-based. The authors attempt to present and evaluate a proposed model using Fast Fourier Transform (FFT) techniques available in MATLAB to ascertain the effectiveness and environmental impacts of the different charging modes. It demonstrates that while EVs seem to be the green solution, their charging often comes at an environmental cost as the energy is mainly obtained from the grids. Wind and solar charging systems offer the best solutions for sustainable inclusion of electric vehicles in the future, but because of the varying weather conditions, the amount of power that can be produced from those sources can vary significantly. This research aims to identify the sources of THD challenges and other factors that may influence the effectiveness and stability of electric vehicle charging.

Keywords: Charging efficiency, CO2, EV, FFT, grid, integration, MATLAB, renewable energy, stability, THD

Introduction

Sustainable energy sources are being harnessed worldwide in order to ensure continuous availability and energy security. With the increasing popularity of electric vehicles (EVs), a reliable and effective charging network becomes paramount. Alternative energy-based charging systems such as photovoltaic (PV) and wind energy-based charging provide opportunities to minimize greenhouse emissions and enhance sustainability. The study attempts to carry out a detailed comparative performance assessment of electric vehicle charging using conventional electricity from the grid, solar-photovoltaic energy, and wind energy. MATLAB-based FFT analysis in this work has been carried out to assess the efficiency and reliability, as well as the overall performance of each method under different operating conditions. The novel work aims to comprehensively understand each charging system, including an accurate assessment of its strengths and weaknesses. The evaluations are also done by comparing performed indicator values with simulation data in order to create a proper baseline for evaluation. The purpose is to guide the designs of EV charging systems by evaluating the economic viability and technological and ecological aspects of solar photovoltaic, wind energy, and conventional grid charging. In this way, it is expected that the findings of this research will facilitate the transition to greener transport systems across the globe, support the deployment of renewable resources, and strengthen them. This comparison analysis aims to determine the most sustainable, reliable, and efficient charging techniques, thereby enhancing the deployment and operation of electric vehicles. Significantly. It has been observed that very little work has been done on the performance study of EV charging stations. By understanding the gravity and necessity of the topic, the authors are highly motivated to contribute significantly to the progress. This study will serve as a valuable resource for novice researchers, enabling them to embark on their research journey and continue the legacy of sustainable innovation.

Literature Review

The emerging reputation of EVs has led to increasing interest in the development of green and reliable charging infrastructure. Consequently, several studies were conducted to evaluate distinctive charging systems and investigate their overall performance in diverse

[a]tapash@gkciet.ac.in, [b]ayan@gkciet.ac.in, [c]bikashgopaltewary@gamil.com, [d]sudarshandas452@gmail.com, [e]ayandassarkar@gmail.com

DOI: 10.1201/9781003663348-4

aspects. This literature assessment targets to integrate and synthesize the findings from that research, perceive know-how gaps, and propose capability future studies directions in electric vehicle charging systems. Zheng et al. (2014) carried out a contrast examine of electric vehicle battery charging/switch stations in distribution systems [13]. The take a look at discovered that battery change stations are more suitable for public transportation in distribution systems. This insight shows the significance of thinking about specific use cases and alertness eventualities whilst designing and making plans EV charging infrastructure. Pandžić et al. [1] assessed the troubles associated with the locations, power, and availability of charging factors for electric vehicles. The observation has first-hand stories and recommendations for the destiny improvement of EV charging infrastructure [2]. These locating highlights the need for addressing sensible challenges and optimizing the deployment of charging stations to fulfill the growing call for electric automobile charging. This research finding contributes to the understanding of wireless charging technology and their capacity for high-performance electricity transfer in electric powered car applications [3].

Proposed Model of Battery Charging using Wind Energy

The use of renewable energy to reduce reliance on traditional charging stations. It solves critical challenges in the EV industry, such as range problems and long charging times, making EVs more practical for everyday use. The adoption of wind energy for EV charging is also in line with global sustainability initiatives. By reducing reliance on fossil fuels and reducing carbon emissions, the technology helps combat climate change. Figure 4.1 illustrates the process of EV battery charging using wind power, showcasing the integration of renewable energy sources into electric vehicle charging systems. The simulation used in the paper are using MATLAB Simulink Model [4].

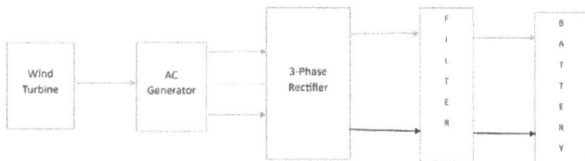

Figure 4.1 EV battery charging using wind power
Source: Author

A permanent magnet alternator is driven by a wind turbine in a wind-fed charger, which runs at a variable speed. A rectifier is used to link the alternator to a battery bank. The wind turbine, alternator, and system configuration can affect the system's characteristics. Depending on the wind speed, a rotor powers a permanent magnet synchronous generator at varying speeds. After being corrected, the generator's output is sent to a bank of batteries. Lithium-ion batteries are generally the most popular kind of battery utilized in grid energy storage systems. Figure 4.2 presents the relationship between battery voltage and charging time, offering valuable insights into how voltage fluctuates during the charging process. [5]

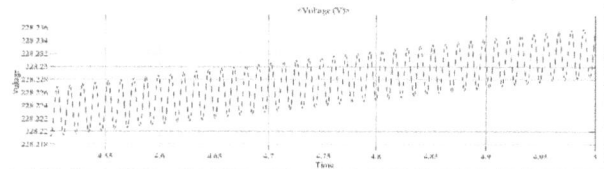

Figure 4.2 Battery voltage vs charging time
Source: Author

Below is the Figure 4.3 that attemt to demonstrates the state of charge (SOC) of the battery over time, providing a visual representation of the battery's charge level throughout the charging cycle.

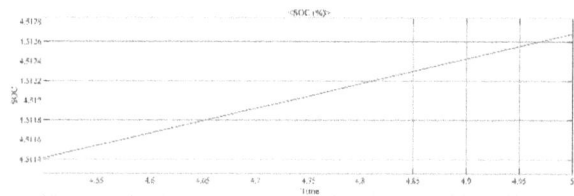

Figure 4.3 Battery SOC w.r.t. time
Source: Author

The nature of SOC graph shows that it is linierly increasing with respect to time which indicates that battery is charging according to the wind speed.

Calculation
Observing the SOC plot in wind fed battery the State of charge (SOC) changes from 4.5001 to 4.5126% in 5 seconds. Thus, 1 sec = 0.0025% SOC in 3600 sec = 9%. Hence the battery takes 0.15 min to charge 1%.

Scheme for Battery Charging using Solar Energy

Solar panels are used to convert solar energy into electricity. The energy generation depends on factors like irradiance, temperature, and panel efficiency [6]. Figure 4.4 highlights the charging of the EV battery through a photovoltaic (PV) array, emphasizing the role of solar energy in sustainable electric vehicle charging.

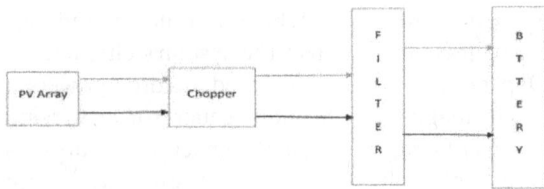

Figure 4.4 EV battery charging using PV array
Source: Author

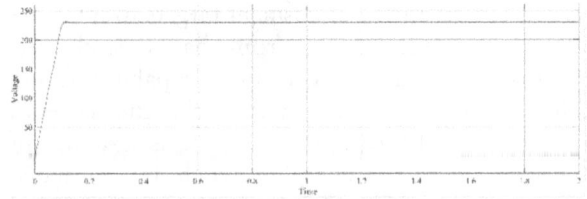

Figure 4.5 PV voltage battery w.r.t time
Source: Author

Figure 4.5 depicts the voltage of the PV array battery in relation to time, revealing the dynamics of solar energy input and battery voltage over the course of the charging period.

Here the plot describes battery voltage w.r.t. Time, when battery is charged by PV Array it shows nearly about 230V which is equivalent to the nominal voltage of the battery.

Figure 4.6 shows the battery SOC w.r.t. when battery is charged by PV Array It is observed that the electric vehicle is charged 100% with solar PV within

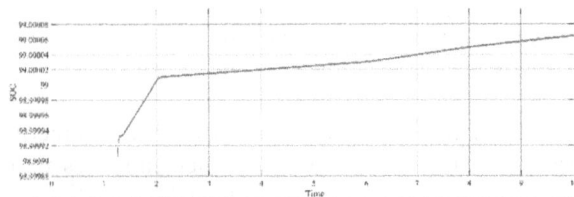

Figure 4.6 X axis shows time % Y axis shows battery SOC
Source: Author

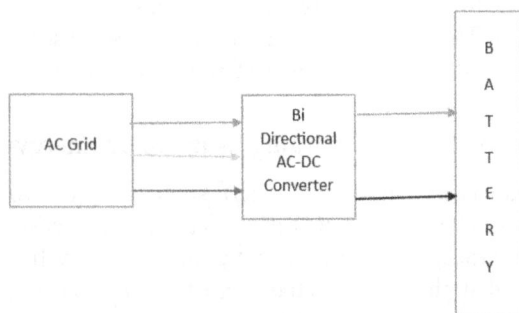

Figure 4.7 EV battery charging using conventional grid
Source: Author

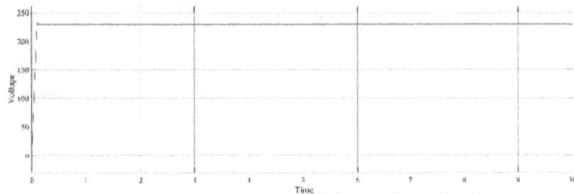

Figure 4.8 Grid connected battery voltage w.r.t time
Source: Author

a short period (10s). And 0 to 1.5 sec.it is charging slowly then later charging speed increases according to temperature irradiance [7].

Battery Charging Topology using Conventional Grid

In this mode of operation, the battery is charged from the grid. This mode operates in the absence of solar power. The battery charging from grid circuit configuration is shown in Figure 4.7.

In Figure 4.8 as demonstrated in below has described the footprint of grid connected battery voltage in the time domain [8].

Figure 4.9 shows the SOC of the battery connected to the grid over time, illustrating the interaction between grid power and battery charge levels during the charging process.

The X axis shows time in seconds. Y axis shows battery voltage. When time increases, battery voltage also increases. This graph is increasing in nature. The figure given below shows a graphical representation of the percentage of battery charge with respect to time. Where time is shown on the X axis in seconds.

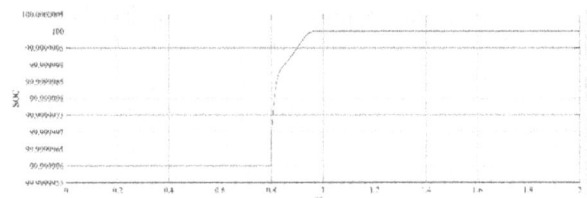

Figure 4.9 SOC of battery connected to grid vs time
Source: Author

The Y axis represents the percentage of battery SOC. When time increases, the percentage of battery SOC also increases. By observing the SOC graph here, we conclude that battery is charge in very short period.

FFT Analysis for Solar, Wind & Grid

Fast Fourier Transform (FFT) analysis is a powerful tool used to analyze the frequency components of

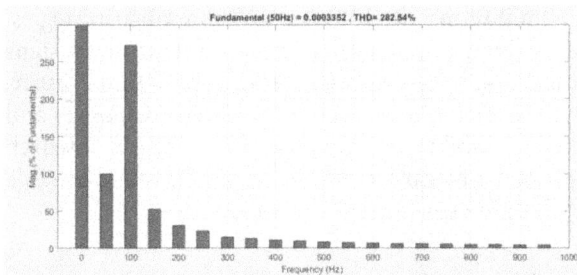

Figure 4.10 Frequency vs THD of wind turbine
Source: Author

signals. In the context of charging systems for solar, wind, and conventional grid-based sources, FFT analysis can help in understanding the harmonic content and overall power quality. This is critical for ensuring the efficient and stable operation of these systems.

$$X(k) = \sum_{j=1}^{N} x(j)\omega_N^{(j-1)(k-1)} \qquad (1)$$

$$x(j) = (1/N) \sum_{K=1}^{N} X(k)\omega_N^{-(j-1)(k-1)} \qquad (2)$$

Where, $\omega_N = e^{(-2\pi i/N)}$ *is an Nth root of unity.*

Wind: FFT can essentially determine the frequency of wind turbine output. These frequencies are associated with variable wind speed and mechanical characteristics of the turbine. Analysis of these frequencies helps to understand the response of the system to wind

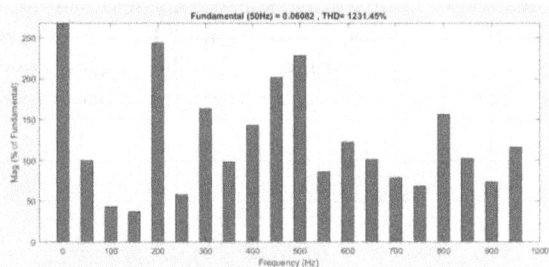

Figure 4.11 Frequency vs THD of wind
Source: Author

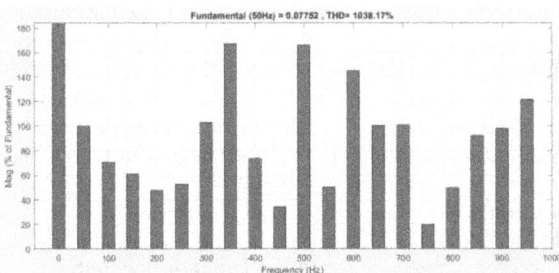

Figure 4.12 Frequency vs THD of conventional grid
Source: Author

fluctuations and the possible mechanisms of power control for attenuation [9].

Figure 4.10 shows the harmonics components present in battery voltage connected to the wind turbine. ie 282.54% of fundamental frequency. Here the magnitude of DC harmonics is more than 250, magnitude of 1st order harmonics is 100, 2nd order harmonics magnitude is slightly greater than 230 and rest all the harmonics is gradually decreasing in nature.

Solar: Like wind, FFT can reveal the cyclical nature of solar output based on daily variations in solar radiation. By analyzing these frequencies, engineers can predict the performance of power generation systems and optimize power connections.

Figure 4.11 shows the harmonics components present in battery voltage connected to the PV Array.ie 1231.45% of fundamental frequency. Here magnitude of DC harmonics is more than 250 and 1st order harmonics magnitude is 100 and all the rest harmonics are of varying magnitude due to the different nature of wind speed.

Conventional grid: FFT analysis of conventional sources typically yields a more constant frequency spectrum compared to wind and solar. However, FFT can still be used to detect potential anomalies or equipment faults that might cause variations in power output [10].

Figure 4.12 shows the harmonics components present in battery voltage connected to the conventional grid.ie 1038.17% of fundamental frequency. Here magnitude of DC harmonics is slightly greater than 180, magnitude of fundamental harmonics is 100, 1st order harmonics magnitude is approx. 70 and the rest orders are varying in nature due to grid stability [11].

Factors under Consideration

(i) Grid effects: Examine how the harmonic profile of the charging system affects the overall network stability and potential interference with other associated loads.
Electromagnetic interference (EMI): Determine the potential for EMI generated by the charging system in a specific range of frequencies that can interfere with nearby electronic devices.

(ii) Future-proofing: Consider developing EV charging technologies and possible future standards when designing the FFT evaluation method.

(iii) FFT-based comparative performance analysis with careful consideration of these factors can provide valuable insights into the strengths and weaknesses of EV charging systems. Use this information to optimize charging system design, get the energy moving effective, making electric vehicles more efficient and reliable It can also be done to ensure cost savings.

(iv) Battery chemistry: Battery type (e.g., lithium-ion, lead-acid) can affect frequency response during charging due to its internal impedance and charging behavior [12].

(v) Converter topology: Consider the type of converter used in the charging system (e.g. bi-directional converter, Vienna rectifier). Different converter topologies introduce characteristic harmonics in the current and voltage waveform.

Proposed Algorithm

i.   Study Configured for EV Charging System.
ii. Consideration of solar, wind and conventional grid.
iii. Implementation of various sources to the proposed scheme.
iv. Analyze the performance of different sources using FFT model.
v.   The categorization of sources has been identified.
vi. We have prepared and proposed a specific outcome to enhance the clarity of the study for the convenience of the reader [13].

Specific Outcome

The paper entitled "A Fast Fourier Transform (FFT) based comparative performance analysis of electric vehicle (EV) charging topologies" offers valuable insights into different scientific parameters, including sub-harmonics, total harmonic distortions and transiency behavior [14]. This review study may provide an in-depth understanding of the diverse advantages and limitations of different EV charging strategies. The findings can be used to enhance structure designs to increase energy gain, reduce harmonic distortion, create improvements in transient response, and enhance stability and reliability. Finally, our research contributes toward establishing a robust and reliable EV charging sustainable infrastructure [15].

Conclusion

A comparative performance evaluation of electric vehicle (EV) charging systems using the Fast Fourier Transform (FFT) may elucidate particular consequences concerning efficiency, harmonic distortion, and transient behavior. Comparison research using FFT offers significant insights into the efficacy of several EV charging schemes. This information may optimize designs for enhanced efficiency, less harmonic distortion, and superior management of transient behavior, resulting in a more resilient and effective EV charging infrastructure. In summary, comparative FFT-based performance analysis serves as an essential instrument for assessing and enhancing EV charging systems. This paper attempts to covers significant processes, which includes specific evaluation of subharmonic components, the presence of total harmonic distortion (THD), and the dc component for quality, reliability, and efficiency across different frequencies. The new research is further advanced by considering parameters like charging mode, chemistry, converter design, and types of batteries used. Ultimately, the application of comparative spectrum assessment aims in the design and implementation of EV battery systems, ensuring their efficiency, reliability, and compatibility with both current and future grid environments.

Table 4.1 Inter-comparative studies between various proposed charging topologies [13]

Source used	Advantages	Disadvantages	FFT Analysis
Wind turbine	1. Renewable energy source 2. Reduces reliance on fossil fuels	1. Variable power output depending on wind speed 2. Requires additional infrastructure	Analyze dominant frequencies to understand wind fluctuations and optimize charging based on wind patterns.
Solar PV array	1. Renewable energy source Environmentally friendly 2. Can be decentralized for home charging	1. Intermittent power output depending on sunlight 2. Lower power output compared to wind turbines	Analyze cyclical frequencies to predict power generation and optimize charging during peak sunlight hours.
Conventional Grid	1. Consistent and reliable power source 2. Widely available infrastructure	1. Relies on fossil fuels (may not be sustainable) 2. Potential for power outages	Analyze for presence of harmonics and potential grid stability issues during high EV charging loads.

Source: Author

References

[1] Pandžić, H., Franc, B., Stipetić, S., Pandžić, F., Mesar, M., Miletić, M., et al. (2023). Electric vehicle charging infrastructure in croatia – first-hand experiences and recommendations for future development. *Journal of Energy - Energija*, 71(3), 16–23. http://doi.org/10.37798/2022713414.

[2] Kan, T., Nguyen, T., White, J., Malhan, R. K., & Mi, C. (2017). A new integration method for an electric vehicle wireless charging system using LCC compensation topology: analysis and design. *IEEE Transactions on Power Electronics*, 32, 1638–1650. http://doi.org/10.1109/TPEL.2016.2552060.

[3] Mehta, R., Srinivasan, D., Khambadkone, A., Yang, J., & Trivedi, A. (2018). Smart charging strategies for optimal integration of plug-in electric vehicles within existing distribution system infrastructure. *IEEE Transactions on Smart Grid*, 9, 299–312. http://doi.org/10.1109/TSG.2016.2550559.

[4] Das, T. K., Banik, A., Chattopadhyay, S., & Das, A. (2017). Sub-harmonics based string fault assessment in solar PV arrays. In Modelling and Simulation in Science, Technology and Engineering Mathematics – Proceedings of International Conference on Modeling and Simulation(MS-17), ISBN-9783319748078, Computer Science, ISSN-21945357, paper ID-132, 4th-5th November, 2017, Kolkata, India.

[5] Dutta, S., Debnath, D., & Chatterjee, K. (2018). A grid-connected single-phase transformerless inverter controlling two solar PV arrays operating under different atmospheric conditions. *IEEE Transactions on Industrial Electronics*, 65(1), 374–385.

[6] Sangwongwanich, A., Yang, Y., Sera, D., Blaabjerg, F., & Zhou, D. (2018). On the impacts of PV array sizing on the inverter reliability and lifetime. *IEEE Transactions on Industry Applications*, 54(4), 3656–3667.

[7] Rahimi, R., Farhangi, S., Farhangi, B., Moradi, G. R., Afshari, E., & Blaabjerg, F. (2018). H8 inverter to reduce leakage current in transformerless three-phase grid-connected photovoltaic systems. *IEEE Journal of Emerging and Selected Topics in Power Electronics*, 6(2), 910–918.

[8] Hu, K., Li, W., Wang, L., Zhu, F., & Shou, Z. (2018). Topology and control strategy of power optimization for photovoltaic arrays and inverters during partial shading. *IET Generation, Transmission and Distribution*, 12(1), 62–71.

[9] Iman-Eini, H., Bacha, S., & Frey, D. (2018). Improved control algorithm for grid-connected cascaded H-bridge photovoltaic inverters under asymmetric operating conditions. *IET Power Electronics*, 11(3), 407–415.

[10] Leonard, J., Hadidi, R., Fox, J. C., Salem, T., Gislason, B., & McKinney, M. H. (2018). Evaluating megawatt-scale smart solar inverters: a commissioned 2.5-MW dc supply for testing grid-tie inverters. *IEEE Industry Applications Magazine*, 24(5), 52–61.

[11] Kar Ray, D., Chattopadhyay, S., & Sengupta, S. (2020). Multi resolution analysis based line to ground fault detection in a VSC based HVDC system. *IETE Journal of Research*, DOI: 10.1080/03772063.2018.1502626. 66(4), 491–504.

[12] Cochran, W. T., Cooley, J. W., Favin, D. L., Helms, H. D., Kaenel, R. A., Lang, W. W., et al. (1967). What is the fast fourier transforms? *Proceedings of the IEEE*, 55(10), 1664–1674. DOI: 10.1109/PROC.1967.5957.

[13] Zheng, Jiantao, Yan, Junjie, Pei, Jie, Liu, Guanjie, Solar Tracking Error Analysis of Fresnel Reflector, *The Scientific World Journal*, 2014, 834392, 6 pages, 2014. https://doi.org/10.1155/2014/834392

[14] S. Thangamayan, M. D. Walunjkar, D. K. Ray, M. Venkatesan, A. Banik, and K. P. Amrutkar, "5G modulation technique comparisons using simulation approach," in 2022 3rd International Conference on Intelligent Engineering and Management (ICIEM), London, United Kingdom, 2022, pp. 848–856, doi: 10.1109/ICIEM54221.2022.9853137.

[15] S. B. B, H. Bhat, S. Poornima, R. Bharanidharan, M. Sridharan, and A. Banik, "Conventional protection of power transformers at distribution grid side using artificial neural network," in 2023 Second International Conference on Electronics and Renewable Systems (ICEARS), Tuticorin, India, 2023, pp. 901–906, doi: 10.1109/ICEARS56392.2023.10085631.

5 Real time detection and classification of dust and shading on solar panels using multi agent system and hybrid machine learning

Sudeep Samanta[1,a], Samrat Hazra[2,b], Sucharita Saha[2,c], Subarna Pal[2,d] and Sumangal Bhoumik[3,e]

[1]Assistant Professor, Department of EE, MCKV Institute of Engineering, Howrah, West Bengal, India

[2]B. Tech Student, Department of EE, MCKV Institute of Engineering, Howrah, West Bengal, India

[3]Assistant Professor, Department of EE, Abacus Institute of Engineering and Management, West Bengal, India

Abstract

Nowadays, solar energy is a viable alternative to traditional energy sources. Due to dust accumulation and shading on solar panels, its efficiency is reduced significantly. Hence, regular cleaning is a prime issue for maintaining optimal photovoltaic performance. Many times, manual cleaning can be challenging under this condition. This paper presents a novel multi agent system detecting and classifying the dust and shade level on solar panels. This system employs three distinct agents, each specializing in a particular task. Agent 1 detects the state of panels as healthy, dusty or shaded using K- Nearest Neighbor classification. Agent 2 classifies the dust level and agent 3 classifies the shading level using Deep Neural Network. By distributing tasks to across specialized agents, the system offers an efficient and scalable solution for real time monitoring and maintenance of solar panels. The experimental results on 5 KW solar panel demonstrate that the proposed system can accurately classify the condition of solar panels and offers an efficient solution for real time monitoring and maintenance of solar panels, thereby maximizing energy output and finally improving the operational longevity of the panels.

Keywords: Confusion matrix, data pre-processing, deep neural network, dust detection, K nearest neighbor, model training, multi agent system, solar panels

Introduction

Solar energy is increasingly recognized as an alternative to traditional energy sources due to its renewability, long term cost efficiency and low carbon emission [1, 2]. The solar panel efficiency can be significantly reduced by different environmental factors like dust accumulation, shading, which significantly reduces the amount of sunlight reaching the panels as well as panel output. Studies have demonstrated that very light dust accumulation can also reduce efficiency to 20%, while shading can lead to substantial power losses mainly in partially shaded condition [3]. For maintaining optimal performance, a periodic monitoring and maintenance is necessary. Traditionally, it involves regular manual inspections which can involve intensive labor, time consuming and cost. As solar plants expanding in India, there is an increasing need for automated solutions for continuous monitoring and assessing the health of solar panels.

Recent advancements in artificial intelligence and machine learning offer innovative solutions to these challenges. Many automated systems use data from sensors which measure voltage, current, light intensity, temperature and continuously monitor solar panel conditions and initiate maintenance actions if needed [4]. Undoubtedly multi agent system has proven to be a promising approach, as it distributes tasks among all specialized agent, thereby enhancing the scalability and efficiency of monitoring process. The proposed work uses multi agent control system for detecting and classifying dust and shading levels of solar panels. The system utilizes three agents: agent 1 detects whether a panel is healthy, dusty, or shaded using K-nearest neighbors (KNN) classification. If the panel is detected as dusty, agent 2 classifies the dust level using a deep neural network (DNN) model. Similarly, if the panel is detected as shaded, agent 3 classifies the shading intensity using another DNN. This modular approach allows for efficient and real-time classification, ensuring that solar panels are monitored and maintained with precision.

The rest of the paper is organized as follows: Section 2 represents literature review. Section 3 details

[a]sudeep0809@gmail.com, [b]samrathazra18092005@gmail.com, [c]rai.email2006@gmail.com, [d]subarnapal2838@gmail.com, [e]sumangalstar26@gmail.com

DOI: 10.1201/9781003663348-5

the methodology of the proposed multi-agent control system, including the design of each agent and their respective classification models. Section 4 presents the experimental setup and results, demonstrating the system's effectiveness in real-world scenarios. Finally, Section 5 concludes the paper.

Literature Review

Various techniques have been explored in the literature for detecting dust on solar panels. Some methods involve image processing-based analysis [5]. Additionally, artificial neural networks (ANNs) have been used to predict solar module power output. ANNs consist of an input layer, hidden layers, and an output layer, with neurons and their transfer functions playing a crucial role. Data collection involves monitoring dust and irradiance as inputs, and voltage and current as outputs, with training performed through back propagation [6]. Other approaches include the KNN classification algorithm for developing intrusion detection systems in wireless sensor networks. Another technique focuses on mitigating the impact of dust on solar panel efficiency by integrating a cleaning system. This system features three main components: a dust density sensor that sends SMS alerts, real-time output measurement using an Arduino Uno with current and voltage sensors, and an automated windshield wiper activated when output power drops below 50% of its rated value. A light dependent resistor (LDR) sensor distinguishes between day and night [7].

High-resolution images for measuring dust particle dimensions using computer vision are discussed [8]. Particle identification through Random Forest classification for detecting dusty panels is presented [9]. Deep learning models have also been employed for this purpose. For instance, Saquib et al. [10] developed an ANN model, with outputs recorded as voltage and current. Parameters such as irradiance, current, and voltage were measured using an LDR and a multimeter. The neural network had a single hidden layer with nine neurons, forecasting panel output voltage based on irradiance and dust levels. Similarly, Mehta et al. [11] proposed using CNNs for this classification task.

Detailed Methodology

This section describes the detailed architecture and functioning of the multi-agent control system for dust and shading level detection on solar panels. The system consists of three distinct agents and each agent assigned to a specific task. Agent 1 is responsible for classification of the panel's health status (healthy, dusty, shaded or both). Agent 2 triggered when the panel is classified as dusty, and it further classifies dust levels and agent 3 triggered when the panel is classified as shaded and it classifies shading intensity. Agents 2 and 3 are activated according to the output of agent 1. By distributing tasks across these agents, the system ensures efficient and accurate real-time monitoring. The data flow between these agents follows a conditional workflow, as shown in Figure 5.1.

Data pre-processing

For training of different agents, sufficient data for solar panel output is required from healthy, dusty and shady environment. Solar panel output voltage, LDR output voltage and output from temperature sensors are used as input to proposed system. Initially, the input data is normalized using Z score normalization following equation (1). The normalized dataset is used for training.

$$X_i^{Norm} = \frac{X_i - \mu}{\sigma} \qquad (1)$$

Where, X_i is i^{th} instance of dataset, σ and μ are the standard deviation and mean of the dataset, respectively.

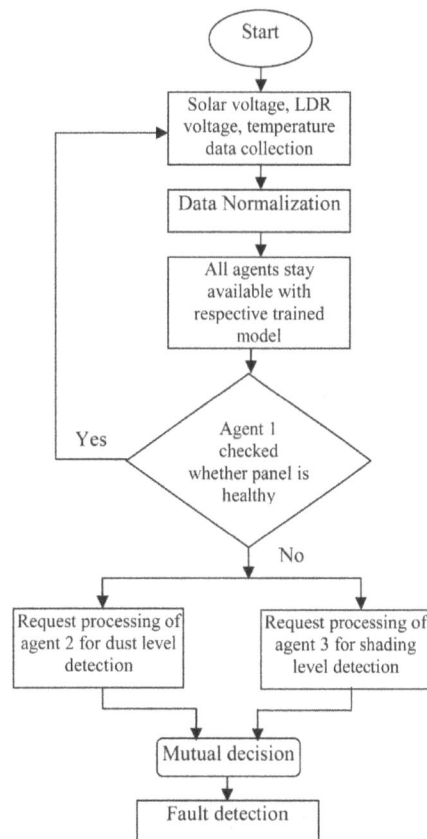

Figure 5.1 Flow chart of proposed methodology
Source: Author

Agent 1 operation

Agent 1 receives output data from a solar panel, LDR voltage sensor and temperature sensor. This data is preprocessed, and features are extracted to capture relevant patterns indicative of the panel's condition. The KNN algorithm is used by Agent 1 to classify the panel's condition into one of four categories: Healthy (*C1*), dusty (*C2*), shaded (*C3*) and both (*C4*). Initially, the KNN classifier is trained using labeled data collected under different states (healthy, dusty and shaded) with different climate conditions. When new data is fed into the system, Agent 1 compares it to the stored training examples and classifies the panel's condition based on the majority class among the k nearest neighbors. If the panel is "Healthy", no further action is required but if the panel is "Dusty", agent 2 is activated to classify the dust level and if "Shaded", agent 3 is activated to classify the shading intensity.

Agent 2 operation

Once Agent 1 determines that a panel is dusty, agent 2 is triggered to perform further analysis. The input to agent 2 is similar to that of agent 1 but focuses on features indicative of dust accumulation. Agent 2 utilizes DNN for classifying the dust levels. The DNN is designed to learn the exact hierarchical features from input data automatically and make the system well suited for more complex tasks like multi-level classification.

The DNN developed with one input layer consisting three input neurons, three hidden layers with 64, 32 and 16 neurons respectively, determined by Grid Search method, and one output layer with three neurons identifying three different dust levels (*D1*: Low, *D2*: Moderate, *D3*: High). The weights and bias of each hidden layer are initialized by Xavier optimization method. The hidden layer uses ReLU activation function, and the output layer consists of SoftMax activation function. The 10-fold cross validation is used to improve DNN performance. The DNN model is trained on the labeled dataset of panels with varying levels of dust accumulation. Training involves minimizing a categorical cross-entropy loss function.

The number of trainable parameters from input to hidden layer 1 are $(3 \times 64) + 64$ bias = 256 parameters. Again the number of trainable parameters from hidden layer 1 to hidden layer 2, hidden layer 2 to hidden layer 3 and hidden layer 3 to output layer are $(64 \times 32) + 32$ bias = 2080, $(32 \times 16) + 16$ bias = 528 and $(16 \times 3) + 3$ bias = 51 parameters. So, a total of 2916 numbers of weights and bias are being updated at each epoch through back propagation.

Agent 3 operation

If Agent 1 detects that the panel is shaded, Agent 3 is activated. Similar to agent 2, it receives normalized

sensor data and again uses DNN model to classify the shading intensity into five levels: 20% (*S1*), 40% (*S2*), 60% (*S3*), 80% (*S4*), and 100% (*S5*). The DNN architecture for Agent 3 contains similar input layer as agent 2, three hidden layers with 128, 64 and 28 neurons respectively. The output layers have 5 neurons which represent shading level. The DNN for Agent 3 is trained on labeled data consisting of examples where panels experience various degrees of shading.

Total number of trainable parameters are $\{(3 \times 128) + 128$ bias$\} + \{(128 \times 64) + 64$ bias$\} + \{(64 \times 28) + 28$ bias$\} + \{(28 \times 5) + 5$ bias$\} = 10733$ parameters are being updated at every epoch through back propagation. The training process follows the same optimization steps as Agent 2, using 0.2 dropout rate for regularization to prevent over fitting.

System integration and decision making

The agents work together in a sequential and conditional manner. Initially, agent 1 performs an initial health check of the panel and classifies it as healthy, dusty, or shaded. If dust is detected, Agent 2 is activated to further classify the dust level, and if shading is detected, Agent 3 is activated to classify the shading intensity. The outputs from Agents 2 and 3 are used to trigger specific maintenance actions, such as cleaning or repositioning the panels. By utilizing the Multi Agent approach, the system ensures that each task is completed by a specific agent, optimizing both resource utilization and performance. This also allows for adoptability and easy scalability in different operational environments.

Experimentation with Real System

Dataset preparation

For experimental verification, the dataset of solar voltage, LDR voltage and temperature was prepared from a 5 KW solar plant located on the rooftop of MCKV Institute of Engineering, West Bengal, India, as shown in Figure 5.2. Data have been collected over the entire year from 1st June, 2023 to 31st May, 2024 and saved in .csv format and next applied for z score normalization, as in equation (1). Total 2000 numbers of data collected throughout the year with different dust and shading level at various climate conditions. So, 2000 × 3 dimension data is prepared for solar panel condition monitoring, among which 400 data for healthy condition, 600 data with dusty level, 600 data with shading and other 400 data for both shading and dusty condition.

Results and discussion

Five different classification methods were utilized to gather data for defining the optimal behavior of

Figure 5.2 5 KW rooftop solar plant
Source: Author

each agent tasked with classification in the proposed system. All methods were tested through ten-fold cross-validation method. Based on obtained result, it was observed that agent 1 performance is best for KNN and other two agent performance are best in DNN classifier. In agent 1, from 2000 data, 80% data were used for training and 20% data for testing KNN classifier. The value of K was considered as 3 and Euclidian distance is used for search algorithm. From the confusion matrix shown in Figure 5.3, it can be observed overall accuracy of agent 1 becomes 98.5%.

Next, for agent 2, among 1000 data, 800 data (80% of 1000) used to train, and rest was used to test the DNN model. During training, mini batches of size 50 were used. The number of epochs was defined as 50. The confusion matrix for agent 2 is shown in Figure 5.4.

For agent 3, also from 1000 data, 80% data was used for training and 20% data used for testing the DNN model with mini batch size 50 and total 100 number of epochs have been used. For both agent 2

and 3, Xavier initialization is used to initialize the weight and bias.

For acceleration of training process, a momentum hyper parameter of 0.9 is used to update the weight and bias. An initial learning rate of 0.01 is selected for ensuring proper convergence of gradient descent algorithm. The confusion matrix for agent 3 is shown in Figure 5.5.

The total model is trained in Python using Keras with Tensor Flow as the backend in a system with 8GB RAM and a clock speed of 2.70 GHz. The time taken for training of agent 2 and 3 are 8.16 sec and 10.54 sec respectively.

The overall result obtained from three different classifiers using real solar plant data samples are analyzed by five standard performance metrics such as accuracy, sensitivity, specificity, precision, F_ measure, as given in Table 5.1.

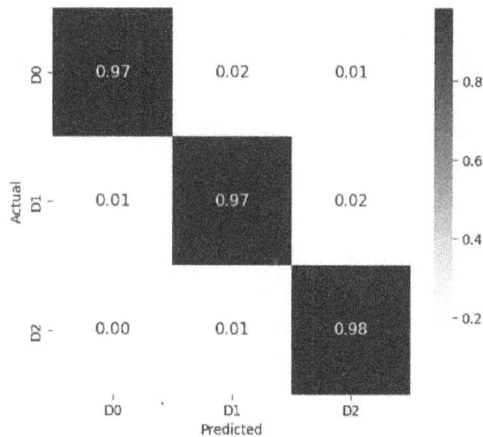

Figure 5.4 Confusion matrix for agent 2
Source: Author

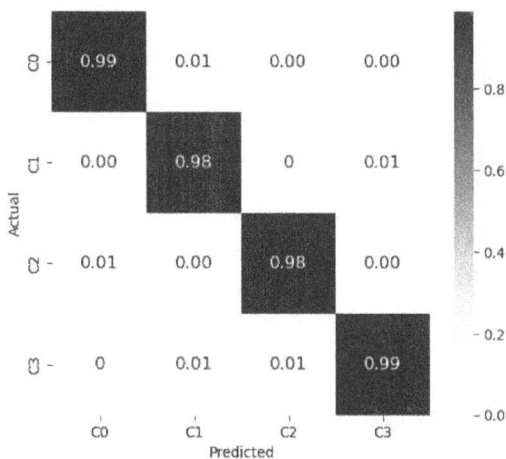

Figure 5.3 Confusion matrix for agent 1
Source: Author

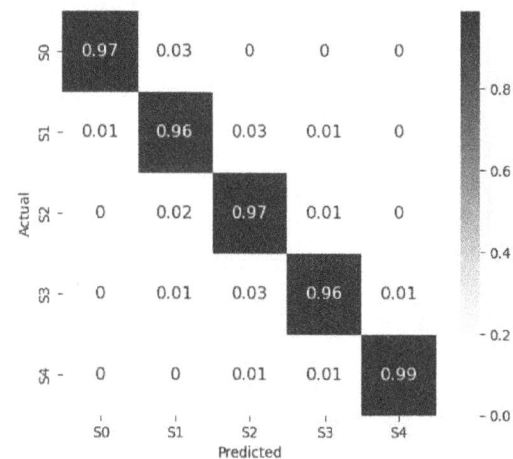

Figure 5.5 Confusion matrix for agent 3
Source: Author

Table 5.1 Performance matrices for different classifier

Agent	Class	Accuracy	Sensitivity	Specificity	Precision	F_measure
Agent 1	C1	98.5%	99%	100%	98%	98%
	C2	98.5%	98%	100%	99%	99%
	C3	98.5%	98%	100%	99%	99%
	C4	98.5%	99%	99%	97%	98%
Agent 2	D1	97.33%	97%	99%	98%	98%
	D2	97.33%	97%	98%	97%	97%
	D3	97.33%	98%	99%	97%	98%
Agent 3	S1	97%	97%	100%	99%	98%
	S2	97%	96%	99%	95%	95%
	S3	97%	97%	99%	95%	96%
	S4	97%	96%	99%	97%	97%
	S5	97%	99%	100%	99%	99%

Source: Table generated from develop DNN model by author

Conclusion

This work developed a Multi Agent control system for accurate detection and classification of dust and Shading level on solar panels. The total system is developed to increase the efficiency and longevity of Solar PV systems by adopting real-time monitoring and automated maintenance alerts. The proposed system offers a modular and scalable solution for detecting dust accumulation and shading by distribution relevant tasks among all specialized agents and significantly improves plant efficiency. The experimental results demonstrate that the proposed multi-agent system is highly effective in classifying the condition of solar panels with a high degree of accuracy. This enables more targeted and timely maintenance interventions, ultimately improving the overall energy output and reducing operational costs.

References

[1] Cipriani, G. (2020). Convolutional neural network for dust and hotspot classification in PV modules. *Energies*, 13(23), 342–351.

[2] Elminir, H. K. (2006). Effect of dust on the transparent cover of solar collectors. *Energy Conversion and Management*, 47(18–19), 3192–3203.

[3] Lee, J. (2021). Machine learning based algorithms for global dust aerosol detection from satellite images: inter comparisons and evaluation. *Remote Sensing*, 13(3), 1–24.

[4] Al-dahoud, A., Fezari, M., & Aldahoud, A. L. I. (2023). Machine learning in renewable energy application: intelligence system for solar panel cleaning. *WSEAS Transactions on Environment and Development*, 19, 472–478.

[5] Onim, M. S. H. (2023). SolNet: a convolutional neural network for detecting dust on solar panels. *Energies*, 16(1), 1–19.

[6] Saquib, D., Nasser, M. N., & Ramaswamy, S. (2020). Image processing based dust detection and prediction of power using ANN in PV systems. In Proceedings of the 3rd International Conference on Smart Systems and Inventive Technology. ICSSIT 2020, (pp. 1286–1292).

[7] Li, W. (2014). A new intrusion detection system based on KNN classification algorithm in wireless sensor network. *Journal of Electrical and Computer Engineering*, 1, 536–544.

[8] Igathinathane, C., Melin, S., & Sokhansanj, S. (2009). Machine visions based particle size and size distribution determination of airborne dust particles of wood and bark pellets. *Powder Technology*, 196, 202–212.

[9] Maitre, J., Bouchard, K., & Bedard, L. P. (2019). Mineral grains recognition using computer vision and machine learning. *Computers and Geosciences*, 130, 84–93.

[10] Saquib, D., Nasser, M. N., & Ramaswamy, S. (2020). Image processing based dust detection and prediction of power using ANN in PV systems. In Proceedings of the 2020 Third International Conference on Smart Systems and Inventive Technology (ICSSIT). Tirunelveli. India. 20–22 August 2020, (pp. 1286–1292).

[11] Mehta, S., Azad, A. P., Chemmengath, S. A., Raykar, V., & Kalyanaraman, S. (2018). Power loss prediction and weakly supervised soiling localization via fully convolutional networks for solar panels. In Proceedings of the 2018 IEEE Winter Conference on Applications of Computer Vision (WACV), Lake Tahoe, NV, USA, 12–15 March 2018 (pp. 333–342).

6 A stochastic approach for optimal scheduling in mixed generation

Reshmi Chandra[1,a], Kamalika Banerjee[2,b], Pallav Dutta[2,c], Suman Moitra[3,d] and Bishaljit Paul[2,e]

[1]Assistant Professor, Abacus Institute of Engineering and Management, West Bengal, India

[2]Assistant Professor, Narula Institute of Technology, West Bengal, India

[3]Assistant Professor, Elitte College of Engineering, West Bengal, India

Abstract

The permeation of renewable units with minimum volume of conventional sources has improved immensely over past ten years. The prime reason of this scenario is due to significant rise of interest of energy sources with renewable features for the need with environmental sustainability in addition to balance the decreasing availability of fossil fuels. The assembly of the renewable units are indeterminate and flexible which is governed by meteorological phenomena. Their contribution to make the most of the profit of the owners, necessitates the optimal scheduling of the plant. Furthermore, a potential elucidation to eliminate the harmful impact of the uncertain performance of the assembly reserve market having volume and energy prices are amalgamated in this research work. This indication is to analyse a prototypical model for a decision making observation through mixed integer linear programming optimization. The stochastic approach regulates the optimum result beneath a distinct future comprehension of the uncertain behaviour. Here the optimization variables have their discrete values. The energy market with day-ahead forecasting is considered here. In this paper it has been revealed through a stochastic scheduling problem with a three-hour planning horizon for the explicit volume of power to be merchandized in the market. Here, the energy charges are measured for the up and down reserve with their corresponding confidence bounds, along with the technical and monetary data of the plants. This optimization mathematics is answered through an algorithm of branch and bound in MATLAB version 2015a. The uniqueness is concentrated through the design of up and down reserves and distribution requests which emphases on here-and-now decisions. This paper lastly concludes with practice of flexibility as reserve markets, the whole power production is organized and merchandized to produce substantial profit.

Keywords: Decision maker, here-and-now decisions, uncertain components

Introduction

The virtual power plants (VPPs) have their own expansion planning and operation decisions that works under uncertain situation. The uncertainties intricate in decision- management complications of VPPs, are specifically addressed with stochastic programming along with robust optimization. The robust optimization and stochastic programming are two substitute optimization models typically reflected very effective with the purpose of supervision of the uncertainties in the decision-management complications. The stochastic programming approach which is scenario-based, implies a determinate quantity of scenarios signifying dissimilar comprehensions of the uncertain constraints. Furthermore, every scenario has mathematically concomitant a probability of occurrence. The mathematical modelling handles the interpretation about the mathematical constrictions of the problematic scenario for every consequence apprehension. After that, the programming methodology with stochastic tactic regulates the variables with decisions which minimalize the estimated price or make the most of the probable turnover respectively for all the circumstances. Primarily, the two important problems being raised whenever to objectify the precise scenarios. The first problem is that information is required about uncertain parameters related to probability distribution functions. The second problem is that the problem practicability is only addressed for specifically taken scenarios and a bulk set of scenarios are required to examine. The robust optimization is the substitute method for stochastic methodology. In that case, certain preselected scenarios are not applied, rather some set of mathematical uncertainty constructed on confidence bounds, are applied. The poorest case for the objective function is optimized

[a]reshmichandra.2024@gmail.com, [b]kamalikabanerjee09@gmail.com, [c]pallav.dutta@nit.ac.in, [d]sumu2009@gmail.com, [e]bishaljit.paul@nit.ac.in

DOI: 10.1201/9781003663348-6

with robust optimization methodology where the parameters of uncertainty mathematics may have any data within confidence bounds of uncertainty set. The practical implementation ability of the mathematical approaches are assured for all considerations of the uncertain data within these set of uncertainties.

Literature Review

Uzunoglu et al. have projected the revolutionary disciplined tactic connected to load dispatching further forecasting structure in a computed power system. An extraordinary extension in conclusion was achieved linked to the temperature, speed and solar radiation of the wind from their research innovations. The union of the virtual power plant carries optimum monetary data directed at energy demand dealings. Zhang et al. predicted a dissimilar workable bidding awareness in their research inferences to squeeze with Virtual power plant scheme machinery and acclimatization of renewable energy. A self-contradictory virtual power plant machinery, flexible demand situation and renewed market planning might lead to moderate energy assessing in power marketplace [1–4].

Merlin and Sandrin focused their research exertion with a radical Lagrangian relaxation manner to adopt the straight unit commitment development. The heuristic strategy approach licenses suitable path of pumping storage points. The probabilistic forecasting route was functional for undertaking the spinning reserve difficulty in their investigation effort. The chief deficiency in their sightings annoyed due to quicker convergence and computational intermission in the optimization for the investigation system [5]. Zhai et al. suggested in their examination work, a dynamic programming direction with constructive inequality. They did appraise optimum power making size, deficient generation discretization. A brilliant style for dissimilar commitment statuses with the backing of Lagrangian relaxation procedure was well-arranged in their pioneering purpose [6]. Osório et al. navigated a research exertion for the assessment of unit commitment development linked with renewable sources amalgamation. The wind energy creation and its self-correlated fitting property, meeting error assessment were planned in the research struggle of them. The scientific style with prophetic UC problem was deliberate with probabilistic survey system [7]. Jun et al. approved a dissimilar exercise in their scrutiny findings intersected with control and bidding structure in day-ahead real time marketplace collected with energy storage summary in totalling to demand response. A prominent progress in computational precision was formulated. The indication of bi-level multi agent model was straight in the practicable development of

power marketplace [8]. Jadid et al. agreed an unlike probabilistic stylishness allied to scheduling of energy inevitability and spinning reserve in market hesitancy. An active methodology to pact with governing penalty incited due to insecurity and policy for cheap bidding was presented in their research inquiry [9]. Shen et al. did commendation a hi-tech bi-level stochastic optimization pointed at falling VPP system charge technology in congruent with power marketplace. The local search algorithm was realistic to pursuit supply side and demand side approximating of load observation in market hesitancies. The survey ensuing with wide-ranging conclusions to complete with scope accumulation of elastic demand and fudging of the surplus assessing [10]. Paul et al. carried out several research effort to scrutinize the bottleneck development for the stream of power in the transmission grid which outcomes in price surge in the power system. Here is positive restriction in the transmission proficiencies that prerequisites congestion regulation else instigating bottleneck setting in dissimilar buses in organized power system. The energy marketplace turn out to be identical low-priced due to presence of buyers and sellers in enthusiastic state. The power stability inequalities and unhinged load demand produces the system convoluted to analyses [11–16].

Case Study

This circumstance is a scenario based stochastic programming model. In this scientific commentary a two stage programming model with stochastic tactic has been considered. It can also be applied to Multi stage stochastic model if it is to be extended further in future. The parameters which are seems to be uncertain are modelled using a number of scenarios. These uncertainties in the market are presented by a number of scenarios as ω and the set of scenarios are as given by Π. The probability of occurrence of these scenarios are given by P_ω. Here ω^I and ω^{II} are the classification of optimisation variables depending on the decision variables of first and second stage.

There is a power facilitator of a conventionally operated power plant which has a capability of 50 MW and a inconstant price of \$18/MWh. In this first stage of two stage scheduling problem, the power facilitator sets that the electricity magnitude x^I which may be commercialised has a upper bound power limit with 30 MW and a static price of \$18/MWh. Similarly in the second phase the control facilitator produce the commercialised power quantity x^{II} with an uncertain price using conventional power plant. These uncertainties are represented by set Π and the corresponding probability of occurrences as P.

Methodology and Model Specifications

The set of scenarios are as given by
$\Pi = \{ 15.0, 20.0, 8.0, 12.0, 21.0 \}$ \$/MWh.

The probability of incidence of these circumstances are
$P = \{0.20, 0.40, 0.10, 0.10, 0.20\}$

Objective function for two stage stochastic problem is given by

$$\max_{\emptyset_{I,II}} \qquad f^I(\emptyset)^I$$
$$E_\pi[f^{II}_\omega(\omega^{II}_\omega) \qquad (1)$$

Such that

$$h^I_k(\emptyset)^I = 0,$$

$$g^I_i(\emptyset)^I \le 0$$

$$h^{II}_{k,\omega}(\emptyset^I, \emptyset^{II}_\omega) = 0$$

$$g^{II}_{i,\omega}(\emptyset^I, \emptyset^{II}_\omega) \le 0, \forall i \in \Omega^{I,II}_\omega$$

$$\emptyset^{II}_\omega \in R^M$$

Here the objective function comprises with of two fragments, first fragment is revenue and second part is expected profit. It is given by

$$Z = 18x^I + 0.2(15x^{II}_{11} - 18x^{II}_{21}) + 0.4(20x^{II}_{12} - 18x^{II}_{22}) + 0.1(8x^{II}_{13} - 18x^{II}_{23}) + 0.1(12x^{II}_{14} - 18x^{II}_{24}) + 0.2(21x^{II}_{15} - 18x^{II}_{25}) \qquad (2)$$

The constraints are gives by:
$0 \le x^I \le 30$, this is inequality constraints for first stage

$$x^I + x^{II}_{11} - x^{II}_{21} = 0$$
$$x^I + x^{II}_{12} - x^{II}_{22} = 0$$
$$x^I + x^{II}_{13} - x^{II}_{23} = 0$$
$$x^I + x^{II}_{14} - x^{II}_{24} = 0$$
$$x^I + x^{II}_{15} - x^{II}_{25} = 0,$$

these are equality constraints for second stage
$$0 \le x^{II}_{21} \le 50$$
$$0 \le x^{II}_{22} \le 50$$
$$0 \le x^{II}_{23} \le 50$$
$$0 \le x^{II}_{24} \le 50$$
$$0 \le x^{II}_{25} \le 50,$$

these are inequality constraints for second stage

Results

The two mathematical sets given by:
$$\phi^I = \{x^I\}$$
$$\phi^{II} = \{x^{II}_{11}, x^{II}_{12}, x^{II}_{13}, x^{II}_{14}, x^{II}_{15}, x^{II}_{21}, x^{II}_{22}, x^{II}_{23}, x^{II}_{24}, x^{II}_{25}\}$$

are considered here as the variables originated from optimization in the initial stage problem also with the subsequent stage problem correspondingly. The scenario realization addresses the second subscript for the set of the variables.

The profit associated with the first stage is \$540 and for the subsequent stage is negative (–\$446). The summation of the profit from the initial stage and the subsequent stage is \$94. This \$94 is the finest assessment of the objective function calculated with software version MATLAB 2015a.

Table 6.1 Optimal solutions [MW]

x^I	x^{II}_{11}	x^{II}_{12}	x^{II}_{13}	x^{II}_{14}	x^{II}_{15}	x^{II}_{21}	x^{II}_{22}	x^{II}_{23}	x^{II}_{24}	x^{II}_{25}
30	-30	20	-30	-30	20	0	50	0	0	50

Source: Estimated arbitrary values are taken so that result converges

x^I = Revenue obtained for the First-stage in Day-Ahead Electricity Market.

x^{II}_{11} = The expected profit obtained for the second stage correspond to first conventional generator with first scenario realization in Real Time Electricity Market.

In this paper it has been revealed through a stochastic scheduling problem with a three-hour planning horizon for the explicit volume of power to be merchandized in the market. Here, the energy charges are measured for the up and down reserve with their

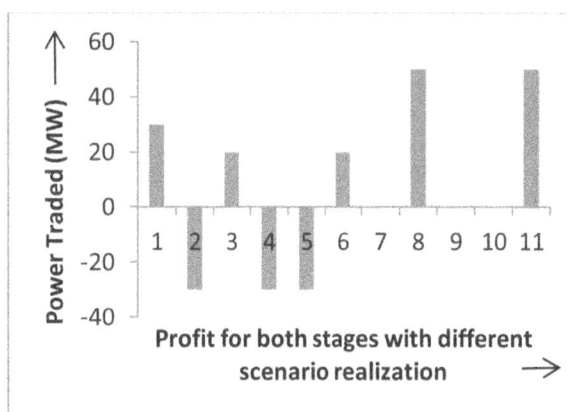

Figure 6.1 Optimal power transacted in day-ahead forecasting and real-time electricity market with conventional generators and different scenario realizations
Source: This is our result analysis from Matlab simulation

corresponding confidence bounds, along with the methodological and monetary records of the plants.

Conclusion

This optimal solution is presented in this research work. The mathematical model with stochastic approach intends to optimize the estimated value. The problem connected along with uncertain data is neglected for the subsequent stage. In actual statement, the objective function dealing with uncertainty phenomena is an arbitrary variable. Consequently, the result is dissimilar from the estimated value. The inconsistency for the objective function may possibly consequence in numbers lesser comparing the estimated numbers with greater probability. A risk-indifferent decision-strategist is neutral to such type of risks. The decision-maker merely improves the estimated value in view of mathematical modelling.

References

[1] Tascikaraoglu, A., Erdinc, O., Uzunoglu, M., & Karakas, A. (2014). An adaptive load dispatching and forecasting strategy for a virtual power plant including renewable energy conversion units, *Applied Energy*, 119, 445–453. http://dx.doi.org/10.1016/j.apenergy.2014.01.020.

[2] Zhang, G., Jiang, C., & Wang, X. (2019). Comprehensive review on structure and operation of virtual power plant in electrical system. *IET Generation, Transmission and Distribution*, 13(2), 145–156. The Institution of Engineering and Technology 2018, doi: 10.1049/iet-gtd.2018.5880.

[3] Conejo, A. J., & Baringo, L. (2018). Power Electronics and Power Systems. Springer International Publishing AG. https://doi.org/10.1007/978-3-319-69407-8.

[4] Baringo, L., & Rahimiyan, M. (2020). Virtual Power Plants and Electricity Markets, Decision Making Under Uncertainty. Switzerland AG: Springer Nature. https://doi.org/10.1007/978-3-030-47602-1.

[5] Merlin, A., & Sandrin, P. (1983). A new method for unit commitment at electricite de France. *IEEE Transactions on Power Apparatus and Systems*, PAS-102(5), 1218–1225. 0018-9510/83/0500-1218$01.00 © 1983 IEEE.

[6] Fan, W., Guan, X., & Zhai, Q. (2002). A new method for unit commitment with ramping constraints. *Electric Power Systems Research*, 62, 215–224.

[7] Osório, G. J., Lujano-Rojas, J. M., Matias, J. C. O., & Catalão, J. P. S. (2015). A new scenario generation-based method to solve the unit commitment problem with high penetration of renewable energies. *Electrical Power and Energy Systems*, 64, 1063–1072. http://dx.doi.org/10.1016/j.ijepes.2014.09.010.

[8] Tang, W. J., & Yang, H. T. (2019). Optimal operation and bidding strategy of a virtual power plant integrated with energy storage systems and elasticity demand response. *IEEE Access*, 7, 2169–3536. DOI 10.1109/ACCESS.2019.2922700.

[9] Zamani, A. G., Zakariazadeh, A., & Jadid, S. (2016). Day-ahead resource scheduling of a renewable energy based virtual power plant. *Applied Energy*, 169, 324–340. http://dx.doi.org/10.1016/j.apenergy.2016.02.011.

[10] Zhao, Q., Shen, Y., & Li, M. (2016). Control and bidding strategy for virtual power plants with renewable generation and inelastic demand in electricity markets. *IEEE Transactions on Sustainable Energy*, IEEE, 7(2), 1949–3029. Digital Object Identifier 10.1109/TSTE.2015.2504561.

[11] Chanda, C. K., Pal, J., Paul, B., & Pathak, M. K. (2017). On transmission congestion management strategies and forecasting locational marginal prices in a deregulated competitive power market. In IEEE 2017 Australasian Universities Power Engineering Conference (AUPEC) - Melbourne, VIC (2017.11.19-2017.11.22), DOI: 10.1109/AUPEC.2017.8282418.

[12] Paul, B., Chanda, C. K., Pal, J., & Pathak, M. K. (2020). Congested power transmission system in a deregulated power market, computational advancement in communication circuits and systems. *Lecture Notes in Electrical Engineering*, 575. Springer Nature Singapore Pte Ltd., 3–14. https://doi.org/10.1007/978-981-13-8687-9_1

[13] Paul, B., Chanda, C. K., Pathak, M. K., & Pal, J. (2018). Effective scheduling of spinning reserve services and cost of energy in a deregulated power market. In IEEE 2018, 2nd International Conference on Energy, Power and Environment: Towards Smart Technology (ICEPE) - Shillong, India (2018.6.1-2018.6.2). DOI: 10.1109/EPETSG.2018.8658497.

[14] Paul, B., Pathak, M. K., Chanda, C. K., & Pal, J. (2017). Pricing of energy in a deregulated power market. In IEEE 2017, International Conference on Computer, Electrical & Communication Engineering (ICCECE) - Kolkata, India (2017.12.22-2017.12.23), DOI: 10.1109/ICCECE.2017.8526200.

[15] Paul, B., Pathak, M. K., Pal, J., & Chanda, C. K. (2017). A comparison of locational marginal prices and locational load shedding marginal prices in a deregulated competitive power market. In IEEE 2017, Calcutta Conference (CALCON) - Kolkata (2017.12.2-2017.12.3), DOI: 10.1109/CALCON.2017.8280693.

[16] Moitra, S., Saha, P., Paul, B., & Chanda, C. K. (2022). Evaluation of azimuth angle profile for solar photovoltaic system in humid subtropical climate of Varanasi City. *Lecture Notes in Electrical Engineering*, 786, 233–242. Online ISBN: 978-981-16-4035-3, Springer, Singapore, DOI: https://doi.org/10.1007/978-981-16-4035-3_21.

7 Development of accurate fault diagnosis system for electric vehicle pertained lithium-ion batteries using deep neural network

Sudeep Samanta[1], Titas Bhaumik[2,a], Sumangal Bhaumik[2,b] and Arpita Sarkar[3]

[1]Assistant Professor, MCKV Institute of Engineering, Howrah, West Bengal, India

[2]Assistant Professor, Abacus Institute of Engineering and Management, West Bengal, India

[3]B. Tech Student, MCKV Institute of Engineering, Howrah, West Bengal, India

Abstract

This paper introduces a novel fault diagnosis and prognosis method for lithium-ion batteries applicable for electric vehicles using back propagation deep neural network. The frequent occurrence of electric vehicle fires highlights the safety risks associated with batteries. When a battery malfunctions, its stability is compromised, leading to a rapid decline in safety performance and posing significant threats to electric vehicles. Consequently, accurate fault diagnosis and prognosis of batteries are crucial for ensuring the safe and reliable operation of electric vehicles. Lithium-ion battery packs for electric vehicles present complex fault types, making their management challenging. In this work, four different faults like thermal runway, battery life degradation, faults occurring from either over voltage condition or under voltage condition etc. Initially, a hardware set up of three lithium-ion battery pack was developed through which fault data were collected using voltage and current sensing unit. By supervised learning, a deep neural network model with multi layered perceptron is trained using back propagation to enable the system for effective classification of lithium-ion battery faults at very early stage. The validation result of trained model demonstrates that the proposed technique could effectively identify fault information in lithium-ion battery packs, achieving a diagnosis accuracy of approximately 99%.

Keywords: Back propagation, deep neural network, electric vehicles, fault classification, lithium-ion batteries

Introduction

Electric vehicles (EVs) utilizing lithium-ion batteries are picking up worldwide consideration due to their effectiveness, maintainability, and capacity to address vitality emergencies and natural contamination. These batteries offer tall vitality thickness, control thickness, cell voltage, long cycle life, and lightweight characteristics, making them basic for EVs and other vitality capacity frameworks. In any case, fabricating absconds, abuse, maturing, and auxiliary asymmetries can cause flaws in person cells, posturing noteworthy security dangers, such as warm runaway, which imperils vehicle operation and driver security. Voltage peculiarities are key pointers of such flaws. Profound neural systems (DNNs), a subset of counterfeit neural systems (ANNs), exceed expectations at learning complex information representations. This paper presents a novel strategy to extricate key highlights from a dataset created by a lithium-ion battery equipment show, which are at that point normalized and utilized as inputs for a DNN prepared by means of backpropagation with cross-validation folds. The paper outlines this approach, its methodology, results, and conclusions [1–6].

Literature Review

Current investigation centers on condition observing of battery packs in EVs, which can be categorized into model-based, threshold-based, and data-driven approaches [7]. Neural systems (NNs) are especially successful in foreseeing different information designs, making them profitable for tending to building challenges, particularly in vehicle designing [8]. Machine learning strategies have been utilized to anticipate remaining valuable life and analyze issues in batteries [9]. Long short-term memory (LSTM) neural systems have been utilized for battery blame determination [10]. A cross-breed approach combining convolutional neural systems (CNN), LSTM, and molecule swarm optimization (PSO) calculations has been proposed to demonstrate the nonlinear characteristics of lithium-ion batteries precisely [11]. Whereas LSTM

[a]titasbhaumik@gmail.com, [b]sumangalstar26@gmail.com

DOI: 10.1201/9781003663348-7

systems appear guarantee in blame conclusion, they regularly confront issues with complex structures and low forecast effectiveness. A wavelet-neural organized approach has been utilized to recognize deficiencies in lithium-ion batteries, emphasizing the critical effect of voltage contrasts on symptomatic precision [12]. The wavelet-Markov stack investigation strategy has been connected to anticipating the control state of lithium-ion batteries [13], whereas back engendering (BP) NN has been utilized to analyze battery conditions based on nine parameters [14]. Bolster vector machine (SVM) strategies have too been utilized for battery distinguishing proof [15]. In spite of the developing notoriety of EVs and the significance of lithium-ion battery packs, there remains a requirement for more broad investigate on testing and diagnosing these battery packs.

Mathematical Modelling of Li-ion Battery

Lithium-ion battery modeling can be done through electrochemical or comparable circuit strategies. This paper employments the comparable circuit approach, as appeared in Figure 7.1, for battery modeling [16]. In this model, V_{oc} stands for OC voltage, R_s denotes the resistance in series, whereas the diffusion resistance & diffusion capacitance are denoted by R_d & C_d respectively. Also, $V_b(t)$ is the battery's output voltage, $i_b(t)$ is the current flowing through the circuit. V_c is the voltage across the diffusion capacitor.

The equation of lithium-ion battery output voltage at terminal is represented by the following equation 1.

$$V_b = V_{oc}(SoC) - i_b * R_s - V_c \qquad (1)$$

When the system is in steady state condition, the equation 1 becomes,

$$V_b = V_{oc}(SoC) - i_b * (R_s + R_d) \qquad (2)$$

State of charge (SoC) may be a key highlight for blame classification. A common strategy for calculating SoC is coulomb checking, which coordinates the current streaming into and out of the battery over time. This approach permits SoC to be decided employing a particular condition.

$$SoC_{new} = SoC_{old} + \frac{\Delta Q}{Ah} \qquad (3)$$

Where, $\Delta Q = I \times \Delta t$, Δt is the time interval during which current is measured. Ah is the nominal battery capacity. The state of health (SoH) of a lithium-ion (Li-ion) battery reflects its overall condition and performance relative to its original state. While SoC provides information about how much energy is left in the battery, SoH indicates how much capacity the battery has lost over time due to aging and degradation. SoH can be calculated as,

$$SoH (\%) = \frac{current\ Ah}{Nominal\ Ah} x\ 100\% \qquad (4)$$

Deep Neural Network

The DNN is trained using a supervised learning algorithm, and it involves multiple layers of neurons. The mathematical modeling includes the forward propagation to compute the output and the backward propagation to update the weights and biases. The typical architecture of a DNN is shown in Figure 7.2. Let us consider, in DNN architecture, *n and m* denotes no. of neurons present in the input and output layers. *L* is the total number of hidden layers in the network.

Forward propagation: For each hidden layer *l*, the inputs are weighted and biased, then passed through an activation function *(ReLU)* to produce the output of that layer. o^l denotes the activated output of layer *l*, which is forward propagated as input to the next layer. Hence, o^{l-1} is the input to layer *l*. Then output of layer *l* can be represented as,

$$z^l = W^l o^{l-1} + b^l \qquad (5)$$

Where, W^l is the weight matrix for layer *l*, o^{l-1} is the activation from the previous layer, b^l is the bias vector for layer *l*.

z^l is further processed by ReLU activation function Ɛ, to enhance the feature learning ability. So,

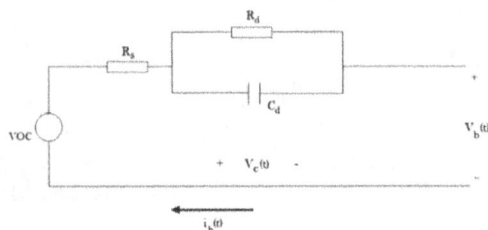

Figure 7.1 Mathematical model of Li ion battery
Source: Author

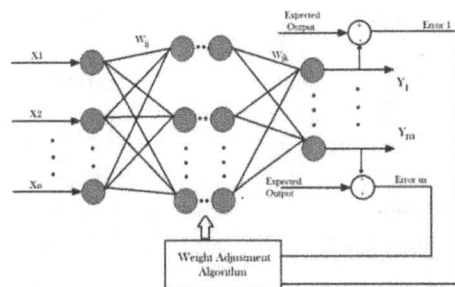

Figure 7.2 Typical architecture of DNN
Source: Author

$$o^l = \mathcal{E}^l(z^l) \qquad (6)$$

Where, $\mathcal{E}^l$ is the Relu activation function applied element wise. If there are j number of neurons in layer l, then

$$\mathcal{E}^l(z_j{}^l) = z_j{}^l \ , \text{for } z_j{}^l \geq 0$$
$$= 0 \ , \text{for } z_i{}^l < 0 \qquad (7)$$

The activated output of layer l is fed to the next hidden layer as input. The final output is calculated using similar approach. The output layer represented as (L+1). In output layer SoftMax activation function is used to predict the probabilities of various categorical classes. It can be denoted as *f(.)* and represented by the following equation,

$$\hat{Y}_j = f(z_j{}^{L+1}) = \frac{e^{z_j{}^{L+1}}}{\sum_{j=1}^m z_j{}^{L+1}} \qquad (8)$$

To quantify the error between the predicted output y and the actual target y, the cross entropy loss function L can be represented as,

$$L(y, \hat{y}) = -\sum_{j=1}^m y_j \ln(\hat{y}_j) \qquad (9)$$

Where, m is the number of output classes.

Back propagation: The back Propagation algorithm obtains the gradient of the weight associated with the loss function, and the weights and bias are updated by the gradient descent algorithm. The gradient of the loss with respect to the output Z^{L+1} is obtained by chain rule,

$$\frac{\partial L}{\partial z^{L+1}} = \frac{\partial L}{\partial \hat{y}_j} \cdot \frac{\partial \hat{y}_j}{\partial z^{L+1}} = \hat{y} - y \qquad (10)$$

For each hidden layer l from L to 1, the gradient of the loss,

$$\frac{\partial L}{\partial z^l} = [W^{(l+1)}]^T \frac{\partial L}{\partial z^{l+1}} \odot \mathcal{E}^{l'}(Z^l) \qquad (11)$$

Where, $\odot$ denotes element wise multiplication and $\mathcal{E}^{l'}$ is derivative of ReLU activation function in l^{th} hidden layer. Hence gradient of weights and bias can be represented as,

$$\frac{\partial L}{\partial w^l} = \frac{\partial L}{\partial z^l} [o^{l-1}]^T \qquad (12)$$

$$\frac{\partial L}{\partial b^l} = \frac{\partial L}{\partial z^l} \qquad (13)$$

After applying chain rule, finally weights and bias are updated using gradient descent algorithm as in equation (14, 15),

$$W^l \leftarrow W^l - \eta \frac{\partial L}{\partial w^l} \qquad (14)$$

$$b^l \leftarrow b^l - \eta \frac{\partial L}{\partial b^l} \qquad (15)$$

η is termed as learning rate.

Detailed Methodology

The primary stage of blame classification includes getting solid framework information, taken after by precisely classifying the data, as appeared within the flowchart in Figure 7.3.

Development of hardware setup

A commercially available battery pack, consisting of three lithium-ion cells connected in series, each with a capacity of 4.2V and 2 Ah, was used to develop the model. The normal operating voltage for the battery is 2.7 V at minimum to a maximum 4.2 V, as required by most commercial applications. Charging was performed at 1.5A, and discharging at 2A. Deviations from this voltage range could result in operational faults and potentially severe downtime. Charging was conducted in constant current mode up to 4.2V, after which it switched to constant voltage mode until the current decreased to 20mA.

During discharge, a constant current mode was maintained with a stable current of 2A. This current is maintained at the value till voltage becomes at least 2.7V. The testing was carried out at an ambient temperature range of 23°C to 26°C. The hardware setup for this development is shown in Figure 7.4.

Preparation of dataset

For data acquisition, a voltage divider circuit was used to measure the voltage across each battery cell, while an ACS712 current sensor was employed to monitor the current of the battery pack. Data collection was performed during both charging and discharging

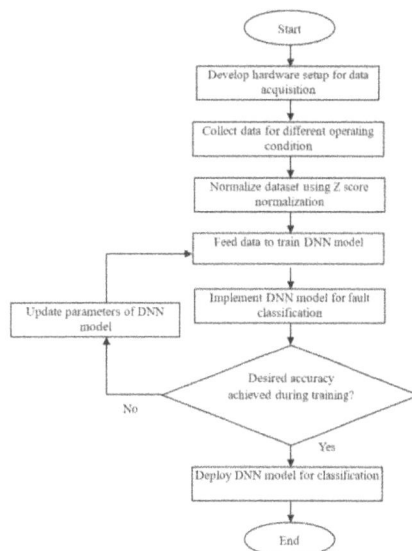

Figure 7.3 Detailed methodology of proposed work
Source: Author

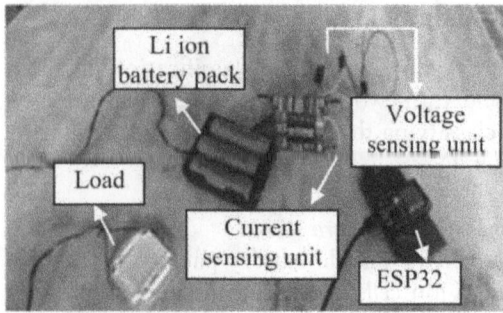

Figure 7.4 Developed hardware setup for data acquisition
Source: Author

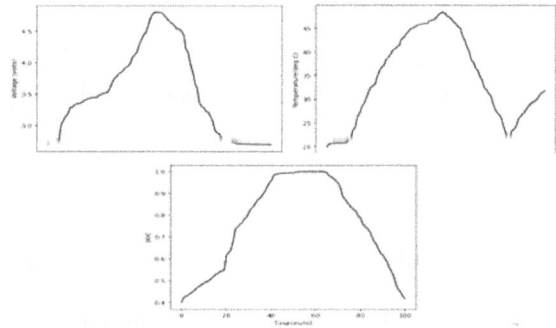

Figure 7.5 Voltage, temperature and SoC profile of cell 1
Source: Author

processes, as well as under various operating conditions. For instance, scenarios included high energy consumption on highways, which leads to higher charge outputs, compared to moderate and delicate loads in rural and urban driving conditions. The three key parameters like cell voltage, current, and temperature were monitored using the hardware setup. State of charge (SoC) and State of Health (SoH) were estimated using equations 3 and 4. Temperature was measured with an LM35 temperature sensor. The parameters were tracked during high-frequency and low-frequency charging and discharging cycles. The voltage profile during high-frequency cycles showed a discharge drop to 3.18V and a charge rise to 4.21V, reflecting both theoretical and practical values for lithium-ion batteries. Fault identification was further supported by monitoring battery temperature and SoC. Figure 7.5 illustrates the voltage, temperature profile and SoC of Battery Cell 1. The data indicates that battery temperature fluctuates with SoC, which is directly related to the voltage. During charging, SoC reaches a maximum value of 1.0, and continued charging beyond this point causes overcharging stress and a rise in battery temperature. This stress is alleviated during discharge, leading to a decrease in temperature. Temperature rises again later in the graph which represents load characteristics. The data set was created under diverse operational conditions.

Experiments were conducted across multiple instances and fault scenarios, with different charge/discharge cycles both uniform and non-uniform, all at different frequencies. Data was collected using an ESP 32 microcontroller, results in a data set, containing 2080 data points with five inputs features (cell voltages, currents, temperature, SoC and SoH) for each cell, capturing a broad range of operational conditions. Next all the data points are normalized using Z score normalization following equation (16). The normalized dataset is used for training the proposed DNN model.

$$d_i^{Norm} = \left(\frac{d_i - \mu}{\sigma}\right) \qquad (16)$$

Where, d_i is i^{th} instance of dataset, μ and σ are the mean and standard deviation for the dataset, respectively. The dimension of normalized data set is 2080 × 5.

Fault classification

The five normalized highlights are input to a profound neural arrange (DNN) with an engineering of three covered up layers: 64, 32, and 16 neurons, decided utilizing network look. The yield layer contains five neurons speaking to four blame classes—thermal runaway (FC1), battery life debasement (FC2), beneath voltage (FC3), and over voltage (FC4)—plus one solid lesson (FC0). ReLU actuation is utilized within the

Table 7.1 Performance matrices for different output class

Class	% Accuracy	% Sensitivity	% Specificity	% Precision	F measure
C_0	98.07	100	100	100	100
C_1	98.07	97.1	99.3	97.1	97
C_2	98.07	96.2	99.3	96.2	96
C_3	98.07	97.1	99.3	97.1	97
C_4	98.07	100	100	100	100

Source: Author

covered-up layers, while SoftMax is connected within the yield layer. A ten-fold cross-validation is actualized to diminish execution metric change and upgrade DNN execution. Of 2080 information focuses, 1560 (75%) are utilized for preparing, and 520 (25%) for testing.

The demonstration is created in Python utilizing Keras with TensorFlow as the backend. Preparing employments mini-batches of measure 100, with a force hyperparameter of 0.9 for parameter overhauls and an introductory learning rate of 0.01 for merging. They demonstrate trains for 200 ages. The number of trainable parameters incorporates 384 for the input to covered up layer 1, 2080 for covered up layer 1 to 2, 528 for covered up layer 2 to 3, and 85 for covered up layer 3 to yield, totaling 3077 parameters overhauled per age. Preparing on a framework with 8GB Smash and a 2.70 GHz clock speed takes 10.58 seconds. The overall result obtained from DNN classifier using real motor data samples are studied with four standard performance metrics such as accuracy, sensitivity, specificity, precision, F_ measure. These metrics are defined as,

$$Accuracy = \frac{TP+TN}{TP+TN+FP+FN} \qquad (17)$$

$$Sensitivity = \frac{TP}{TP+FN} \qquad (18)$$

$$Specificity = \frac{TN}{TN+FP} \qquad (19)$$

$$Precision = \frac{TP}{TP+FP} \qquad (20)$$

$$F\_measure = \frac{2TP}{2TP+FP+FN} \qquad (21)$$

Here, TP stands for correctly predicted positive cases, FP refers to incorrectly predicted positive cases, TN denotes correctly predicted negative cases, and FN indicates incorrectly predicted negative cases. The overall performance of the fault detection system across all five classes displayed in Table 7.1. From the table and Figure 7.6, it can be observed that, the accuracy is 98.07% for all classes. The achieved sensitivity is 100% for classes C_0 and C_4, 97.1% for other two classes C_1 and C_3. The specificity achieved for C_1 and C_4 is 100% and becomes 99.3% for all other classes. Precision is 97.1% for classes C_1 and C_3, 96.2% for class C_2 and 100% for all other classes. Finally, F measure becomes 100% for C_0, C_4, 97% for C_1, C_3 and 96% for C_2. The classification model performance is illustrated by the ROC plot in Figure 7.7. The X axis represents False Positive Rate (FPR) and Y axis shows true positive rate (TPR). The dashed line from (0,0) to (1,1) represents the performance of a random classified. If a classifier's ROC curve aligns with this line, it indicates performance equivalent to random guessing. The area under the curve (AUC) quantifies the model's ability to distinguish between classes. An AUC value closer to 1 indicates better performance, while a value around 0.5 suggests performance akin to random guessing. Each class's ROC curve is associated with an AUC value. Classes C0 and C4, with AUC values of 1.00, demonstrate superior discriminative performance compared to the other three classes, which have AUC values of 0.98.

Conclusion

In the current electric vehicle market lithium-ion batteries are essential. A method for identifying and categorizing observable flaws in lithium-ion batteries is presented in this paper. Because of its capacity for self-learning the multi-layer perceptron (MLP) exemplifies superior performance in terms of output asset utilization and classification accuracy. This approach presents a promising arrangement in the developing field of blame classification in lithium-ion batteries for electric vehicles. According to comparative analyses MLP is incredibly adaptable and effective for comparative datasets confirming its sufficiency. For thorough

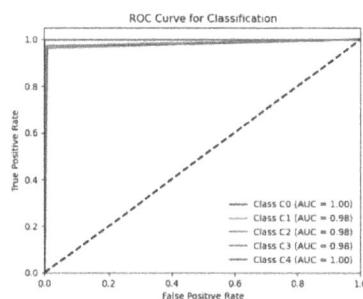

Figure 7.6 Confusion matrix for fault detection system
Source: Author

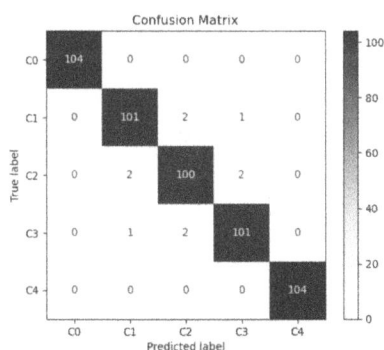

Figure 7.7 ROC plot for proposed fault detection system
Source: Author

blame administration future work should amplify the calculation to handle combinations of deficiencies and address more minor issues.

References

[1] Zhang, X., Wang, Y., & Xia, L. (2020). Data-driven methods for lithium-ion battery health monitoring and management: a review. *Renewable and Sustainable Energy Reviews*, 127, 109852. https://doi.org/10.1016/j.rser.2020.109852.

[2] Eddahech, A., Briat, O., & Vinassa, J.-M. (2012). Performance comparison of four lithium-ion battery technologies under calendar aging. *Energy*, 42(1), 229–237. https://doi.org/10.1016/j.energy.2012.04.029.

[3] Li, W., Zhang, G., Luo, Y., & Xu, Y. (2020). A hybrid machine learning-based fault diagnosis approach for lithium-ion batteries. *Journal of Power Sources*, 451, 227763. https://doi.org/10.1016/j.jpowsour.2019.227763.

[4] Liu, C., & Peng, Z. (2021). Deep learning-based battery health management: a review. *IEEE Access*, 9, 14698–14711. https://doi.org/10.1109/ACCESS.2021.3052277.

[5] Xia, B., Cao, J., & Xiao, F. (2017). Fault diagnosis and prognosis for lithium-ion battery systems: a review. *Journal of Power Sources*, 367, 126–139. https://doi.org/10.1016/j.jpowsour.2017.09.009.

[6] Feng, X., Ouyang, M., & Lu, L. (2018). Online battery state of health estimation based on short-term degradation data: a comparative study. *Journal of Power Sources*, 379, 1–12. https://doi.org/10.1016/j.jpowsour.2018.01.023.

[7] Zhang, W., & Wang, C. (2018). Deep learning-based state of health estimation and remaining useful life prediction for lithium-ion battery. *Journal of Energy Storage*, 17, 299–308. https://doi.org/10.1016/j.est.2018.03.019.

[8] Ma, X., Li, H., & Xiong, R. (2019). A deep learning method for battery state-of-health estimation based on an improved gated recurrent unit. *Applied Energy*, 253, 113603. https://doi.org/10.1016/j.apenergy.2019.113603.

[9] Tian, H., Wei, Z., & Wang, D. (2020). Fault diagnosis of lithium-ion battery based on deep learning and early warning system. *Energy*, 199, 117421. https://doi.org/10.1016/j.energy.2020.117421.

[10] Hu, X., Li, S., & Peng, H. (2012). A comparative study of equivalent circuit models for lithium-ion batteries. *Journal of Power Sources*, 198, 359–367. https://doi.org/10.1016/j.jpowsour.2011.10.013.

[11] Berecibar, M., Gandiaga, I., Villarreal, I., Omar, N., Van Mierlo, J., & Van den Bossche, P. (2016). Critical review of state of health estimation methods of Li-ion batteries for real applications. *Renewable and Sustainable Energy Reviews*, 56, 572–587. https://doi.org/10.1016/j.rser.2015.11.042.

[12] Zhang, Y., & Lee, J. (2011). A review on prognostics and health monitoring of Li-ion battery. *Journal of Power Sources*, 196(15), 6007–6014. https://doi.org/10.1016/j.jpowsour.2011.03.101.

[13] Wu, B., Zhang, W., & Cao, Y. (2019). Machine learning techniques for real-time state of charge estimation for electric vehicle batteries. *IEEE Transactions on Vehicular Technology*, 68(5), 4188–4196. https://doi.org/10.1109/TVT.2019.2905731.

[14] Li, Y., Shi, J., & Wu, L. (2021). An improved convolutional neural network for state-of-charge estimation of lithium-ion batteries. *IEEE Access*, 9, 32114–32123. https://doi.org/10.1109/ACCESS.2021.3061044.

[15] He, W., Williard, N., Chen, C., & Pecht, M. (2014). State of charge estimation for Li-ion batteries using neural networks modeling and unscented Kalman filter. *Applied Energy*, 121, 20–27. https://doi.org/10.1016/j.apenergy.2014.01.019.

[16] S. Li and B. Ke, 2011 "Study of battery modeling using mathematical and circuit oriented approaches," *IEEE Power and Energy Society General Meeting, Detroit*, MI, USA, 2011, 1–8, doi: 10.1109/PES.2011.6039230.

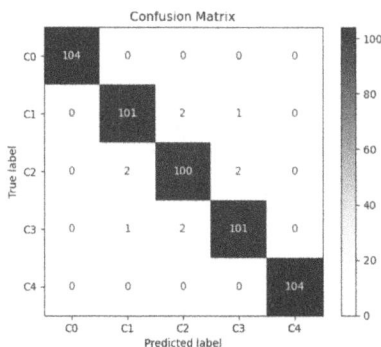

8 Development of sensor-less speed estimation method of IFOC based IM drive using genetic algorithm

Tista Banerjee[1,a], Sumangal Bhaumik[2,b], Titas Bhaumik[2,c], Jinia Datta[2,d] and Debasish Mitra[3,e]

[1]Institute of Engineering and Management (IEM), University of Engineering and Management Kolkata, West Bengal, India

[2]Abacus Institute of Engineering and Management, Hooghly, West Bengal, India

[3]Hooghly Engineering & Technology College, Hooghly, West Bengal, India

Abstract

This paper describes the development of a modern controller for an indirect field drive of an asynchronous motor for sensor-less speed estimation based on reactive power measured by voltage and current sensors. This paper solves the problems of sensor-less speed estimation. The challenges for estimation of speed without a sensor are achieved in this work. The proposed method has been tested here both using SIMULINK environment and also using a prototype model. It has been observed that this method satisfactorily estimates the speed and whose value is compatible with that measured using an optical sensor.

Keywords: AI, ASD, GA, IFOC, IM, sensor-less

Introduction

Sensor-less adjustable speed drives (ASD) refers to the application of vector or field-oriented control without the need for a speed sensor [1]. In advanced ASD systems with induction motor (IM) drives, sensor-less speed control combined with a proportional-integral (PI) controller offers improved efficiency and dynamic performance [1–11]. This method utilizes control with space vector modulation, eliminating the need for costly and complex mechanical speed sensors [1]. Various schemes or method can be used to estimate the speed in Sensor-less ASD based systems, such as the slip estimation method, model reference adaptive systems, extended Kalman filter, slot harmonics, and injection of auxiliary signals. Proper equivalent motor parameter estimation is crucial for the accuracy of these methods [17–21].

One effective and reliable method for speed estimation is the reactive power calculation method [1]. This method has shown satisfactory results compared to the traditional optical speed sensor mounted on the motor, eliminating the need for additional sensors.

Conventional control methods, for tuning PI or PID controllers, like the Ziegler-Nichols (Z-N) method, have limitations [12, 13]. These methods need mathematical model of the system, which may not always accurately reflect real-world conditions. The performance of classical linear control can be affected by load disturbances, motor saturation, and thermal variations.

Artificial intelligence (AI) algorithms are increasingly being used for controlling the speed of induction motor drives. Heuristic algorithms like Genetic Algorithms (GA), swarm intelligence like particle swarm optimization (PSO), knowledge-based systems like fuzzy logic, and machine learning methods such as artificial neural networks (ANN) are utilized for better accuracy [14–16]. A combination of AI algorithms is often employed to tune the PI or PID gain parameters, optimizing the speed control of electrical machines.

In this work, reactive power based senseless speed method is used where the PI gain parameters present in the path of slip speed calculation of IFOC IM path are tuned using GA algorithm. The performance of the GA algorithm is evaluated to determine its effectiveness in achieving optimal speed control.

The integration of Sensor-less speed control with PI controller in induction motor drives represents the future of efficient and dynamic motor control systems. By leveraging AI algorithms and innovative speed estimation methods, industries can achieve significant improvements in performance and reliability without the need for costly mechanical sensors.

Here, in this work the IFOC-IM ASD is modeled in MATLAB SIMULINK first where the proposed Sensor-less scheme has been implemented and then the

[a]tista.banerjee@gmail.com, [b]sumangalstar26@gmail.com, [c]titasbhaumik@gmail.com, [d]dattajinia@gmail.com, [e]debasish9999@yahoo.in

DOI: 10.1201/9781003663348-8

prototype model is built. The prototype model also contains sensor so that a comparative study between the speed measured using sensor and sensor-less method can be studied. The results show satisfactory result in support of the proposed Sensor-less proposed method.

Mathematical Model of IFOC Control Based ASD

In ASD, Indirect Field Oriented Control (IFOC) is recognized as a potent and effective technique. By employing IFOC, engineers and researchers can get precise control over flux and torque, leading to enhanced performance and efficiency in various industrial applications. This advanced controller, illustrated in the diagram provided, is meticulously designed based on a set of equations ranging from (1) to (4) [1–11]. One of the key components of IFOC is the manipulation of d-q stator currents to optimize flux and torque. These stator current reference values are derived from the flux and torque flow paths [1] using equations (1) and (4). Equation (1) encapsulates the flux flow path, while equation (2) delineates the torque flow path.

$$\psi_r = L_m i_{ds}^*, i_{ds}^* \triangleq \frac{\psi_r}{L_m}, G_1 = \frac{1}{L_m} \quad (1)$$

$$T_e \triangleq \frac{3}{2}\left(\frac{p}{2}\right)\frac{L_m}{L_r}i_{ds}^* i_{qs}^*, G_2 = \frac{4L_r}{3pL_m^2}, i_{qs}^* = \frac{T_e G_2}{i_{ds}^*} \quad (2)$$

In a field-oriented control system (IFOC), the rotor flux denoted by ψ_r and the electromechanical torque T_e, play an important role [1,2]. The number of poles, p, influences these parameters. Within the system, G_1 represents the ψ_r flow path gain, while G_2 signifies the gain of T_e flow path [1]. Specifically, in IFOC, ψ_{qr} is set to 0 and ψ_{dr} equals ψ_r. The i_{ds}^*, i_{qs}^*, v_{ds}^* and v_{qs}^* are evaluated from the motor nameplate data in feedforward manner. To further improve the control, we use the gain G_3 to estimate the slip ω_{sl} as in equation (3). This estimation is pivotal as it enables the generation of the unit vector for axis transformation, defined in equation (4). This transformation process is essential for accurate and efficient control of flux and torque in electromechanical systems.

$$G_3 = \omega_{sl} = \frac{1}{T_r}\frac{i_{qs}}{i_{ds}}, T_r = \frac{L_r}{R_r} \quad (3)$$

$$\theta_s \triangleq \int \omega_s dt \quad (4)$$

There are three PI controllers present as shown the Figure 8.1. One of them is the way of the torque flow, and the other two are used to evaluate the voltage of the d-q axis from the current. The generated electromagnetic torque dynamically affects the speed, so torque control can be performed from the speed error

[11]. The torque T_e can be represented as shown in equation (5).

$$\left.\begin{array}{l} T_e \triangleq \left(K_{pls} + \frac{K_{ils}}{s}\right)(\omega_r^* - \omega_r), \\[2mm] i_{qs}^* = \frac{\left(K_{pls} + \frac{K_{ils}}{s}\right)(\omega_r^* - \omega_r)G_2}{i_{ds}^*} \end{array}\right\} \quad (5)$$

The error between the reference values and actual value of i_{ds} and i_{qs} are provided as input to the respective PI controllers (PI$_f$ and PI$_T$) to generate equivalent d-q axis voltages (v_{ds}^*, v_{qs}^*).

$$v_{ds}^* = \left(K_{plf} + \frac{K_{ilf}}{s}\right)(i_{ds}^* - i_{ds}), v_{qs}^* = \left(K_{plt} + \frac{K_{ilt}}{s}\right)(i_{qs}^* - i_{qs})\} \quad (6)$$

The v_{ds}^* and v_{qs}^{**} are converted to (a-b-c) axis variables via space vector modulation method. To implement the SVM scheme, the calculated flux ψ_r from the stator voltage V_s, ω_r^* is estimated from nameplate information both before and after the engine starts. The SVM generates vsvm to produce PWM pulses for the inverter after converting to vα and vβ [1]. The following formula is used to determine vsvm and its angle α:

$$v_{svm} = \sqrt{v_\alpha^2 + v_\beta^2}, \alpha = tan^{-1}\frac{v_{qs}}{v_{ds}} = tan^{-1}\frac{v_\alpha}{v_\beta} \quad (7)$$

To implement the SVM scheme, the flux ψ_r is estimated from the stator voltage V_s, ω_r^* is estimated from nameplate data before engine start and remains approximately constant during operation, and ids also remains constant.

$$V_{an} = \frac{V_s}{2T_s}(t_1 + t_2), V_{bn} = \frac{V_s}{2T_s}(t_1 + t_2),$$

$$V_{cn} = \frac{V_s}{2T_s}(t_1 - t_2) \quad (8)$$

Where t_1 and t_2 are given by equation (9) and M is the modulation index.

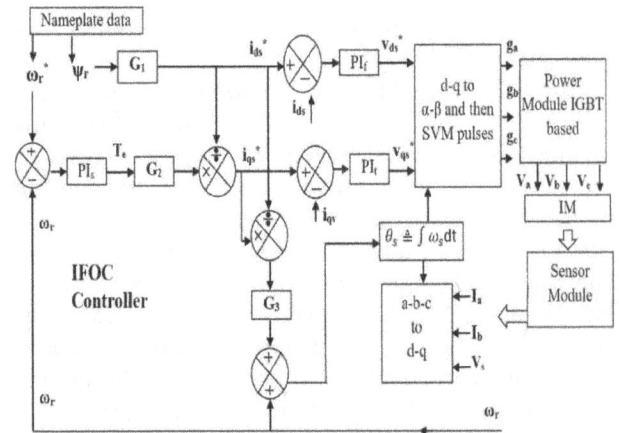

Figure 8.1 IFOC based controller design for VSI-IM
Source: Author

$$t_1 = \frac{2\sqrt{3}}{\pi} MT_s \, Sin\left(\frac{\pi}{3} - \alpha\right)$$

$$t_2 = \frac{2\sqrt{3}}{\pi} MT_s \, Sin(\alpha) \text{, where M=} \frac{V_c}{\frac{2}{\pi}V_{dc}} \tag{9}$$

Basic Plan of the Proposed Work

The effectiveness of Sensor-less speed estimation using the indirect field-oriented control (IFOC) method is significantly affected by factors such as temperature fluctuations due to changes in weather condition, presence of harmonics, and overloads. The motor performance is impacted by these fluctuations in both the rotor resistance (Rr) and stator resistance (Rs). The schematic has been shown in Figure 8.2.

Fluctuations in IM speed can occur as the load changes and alterations occur in time constant resulting from temperature variation. When the rotor resistance (R_r) changes, the number of times ($\tau_r = L_r / R_r$) changes in reverse. As the temperature increases, the rotor and stator copper losses (P_{rcl}, P_{scl}) rise, leading to an overall increase in total losses (P_{loss}). Under normal operating conditions, the motor accelerates gradually, and the torque T_e reaches a maximum value at slip s_m and then stabilizes at the operating point as per equation (13). The rotor speed is directly impacted by variations in equivalent circuit parameters (ECP) estimated using a model reference adaptive system (MRAS), where G_1, G_2, and G_3 are calculated.

$$\omega_r = \omega_s\left(1 - \frac{s_m T_e}{2T_{em}}\right), s_m = \frac{R_r}{L_s + L_r} \tag{10}$$

At steady-state, considering the constant magnetic flux in the motor, the d-q axis voltages are determined by equations (11) and (12). The reactive power, a crucial parameter for effective motor control, is expressed through a set equation. By substituting the reference values of voltage and current into the equations (13) and (14), the reactive power is accurately estimated, assuming minimal measurement error.

$$v_{ds}{}^* = R_s i_{ds}{}^* - \omega_s \sigma L_s i_{qs}{}^*, \sigma_1 = 1 - \frac{L_m{}^2}{L_s L_r} \tag{11}$$

$$v_{qs}{}^* = R_s i_{qs}{}^* + \omega_s \sigma_1 L_s i_{ds}{}^* \tag{12}$$

One way to express the reactive power is as

$$Q_1 = v_{qs} i_{ds} - v_{ds} i_{qs} \tag{13}$$

The reference value of the reactive power is obtained by entering the values of $v_{ds}{}^*$ and $v_{qs}{}^*$ into the equation above.

$$Q_1{}^* = \omega_s(L_s i_{ds}{}^{*2} + \sigma_1 L_s i_{qs}{}^{*2}) \tag{14}$$

$$\omega_s = \frac{v_{qs} i_{ds} - v_{ds} i_{qs}}{(L_s i_{ds}{}^{*2} + \sigma_1 L_s i_{qs}{}^{*2})} \tag{15}$$

To accurately estimate the shaft speed (ω_r), an equation (16) is derived involving a PI controller to ensure precise speed estimation. The gain parameters of PI controller is first tuned using the Ziegler-Nichols method and then optimized using Genetic algorithm (GA) to improve performance. The values of gain parameters, G_1, G_2, and G_3, are obtained through parameter estimation methods discussed in previous studies.

$$\omega_r = \omega_s - (K_{p\omega sl} + \frac{K_{i\omega sl}}{s})G_3 \tag{16}$$

Where ω_{sl} can be expressed as in equation (16).

To guarantee precise assessment of the ω_{sl}, a PI controller with gain $kp_{\omega sl}$, $ki_{\omega sl}$ is provided as shown in equation (16). The speed estimation method described is crucial to achieve accurate speed control and reactive power management in IFOC IM drives. Using modern control techniques such as sensor less speed estimation and reactive power control can improve motor performance and energy efficiency. The motor equivalent circuit parameters are estimated as in [22]. The fitness function used here is

$$\int e^2(t)dt, e(t) = (\omega_{sl}^* - \omega_{sl}) \tag{17}$$

Where ω_r^* is the demand or the reference speed and ω_r is the speed.

Genetic Algorithm used for Estimation of PI Gain Parameters of Slip Speed Path

Step 1: Initialize the values of $K_{p\omega sl}$, $K_{i\omega sl}$ using the ZN method

Step 2: Choose ten people to create the population size. close to $K_{p\omega sl}$, $K_{i\omega sl}$

Step 3: Change each set of $K_{p\omega sl}$, $K_{i\omega sl}$ from decimal to binary.

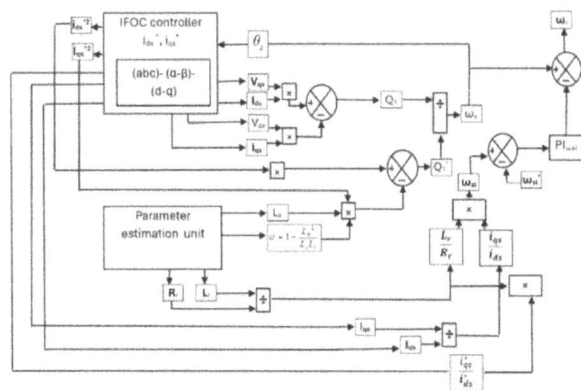

Figure 8.2 Schematic of speed estimation method
Source: Author

Step 4: Put each pair through a single-point crossover.

Step 5: Put each pair through a single-point mutation.

Step 6: Change each set of $K_{p\omega sl}$, $K_{i\omega sl}$ from binary to decimal.

Step 7: Simulate the machine model using the new sets of $K_{p\omega sl}$, $K_{i\omega sl}$

Step 8: Calculate the error e(t) between the refrerence speed and the actual speed.

Step 9: Check whether e(t) < 0.0001. Set the values of $K_{p\omega sl}$ and $K_{i\omega sl}$

Step 10: Declare the obtained values of $K_{p\omega sl}$, $K_{i\omega sl}$ as parent values and repeat the process from step 2

Results

In the context of validation, a 3 kW, 3-φ, 415V, 1415 rpm ,6.9A 50Hz, and pf=0.8 system is utilized. First, IFOC-IM is modeled and simulated, as illustrated in Figure 8.1. A detailed comparative analysis is conducted, focusing on the reference speed, the speed estimated through PI controller gain tuning using the ZN Method, and the comparison with GA optimization, as outlined in Figure 8.3, 8.4 and Table 8.1a. performance under a wide range of conditions, and the integration of hardware components further improves the accuracy and reliability of the results obtained. The value of PI gain parameters obtained using ZN method and for GA gas been shown in Table 8.1c for the reference speed 1000 rpm. Another SCIM 3.7 kW, 415 V, 7.9 A, 1420 rpm, 4pole, and 50 Hz is used for validation of the proposed algorithm as shown in Table 8.1b. After completing the simulation process, a hardware setup has been meticulously designed. This setup includes a power unit constructed with H-Bridge configuration IGBTs. Additionally, a control unit has been implemented with a DSP based microcontroller. This microcontroller contains the firmware of the proposed software. The speed estimation method and the IFOC algorithm. The hardware unit has a sensor unit with Hall sensors, an optical encoder and a temperature sensor. These sensors are responsible for detecting stator voltage and current, stator speed and temperature.

The required gate trigger SVM pulses are generated by the DSPIC controller as described in Equation (11) and Equation (12). The SVM with a frequency of 5 kHz is used to obtain the required voltage and frequency from the power module, which corresponds to the required speed and torque of the motor. The firmware has been carefully designed to handle all the necessary measurements, error calculations and corrective actions.

According to the proposed algorithm, the desired PWM signal can be generated. The PWM duty cycle of SVM is updated in each cycle with a control time corresponding to the frequency of the PWM. The firmware has been designed in such a way that for online monitoring purposes, the evaluated parameters are transmitted to the PC via serial communication at a speed of 76.8 kbps. This data is displayed on the developed graphical interface unit, allowing the user to easily monitor the system in real time. The firmware also includes functionality for adjusting the engine operating conditions through a graphical user interface (GUI). The GUI allows the user to directly input commands that the firmware can receive and execute. Additionally, the firmware collects feedback signals from Hall-voltage, Hall-current, optical speed and temperature sensors to optimize the system.

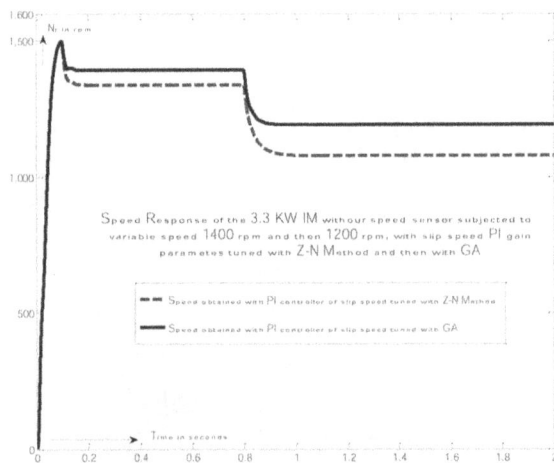

Figure 8.3 Speed response of 3.3KW IM without using sensor subjected to a unit reference speed of 1400 rpm
Source: Author

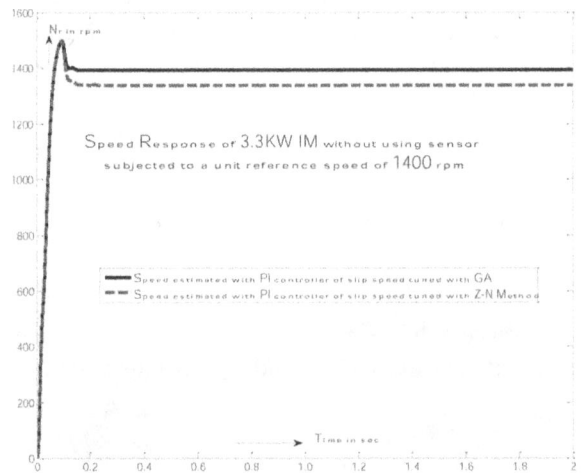

Figure 8.4 Speed response of 3.3KW IM without using sensor subjected to variable speed
Source: Author

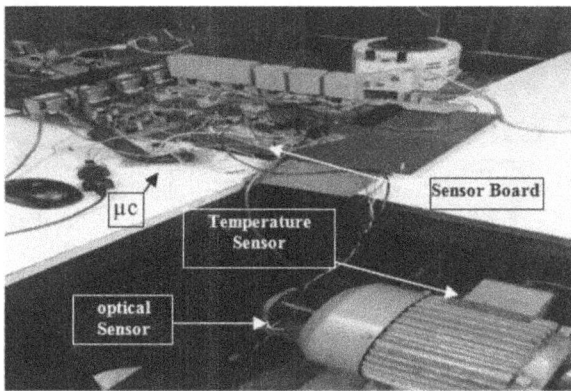

Figure 8.5 Hardware representation of the IFOC based VSI-IM prototype model
Source: Author

By integrating advanced hardware components with efficient firmware and an easy-to-use graphical user interface, this setup provides a comprehensive solution for speed estimation and motor control applications. The speed estimated from the optical sensor has been compared with the sensor less speed estimation proposed system as shown in Table 8.2a.

Where, (Nr^*) represents Demand Speed in rpm, (N_{r1}) represents speed calculated with PI gains estimated by ZN method and (N_{r2}) represents Speed calculated by gain parameters tuned with GA.

There is just one speed value displayed in Table 8.2a. While (Nr_2) estimates speed as measured by an optical sensor in rpm, (Nr_1) represents speed assessed using the suggested method in rpm [23].

Conclusion

From the work it becomes clear that speed can efficiently be sensed by the proposed method if the gain parameters are tuned using AI based algorithm like

Table 8.1a Comparison between estimated speed by the proposed method, where gain parameters of the slip speed tuned using Z- N method and then with GA for 3.3KW IM

Nr*	N_{r1}	N_{r2}	N_r-N_{r1}	N_r-N_{r2}
400	397.2	399.4	-0.7	-0.15
600	595.8	598.7	-0.7	-0.2166
800	798.1	799.3	-0.237	-0.0875
1000	995.8	1000.4	-0.42	0.04
1200	1193.09	1199.5	-0.575	-0.041
1400	1396.2	1398.1	-0.271	-0.135

Source: Author

Table 8.1b Comparison between estimated speed by the proposed method, where gain parameters of the slip speed tuned using Z- N method and then with GA for 3.7KW IM

(Nr^*)	(N_{r1})	(N_{r2})	%Error $(N_r$-$N_{r1})$	%Error $(N_r$-$N_{r2})$
300	298.1	299.7	-0.633	-0.1
500	495.8	498.63	-0.84	-0.274
700	697.651	699.813	-0.335	-0.0875
900	896.8	899.4	-0.355	0.026
1100	1095.09	1099.93	-0.446	-0.06
1300	1298.46	1299.1	-0.118	-0.118

Source: Author

Table 8.1c Comparison between estimated speed with the proposed sensor-less scheme and the PI gain parameters tuned with Z- N method and then with GA

(Nr^*)	(Nr_1)	(Nr_2)	%Error $(N_r$-$N_{r1})$	%Error $(N_r$-$N_{r2})$
400	399.86	400	0.035	0
600	600.2	601	-0.033	0.1
800	799.71	800	0.03	0
1000	1000.2	1000	0.02	0
1200	1199.5	1201	0.041	0.1
1400	1400.3	1400	-0.02	0

Source: Author

Table 8.2a Comparison of speed estimation by the proposed method and by the optical sensor using 3.3 KW IM within the rated speed 1000 rpm

(N_r)	slip speed estimated by ZN method			slip speed estimated by GA method			$(N_r$-$N_{r1})$	$(N_r$-$N_{r2})$
	Nr_1	P	I	Nr_2	P	I		
1000	995.8	8.48	11.7	1000.4	6.13	10.4	-0.42	0.04

Source: Author

Table 8.2b Comparison of speed estimation by the proposed method and by the optical sensor using SCIM of 3.7KW above the rated speed. The SCIM 3.7KW is made to operate at the field weakening mode using the same proposed method of speed estimation as shown in Table 3a

(Nr^*)	(Nr_1)	(Nr_2)	$(N_r$-$N_{r1})$	$(N_r$-$N_{r2})$
1520	1519.543	1520	-0.030	0
1535	1534.932	1536.3	-0.0044	0.08
1550	1549.71	1549.8	-0.018	-0.01
1570	1570.2	1571	0.02	0

Source: Author

GA. But for optimum operation of the proposed method, the equivalent circuit parameters are necessary which have been done by using an MRAS as described in paper [18]. The results obtained from the simulation method and from the prototype method are quite satisfactory hence can be used in IFOC based ASD [24].

References

[1] Bose, B. K. (??). Power Electronics & Drives. CRC Press, 391–450.

[2] Rashid, M. H. (??). Power Electronic Circuits, Devices, and Applications. Low price edition, 250–300. ISBN: 978-81-317-0246-8.

[3] Finch, J. W., & Giaouris, D. (2008). Controlled AC electrical drives. *IEEE Transactions on Industrial Electronics*, 55(2), 481–491. doi:10.1109/tie.2007.911209.

[4] (1998). Field orientated control of 3-Phase AC - motors. Literature Number: BPRA07, Texas Instruments Europe.

[5] Ross, D., Theys, J., & Blowing, S. (2004). Using the DSPIC30F for Vector Control of an ACIM. Microchip Technology Inc, AN908.

[6] Masoudi, S., Feyzi, M. R., & Sharifian, M. B. B. (2009). Speed control in vector controlled induction motors. In 2009 44th International Universities Power Engineering Conference (UPEC), (pp. 1–5). IEEE. 978-0-947649-44-9/09.

[7] Hiware, R. S., & Chaudhari, J. G. (2011). Indirect field oriented control for induction motor. In Fourth International Conference on Emerging Trends in Engineering & Technology.

[8] Bhimbrah, P. S. (2021). Generalized Theory of Electrical Machines. Khanna Publishing.

[9] Haweel, T. I. (2012). Modeling of Induction Motors. *International Journal on Electrical Engineering and Informatics*, 4(2), 361.

[10] Karrer, S. (1989). Measurement and simulation of induction motor characteristics. *Measuremnt*, 7(3), 134–140.

[11] Banerjee, T., & Bera, J. N. (2019). Indirect field oriented control for three-phase induction motor drive using DSP controller. In Advances in Communication, Devices and Networking: Proceedings of ICCDN 2018, (pp. 435–443). Springer Singapore.

[12] Ang, K. H., Chong, G., & Li, Y. (2005). PID control system analysis, design, and technology. *IEEE Transactions on Control Systems Technology*, 13(4), 559–576.

[13] Tzafestas, S., & Kapsiotis, G. (1994). PID self-tuning control combining pole-placement and parameter optimization features. *Mathematics and Computers in Simulation*, 37(2-3), 133–142.

[14] Mirjalili, S. (2019). Genetic algorithm. In Evolutionary Algorithms and Neural Networks. (pp. 43–55), Cham: Springer.

[15] Banerjee, T., Choudhury, S., & Bera, J. N. (2010). Offline optimization of PI/PID controller for a vector controlled induction motor drive using genetic algorithm. In International Conference on. Electrical Power and Energy Systems, ICEPES.

[16] Banerjee, T., Chowdhuri, S., Bera, J. N., & Sarkar, G. S. (2012). Performance comparison between ga and pso for optimization of PI and PID controller of FOC induction motor drive. *International Journal of Scientific and Research Publications*, 2(7), 1–8.

[17] Chen, J., Huang, J., & Sun, Y. (2019). Resistances and speed estimation in sensor-less induction motor drives using a model with known regressors. *IEEE Transactions on Industrial Electronics*, 66(4), 2659–2667.

[18] Kumar, A., & Ramesh, T. (2015). MRAS speed estimator for speed sensor-less IFOC of an induction motor drive using fuzzy logic controller. In 2015 International Conference on Energy, Power and Environment: Towards Sustainable Growth (ICEPE), (pp. 1–6). IEEE.

[19] Zaky, M. S., Metwaly, M. K., Azazi, H. Z., & Deraz, S. A. (2018). A new adaptive SMO for speed estimation of Sensor-less induction motor drives at zero and very low frequencies. *IEEE Transactions on Industrial Electronics*, 65(9), 6901–6911.

[20] Faten, G., & Lassaâd, S. (2009). Speed sensor-less IFOC of PMSM based on adaptive Luenberger observer. *International Journal of Electrical and Electronics Engineering*, 2(1), 7–13.

[21] Pinheiro, D. D., Carati, E. G., Del Sant, F. S., da Costa, J. P., Cardoso, R., & de Stein, C. M. P. (2018). Improved sliding mode and PLL speed estimators for sensor-less vector control of induction motors. In 2018 13th IEEE International Conference on Industry Applications (INDUSCON), (pp. 1030–1037). IEEE.

[22] Banerjee, T., & Bera, J. N. (2022). Online equivalent parameter estimation using BPANN controller with low-frequency signal injection for a sensor-less induction motor drive. *Journal of Electrical Systems and Information Technology*, 9(1), 20.

[23] Banerjee, T., & Bera, J. N. (2023). An improved torque ripple reduction controller for smooth operation of induction motor drive. *Journal of Control, Automation and Electrical Systems*, 34(2), 247–264.

[24] Banerjee, T., & Bera, J. N. (2022). Adadelta-based BPANN controller for online equivalent parameter estimation of squirrel cage induction motor drive. *Transactions of the Indian National Academy of Engineering*, 7(4), 1291–1310.

9 Comparative performance analysis of novel hybrid approach in sustainable PV MPPT architecture

Soumyendu Bhattacharjee[1,a], Debabrata Mazumdar[2,b] and Shounak Bandyopadhyay[1,c]

[1]Department of Electronics & Communication Engineering, Abacus Institute of Engineering and Management, Hooghly, West Bengal, India

[2]Department of Electrical Engineering, Abacus Institute of Engineering and Management, Hooghly, West Bengal, India

Abstract

Large-scale solar photovoltaic (PV) systems are vulnerable to conditions known as partial shadowing (PSCS). PSCS may degrade the efficiency of the PV system by producing recurrent peaks in the power-voltage characteristic. The PV system must be operated at global maximum power point (GMPP) in order to maximize its potential. This paper develops a hybrid approach for GMPP tracking that relies on both Drone Squadron optimization (DSO) and the particle swarm optimizations (PSO). This study proposes a unique MPPT algorithm that combines the DSO and PSO algorithms to benefit from the advantages of both approaches. To validate the proposed method, it is tested under the uniform irradiance condition and with different occurrences of partial shadowing. There is a noticeable improvement in performance when comparing the results of the simulation and experiments to the original PSO technique and DSO algorithm.

Keywords: Drone squadron optimization, maximum power point, partial shading condition

Introduction

In recent decades, concerns about global warming and environmental damage have grown significantly. By using pollution free, renewable energy resources, researchers are currently looking at ways to lower the quantity of greenhouse gas emissions and the cost of power per unit. Their integration into a power-producing system is the most effective approach to rectify the issue, since their attributes surpass those of singular sources [1]. Numerous generation technologies fall under the broad category of renewable energy resources (RERS). The most well-known of these, which may be seen as dependable substitutes for traditional power plants that burn fossil fuels, are solar and wind energy sources in the form of photovoltaic (PV) panels, hydro and tidal energy, and wind energy [2]. Although PV systems offer plenty of advantages to the electrical network, but some significant drawbacks must be properly taken into account. Several research investigations have looked into the power inefficiencies of PV solar panels in light of the aforementioned challenges, and several maximum power point tracking (MPPT) methods developed to solve previously mentioned issues with power output accuracy. Determining the conditions in which a photovoltaic system produces its maximum electricity is crucial. In order to maximize its usefulness, MPPT algorithms are typically utilized to extract the most power. Results from PV systems are significantly impacted by abrupt changes in climate, particularly when there is partial shadowing (PSCS). Finding the global maximum power point (GMPP) in a PV array's power-voltage graph can be difficult since there are several local maximum power peaks (LMPP).

Boost converters are usually used to achieve the optimal results at a specific intensity. It is possible to alter the output load by adjusting the duty cycle that the converter provides. Conventional MPPT algorithms are occasionally modified to enhance performance in a variety of partial shade and irradiance scenarios. Adaptive P&O algorithm [3], and the incremental conductance method [4] are some MPPT techniques that were thoroughly discussed already. Using the harmony search (HS) method, an adaptive MPPT with parameters taken from a PI controller was optimized [5]. These techniques were desirable and distinct for their simple structure, ease of implementation, and quick convergence. However, the distinction between the LMPP and the GMPP cannot be made using these methods [6].

[a]s.microwave@gmail.com, [b]mazumdardeba@gmail.com, [c]aakash2690@gmail.com

DOI: 10.1201/9781003663348-9

Due to certain changing atmospheric conditions, two different plots, V-P and V-I are not linear in photovoltaic cell. In the last few years, several MPPT techniques solicited by scientists. Two customary ways adopt to reach MPP are: 1) Traditional methods (TMPPT) and 2) Soft Computing methods. Over the past ten years, a number of soft computing-based optimization techniques, such as fuzzy logic controller [7], and neural network [8] have been used to get around the drawbacks of conventional methods. High efficiency and superior performance are achieved with optimization techniques in both partial shade and constant irradiance scenarios. With relatively little oscillation, these algorithms are highly helpful in rapidly arriving at GMPP. Grey Wolf Optimization (GWO) [9] and genetic algorithms (GA) [10] and are examples of evolutionary algorithms. Swarm-based algorithms, like particle swarm optimization [11], cuckoo search [12], and Drone Squadron Optimization (DSO) [13] have shown their acceptability very well.

In light of the arguments above, it seems sense to claim every beneficial index using a single algorithm. Consequently, to harvest the highest power in the shortest amount of settling time, a hybrid strategy combining soft computing techniques is required. This work first analyses the performance of two promising soft computing-based algorithms, DSO and PSO, and then introduces a novel hybrid technique based on these two algorithms. The newly created method's performance was assessed in a MATLAB Simulink environment using a boost converter integrated with PV systems under both uniform and PSCS circumstances. Simulation findings showed that the suggested technique produces relatively large power without sacrificing efficiency or settling time, nor does it fail to fulfil GMPP promptly and efficiently.

The rest of the work is structured as: A quick introduction to PV system modelling with numerous peaks occurring in the P-V curve is provided in section 2. Conventional DSO and PSO based MPPT are reviewed in part 3, after which the performance of the newly developed approach is investigated under various atmospheric circumstances. Section 4 presents experimental findings that demonstrate the superiority of the hybrid technique over DSO and PSO. The last thoughts expressed in section 5.

PV System Mathematical Approach

The components of a PV cell are a series array of resistors, a current source, and a diode, shown in Figure 9.1. The current output of PV cell is:

$$I_{PVL} = I_R - I_{D1} - I_{PAL} \tag{1}$$

I_{PAL}, I_{D1}, I_R stand for parallel resistance (RPAL), diode, and photovoltaic current. R_{SERIES} stands for series resistance

$$I_{D1} = I_{RSC} + e^{q\frac{V_{PVOLTAGE} + I_{PVOLTAGE} * R_{SERIES}}{nKT} - 1} \tag{2}$$

I_{SCN} is for short circuit current, and T_0 and W_0 stand for reference irradiance and temperature, respectively.

$$I_{PVOLTAGE} = I_R - I_{RSC} + [e^{q\frac{V_{PVOLTAGE} + I_{PVOLTAGE} * R_{SER}}{nKT}}$$
$$- 1] - \frac{V_{PVOLTAGE} + I_{PVOLTAGE} * R_{SERIES}}{R_{PAL}} \tag{3}$$

V-I and P-V curve of PV array is shown in Figure 9.2. A steady output cannot be generated when parts of the PV cells are blocked by clouds, tall buildings, birds, etc. Figure 9.3. includes a photovoltaic module in a shaded array. Because almost every diode receives a uniform solar irradiance, or 1000 W/m² under uniform irradiance circumstances, there won't be a voltage drop.

PV modules must withstand a range of irradiation intensities in PSCS settings. When PV cells receive different levels of irradiance, the power-voltage curve shows several crests. The peak associated with the maximum power is known as GMPP, while other peaks are known as LMPP.

System Description

In the PV system chain, the MPPT control procedure is an essential link. The DC/DC converter generates a variable duty cycle that is sufficient to pulse the switch.

Particle swarm optimization

Kennedy and Ebrahat created PSO, a robust optimization technique, in 1995. The development of this algorithm relies heavily on the collective behavior of schooling fish and flocking birds. In this procedure, many community agents are deployed to exchange information. Each particle is directed towards an ideal solution, or one that is nearly optimal, at the conclusion of this process [14]. PSO may be said to provide better class solutions in a shorter amount of time when compared to another optimizations method.

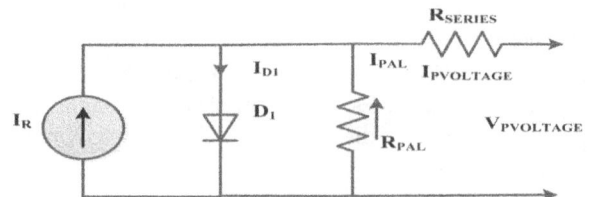

Figure 9.1 Analogous schematic of PV cell
Source: Author

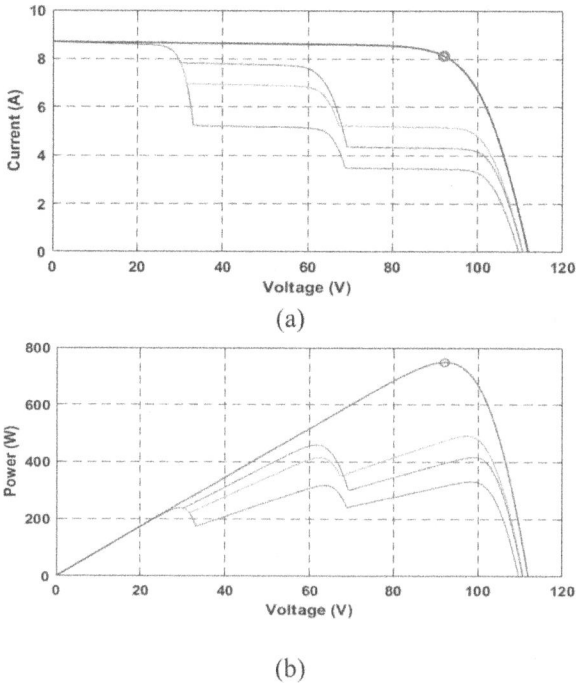

Figure 9.2 Characteristics of a solar cell for (a) V-I and (b) P-V curve
Source: Author

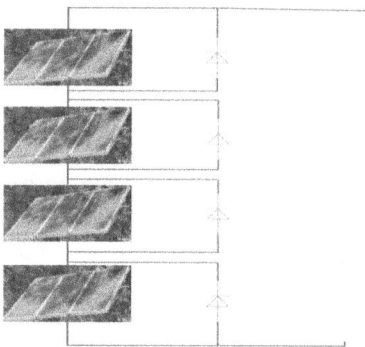

Figure 9.3 Solar PV cells in partial shading states
Source: Author

This algorithm's three primary phases are allocation, gesture, and judgement.

In this current article, output power is considered an objective function and output voltage or current may behave as particles. In order to attain a globally optimal solution, the main objective of implementing this approach in MPPT is to supply researchers with the microscopic particles found using PSO. PSO has been included to track GMPP in the shortest settling time and boost the overall efficiency of the PV module. PSO performance is heavily impacted by changes in the environment and the formation of the group.

Drone squadron optimization
One unique characteristic of airplanes, other flying vehicles like helicopters, is their ability to move autonomously. They are also able to use solar power and establish communication over long distances with the aid of sensors. However, their greatest quality is that they are able to enhance or upgrade both the relevant software and hardware.

The DSO method described in this paper is equipped with a command center and many teams. The command center uses the data that the drones collect to carry out two tasks: 1) To manage a segment of the search, 2) to set up a different firmware to operate drones as shown in Figure 9.4. A command center is often a central location from which drone control actions may be carried out. It gathers inputs, processes information, and produces results. When it's all over, the command center may update the fleet of drones.

In actual use, one can see that drones have an objective function—a hunt assignment—to find a predetermined target in the landscape. The sensors on the drone save the values of this action. The teams do not need to locate a specific, far-off section of the terrain. In actuality, everyone starts from predetermined starting places, which may not change for some teams. The squads may reach different coordinates but may overlap in the exploring region since they have different firmware. Additionally, until the command center crypts such a character in firmware, one squad may be restricted from accompanying the other squad.

A non-nature-motivated evolutionary metaheuristic called Drone Squadron Optimization imitates the actions of drones that are piloted by a central command center and fly over an area to conduct exploration. Since the duty cycle (D) is linked to the PV module and may be controlled to reduce MPP, it is regarded as an optimistic variable in soft computing optimization issues. In a matter of 2-3 repetitions, all the particles reach their maximum value and the PV module produces the maximum power.

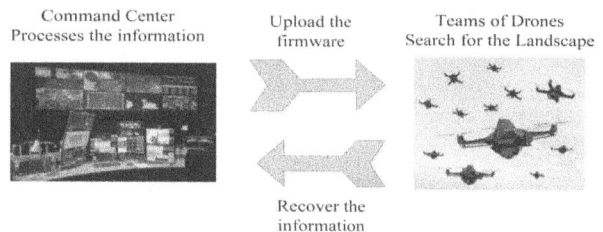

Figure 9.4 DSO based approach: with its associate functions
Source: Author

Proposed method

The DSO is an optimization technique that is commonly used to track MPP in solar PV systems. In several application domains, DSO has shown better convergence and compatibility when compared to alternative optimization techniques. We can contend that our approach outperforms other algorithms in terms of PV output power, however it does not always track MPP in the smallest period of time. Most of the time, especially when PSCS is in place, it will swing around MPP. PSO does, however, have a much visible settling period and a much lower PV output power than other optimization techniques.

In order to address these drawbacks, a unique strategy that effectively combines DSO and PSO is proposed in this study. The suggested method not only yields the maximum power output from a PV panel but also settles faster than DSO.

In the circuit three PV panels were used. Bypass diodes are incorporated in series between the PV modules, which are connected in parallel. Table 9.1 lists the exact irradiances for PV modules in each of the scenarios. The load is resistive, and the converter is specifically a boost converter. PWM generator then feeds the value to the converter once the duty cycle determined by the algorithm is saturated to a least value from 0 to 1. By giving the boost converter a duty cycle, the converter efficiently becomes a viable. Load connected to the output of PV array. Here the PWM receives the duty cycle and transforms it into modulated pulse. The system is given enough time for all three algorithms to settle certain power. Each algorithm is tested in three different scenarios. PV module specification is given in Table 9.2. The MPPT controller connected to a resistive load and coupled with boost converter makes up the entire system. Parameters of aforesaid converter is stated as: C_{IN} = 100 µF, C_{OUT} = 100 µF, L = 3mH. The resistive load used here is 30 ohms. The switching frequency is considered as 10 kHz. Here the two different case studies have been conducted. Initially, performance of solar PV panel was tested under three different partial shading conditions while in another case it receives constant irradiance of 1000 w/m². Under diverse climate conditions, the developed approach proves its superiority. Corresponding comparison is also shown in Table 9.3. Output PV power and power across load has shown in Figure 9.5.

Results Analysis and Case Comparison

The MATLAB/Simulink program utilized for simulations, and developed MPPT approach's performance is examined. The enhanced performance of the suggested technique in tracking GMPP under both normal and

Table 9.1 Details of irradiances and GMPP power and voltages

Case No	Irradiance (W/m²)	Power at GMPP(W)	Voltage at GMPP(V)
UNIFORM	[1000-1000-1000]	66.92	12.51
PSCS 1	[1000-800-600]	155.24	25.21
PSCS 2	[1000-900-500]	188.53	49.71

Source: Author

partial shading circumstances was examined using MATLAB 2018 version. Subsequently, a thorough comparison was conducted between the hybrid MPPT methodology and two other algorithm-based MPPT methods, namely PSO and DSO.

Here, two different kinds of case studies have been performed. The solar photovoltaic panel is first configured with a constant irradiation of 1000 W/m² and a set temperature of 25°C. From Figure 9.5a and 9.5b it can be concluded that the proposed method outperforms in comparison with the other two techniques in terms of output power, oscillation etc.

In second instant, again three PV panels were employed, and the irradiance is kept as Table 9.1. Detailed results are well elaborated in Figure 9.6. In this instant also, developed hybrid method proved its effectiveness compared with other techniques. Table 9.4 discussed the MPPT results in PSCS-1 condition.

In the third scenario, irradiance is maintained at [1000-900-500] w/m². In Figure 9.7 comparative results are discussed. Here also, our proposed methods performance is outstanding, and we are getting more output power compare with other ones. In Table 9.5, the performance of three techniques is elaborated.

Table 9.2 Specification of photovoltaic module used in simulation

Parameters	Value
Open circuit voltage	12.64 Volt
Short circuit current	8.62 Amp
Coefficient of Temperature	-0.33969(V/°C)
Coefficient of Temperature	0.063701(A/°C)
Cells per module	20
Maximum Power	83.2824(W)
Series connected module per string	1
Parallel strings	1

Source: Author

Table 9.3 Performance comparison in standard conditions

Observations	Algorithm	Power at GMPP (W)	Power received (W)	Settling time (Sec)	Efficiency
Uniform irradiance (1000W/m²)	DSO	175.2	133.6	1.22	76.25
	PSO		158.2	1.10	90.29
	Proposed method		164.8	1.11	94.06

Source: Author

Table 9.4 Performance comparison under PSCS-1

Observations	Algorithm	Power at GMPP (W)	Power received (W)	Settling time (Sec)	Efficiency
Uniform irradiance (1000W/m²)	DSO	250	248.6	0.83	99.44
	PSO		247.7	0.92	99.08
	Proposed method		249.8	0.72	99.99

Source: Author

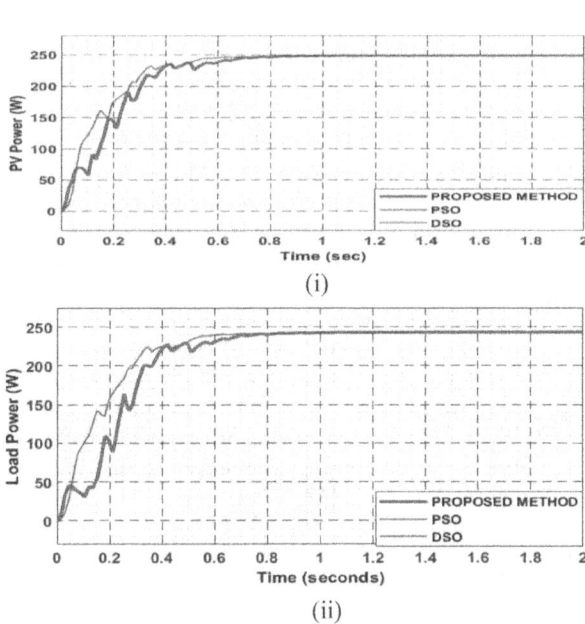

Figure 9.5 Parity of (i) PV power (ii) load power in invariable irradiance
Source: Author

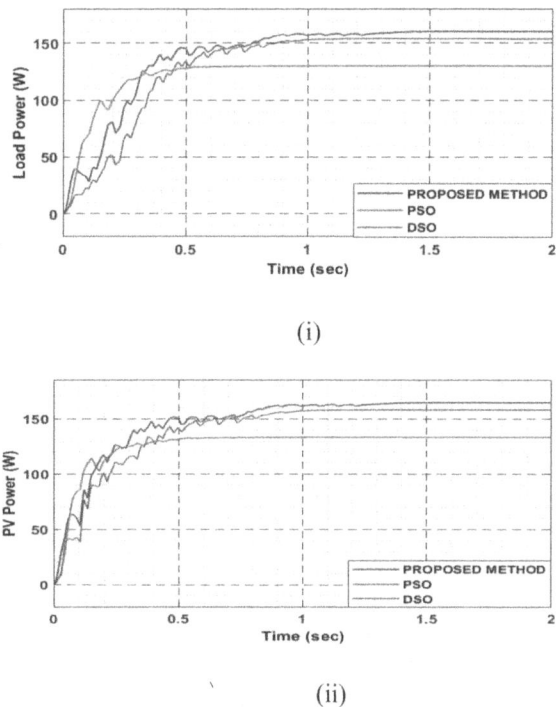

Figure 9.6 (i) PV power and (ii) Load power comparative analysis for PSCS-1
Source: Author

Conclusion

This work has examined the global maximum power point (GMPP)-imitating capabilities of a unique hybrid model. The outcome of simulation has been compared with those of particle swarm optimizations (PSO) and Drone Squadron Optimization (DSO), two of the most popular and often used algorithms, which are also simulated using the same circuit. The time complexity to reach the MPP and power generation efficiency of the method are evaluated under

Table 9.5 Performance comparison in PSCS-2

Observations	Algorithm	Power at GMPP (W)	Power received (W)	Settling Time (Sec)	Efficiency
Uniform Irradiance (1000W/m²)	DSO		148.3	0.90	92.68
	PSO	160	135.8	0.95	84.87
	Proposed Method		149.7	0.88	93.56

Source: Author

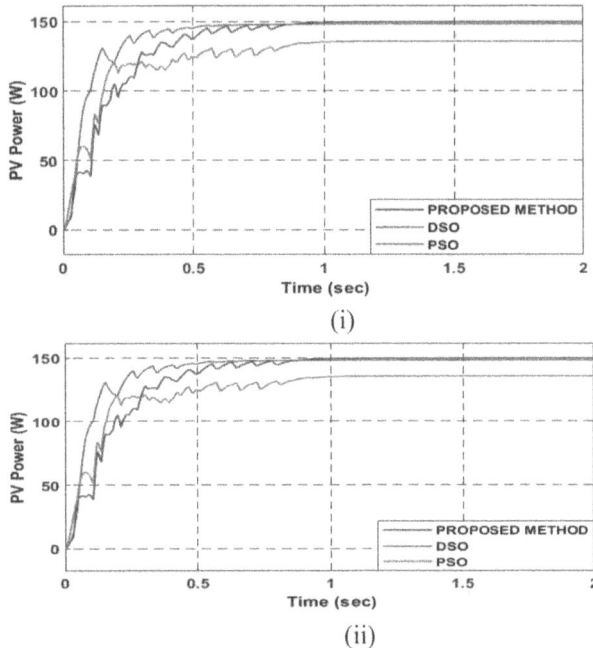

Figure 9.7 Parity of (i) PV power (ii) load power in invariable irradiance
Source: Author

various atmospheric scenarios. The suggested method is more reliably successful in both partial and uniform shadows, according to the data comparison. Under uncertain scenarios, the algorithm's efficacy, the time it takes to converge with the MPP, and their settling times have all been examined. In terms of PV power and load power output, recently developed hybrid approaches perform better than PSO and DSO algorithms. Hybrid techniques provide fast and precise monitoring with the option to boost power output later to compensate for excessive oscillations at first. The use of this technology of solar-powered vehicles, which require fast MPPT owing to rapidly changing irradiance, may expand the scope of this work. Using the hardware configuration, we want to evaluate the developed hybrid approach technique of MPPT in later research on an experimental platform.

References

[1] Nurunnabi, M., Roy, N. K., & Pota, H. R. (2019). Optimal sizing of grid-tied hybrid renewable energy systems considering inverter to PV ratio - a case study. *Journal of Renewable and Sustainable Energy*, 11. http://dx.doi.org/10.1063/1.5052492.

[2] Mazumdar, D., Sain, C., Biswas, P. K., Sanjeevkumar, P., & Khan, B. (2024). Overview of solar photovoltaic MPPT methods: a state of Art on conventional and artificial intelligence control techniques. *International Transactions on Electrical Energy Systems*, 2024(1), 8363342. Doi: https://doi.org/10.1155/2024/8363342.

[3] Mazumdar, D., Biswas, P. K., Sain, C., & Ustun, T. S. (2024). GAO optimized sliding mode based reconfigurable step Size Pb&O MPPT controller with grid integrated EV charging station. *IEEE Access*, 12, 10608–10620. Doi: 10.1109/ACCESS.2023.3344275.

[4] Bouksaim, M., Mekhfioui, M. M., & Sirfi, M. N. (2021). Design and implementation of modified INC, conventional INC and fuzzy logic controllers applied to a PV system under variable weather condition. *Design*, 5, 71.

[5] Mazumdar, D., Sain, C., & Biswas, P. K. (2022). A comprehensive review based on conventional and artificial intelligence strategies for MPPT-based solar PV system. In Sikander, A., Zurek-Mortka, M., Chanda, C. K., & Mondal, P. K. (Eds.), Advances in Energy and Control Systems. ESDA. Lecture Notes in Electrical Engineering, (Vol. 1148). Singapore: Springer. https://doi.org/10.1007/978-981-97-0154-4_3.

[6] Mazumdar, D., Biswas, P. K., & Sain, C. (2023). Introduction to new hybrid approach in sustainable PV MPPT architecture. In 2023 IEEE 20th India Council International Conference (INDICON), Hyderabad, India, (pp. 989–994). doi: 10.1109/INDICON59947.2023.10440825.

[7] Ge, X., Ahmed, F. W., Rezvani, A., Aljojo, N., Samad, S., & Foong, L. K. (2020). Implementation of a novel hybrid BAT-Fuzzy controller based MPPT for grid-connected PV-battery system. *Control Engineering Practice*, 98, 104380.

[8] Issaadi, S., Issaadi, W., & Khireddine, A. (2019). New intelligent control strategy by robust neural network algorithm for real time detection of an optimized maximum power tracking control in photovoltaic systems. *Energy*, 187, 115881.

[9] Mazumdar, D., Biswas, P. K., Sain, C., Ahmad, F., & Fagih, L. A. (2024). A comprehensive analysis of the adaptive FOPID MPPT controller for grid-tied photovoltaics system under atmospheric uncertainty. *Energy Reports*, 12, 1921–1935. Doi: https://doi.org/10.1016/j.egyr.2024.08.013.

[10] Kulaksız, A. A., & Akkaya, R. (2012). A genetic algorithm optimized ANN-based MPPT algorithm for a stand-alone PV system with induction motor drive. *Solar Energy*, 86, 2366–75.

[11] Ishaque, K., Salam, Z., Amjad, M., & Mekhilef, S. (2012). An improved particle swarm optimization (PSO)–based MPPT for PV with reduced steady-state oscillation. *IEEE Transactions on Power Electronics*, 27(8), 3627–3638. doi: 10.1109/TPEL.2012.2185713.

[12] Mazumdar, D., Biswas, P. K., Sain, C., Ahmad, F., & Fagih, L. A. (2024). An enhance approach for solar PV-based grid integrated hybrid electric vehicle charging station. *International Journal of Energy Research*, 2024, 7095461.

[13] Mazumdar, D., Biswas, P. K., Sain, C., Ahmad, F., Sarker, R., & Ustun, T. S. (2024). Optimizing MPPT control for enhanced efficiency in sustainable photovoltaic microgrids: a DSO-based approach. *International Transactions on Electrical Energy Systems*, 2024(1), 5525066.

[14] Yi-Hwa, L., Shyh-Ching, H., Jia-Wei, H., & Wen-Cheng, L. (2012). A particle swarm optimization-based maximum power point tracking algorithm for PV systems operating under partially shaded conditions. *IEEE Transactions on Energy Conversion*, 27, 1027–1035.

10 Evolution of linear electrical model of non-branded white LEDs and its application in the field of dimmable driver design

Shibangi Bhattacharya[1,a], Khushi Rao[1,b], Sudhangshu Sarkar[2,c], Pallav Dutta[2,d] and Biswadeep Gupta Bakshi[2,e]

[1]Student, Narula Institute of Technology, Kolkata, West Bengal, India

[2]Assistant Professor, Narula Institute of Technology, Kolkata, West Bengal, India

Abstract

The linear model is the most extensively utilized model to describe the electrical behavior of a light-emitting diode (LED). Usually, the I-V characteristics published in commercial LED reports are used to develop their linear model. However, reliable reports are rarely available for non-certified LEDs, extensively used for display and decorative lighting. This paper presents an experimental procedure towards linear model development for those LEDs. The developed model is experimentally verified for a single LED chip, after which the verified model is applied to simulate an LED module housing multiple LED chips in a series-parallel combination, forming an array. The power vs. forward current characteristics of the considered LEDs are synthesized from the developed model, which is utilized to design a closed-loop buck LED driver having a wide range of dimming facilities. The performance of that designed driver is evaluated in MATLAB-Simulink environment, which shows high dimming accuracy, about 80% circuit efficiency, and satisfactory step response.

Keywords: Light-emitting diode-driver, light emitting diode, linear model, simulation

Introduction

Light-emitting diodes (LEDs) rapidly replace traditional light sources because of low power consumption, long lightning hours, eco-friendliness, and reduced cost [1]. A semiconductor diode, LEDs are current-controlled devices [2]. In order to get the desirable light output, an LED must undergo a regulated forward current facilitated by an electronic circuit, commonly termed the LED driver [3]. For designing an LED driver, the load model of the LED module comprising a number of chips in a series-parallel combination is required. Several modelling approaches for LEDs have been reported in the literature, such as the linear, piecewise-linear, Shockley, and limited electrical models [4–7]. Out of all these, the linear model is the most widely used one because of its simplicity and reliability over wide operating regions [4].

The standard way to obtain the linear model of an LED is to use its I-V characteristics provided in the datasheet for a given nominal ambient temperature [8]. This method is well-established, especially for high-power, branded, surface-mounted LED chips. In the Indian scenario, however, non-certified, generic LEDs of both conventional (through-hole) and SMD varieties are vastly commissioned for signaling, display and decorative purposes. For these LEDs, published manufacturer's reports are hardly available. Because of this reason, LED driver manufacturers rely upon the rated current and forward voltage specifications, leading to under driving or overdriving of the LED load. Consequently, the LED module or string made of non-certified LEDs glows either with insufficient brightness or excessive intensity, undermining its longevity.

The practical problem mentioned above whose solution is addressed in the present work. An alternate experimental approach is taken to develop the linear model for a 3 × 20 LED module made of conventional, non-branded, cool daylight LED chips. Based on only one specified electrical data, i.e., the rated forward current of a single chip, the module's nominal power and voltage are precisely computed. These computed parameters are then used to design a dimmable bulk LED driver. Power versus current characteristics of the module is also utilized in the closed-loop dimming control of the LED driver. Results of the MATLAB-Simulink simulation reveal the satisfactory operation of the designed circuit in terms of (i) high dimming

---

[a]shibangibhattacharya123@gmail.com, [b]17khushirao02@gmail.com, [c]sudhangshu.sarkar@nit.ac.in, [d]pallav.dutta@nit.ac.in, [e]biswadeep.guptabakshi@nit.ac.in

DOI: 10.1201/9781003663348-10

accuracy (max. deviation 0.4%) over 20–100% dimming range; (ii) about 80% circuit efficiency at rated power; (iii) desirable step response.

The original contributions of this work are: (i) development of linear model of non-branded LED, (ii) model parameter estimation from experimental data and (iii) development of a methodology to design and simulate dimmable driver.

Working Model

Experimental setup

White, non-branded 5 mm LEDs, which are considered for the 3 × 20 LED module formation, have a forward current rating of 25 mA as the only specified electrical parameter. Five of the 60 LEDs were randomly chosen, whose I-V characteristics were obtained experimentally by operating them individually from a programmable DC power source. The experimental circuit diagram is given in Figure 10.1. The power supply's output is varied over 0–5V range, and the corresponding forward current is measured. During the test, the mean ambient temperature is kept constant at 26°C, as monitored by a resistance temperature detector (RTD).

Model formulation for single LED chip

The following equation gives the I-V characteristics of an LED [4, 5].

$$V_D = I_D \cdot R_D + V_{FD} \tag{1}$$

V_D and I_D are, respectively, the forward voltage and current of individual LED; R_D is the dynamic resistance of the LED; and VFD is the forward voltage drop of the LED. Measured I-V characteristics of individual LED are imported into the MATLAB curve fitting toolbox [9]. The two constants of the linear model (R_D, V_{FD}) are calculated by fitting the experimental data upon a linear polynomial. The algebraic mean of the mentioned constants retrieved for the five samples is V_{FD} = 2.925 V and R_D = 15.22 Ω.

A sample curve fitting window is shown in Figure 10.2. Measured I-V and simulated I-V characteristics are plotted together in Figure 10.3, revealing the model's validity.

Modelling of LED module

For the modelling of an N × M LED module, the following variables are considered: (i) V_S: applied voltage across the module; (ii) I_S input current fed to the module; (iii) N: number of series-connected LED per string; (iv) M: number of string in parallel. Considering equal potential drop across each LED and equal current sharing in each brunch, (1) can be modified as per (2) [8]

$$V_S = \frac{N}{M} \cdot I_S \cdot R_D + N \cdot V_{FD} \tag{2}$$

The I-V plot of the 3 × 20 module is generated using (2), as depicted in Figure 10.4.

Design of Driver Based on the Developed Model

To design the dimmable buck converter, the power vs current characteristic of the module is synthesized by simulating the developed model given by (2). The mentioned characteristic can be best expressed by the following quadratic polynomial fit (3).

$$P_{LED} = 2.28I_S^2 + 8.79I_S + 1.13 \times 10^{-15} \tag{3}$$

From (3), the rated power corresponding to the module's rated current (25 mA × 20 = 0.5A) is calculated as 4.97W, and the forward voltage is computed as 9.94V.

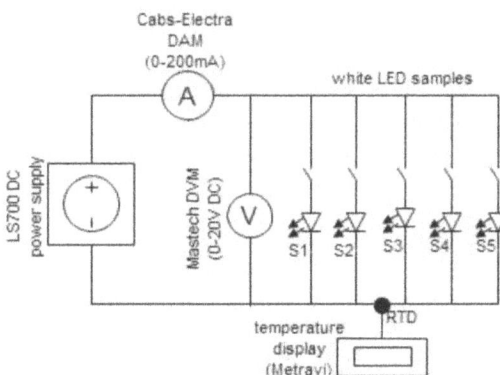

Figure 10.1 Experimental circuit diagram (DAM: Digital Ammeter, DVM: digital voltmeter, S1-S5: White LED under test)

Source: Author

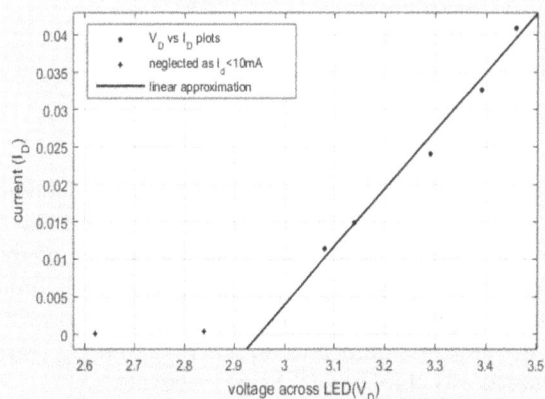

Figure 10.2 Fitting the experimental data upon linear polynomial

Source: Author

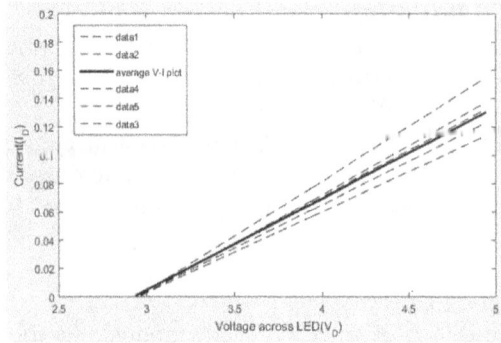

Figure 10.3 Experimental validation: Simulated (Solid Line) and measured I-V characteristics (Dashed Line)
Source: Author

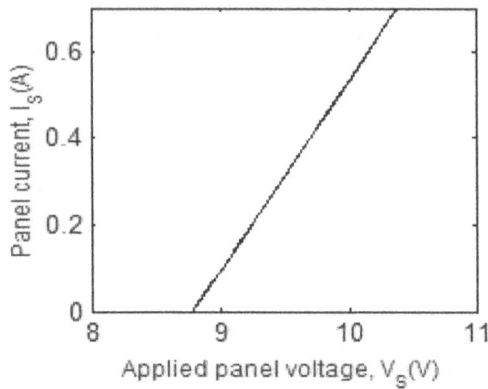

Figure 10.4 I-V characteristics of 3 × 20 LED module
Source: Author

The schematic circuit diagram of the buck LED driver, considered for design, is shown in Figure 10.4, which comprises a power circuit and a control block. Inductor (L_{buck}) and output capacitor (C_{buck}) of the driver are estimated by (4) and (5) respectively for continuous conduction mode.

$$L_{buck} \geq \frac{V_S \cdot (V_i - V_S)}{f_{buck} \cdot \Delta I \cdot V_i} \qquad (4)$$

$$C_{buck} \geq \frac{\Delta I}{8 f_{buck} \cdot \Delta V} \qquad (5)$$

where f_{buck} is the driving frequency, which is taken as 60 kHz, ΔI is the maximum current ripple taken as 3% of rated current; V_i is the input voltage of the driver considered to be 12V, ΔV is the allowable limit of the voltage ripple taken as 0.1V. For these considered parameters, the values of the L_{buck} and C_{buck} are estimated as 1.9 mH and 1 μF, respectively.

The power circuit of the driver shown in Figure 10.5 is implemented in Simulink SimPowerSystems [10] for performance evaluation. The control block is

built logically as shown in Figure 10.6. At a certain dimming level, the current through the LED module is evaluated using (3) and used as a reference signal to produce the triggering pulse of the buck converter. The triggering pulse is generated by the principle of pulse width modulation (PWM). The proportional gain (K_p) and the integral gain (K_i) are obtained by the trial and error method as 1000 and 25, respectively.

Performance Analysis and Simulation Results

The Simulink model of the buck driver-driven LED module is simulated for 20–100% dimming level based on Ode23tb continuous solver. Switching non-linearities are neglected in the present analysis. Voltage (V_S), current (I_S), and consumed power (PLED) of the module are recorded at each dimming level and furnished in Table 10.1. Percentage deviations in module power are computed by (6) as a measure to assess the dimming control accuracy.

$$deviation(\%) = \left| \frac{desired\ power - simulated\ power}{desired\ power} \right| \times 100 \qquad (6)$$

Table 10.1 shows that the maximum deviation is 0.4%, which is negligible and thus indicates high dimming control accuracy. The response of the output

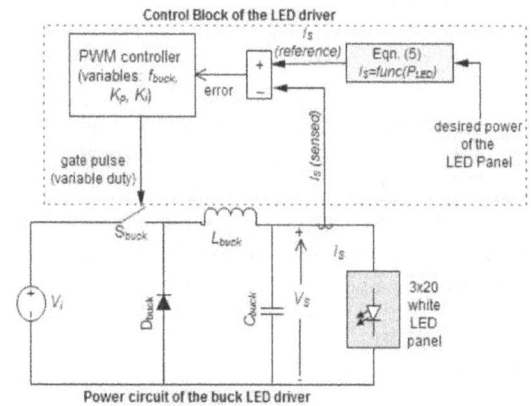

Figure 10.5 Buck LED-driver considered for design and evaluation
Source: Author

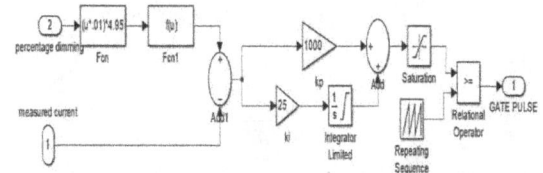

Figure 10.6 Simulink implementation of the control block
Source: Author

power curve under a step-change in the desired dimming level command is shown in Figure 10.7. It is evident that the response of the controllable parameter (i.e. power) is quite fast and accurate as it offers a very short settling period and negligible steady-state deviations. This indicates the speedy and precise operation of the control circuit.

Assuming the internal resistance of the inductor coil as 3Ω and the total switching and conduction loss of the converter switch as 500 mW equal to that of a 2N2222A transistor, the efficiency of the buck driver is calculated to be 79.99% at rated power condition. Since the rated power is about only 5W, this much of efficiency (≈80%) is quite satisfactory. The efficiency can be improved further if a larger LED array is considered [11].

Conclusion

This paper presents the experimental development of the linear light-emitting diode (LED) model of non-branded LEDs and its application to design a dimmable LED driver. Remarkable improvements are

Table 10.1 Simulation result of the buck LED-driver

Dimming level (%)	V_s (V)	I_s (A)	P_{LED} (W)	Devn. in consumed power (%)
20	9.040	0.109	0.998	0.40
40	9.278	0.213	1.984	0.20
60	9.506	0.313	2.975	0.23
80	9.724	0.408	3.976	0.00
100	9.932	0.500	4.972	0.04

Source: Author

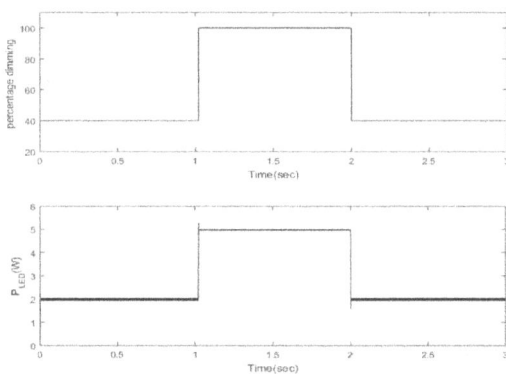

Figure 10.7 Step input (Top) of desired power level and the response of the simulated power of the LED module (Bottom)

Source: Author

reported in this paper, which are: firstly, no datasheet is required to estimate the electrical parameters of an LED chip; secondly, although the sample design is shown for low-power (5W) applications, the same technique can also be applied for higher power. Finally, the designed driver has displayed high dimming accuracy (deviations ≤0.4%), satisfactory circuit efficiency (≈80%) and speedy and accurate step response, all of which have established the satisfactory performance of the proposed design method. However, the effects of the temperature variation and switching non-linearity have been excluded from the present analysis. In the future, the designed circuit can be implemented as hardware. The modelling approach may also be followed for non-branded or non-certified SMD LEDs.

References

[1] Coaton, J. R., & Marsden, A. M. (1996). Lamps and Lighting. 3rd edn. London: Wiley Publication.

[2] Simpson, R. S. (2003). Lighting Control-Technology and Applications. 1st edn. Italy: Focal Press.

[3] Illuminating Engineering Society of North America (IESNA) (2000). IESNA Lighting Handbook: References and Applications. 9th edn. New York (NY): Illuminating Engineering Society of North America (IESNA).

[4] Gacio, D., Alonso, J. M., Calleja, A. J., Garcia, J., & Rico-Secades, M. (2009). A universal-input single-stage high-power-factor power supply for HB-LEDs based on integrated buck-flyback converter. In 2009 Twenty-Fourth Annual IEEE Applied Power Electronics Conference and Exposition, Washington, DC, (pp. 570–576).

[5] Lin, R., Liu, S., Lee, C., & Chang, Y. (2013). Taylor-series-expression-based equivalent circuit models of LED for analysis of LED driver system. *IEEE Transactions on Industry Applications*, 49(4), 1854–1862.

[6] Khanh, T. Q., Bodrogi, P., Vinh, Q. T., & Winkler, H. (Eds.). (2015). LED Lighting: Technology and Perception. 1st Edn. Wiley-VCH.

[7] Li, G., Yu, S., & Shi, J. (2016). Dynamic response analysis of buck driver for LED based on second-order model. *Electronics Letters*, 52(24), 2005–2007.

[8] Lin, R., Tsai, J., Liu, S., & Chiang, H. (2015). Optimal design of LED array combinations for CCM single-loop control LED drivers. *IEEE Journal of Emerging and Selected Topics in Power Electronics*, 3(3), 609–616.

[9] The Mathworks Inc (2020). *Curve Fitting Toolbox (Version 2020a)* [Computer software]. MathWorks. https://in.mathworks.com/products/curvefitting.html.

[10] The Mathworks Inc (2020). *Simscape (Version 2020a)* [Computer software]. MathWorks. https://in.mathworks.com/products/simscape-electrical.htm

[11] Osorio, R., Alonso, J. M., Pinto, S. E., Martínez, G., Vázquez, N., Ponce-Silva, M., et al. (2017). Simplified electrical modeling of power LEDs for DC–DC converter analysis and simulation. *International Journal of Circuit Theory and Applications*, 45(11), 1760–1772.

11 Comprehensive study of encryption algorithms for obtaining privacy protection in smart grids

Rakhi V. Gupta[a] and Parminder Singh[b]

School of Computer Science and Engineering, Lovely Professional University, Phagwara, Punjab, India

Abstract

Everyday advancement in smart grid technologies is provoking an unsafe environment for data security. In order to retain the privacy and secrecy of data operations in order to retain the privacy and secrecy of smart grid infrastructure is a major challenge. Along with optimization, the research is also focusing on the enhancement of encryption algorithms to reinforce the privacy protection mechanisms in the context of smart grids. Specific investigation is made on the performance of widely used encryption algorithms, including AES128, DES, RSA to attain an optimum balance between the robust privacy safeguards and computational efficiency. Thorough comprehensive analysis is done and hence we evaluated the strengths and weakness of each algorithm in context with the smart grid, considering factors such as speed security and resource utilization. The research contributes a novel insight with practical implementation of encryption strategies to match up the inimitable demands of the smart grid environments. The findings aim to develop an enhanced privacy protection model, that ensures the efficiency and resilience of encryption algorithm to cope up the specific challenges laid by the evolving smart grid technologies.

Keywords: Asymmetric key, encryption, performance metrics, symmetric key

Introduction

Emerging smart grid (SG) technologies are transmuting the era to generate, distribute and consume electricity. Integration of the upcoming communication technologies aggravates SGs to enhance proficiency, consistency and sustainability in power systems. Moreover, advanced connectivity produces the unavoidable challenge of making the data private and secured in the smart grid environment. The consumer trust and ensured data transmission can be achieved by prioritizing the privacy protection of the sensitive data.

Encryption algorithms play an important role in achieving secure data transmissions within the SGs, ensuring sensitive information remains confidential and immune to unauthorized access [1]. The research endeavors to address the interplay between computational efficiency and privacy protection by exploring the performance of the prominent encryption algorithms. The scrutinized algorithms include AES128, DES, RSA, AES256, BLOWFISH, RC4, each known for its unique computational attributes and cryptographic properties [3].

The study gives a comprehensive study of the above-mentioned encryption algorithms to determine their efficiency in context of SG privacy-protection. By examining the algorithms on different performance metrics, we aim to determine avenues for enhancing the computational proficiency of mentioned algorithms while safeguarding the preservation of robust privacy measures [2]. The multifaceted nature of SG ecosystem demands nuanced understanding of encryption algorithm performance. The amalgamation of enhanced encryption strategies can give an assurance of fortifying the resilience and efficiency of the smart grid operations, that nurtures a secure exchange of critical information.

Through this research, we aim to contribute valuable acumens that would shed light on the strength and weaknesses of the algorithms. We also try to contribute to developing a tailored privacy protection protocol. As smart grids continue to evolve the findings of this study aspire to inform the ongoing discourse on privacy protection, thereby facilitating the sustainable and secure advancement of smart grid technologies [4, 5].

Comprehensive Study

AES128
Steps of AES128 Encryption:

Key Expansion

- *The 128-bit key is expanded into a set of round keys, which are used in each round of encryption*

Initial Round (Add Round Key)

- *XOR the state with the first- round key.*

[a]rakhig4@gmail.com, [b]parminder.16479@lpu.co.in

DOI: 10.1201/9781003663348-11

Rounds (SubBytes, Shift Rows, MixColumns)
- *SubBytes:* Non- linear substitution of each byte using an S-box.
- ShiftRows: Pemute the bytes in each row.
- MixColumns: Combine the bytes in each column.

Final Round ((SubBytes, Shift Rows, MixColumns)
- Similar to the regular round but without the Mix-Columns step.

Ciphertext
- The final state after the last round represents the encrypted data Figure 11.1 and 11.2.

AES128 is a widely used symmetric-key block cipher that employs a combination of substitution, permutation, and mixing operation in order to provide a high level of security. The key expansion process is crucial for generating a unique subkey for each round. The diagram illustrates the flow of data through the key expansion, initial round, several rounds, and the final round to produce the ciphertext [6, 11].

AES256 encryption indeed follows a structure alike to AES128, but along with larger key size (256 bits) number of rounds also increase. The key expansion ensures that each round uses a unique subkey [10].

RSA
Rivest-Shamir-Adleman (RSA) is an elementary algorithm used for cryptography, that enables secure data communication across the number of digital platforms.

Steps of RSA Encryption:

Key Generation
- Take two prime numbers, say a and b.
- Calculate P = a * b.

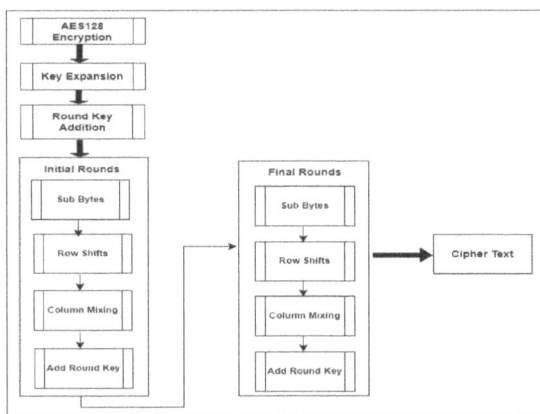

Figure 11.1 AES128 data flow through the key expansion
Source: Author

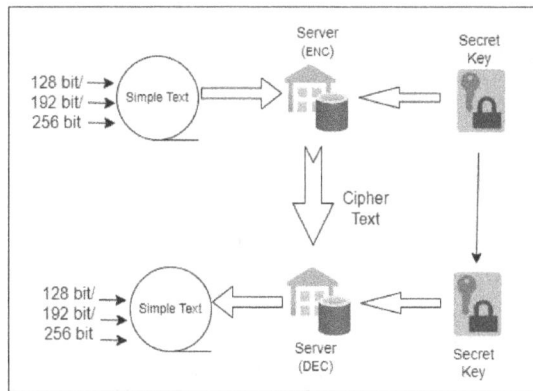

Figure 11.2 Working of AES algorithm
Source: Author

- Totient, $\varphi(P) = (a - 1) * (b - 1)$.
- Pick a public exponent say x, such that, $\varphi(P) > x > 1$.
- Calculate private exponent y, such that, $(y * x) \bmod \varphi(P) = 1$
- Public key = (P, x)
- Private key =(P,y)

Encryption (Ciphertext Calculation)
- *Denoting integer Q as the plaintext message.*
- *Ciphertext C= Q^x(mod P).*

Ciphertext
- *The ciphertext value C can be transmitted securely.*

RSA encryption algorithm makes use of modular exponentiation with public and private key for encryption and decryption. The algorithm's security is fundamentally based on the complexity of factoring large prime products, ensuring that deriving the private key from the public key is computationally impractical. Fig.3 provides a simplified illustration of the key generation and encryption steps in RSA [8].

DES
Data Encryption Standard) is mostly of historical interest rather than practical application in modern cryptography. While it played a significant role in early cryptographic systems, its vulnerabilities have led to the adoption of more secure encryption standards, such as AES. Steps of DES Encryption:

Key Generation
- Generate a 56-bit key(K) from an original 64-bit key by removing every 8th bit.

Initial Permutation
- Permute the 64-bit plaintext block using an initial permutation table.

Rounds (16 rounds)
Each round involves

- **Expansion:** Expand the 32-bit half-block to 84bits.
- **Substitution:** Substitute 48 bits using S-boxes.
- Next permute the result.
- **XOR with Key:** XOR the permuted result with the round key obtained from the original key.

Final Permutation
- Now permute the integrated halves of the last output.
- 64-bit ciphertext is obtained after the last permutation.

The shortened key length can be the reason to compromise DES's security due to which it is not recommended for modern applications. These limitations have led to its outmodedness in favor of more robust algorithms like AES [2].

Comparative Analysis

AES128 is a versatile and widely approved encryption standard, that offers a robust balance of security and performance. Its effectiveness leads to its integration into numerous security protocols and applications, making it a cornerstone of modern cryptography. For most applications, AES128 remains an excellent choice for ensuring data confidentiality and integrity.

AES256 provides a higher-level security compared to AES128, particularly against brute-force attacks. Also, the performance is a bit slower than AES128 due to its longer key length. Used in scenarios where extra security is required, but with a trade-off in performance.

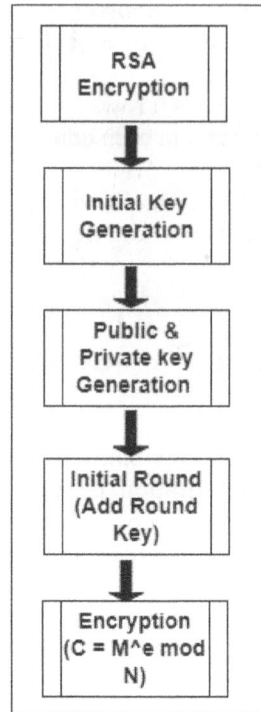

Figure 11.3 Key generation and encryption steps in RSA
Source: Author

DES is considered insecure for modern applications due to its short key length, making it susceptible to brute-force attacks. Rarely used in current applications due to their vulnerabilities. RSA's security is fundamentally tied to the mathematical complexity of factoring large semiprime numbers. Key lengths vary, with 2048 and 3072 bits being common for secure communication. Widely used for secure key exchange, digital signatures, and asymmetric encryption.

Table 11.1 Comparative analysis of the encryption algorithms

Algorithms	Key Length	Security	Performance	Common usage
AES128	128 bits	Secure	Fast	Secure communication, file encryption
AES256	256 bits	Higher Security	Slightly slower than AES128	High-security scenarios, trade-off in speed
DES	56 bits	Insecure (vulnerable to brute force attacks)	Fast	Rarely used due to security vulnerabilities
RSA	Variable (2048, 3072 bits)	Secure for key exchange, digital signatures	Slower for large key sizes	Key exchange, digital signatures, asymmetric encryption
Blowfish	Variable (32 to 448 bits)	Generally considered secure	Moderate speed	Historically used, but less common now
RC4	Variable (128/ 256 bits)	Vulnerable to certain attacks	Fast	Previously used in SSL/ TLS, discouraged now

Source: Author

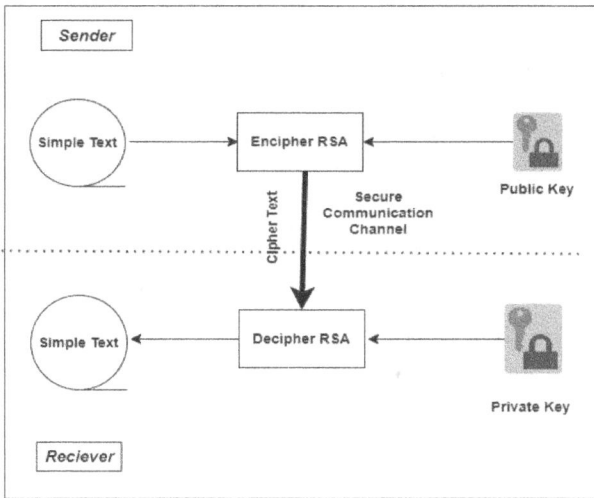

Figure 11.4 Working of RSA algorithm
Source: Author

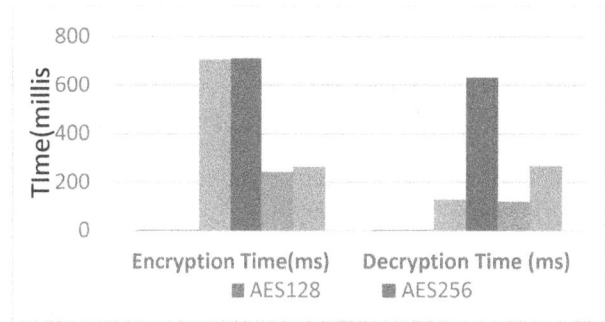

Figure 11.5 Graphical representation of statistical analysis
Source: Author

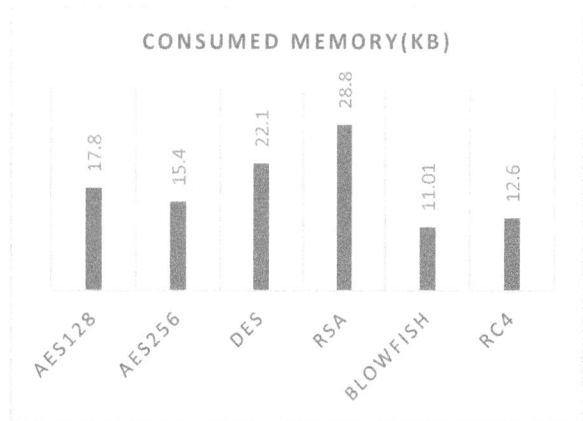

Figure 11.6 Memory consumed for listed algorithms
Source: Author

Implementation and Results

The system has been implemented to compare the different cryptographic algorithms (AES128, AES256, DES, RSA, BLOWFISH and RC4). The implemented system is set around with a pair of ESP328 processors. Out of two processors one is connected to voltage and current sensors. The sensor readings are further converted to wattage per minute. This reading is provided as an input to discussed algorithms. Being more specific, that supports data encryption/decryption, checking the authenticity of data etc.

The algorithms are compared on the parameters such as the encryption time, decryption time and the memory consumed. The table shows the results. Statistical analysis of different algorithms with respect to ascendency parameters have been discussed in Table 11.2. Figure 11.5. shows the graphical representation of the table above. Figure 11.6. shows memory used for unit operations for mentioned algorithms.

Conclusion

In conclusion, for symmetric encryption, AES (especially AES128 and AES256) is widely considered secure and efficient. For asymmetric encryption and digital signatures, RSA is commonly used. DES and RC4 are generally considered insecure for modern applications, and Blowfish is less popular compared to AES. Usage: Historically used in some cryptographic

Table 11.2 Experimental results for various algorithm

Sr. No.	Algorithm	Encryption time (ms)	Decryption time (ms)	Consumed memory (kb)
1	AES128	0.161	0.216	17.8
2	AES256	0.8001	0.1325	15.4
3	DES	705.4	128.4	22.1
4	RSA	708.55	631.8	28.8
5	BLOWFISH	241.001	118.3	11.01
6	RC4	263	267	12.6

Source: Author

applications, but its usage has declined in favor of more modern algorithms. RC4 (Rivest Cipher 4): Key length: Variable (usually 128 or 256 bits). Security: Vulnerable to certain attacks, and its usage is discouraged in modern applications. Usage: Previously used in various applications, especially in SSL/TLS, but its usage has decreased due to security concerns.

References

[1] Bala, T., & Kumar, Y. (2015). Asymmetric algorithms and symmetric algorithms: a review. *International Journal of Computer Applications (ICAET)*, 7(1), 1–4.

[2] Patil, P., Narayankar, P., Narayan, D. G., & Meena S. M. (2016). A comprehensive evaluation of cryptographic algorithms: DES, 3DES, AES, RSA and blowfish. *Procedia Computer Science*, 10(6), 617–624.

[3] Ebrahim, M., Khan, S., & Khalid, U. B. (2013). Symmetric algorithm survey: a comparative analysis. *International Journal of Computer Applications*, 61(20), 12–19.

[4] Kansal, S., & Mittal, M. (2014). Performance evaluation of various symmetric encryption algorithms. In International Conference on Parallel Distributed and Grid Computing, (pp. 105–109). IEEE.

[5] Saini, V., Bangar, P., & Chauhan, H. S. (2014). Study and literature survey of advanced encryption algorithm for wireless application. *International Journal of Emerging Science and Engineering*, 2(6).

[6] Utama, K. D. B., Al-Ghazali, Q. M. R., Mahendra, L. I. B., & Shidik, G. F. (2017). Digital signatureusing MAC address based AES-128 and SHA-2 256-bit. In 2017 International Seminar on (iSemantic), (pp. 72–78). IEEE. doi:10.1109/ISEMANTIC.2017.8251 846.

[7] Nagaraj, S., Raju, G., & Srinadth, V. (2014). Data encryption and authentication using public key approach. In International Conference on Intelligent Computing, Communication and Convergence (ICCC-2014), 48.

[8] Thu, K. M., Hlaing, K. S., & Aung, N. A. (2019). Performance analysis of RSA and elgamal public-key cryptosystems. *International Journal of Trend in Scientific Research and Development (IJTSRD)*, 3(6), 367. e-ISSN: 2456-6470.

[9] Bejo, A., & Adji, T. B. (2017). AES S-box construction using different irreducible polynomial and constant 8-bit vector. In 2017 IEEE Conference on Dependable and Secure Computing, (pp. 366–369). doi:10.1109/DESEC.2017.8073857.

[10] Alyanto, D. (2016). Penerapan algoritma AES: rijndael dalam pengenkripsian data rahasia [Application of AES Algorithm: Rijndael in Encrypting Confidential Data]. (Bachelor's Thesis). Retrieved from https://docplayer.info/36742114-Penerapan-algoritma-aesrijndael-dalam-pengenkripsian-data-rahasia.html.

[11] Anwar, N., Munawwar, M., Abduh, M., & Santosa, N. B. (2018). Komparat if performance model keamanan menggunakan metode algoritma AES 256 bit dan RSA [Model Performance Security Comparative using AES 256 Bit Algorithm Method and RSA]. *Jurnal Rekayasa Sistem dan Teknologi Informasi (RESTI)*, 2(3), 783–791. Retrieved from http://jurnal.iaii.or.id/index.php/RES TI/article/view/606.

[12] Aufa, F. J., Endroyono, E., & Affandi, A. (2018). Security system analysis in combination method: RSA encryption and digital signature algorithm. In Conference: 2018 4th International Conference on Science and Technology (ICST), (pp. 1–5). doi:10.1109/ICSTC.2018.8528584.

[13] Belazi, A., Khan, M., El-Latif, A. A. A., & Belghith, S. (2017). Efficient cryptosystem approaches: S-boxes and permutation–substitution-based encryption. *Nonlinear Dynamics*, 87(1), 337–361. doi:10.1007/s11071-016-3046-0.

[14] Eason, G., Noble, B., & Sneddon, I. N. (1955). On certain integrals of Lipschitz-Hankel type involving products of Bessel functions. *Philosophical Transactions of the Royal Society of London. Series A, Mathematical and Physical Sciences*, A247, 529–551.

[15] Akgün, M., Soykan, E. U., & Soykan, G. (2023). A privacy-preserving scheme for smart grid using trusted execution environment. *IEEE Access*, 11, 9182–9196.

[16] Triantafyllou, A., Jimenez, J. A. P., Torres, A. D. R., Lagkas, T., Rantos, K., & Sarigiannidis, P. (2020). The challenges of privacy and access control as key perspectives for the future electric smart grid. *IEEE Open Journal of the Communications Society*, 1, 1934–1960.

12 Optimal design parameters of a single-phase small transformer by different optimization techniques

Pritish Kumar Ghosh[1], Rakesh Das[2], Raju Basak[3]

[1]Training and Placement Officer, Abacus Institute of Engineering & Management, Mogra, Hooghly, West Bengal, India

[2]Lecturer, Calcutta Institute of Technology, Uluberia, Kolkata, West Bengal, India

[3]Associate Professor, Techno India University, SaltLake, Kolkata, West Bengal, India

Abstract

The entire research community is starting to worry about the problem ofichoosing an appropriate optimization meth od for determining the best designiparameters ofiaicompact, inexpensive transformer, and eachiattempt to create a new algorithm for this purpose has its own drawbacks. Nonlinear objective functions subject to multiple constraints are a feature ofiall engineering optimization problems. Such issues are particularly hard to solve with traditional op timization techniques, and they sometimes fail to produce the global optimaiand end up locked in the local optima. In contrast, nonclassical optimization techniques such as genetic algorithm (GA), particle swarm optimization (PS O), gradient search, pattern search, and so on are known to be more useful since they offer superior flexibility whe n used independently or in concert with one another. Therefore developing such anialgorithm is gaining prominence every day andimaintaining their position as aisignificant area ofistudy as a result ofithe enhanced capability and fle xibility they cooperatively provide. The current effort has been to use GA, simulated annealing, andipattern search t o develop an algorithmito achieve the ideal design parameters foriconstructing a low cost ofimaterial for a 5KVA, 230/115 volt, single phase, coretype, dry transformer. Theipaper's goal is to demonstrate that using nonclassical me thodologies results in aifar better and more acceptable answer. The goal function has been determined to be the tota l cost oficopper and iron.

Keywords: Design optimization, genetic algorithm, pattern search simulated annealing, and objective function

Symbols and abbreviations:

S	Rating,
$V1$, $V2$	Primary and secondary voltage, V
$I1$, $I2$	Primary and secondary current, A
Bm	Maximum value of flux-density, Tesla
δ	Current density, A/sq.mm
Ks, Kw	Stacking factor, window space factor
Hw, Ww, Rw	Window height, width, m, height/width ratio
Et	EMF/turn, V
$T1$, $T2$	No. of turns of the primary and secondary
$a1$, $a2$	C.S. of the primary and secondary conductors, sq. mm
CI, CC, CM	Cost of iron (Rs.); cost of copper (Rs.); total cost (Rs.)

Introduction

In orderito meet the increasing demand for an algorithm that isboth simple to implement and effective in determini ng the glo-bal optimal or quasioptimal point ofioperation, various productive contributions in the recent research era have bee n reported in the development ofiefficient optimization algorithms [3, 5]. The goal ofiminimizingithe initial cost ofia transformer,namey the cost ofithe copper used for windings and the mat erial cost ofistampings, is stated in a paper published in Industrial Technology, ICIT 2006, IEEE International Conference. Two transformer samples are used to test the algorithm. The acquired resultssuggest that a global optimumihas been produced b y the procedure.In order to address the intricate issue ofitransformer design optimization, artificial intelligence tech niques have also been widely employed. Examples ofithese techniques include the use ofiGENETIC

[a]pritishghosh80@gmail.com, [b]rakeshdasbtech@gmail.com, [c]basak.raju@yahoo.com

DOI: 10.1201/9781003663348-12

ALGorithms (GAs) for transformer cost minimization [15], performance optimization oficastresin distribution transformers with s tack core technology [16, 18], and ortoroidal core transformers [2]. Neural network approaches are also utilized for design optimization, as demonstrated. Abdelwanis et al. and Suja and Yuvaraj [1, 13], where they are applied to the prediction ofitransformer losses and reactance, as well as the selection ofiwinding materials. Robust solutions to the transformer design optimization challenge might potentially be obtained through deterministic approaches.The deterministic approach ofigeometric programming has been introduced [10] to address the problem ofidesign optimization for both high frequency and low frequency transformers in this conte xt. Technical literature rarely discusses minimizing manufacturingicosts overall. Instead, most approaches focus on minimizing the costs ofiparticularparts, like magnetic material [6], output power, loadiloss minimization [12, 13], or no-load loss minimization [6].

Selecting the optimal element from a range of accessible choices referred to as optimization. It refers to the process of methodically selecting real or integer variable values from an acceptable set in order to reduce or maximize a real function. There could be several workable answers to a real world issue. The best possible design among numerous workable solutions is known as the optimal design, usually when several inequality restrictions are present. To find the best answer, a variety of optimizing techniques are available, including nonclassical or probabilistic and classical or deterministic methods. These days, nonclassical methods built on soft computing are quite well liked. In this work, three distinct methods pattern search, simulated annealing, and genetic algorithm have been applied to optimize the design of a single-phase transformer [14, 17].

Formation of the cost function

A dry type 230/115 V, 5 KVA, 50 Hz, transformer design has been selected for maximization. The cost function is selected for calculation of the price of the material [1, 18].

Design parameters and limitations
The following two variables have been selected as dynamic Constant (K) = EMF and $Et = K \sqrt{S}$

Window height and width ratio, $Rw = Hw / Ww$

We take standard values for other design and decision variables: Selected core material: CRS;

Wire material: Cu

Maximum flux density, $Bm = 1.4$ Tesla;

Current density, $\delta = 2.4$ A/mm^2

Stepped core has been used and steps No. = 3;

Space factor of Window, $K_w = 0.4$

Stacking factor, $K_s = 0.93$ with CRS core.

Design constraints
The following design constraints considering the practical aspects of standard transformers: Full load ʮ and 0.85 (lag) P. F >96%

Voltage regulation of full load condition and 0.85 (lag) P. F <4%

$I_0 < 2\%$

$T_{FL} < 50° $ C. [4]

The optimizing function [1]
Total material cost (Rs.) = CM =

$$1131(\sqrt{R_w K} + \sqrt{K / R_w}) + 3371 K^{1.5} + 2520 / \sqrt{K} + 803 / (K^{1.5} \sqrt{R_w})$$

Now we have to optimize (minimize) the function

Genetic algorithm

Originally developed by Holland and established by Holland and Dejong, genetic algorithm has been popularized by Goldberg. Several researchers have contributed in many ways to various aspects and applications of genetic algorithms [4, 18]. GA has found its application in almost every branch of engineering and research work. It simulates natural genetics. The flow chart for optimization by genetic algorithm is given in Figure 12.1.

Crossing over, mutation, and reproduction make up the evolutionary computation technique. It begins with a single iteration or class of chromosomes that are haphazardly encrypted, each of them stands for a potential remedy encrypted chromosomes undergo natural selection for recombination through the crossover operator. So, better off springs generate in successive generations. Probabilistic bit mutation incorporates new property to the child chromosomes in search of diversity and better-quality solutions. After a prearranged number of generations, the

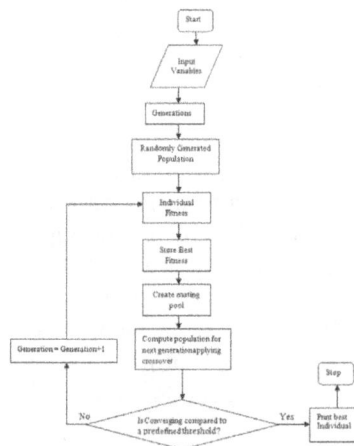

Figure 12.1 Flow chart for genetic algorithm
Source: Author

population naturally evolves to produce a final generation of highly suited chromosomes that represent ideal or nearly effective solutions to the issue.

Simulated annealing

The technique of annealing is used to crystallize metals. Metal atoms that have been heated to a high temperature attain a high energy level and move. Atoms can reach a balanced condition with the least amount of energy when cooling is controlled. Expected value of energy change is $P(\Delta E) = e^{-\Delta E/KT}$ and this process called simulated annealing [8, 9].

In order to obtain global optima, the function that is required to first heated to a high degree and then gradually cooled. First, a random number r, ranging from 0 to 1, is obtained by calculating $e\text{-}\Delta E/KT$. It is kept if $r \leq e\text{-}\Delta E/KT$; if not, it's thrown away and then it is proceeded to the following step.

The main success of this process annealing is determining the primary temperature and the number of repetitions. There will be more iterations when initial temperature is high for convergence; if the initial temperature is low, there won't be enough iterations to thoroughly explore the search space before arriving at actual optima. Maximum number of iterations to reach the quasi-equilibrium stage; however, this will result in a longer calculation time.

Algorithm developed using simulated annealing approach [1]

Step1: Initial point $x(0)$, a termination criterion ε, set T at high temperature, no. of iteration at a particular temperature is n: set t = 0.

Step2: Neighboring point: x (t + 1) = N x (t). neighborhood is created.

Step3: if $\Delta E = E[x (t +1)] - E[x (t)] < 0$, set: If $r \leq e^{-\Delta E/KT}$: set $t = t +1$, else go to step2.

$t = t +1$; create r
in the range (0, 1).

Step4: if $\mid$ x (t+1))-x (t) $\mid$ < ε and T is small, terminate; else lower the value of T according to a cooling schedule: go to step 2;

Step5: end.

Pattern Search Algorithm

Following a series of steps, this process looks for the path. The pattern direction is the name given to the direction [1, 7, 11].

Two points are used to produce a pattern move in the Hooke-Jeeves pattern search technique. Repeatedly developing a set of search parameters operators indicates the pattern of searching directions. By following the flow chart direction of the search will lead to the next location.

Flow Chart for pattern search

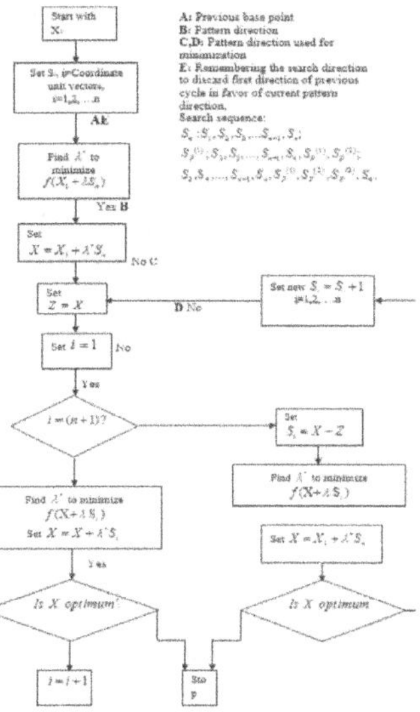

Figure 12.2 Flow chart for pattern search
Source: Author

Case studies:

Three distinct approaches—pattern search, simulated annealing, and genetic algorithms—have been used to create case studies on the design challenge. The outcomes are listed below.

Genetic Algorithm
The outcomes of executing the program using the genetic algorithm are displayed below (Figure 12.3 and Table 12.1).

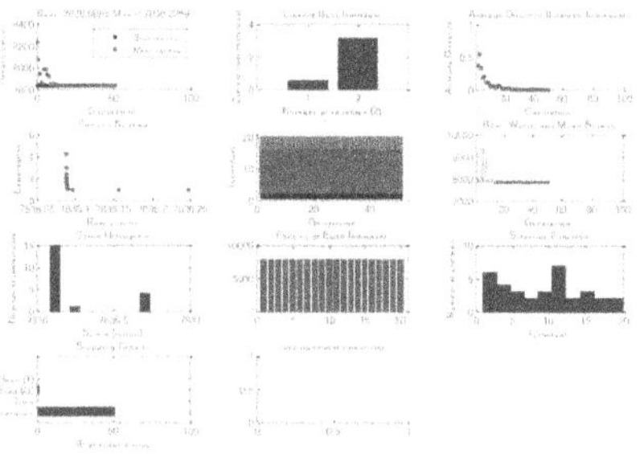

Figure 12.3 Results using genetic algorithm [18]
Source: Author

Simulated Annealing
The results obtained by genetic algorithm are given below (Figure 12.4 and Table 12.2).

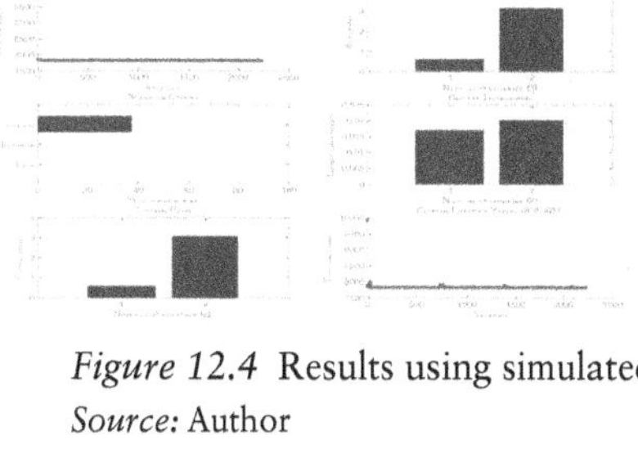

Figure 12.4 Results using simulated annealing [18]
Source: Author

Pattern search

The results obtained by pattern search are given below (Figure 12.5and Table 12.3).

Conclusion

In this work, the material cost of a small, single phase dry transformer has been optimized through the use of three distinct soft computing techniques: pattern

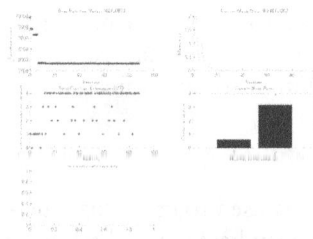

Figure 12.5 Results using pattern search algorithm Table 12.1. Convergence using GA

Source: Author

Table 12.1 Convergence using GA

Generation_f-count	Best f(x)	Variation of mean f(x)	Stall generations	
1	40	7850	8030	0
2	60	7841	7910	0
3-13	80-280	7838	7848-8027	0-4
14-21	300–440	7837	7838-7869	0-5
22-51	460–1040	7836	7836-7838	1-4

Source: Author

Table 12.2 Convergence using SA.

Iteration_f-count	Best f(x)	Current f(x) variation	Mean temperature variation	
0	1	9924.45	9924.45	100
10-30	11-31	7845.23	7859.81-7941.28	56.88
40	41	7842.17	7842.17	12.2087
50	51	7837.17	7837.17	7.30977
60-820	61-825	7836.21	7837.71-7836.2	4.3766-3.805e-007
830-890	835-895	7836.19	7836.19-7836.2	0.14636-0.006743
900	905	7836.17	7836.17	0.00403716
910	915	7836.15	7836.16	0.0024172
920	925	7836.14	7836.14	0.00144726
930	935	7836.13	7836.13	0.000866531
940-1200	945-1207	7836.12	7836.12- 7866.07	35.7863-3.073e-6
1210	1217	7836.11	7836.11	0.00795039
1220	1227	7836.1	7836.1	0.00476019
1230-2230	1237-2243	7836.09	7836.09-7867.44	0.00265-9.51e-7

Source: Author

Table 12.3 Convergence using PSA

Iteration_f-count	f(x)	Mesh size	Method	
0	1	9924.45	1	
3-6	3-6	9146.92	2-0.25	
8-12	8-12	7842.8	0.0625	
13	17	7841.85	0.125	
14-17	21-32	7836.72	0.25-0.03125	Successful polls are applied in all iteration
18-20	34-42	7836.32	0.0625-0.01563	
21-22	45-49	7836.2	0.03125-0.01563	
23-25	51-59	7836.18	0.03125-0.007813	
26-27	63-67	7836.15	0.01563-0.007813	
28-29	68-72	7836.12	0.01563-0.007813	
30-36	76-79	7836.1	0.01563-0.003906	
37-88	99-277	7836.09	0.007813-9.537e-007	

Source: Author

search, simulated annealing, and genetic algorithms. Two design variables, the window height/width ratio, or Rw, and the emf constant K, have been selected as critical variables. The cost of production is directly impacted by these two factors. To attain better performance, copper has been used as the conductor material and CRGOS as the core material. The current density in copper and the flux-density in iron have been adjusted to make sure that the design constraints and criteria are met. The objective function, which is the total cost of iron and copper, was formulated using the design variables. The lowest cost has been established using the three previously mentioned methods.

After a small number of iterations, all three ways have produced the same outcome. An auxiliary program has calculated the dimensions of the optimized transformer based on the value of the design variables selected by the GA, SA, and PS algorithms (the cost has somewhat for substituting nearest integers for real variables at

the relevant places). Calculations have also been made for the performance variables. It should be mentioned that no design restrictions have been broken. This research aims to demonstrate that nonclassical or probabilistic optimization algorithms yield significantly better results than traditional techniques such as the gradient search methodology. Additionally, a comparison of several probabilistic optimization strategies was done, and the best outcome was discovered to be achieved using simulated annealing. However, this outcome might be further enhanced by applying a skillful fusion of many nonclassical soft computing techniques to have a synergistic impact without using them

separately, which is probably what we will try to do in the future.

References

[1] Abdelwanis, M. I., Abaza, A. El-Sehiemy, R. A., Mohamed, I., & Hegazy, R. (2020). Parameter estimation of electric power transformers using coyote optimization algorithm with experimental verification. *IEEE Access*, 8, 50036–50044. https://doi.org/10.1109/ACCESS.2020.2978398

[2] Abu-Siada, A., Mosaad, M. I., Kim, D., & El-Naggar, M. F. (2020). Estimating power transformer high frequency model parameters using frequency response analysis. *IEEE Transactions on Power Delivery*, 35(3), 1267–1277.

[3] Aghmasheh, R., Rashtchi, V. & Rahimpour, E., (2018). Gray box modeling of power transformer windings based on design geometry and particle swarm optimization algorithm. *IEEE Transactions on Power Delivery*, 33(5), 2384–2393.

[4] Basak, R., Yahoui, H., & Siauve, N. (2017). Design of single-phase transformer through different optimization techniques. *International Journal of Information and Communication Sciences*, 2(3), 30–34

[5] Bhowmick, D., Manna, M. & Chowdhury, S. K. (2018). Estimation of equivalent circuit parameters of transformer and induction motor from load data. *IEEE Transactions on Industry Applications*, 54(3), 2784– 2791.

[6] Ćalasan, M. P., Jovanović, A., Rubežić, V., Mujičić, D. & Deriszadeh, (2020). Notes on parameter estimation for single-phase transformer. *IEEE Transactions on Industry Applications*, 56(4), 3710–3718. https://doi.org/10.1109/TIA.2020.2992667

[7] El-Fergany, A. A., Hasanien, H. M., & Agwa, A. M. (2019). Semi-empirical PEM fuel cells model using whale optimization algorithm. *Energy Conversion and Management*, 112197. https://doi.org/10.1016/j.enconman.2019.112197

[8] Illias, H. A., Mou K. J., & Bakar, A. H. A. (2017). Estimation of transformer parameters from nameplate data by imperialist competitive and gravitational search algorithms. *Swarm and Evolutionary Computation*, 36, 18–26.

[9] Kazemi, R., Jazebi, S., Deswal, D. & León, F. D. (2017). Estimation of design parameters of single-phase distribution transformers from terminal measurements. *IEEE Transactions on Power Delivery*, 32(4), 2031–2039.

[10] Li, S., Chen, H., Wang, M., Heidari A. A., & Mirjalili, S. (2020). Slime mould algorithm: a new method for stochastic optimization. *Future Generation Computer Systems*, 111, 300–323.

[11] Martin, P. C. A., Vesna, R., & Adel, D. (2020). Notes on parameter estimation for single-phase transformer. *IEEE Transactions on Power Delivery*, 29(3), 141–149.

[12] Orosz, T., Borbély, B., & Tamus, Z. Á. (2017). Performance comparison of multi design method and meta-heuristic methods for optimal preliminary design of core-form power transformers. *Periodica Polytechnica Electrical Engineering and Computer Science*, 61(1), 69–76.

[13] Suja, K., & Yuvaraj, T. (2021). Transformer health monitoring system using android device. *In 2021 7th International Conference on Electrical Energy Systems (ICEES)*, pp. 460–462

[14] Toren, M. (2023), Optimization of transformer parameters at distribution and power levels with hybrid Grey wolf-whale optimization algorithm. *Engineering Science and Technology*, 43, 101439.

[15] Yilmaz, Z., Okşar, M. & Başçiftçi, F. (2017). Multi-objective artificial bee colony algorithm to estimate transformer equivalent circuit parameters. *Periodicals of Engineering and Natural Sciences*, 5 (3), 271– 277.

[16] Zhao, W., Wang, L., & Zhang, Z. (2019). A novel atom search optimization for dispersion coefficient estimation in groundwater. *Future Generation Computer Systems*, 91, 601–610.

[17] Zhao, X., & Gao, X. S. (2007). Affinity genetic algorithm. *Journal of Heuristics*, 13, 133–150. https://doi.org/10.1007/s10732-006-9005-z

13 Torque ripple control using hybrid converter in light weight electric vehicle

Sunam Saha[1,a], Madhurima Chattopadhyay[2,b] and Moumita Mukherjee[3,c]

[1]Assistant Professor, Department of EEE, Adamas University, Kolkata, West Bengal, India

[2]Professor and H.O.D., AEIE, Heritage Institute of Technology, Kolkata, West Bengal, India

[3]Professor and Dean, R&D, Adamas University, Kolkata, West Bengal, India

Abstract

Electric vehicle (EVs) have become an attractive option and are rapidly growing as an electric powered vehicle now-a-days due to their ability to reduce environmental pollution and its advantages over fossil fuel-based transportation. The major focus on research in EVs is to increase the range of driving and optimize the consumption of energy. Both mentioned research aims can be obtained by improving the performance of the motor used in the electric vehicle. The Brushless direct current motor is one of the popular choices of motor when considering dynamic performance for light weight EVs. In this paper, a hybrid boost converter is proposed to control electromagnetic torque in light weight EVs. The torque control is obtained by controlling the pulse width modulation of the proposed converter. Along with torque control, torque ripple suppression is obtained with the help of the proposed model in Simulink platform. It has been observed from the electromagnetic torque graph obtained in Simulink model, that the torque ripple gets reduced to 29% with the application of the proposed converter in light weight electric vehicle when compared with traditional boost converter fed electric vehicle drive.

Keywords: Boost converter, DC-DC converter, electric vehicle, light weight EV, torque control

Introduction

Electric vehicle (EV) has been considered as a potential option to address the major global issue of environmental pollution and energy saving [1, 2]. Electric vehicle will enable us to electrify vehicle technology and address the global challenge. Electrification of automobiles includes the usage of efficient motor, electronics controllers for motor and advances in power converter technologies [3, 4]. Even though the usage of EV offers many advantages including simplicity, cheaper cost of maintenance and higher efficiency, there are certain issues which are to be addressed for better performance. The major issue is the driving range and smooth operating torque in EV when compared to the IC engine vehicles [5–7].

Electric motors play an important role in improving the performance and power management of EV [8]. Brushless DC motor has become a popular choice in light weight electric vehicle as compared to other motors due to different benefits offered by BLDC including high starting torque, lower noise, and wide range of speed [9]. However, the application of BLDC introduces unwanted torque ripple that prevent smooth operation of the drive [10]. Therefore, different controllers are being widely researched to offer smooth operation of BLDC based EV.

One of the important controllers in BLDC based EV is the motor controller. Motor controllers help in the control of speed, torque, and rotor position at different input voltages. For better performance of EV, proper control of the BLDC motor is required. To control the rotating system with minimized torque ripple, the development of motor controller is a challenging task, which can be accomplished with the help of advanced power converter application [4, 5]. In this article a hybrid power converter is proposed to be utilized for light weigh EV operation.

The basic block diagram of the BLDC based EV is presented in Figure 13.1. The proposed control is an open loop control, which makes it simple in operation. In this work, we propose to use a DC-DC power modulator as a connector from battery to inverter for controlling the torque output generated of the EV drive by modulating the duty ratio or the switch on time of the DC power modulator switches.

In this work, Section 1 introduces the work. The BLDC based light weight EV is elaborated in Section 2. Section 2 discusses a hybrid converter operation along with BLDC motor. Section 3 describes the result

[a]sunamsaha.ee@gmail.com, [b]madhurima.chattopadhyay@heritage.edu, [c]moumita.mukherjee@adamasuniversity.ac.in

DOI: 10.1201/9781003663348-13

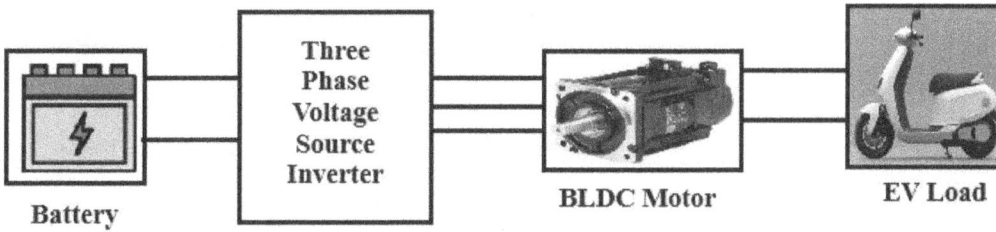

Figure 13.1 Basic block diagram of BLDC based light weight electric vehicle
Source: The figure 13.1 is the basic block diagram representation designed by the author

obtained in Simulink platform after simulation of the proposed drive. Section 4 concludes the paper.

BLDC Based Light Weight Electric Vehicle

The BLDC motors are mostly popular for electric scooter and e-rickshaw application. A basic BLDC based EV is presented in Figure 13.1. The basic block diagram representation of the drive shows that EV is operated with BLDC motor. The motor is operated using the output from a three-phase voltage source inverter. The inverter is given input supply from a battery source. In this work, we presented the application of a hybrid boost converter to provide the DC input voltage to the inverter. The DC is a controllable DC, which is obtained by modulating the duty ratio or the pulse width modulation provided to the switch of the hybrid boost converter.

Hybrid boost converter
Traditional boost converters help in obtaining a gain in voltage. However, the voltage gain obtained using conventional boost is limited due to the limitation of switching stress with increased duty ratio [10, 11]. The Hybrid boost converter comprises of a SEPIC converter along with boost converter. It consists of one MOSFET (M), with two inductor coils (La and Lb). Figure 13.2 presents the converter circuit. The operation of the power modulator can be segregated into two modes based on the current load, i.e. continuous current conduction mode and discontinuous current conduction. Here, the circuit is considered in its continuous conduction mode. The continuous conduction mode can be discussed by dividing the operation further into two more modes, depending on the switching condition of the MOSFET. The circuit equations based on the switching condition of the MOSFET is presented in the Table 13.1.

The value for the circuit component can be obtained with the help of the inductance and capacitance equation presented in Table 13.1 by assuming a small value of current ripple and voltage ripple [12–14]. Also, the

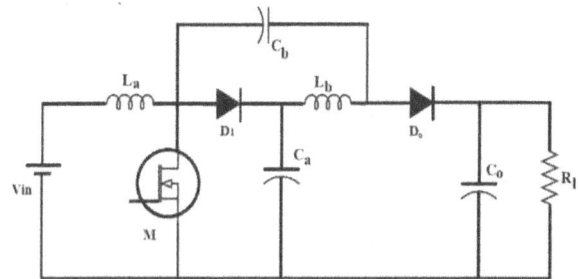

Figure 13.2 Circuit diagram of hybrid boost converter
Source: The circuit is drawn by the author

Table 13.1 Operation of the hybrid boost converter

Parameters	MOSFET is turned on	MOSFET is turned off
Voltage across inductor (L_a)	$V_a = V_{in}$	$V_a = V_{in} - V_{c1}$
Voltage across inductor (L_b)	$V_b = V_{cb} - V_{ca}$	$V_b = V_{c2}$
Inductance	$L_a = L_b = \dfrac{V_{in}DT}{\Delta i_L}$	
Capacitance	$C_a = C_b = \dfrac{I_o T}{\Delta V_c}$	
Relation between Input and output voltage	$\dfrac{V_o}{V_{in}} = \dfrac{1+D}{1-D}$	

Source: The table 13.1 is obtained by applying KCLand KVL(kirchoff current law and kirchoff voltage law in the circuit presented in figure 13.2

voltage gain is obtained by applying voltage second balance across the inductor in both the switching state. The circuit for hybrid boost converter is designed in a Simulink platform using the parameter specifications presented in Table 13.2.

Table 13.2 Parameter specification of the hybrid boost converter

Parameter	Value
Inductor (La)	40µH
Inductor (Lb)	40µH
Capacitor (Ca)	0.3µF
Capacitor (Cb)	10µF
Capacitor (C_0)	10µF
Resistor (R_0)	350Ω
Duty Ratio (D)	0.8
Frequency	50kHz

Source: The table 13.2 data is obtained using Table 13.1 for a frequency of 50kHz and assuming 1% ripple content

Brushless DC motor (BLDCM)

Brushless DC motors have gained a lot of attention-recentcent years in different industrial and domestic applications. It is also gaining importance in the automobile sector owing to its many advantages like light weight, lower noise, and high range of speed. BLDC motors are popular choice for the application in light weight EV like e-scooter and e-rickshaw. The BLDC hub motors are popular with electric scooter applications because of the advantage of retrofitting [14, 15]. The major disadvantage in using BLDC motor for EV is that it becomes difficult to deliver uniform torque, which is a major concern. Hence, different control techniques are presented in the literature to reduce the ripple in torque of BLDC motor [4]. The BLDC motor in light weight EV is operated through a three-phase voltage source inverter by feeding an input from a fixed battery source of voltage ranging from 24 Volt to 48 Volt [16, 6]. The electromagnetic torque ripples can be controlled by modulating the DC input of the inverter. The control of input voltage is obtained using the hybrid boost converter.

Voltage variation for torque ripple reduction

Torque ripple reduction has been a trending research area in electric vehicle applications. To determine the optimum torque ripple reduction method, the sources of torque ripple are required to be understood properly. The sources of torque ripples may be classified based on the following categories: (a) motor construction (b) motor parameter (c) motor control strategies. The torque ripple that exists due to the motor construction and fixed motor parameters like air gap length and mutual flux are difficult to control and suppress. However, the torque ripple that exists due to the motor control strategies like pulse width modulation, non-ideal back emf and current commutation cab be controlled by various control mechanism [8, 17, 18].

A simple and easiest way to minimize the ripple content in torque is by varying the supply voltage to the inverter. This is obtained by connecting a DC power modulator. Here, we have implemented a hybrid DC-DC converter. Therefore, by varying the PWM signal provided to the MOSFET of the DC-DC hybrid boost converter, torque ripple control can be obtained.

Result and Discussion

In the proposed electric vehicle Simulink model, a load torque of 2 Nm is connected. The electromagnetic torque of the proposed EV is obtained by varying the pulse width modulation from 55% to 75%. The electromagnetic torque waveform obtained is presented in the Figure 13.3. The torque ripple is then calculated using the maximum and minimum electromagnetic torque and average torque values. The ripple so obtained is presented in the Table 13.3. It can be observed that the minimum value of torque ripple obtained is 29%, which is obtained at a duty ratio of 65%. Also, the torque ripple is calculated for a BLDC based EV, where inverter is supplied using a conventional boost converter. The ripple obtained in this case is 48%. Thus, it can be stated that with the application of hybrid boost converter as a supply for inverter, the torque ripple in BLDC motor reduces from 48% to 29%. This reduction is obtained without using any difficult controlling mechanism.

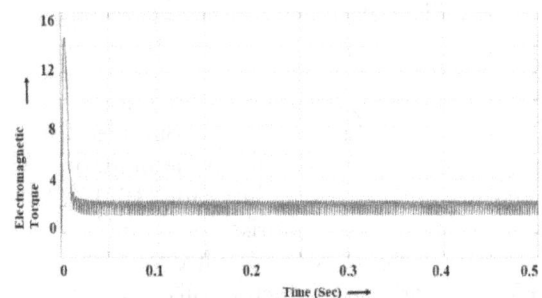

Figure 13.3 Electromagnetic torque of BLDC motor

Source: The graph is obtained in MATLAB simulink on simulation of the circuit presented in figure 13.2 using table data of 13.2

Table 13.3 Torque ripple variation with voltage variation

Duty ratio	Torque ripple (%)
0.55	32.87
0.6	33.85
0.65	29.19
0.7	31.11
0.75	32.36

Source: The data is obtained from simulation of the proposed circuit in MATLAB

Conclusion

The control using varying supply voltage is the easiest and most effective way of control for torque and speed of the electric vehicle. It has been observed that on application of the hybrid boost converter the ripple content in electromagnetic torque reduces to 29% as compared to electric vehicle, where the voltage source inverter is fed from a conventional boost converter. The method proposed does not use any complex control technique and can be considered as a simple control method. However, the proposed control is undesirable for electric vehicle application in hilly region. Therefore, further research may be continued in torque ripple control using V/F method by keeping voltage (V) constant and varying frequency (F).

References

[1] Torres-Sanz, J. A., Garrido, V., Martinez, F. J., & Marquez-Barja, J. M. (2021). A review on electric vehicles: Technologies and challenges. *Smart Cities*, 4(1), 372–404. doi:10.3390/smartcities4010022.

[2] Men X., Guo, Y., Wu, G., Chen, S., & Shi, C. (2022). Implementation of an improved motor control for electric vehicles. *Energies*, 15(13), 4833. doi: 10.3390/en15134833.

[3] Shen, J. X., & Tseng, K. J. (2003). Analyses and compensation of rotor position detection error in sensorless PM brushless DC motor drives. *IEEE Transactions on Energy Conversion*, 18(1), 87–93. doi:10.1109/TEC.2002.808339.

[4] Shi, T., Guo, Y., Song, P., & Xia, C. (2010). A new approach of minimizing commutation torque ripple for brushless DC motor based on DC–DC converter. *IEEE Transactions on Industrial Electronics*, 57(10), 3483–3490. doi: 10.1109/TIE.2009.2038335.

[5] Hais, Y. R., Rameli, M., & Effendie, R. (2018). Design of torque control strategy for hybrid electric vehicle (HEV) with maximum work of self - commutation brushless DC motor using fuzzy-PI. *IPTEK Journal of Proceedings Series*, 4(1), 39. doi: 10.12962/j23546026.y2018i1.3504.

[6] Banaei, M. R., & Sani, S. G. (2018). Analysis and implementation of a new SEPIC-based single-switch buck-boost DC-DC converter with continuous input current. *IEEE Transactions on Power Electronics*, 33(12), 10317–10325.

[7] Saha, S., Chowdhury, D., Chattopadhyay, M., & Mukherjee, M. (2024). A novel technique for torque ripple suppression in BLDC motor drive using switched capacitor-based SEPIC converter. *International Journal of Power Electronics and Drive Systems*, 15(1), 27–34.

[8] Mohanraj, D., Aruldavid, R., Verma, R., Sathiyasekar, K., Barnawi, A. B., Chokkalingam, B., et al. (2022). A review of BLDC motor: state of art, advanced control techniques, and applications. *IEEE Access*, 10, 54833–54869. doi: 10.1109/ACCESS.2022.3175011.

[9] Saha, S., Modal, S., Chattopadhyay, M., & Mukherjee, M. (2022). A scheme for torque ripple minimization in BLDC drive using two-inductor boost converter. In 2022 IEEE International Conference of Electron Devices Society Kolkata Chapter (EDKCON), Kolkata, India, (pp. 1–6). doi: 10.1109/EDKCON56221.2022.10032907.

[10] Lakshmi, M., & Hemamalini, S. (2018). Nonisolated high gain DC-DC converter for DC microgrids. *IEEE Transactions on Industrial Electronics*, 65(2), 1205–1212.

[11] Cao, Y., Samavatian, V., Kaskani, K., & Eshraghi, H. (2017). A novel nonisolated ultra-high-voltage-gain DC-DC converter with low voltage stress. *IEEE Transactions on Industrial Electronics*, 64(4), 2809–2819.

[12] Mamun, A., Arfin, S., & Khan, S. (2021). High gain DC-DC converter for three- wheeler electric vehicles. In 2021 IEEE International IOT, Electronics and Mecha-tronics Conference, Toronto, ON, Canada, (pp. 1–6). doi: 10.1109/IEMTRONICS52119.2021.9422579.

[13] Wu, B., Li, S., Liu, Y., & Smedley, K. M. (2016). A new hybrid boosting converter for renewable energy applications. *IEEE Transactions on Power Electronics*, 31(2), 1203–1215.

[14] Mumtaz, F., Zaihar, N. Y., Sheikh, T. M., Singh, B., Kannan, R., & Ibrahim, O. (2021). Review on non-isolated DC-DC converters and their control techniques for renewable energy applications. *Ain Shams Engineering Journal*, 12, 3747–3763.

[15] Song, J. H., & Choy, I. (2004). Commutation torque ripple reduction in brushless dc motor drives using a single dc current sensor. *IEEE Transactions on Power Electronics*, 19(2), 312–319.

[16] Huang, C. L., Lee, F. C., Liu, C. J., Chen, J. Y., Lin, Y. J., & Yang, S. C. (2022). Torque ripple reduction for BLDC permanent magnet motor drive using DC link voltage and current modulation. *IEEE Access*, 10, 51272–51284. doi: 10.1109/access.2022.3173325.

[17] Weidong, J., & Jinping, W. (2017). Improved control of BLDCM considering commutation torque ripple and commutation time in full speed range. *IEEE Transactions on Power Electronics*, 33(5), 4249–4260.

[18] Arathy Rajeev, V. K., & Prasad, V. (2023). Online adaptive gain for passivity based control for sensorless BLDC motor coupled with DC motor for EV application. *IEEE Transactions on Power Electronics*, 38(11), 13625–13634. doi: 10.1109/TPEL.2023.3288939.

14 Optimizing electric vehicle performance by using MATLAB for NEDC urban and suburban drive cycle

Goutam Kumar Ghorai[1,a], Rohit Nayak[2,b], Chandra Kumar Adak[3,c], Avijit Mondal[4,d] and Soujannya Ballav[5,e]

Ghani Khan Choudhury Institute of Engineering & Technology, Malda, West Bengal, India

Abstract

The objective of this paper is to study the driving range and responsible parameters of an Electrical vehicle (EV). To study the driving range of EV urban and sub-urban driving cycles of the New European driving cycle (NEDC) is taken as a standard driving cycle. In a detailed investigation of NEDC, we may gain a rudimentary understanding of the state of a battery-powered automobile as it progresses through phases such as steady speed, acceleration, and deceleration, which affect driving range and the use of energy. To improve EV performance and provide realistic range estimates, we look at various methods for evaluating parameters such as the energy requirement of EV's, Battery requirement and overall power train efficiency. The studies in many driving conditions provide substantial insights into EVs' performance and driving range during various driving situations. This analysis highlights the real-time parameter estimation and driving range calculation of EVs, which can build up the confidence of users before traveling.

Keywords: Battery capacity, driving cycle, driving range, electric vehicle, new European driving cycle, urban and suburban driving cycle

Introduction

One essential solution to the urgent environmental and financial issues caused by cars that depend on fossil fuels is the electrification of public transportation [1]. As electric vehicles use electrical energy instead of fossil fuel, they have become cost-effective and environmentally friendly. This study article examines electric vehicle operation, range assessment, and parameter assessment, focusing on NEDC. The NEDC is an important way to test how well electric vehicles (EVs) work in real-life driving situations in urban areas and suburbs [2, 3]. We learn crucial pieces of information about when an EV transitions between the three stages of driving (uniform velocity, acceleration, and deceleration) by analyzing these drive cycles. Electric vehicles' range and efficiency are greatly affected by the fluctuations in power usage that occur during these transitions [4, 5]. This study's main goal is to identify important factors that are essential for improving EV functionality [6]. The complex issues of battery capacity, consumption rate of energy, and efficiency of powertrain are explored. To meet Byer's expectations, forecasting the driving range and performance of EVs is an important strategy [7]. We conduct observational experiments in various driving conditions to validate our findings and increase their relevance We stress the importance of analysis of data, modeling with predictive capabilities, and advanced techniques in current time estimation of parameters [8]. These tools can provide accurate and user-friendly range predictions for electric vehicle users, boosting user confidence and supporting a more seamless shift to electric vehicles [9]. EV performance and millage are important factors that lead to market demand throughout the globe. This research enhances these efforts to empower more informed and reliable electric car networks and lead to a greener, more sustainable future.

Working Environment of NEDC Cycle

In NEDC the total running distance is 11.022 km. This driving cycle includes all three stages of driving situations: constant speed, acceleration, and decelerating periods. A total of five driving cycles are depicted in Figure 14.1, four of them being in urban areas and one being in suburban areas. The plot in Figure 14.1 shows the changes in vehicle velocity with time. The physical parameters of vehicle and road parameters are given in Table 14.1 [8].

[a]goutamghorai79@gmail.com, [b]rohitnayak500@gmail.com, [c]chandrakadak04@gmail.com, [d]avijitmondal97212@gmail.com, [e]soujannya01@gmail.com

DOI: 10.1201/9781003663348-14

Table 14.1 Vehicle parameter

Mass of EV (kg)	Coefficient of road rolling resistance	Air drag coefficient	Frontal area (m²)	Radius of tire
1300	0.0131	0.320	2.10	0.285
Factor of rotational mass conversion	Efficiency of Driving System	Efficiency of Battery discharge	Ratio of Transmission	Battery pack voltage
1.0021	0.950	0.90	5.30	320

Source: Author [8]

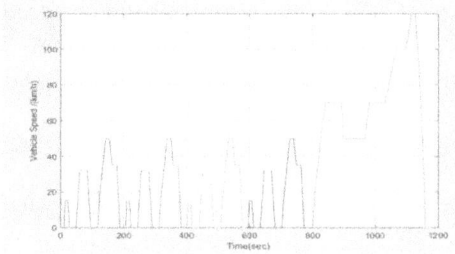

Figure 14.1 Velocity vs time graph for NEDC urban and suburban region
Source: Author

Power Characteristics

Motor power during constant speed
The power demand by a car running on a flat road with constant speed is defined by mathematical equations as

$$P_{m1} = \frac{V_x}{3600\eta}\left(mgf + \frac{C_D A_f V_x^2}{21.15}\right) \qquad (1)$$

Where, Pm_1 = motor power demand. m = mass of the car, $V_{x =}$ constant speed of the car, η = efficiency of driving system, C_D = Coefficient of Air drug, A_f = Frontal surface area, f = Road rolling resistance coefficient, Data in Table 14.2 is used in simulation considering constant speed for NEDC driving cycle.

Table 14.2 Requirement of power for the vehicles in the regions of urban is obtained from equation (1)

Speed (km/h)	Duration (sec)	NEDC range (sec)	Requirement of power (KW)
16	9	16-23	0.7577
31	23	60-85	1.854
36	16	162-178	2.0932
50	13	144-155	3.5827

Source: Author

Table 14.3 Requirement of power for the vehicles in the regions of sub urban is obtained from equation (1)

Speed (km/h)	Duration (sec)	NEDC range (sec)	Requirement of power (KW)
72	50	842-892	6.5764
52	68	898-969	3.5825
72	50	980-1032	6.5766
100	30	1065-1095	14.332
120	10	1115-1125	21.564

Source: Author

Power demand during accelerating period of EV
An electric vehicle's estimated engine power under accelerating and negative acceleration driving conditions on a slope is as follows:

$$P_{M2} = \frac{V}{3600\eta}\left(mgf + \frac{C_D A_f V^2}{21.15} + \delta ma_j\right) \qquad (2)$$

$$V_f(t) = V_0 + 3.6a_j t \qquad (3)$$

Where, Motor power is P_{M2}, Constant speed is V, m = vehicle mass is m, Efficiency is η, Coefficient of rolling resistance is f, Rotating masses conversion factor δ, = Rate of change of speed is a_j, Final speed is V_f, V_0 =Initial speed.

Simulation table for the urban region under acceleration and deceleration using the following parameters [9]

Simulation table for suburban regions under acceleration and deceleration

The power characteristics derived by simulation using data are in Tables 14.2–14.7
The power vs. time graph in Figure 14.2 is plotted according to the data given in the table using equations 1 and 2. This graph shows that NEDC ranges from 1096 sec to 1116 sec. The electric vehicle

Table 14.4 Power requirement in accelerating period of EV for urban region according to equations (2) and (3).

Acceleration (meter/sec²)	Duration (sec)	NEDC range (sec)	Requirement of power (KW)
1.05	5	12-16	6.80
0.69	7	48-54	4.77
0.78	7	54-62	11.65
0.69	7	116-122	4.77
0.50	10	122-133	9.01
0.46	9	133-143	12.5

Source: Author

Table 14.5 Power requirement in the decelerating period of EV for urban regions according to equations (2) and (3)

Deceleration (meter/sec²)	Duration (sec)	NEDC range (sec)	Requirement of power (KW)
-0.82	6	22-27	-4.05
-0.8	10	86-96	-8.2
-0.5	8	155-165	-6.50
-0.97	10	177-178	-11.05

Source: Author

Table 14.6 Power requirement in accelerating period of EV for suburban region according to equations (2) and (3)

Acceleration (meter/sec²)	Duration (sec)	NEDC range (sec)	Requirement of power (KW)
0.69	6	800-808	4.70
0.50	10	808-819	9.01
0.41	10	819-830	11.72
0.39	15	830-840	17.43
0.42	14	965-980	18.24
0.23	35	1030-1065	23.43
0.27	20	1095-1115	34.89

Source: Author

Table 14.7 Power requirement during deceleration period of EV for suburban region according to equations (2) and (3)

Deceleration (meter/sec²)	Duration (sec)	NEDC range (sec)	Requirement of power (KW)
-0.70	9	890-900	-12.15
-0.71	15	1125-1143	-10.23
-1.04	9	1143-1151	-23.62
-1.38	10	1151-1160	-23.36

Source: Author

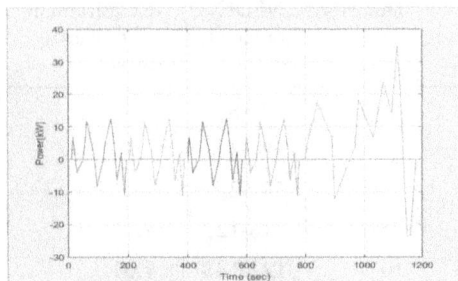

Figure 14.2 Power vs time graph
Source: Author

demand's highest power is 34.89 KW. Therefore, consider a motor with a peak rating of 35Kw.

Power Rating of Motor

Relation between maximum power and rated power of the motor can be written as

Relation of maximum power and rated power of the motor can be written as

$$P_{emax} = \lambda P_e \qquad (4)$$

P_{emax} and P_e are the motor's maximum power and rated power, respectively, overload factor λ. Consider $\lambda = 2.2$. Therefore, the rated power of the driving motor becomes 16 kW. At the topmost running speed of the vehicle, the motor consumes constant power. The maximum speed of driving motor () and maximum speed of the vehicle can be written as

$$\eta_{max} = \frac{V_{xmax} \times i}{0.377r} \qquad (5)$$

Where r is the tire's rolling radius, and i is the transmission ratio.

Motor's maximum speed in driving condition

The relation between rated speed and maximum speed be written as:

$$n_e = \frac{n_{max}}{\beta} \qquad (6)$$

Where n_e is the rated speed of the motor, β is the coefficient of the extended constant power region of the motor [10]. The highest speed of the motor is 5919 rpm, which is approximately equal to 6000 rpm for V_{xmax} = 120km/h. Higher β values result in reduced speed, increased torque, and improved acceleration and climbing capability for electric vehicles [11, 12]. However, the power converter size will increase. β must remain within a reasonable range [13, 14]. Typically, β value ranges from 2 to 4. Assuming β equals 3, the motor's rated speed is 2000 rpm [15, 16].

Determination of Torque

In terms of speed and power, rated torque of the motor is

$$T_e = \frac{9550 P_e}{n_e} \qquad (7)$$

Where T_e, P_e, and n_e are the rated torque, maximum power, and rated torque of the motor, respectively. The motor demands maximum torque as

$$T_{emax} = \frac{9550 P_{emax}}{n_e} \qquad (8)$$

Therefore, the motor's rating should be selected, as its rated torque, 76 N-m, and maximum torque, 167 N-m, are calculated using equations 7 and 8.

Battery Power Capacity

The battery capacity is demanded by the motor running with constant speed on flat road is

$$C_{m1} = \frac{P_{m1}t_{m1}}{3.6V_{batt}} \qquad (9)$$

Where battery capacity is . P_{m1} is the power requirement, V_{batt} is the battery voltage, and t_{m1} is the run time of the motor. At the time of acceleration, the required storage capacity of the battery is

$$C_{m2} = \frac{P_{m2}t_{m2}}{3.6V_{batt}} \qquad (10)$$

Where battery capacity is C_{m2} for power demand P_{m2}, t_{m1} is the run time of the motor. MATLAB simulation is showing the change of batter demand with in Figure 14.3. Mathematical modelling using equation 7,8,9,10 and its simulation results are shown in Table 14.8.

Analysis of urban region

- Battery capacity requirement for Urban Driving cycle when Speed is constant.

Table 14.8 Battery capacity necessary during constant speed. in urban driving cycle, no

Speed in km/h	Duration in sec	NEDC range in sec	Necessary of battery capacity (A.h)
16	9	14-22	0.007
33	23	60-84	0.038
36	14	162-177	0.027
50	13	144-155	0.037

Source: Author

- Battery capacity requirement for Urban Driving cycle during acceleration.

Table 14.9 During acceleration, battery capacity necessary for the urban driving cycle

Acceleration (meter/sec²)	Duration (sec)	NEDC range (sec)	Demand of battery capacity (A.h)
1.04	5	12-16	0.023
0.69	6	50-54	0.024
0.78	7	54-60	0.060
0.69	6	116-125	0.024
0.50	10	125-135	0.086
0.46	10	135-143	0.097

Source: Author

- Battery capacity requirement for Urban Driving cycle during deceleration.

Table 14.10 Battery capacity demand for the urban driving cycle during negative acceleration

Deceleration (meter/s²)	Duration (sec)	NEDC range (sec)	Demand of battery capacity (A.h)
-0.84	6	24-28	-0.011
-0.82	10	86-95	-0.078
-0.53	9	156-165	-0.045
-0.98	11	171-178	-0.096

Source: Author

Analysis of suburban region

- Battery capacity requirement for the suburban Driving cycle during same Speed

Table 14.11 Battery capacity demand for the suburban driving cycle during constant speed

Speed in km/h.	Duration in sec	NEDC range in sec	Necessary of battery capacity in A.h
60	55	840-890	0.285
65	70	900-969	0.214
70	52	980-1032	0.285
100	32	1065-1095	0.368
120	12	1115-1125	0.189

Source: Author

- Battery capacity requirement for suburban Driving cycle during acceleration.

Table 14.12 Battery capacity necessary for the suburban driving cycle during acceleration

Acceleration (meter/sec²)	Duration (sec)	NEDC range (sec)	Demand of battery capacity (A.h)
0.69	5	800-805	0.024
0.50	10	805-816	0.086
0.41	11	816-826	0.101
0.39	13	826-842	0.211
0.42	14	965-980	0.205
0.23	35	1030-1065	0.712
0.27	20	1095-1115	0.605

Source: Author

- Battery capacity requirement for suburban Driving cycle during deceleration.

Table 14.13 Battery capacity demand for the suburban driving cycle during negative acceleration period

Deceleration (meter/s²)	Duration (sec)	NEDC range (sec)	Demand of battery capacity (A.h)
-0.69	8	890-900	-0.084
-0.69	16	1125-1143	-0.142
-1.04	8	1143-1155	-0.164
-1.38	10	1155-1160	-0.202

Source: Author

Variation of Battery demand

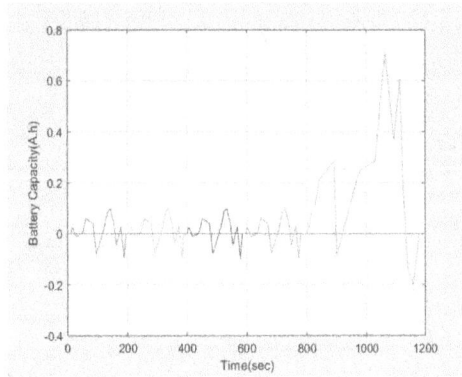

Figure 14.3 Battery capacity vs time obtained from Table 14.9–14.13
Source: Author

Battery capacity demand for urban cycle
The battery capacity vs time graph indicates the fundamental urban cycle needs 0.4281 A.h of battery to maintain consistent and accelerating speeds. During braking, a 20% recovery of brake energy generates 0.0462 A.h. of electrical energy therefore, the predicted battery A.h. limit becomes around 0.3818 A.h [17, 18]. For a typical urban cycle. Over the span of four urban cycles, the total aggregate limit demand amounts to 1.52744 A.h.

Battery capacity demand for suburban cycle
The battery capacity vs. time graph indicates that the suburban cycle needs a battery of 3.29136 A.h. to maintain consistent and accelerating speeds. In addition, 0.593 A.h. of energy is used in the breaking period. Considering a 20% break energy recovery rate, 0.118 A.h. of energy is recovered. Therefore, roughly 3.17 A.h. of battery capacity is required to complete the suburban cycle.

Estimation of Battery Capacity

To complete the total cycle, the battery requirement is approximately 4.7 A.h (1.52 + 3.17 = 4.69 A.h). To complete a 300 km distance, like in the same situation

as NEDC, the power battery limit must be at least 128 A.h. [19]. This is calculated using the formula (4.70011×300)/11.022, which approximates 127.92, rounding up to 128 A.h.

Calculation of Driving Range

The amount of energy used by the EV when driving at a constant speed is

$$E_d = \frac{P_d S_d}{V_x \eta_e} \tag{11}$$

Where energy discharge by the battery is E_d, distance travel by the EV is S_d with uniform speed, and the efficiency of the battery discharge is η_e. Distance travel by the EV is

$$S_d = \frac{V_x t}{3600} \tag{12}$$

Where t and V_x are the run time and speed of the EV.

The amount of energy used by the EV when accelerating is

$$E_j = \frac{P_j S_j}{V_x(t) \eta_e} \tag{13}$$

Where in acceleration of EV the energy discharge by the battery is E_j, distance travelled by the EV is S_j, and efficiency of the battery discharge is η_e.

The distance traveled by the vehicle during acceleration is

$$S_j = \frac{V_{xj}^2 - V_{X0}^2}{25920 a_j} \tag{14}$$

The battery's overall energy load is calculated by

$$E = \frac{Q_m V_{batt}}{1000} \tag{15}$$

Total battery loading is E. Vehicle Capacity Q_m, and battery voltage is V_{batt}. Hence, the travel range of EV considering the situation of NEDC is calculated as

$$S_1 = \sum_{i=1}^{k} S_i \tag{16}$$

Where S_i is the sum of the travelled distance, the intermittent distance traveled by the EV is Si during constant speed, acceleration, and deceleration stages. k is the number of stages completed by the vehicle.

Requirement of Battery Energy

The total energy demand by battery to complete NEDC range is

$$E_1 = \sum_{i=1}^{k} E_i \tag{17}$$

Where total energy demand is E_1, and E_i is the Intermittent energy demand for each stage of NEDC range.

The total distance traveled by the EV is

$$S = \frac{S_1 E}{E_1} \tag{18}$$

Testing the MATLAB code for NEDC driving range reveals that the EV travels the distance S=308.07km. This is consistent with our set criteria [20].

Result

This analysis estimates the following parameters as findings in case of NEDC driving range as

I. Rated power demand = 16 kw
II. Maximum power demand=35 kw
III. Torque demand by the motor= 76 N-m
IV. Peak torque demand by the motor= 167N-m
V. Motor's rated velocity= 2000 rpm
VI. Motor's rated maximum velocity =6000 rpm
VII. Battery capacity demand=128 A.h.
VIII. Total driving range=308 km

Conclusion

This research delves into electric vehicle (EV) driving range during NEDC urban and suburban cycles, focusing on key moments in velocity transitions—constant speed, acceleration, and deceleration. Accurate parameter estimation, covering factors like battery limit, energy utilization, and powertrain effectiveness, is crucial for comprehensive EV performance evaluation. Real-world experiments, incorporating speed variations, validate estimation techniques for diverse driving conditions. The study highlights the importance of predictive modeling, emphasizing constant speed, acceleration, and deceleration dynamics. This contributes to a nuanced understanding of EV behavior and enhances overall range and performance assessment, supporting efficient and sustainable electric mobility in an era of increasing adoption.

References

[1] Mamo, T., Gopal, R., Yoseph, B., Tamirat, S., & Seifu, Y. (2023). Numerical study on low speed electric car for public transportation to predict its aerodynamic performance and stability. *Engineering and Technology Journal*, 8(5), 2183–2190.

[2] Sharmila, B., Srinivasan, K., Devasena, D., Suresh, M., Panchal, H., Ashok Kumar, R., et al. (2022). Modeling and performance analysis of electric vehicle. *International Journal of Ambient Energy*, 43(1), 5034–5040. DOI: 10.1080/01430750.2021.1932587.

[3] Yu, H., & Huang, M. (2008). Potential energy analysis and limit cycle control for dynamics stability of in-wheel driving electric vehicle. In 2008 IEEE Vehicle Power and Propulsion Conference, Harbin, (pp. 1–5).

[4] Rahman, K. M., & Ehsani, M. (1996). Performance analysis of electric motor drives for electric and hybrid electric vehicle applications. In Power Electronics in Transportation, (pp. 49–56). IEEE.

[5] Skuza, A., & Jurecki, R. S. (2022). Analysis of factors affecting the energy consumption of an EV vehicle – a literature study. In IOP Conference Series: Materials Science and Engineering, (Vol. 1247, no. 1, p. 012001). IOP Publishing. DOI 10.1088/1757-899X/1247/1/012001.

[6] Muratori, M., Moran, M. J., Serra, E., & Rizzoni, G. (2013). Highly-resolved modeling of personal transportation energy consumption in the United-States. *Energy*, 58, 168–77.

[7] CarFolio (2021). BMW i4 e-Drive 40 specifications technical data. Available at: www.carfolio.com/bmw-i4-edrive40-728508 (accessed May, 2023).

[8] Tata Motors (2022). Tata Nexon EV specifications. Available at: www.nexonev.tatamotors.com/features/ (accessed June, 2023).

[9] Vempalli, S. K., Ramprabhakar, J., Shankar, S., & Prabhakar, G. (2018). Electric vehicle designing, modeling and simulation. In 2018 4th International Conference for Convergence in Technology (I2CT), (pp. 1–6).

[10] Rahman, K. M., Fahimi, B., Suresh, G., Rajarathnam, A. V., & Ehsani, M. (2000). Advantages of switched reluctance motor applications to EV and HEV: design and control issues. *IEEE Transactions on Industry Applications*, 36(1), 111–121.

[11] Cheng, K. W. E. (2009). Recent development on electric vehicles. In 3rd International Conference on Power Electronics System and it's Application.

[12] Bradley, M. J. (2013). Electric Vehicle Grid Integration in the U.S., Europe, and China. MJ Bradley and Associates, Tech. Rep.

[13] Chan, C. C., Chau, K.T., Jiang, J. Z., Xia, W., Zhu, M., & Zhang, R. (1996). Novel permanent magnet motor drives for electric vehicles. *IEEE Transactions on Industrial Electronics*, 43, 331–339.

[14] Smith, M., & Castellano, J. (2015). "Costs Associated With Non-Residential Electric Vehicle Supply Equipment: Factors to consider in the implementation of electric vehicle charging stations".

[15] Clint, J., Gamboa, B., Henzie, B., & Karasawa. A. (2015). Considerations for corridor direct current fast charging infrastructure in California. California Energy Commission. Available at: http://www. energy. ca. gov/2015publications/CEC-600-2015-015/CEC-600-2015-015. pdf (accessed March 3, 2017).

[16] Chan, C. C., Jiang, J. Z., Chen, G. H., & Chau, K. T. (1993). Computer simulation and analysis of a new polyphase multipole motor drive. *IEEE Transactions on Industrial Electronics*, 40, 570–576.

[17] Zhan, Y. J., Chan, C. C., & Chau K. T. (1999). A novel sliding-mode observer for indirect position sensing of switched reluctance motor drives. *IEEE Transactions on Industrial Electronics*, 46, 390–397.

[18] Nunes, P., Brito, M. C., & Farias, T. (2013). Synergies between electric vehicles and solar electricity penetrations in Portugal. In 2013 World Electric Vehicle Symposium and Exhibition, (pp. 1–8).

[19] Mwasilu, F., Justo, J. J., Kim, E. K., Do, T. D., & Jung, J. W. (2014). Electric vehicles and smart grid interaction: a review on vehicle to grid and renewable energy sources integration. *Renewable and Sustainable Energy Reviews*, 34, 501–516.

[20] Ates, M. N., Gunasekara, I., Mukerjee, S., Plichta, E. J., Hendrickson, M. A., & Abraham, K. M. (2016). In situ formed layered-layered metal oxide as bifunctional catalyst for li-air batteries. *Journal of the Electrochemical Society*, 163(10), A2464–A2474.

15 Safeguarding mechanisms against cyberattacks in smart grids employing grid forming inverters: ensuring stability and resilience

Taha Selim Ustun

Senior Researcher, Fukushima Renewable Energy Institute, (AIST), FREA, Japan

Abstract

Grid forming inverters (GFIs) are becoming more popular because they help manage frequency in the grid. This capability addresses the natural limits of renewable energy systems. Many companies are exploring how GFIs can be used in their networks and are studying their effects on power system operations. Tests in controlled environments show that GFIs can provide the necessary virtual inertia support. However, these devices actively try to adjust system frequency, which could lead to unexpected problems. If a hacker compromises the GFI control, it could result in serious consequences. With the integration of information technologies (IT) and automation in power systems, cybersecurity has become a significant threat. Most research in this area focuses on large power plants connected to the transmission grid. This work analyzes how such attacks affect power system networks. The focus of this work is to provide cybersecurity recommendations to help secure smart grids with grid-forming inverters against cyberattacks.

Keywords: Certificate based authentication, cybersecurity, IEC 61850, IEC 62351, key exchange mechanisms, smartgrids

Introduction

The shift to renewable energy sources is essential for tackling urgent environmental issues. These include climate change and air pollution [1–3]. Unlike fossil fuels, renewable sources like solar, wind, and hydroelectric power provide clean and sustainable alternatives. They significantly reduce greenhouse gas emissions and limit environmental harm [4]. By using renewable energy, societies can lessen the negative effects of climate change. They can also protect natural ecosystems and promote environmental sustainability [5, 6]. Therefore, widespread use of renewable technologies is vital for reaching global climate targets [7–10].

Addition of renewable energy resources to the power systems brings about many issues which further complicates operation [11]. Firstly, these intermittent resources are not dispatchable as traditional generation methods [12–15]. This fact makes it difficult to control voltage and frequency with deep penetration as well as supply-demand matching [16–19]. Novel approaches are required to address these challenges [20].

Researchers have been working on new devices that can not only inject generated power into the system but also provide much needed auxiliary support. Devices such as power conditioning systems, grid forming and smart inverters are to name a few. They take active part in sustaining the two crucial parameters of power systems: system voltage and frequency [21]. Grid forming inverters take this one step further and tries to establish a voltage or frequency reference that can be followed by other generators to form a standalone system [22].

It is obvious from their operation requirements that such devices are very active and dynamic. This necessitates their control algorithm and operation points to be dynamic as well. Unlike traditional systems, these devices need constant interaction with other entities and control systems [23]. These happen at real-time or pseudo real-time and improve system operation a lot [24–27].

Considering the number of devices and the diversity of their manufacturers in a power system, standardized communication protocols, datasets and protocols are a must to achieve a common language. When achieved, real-time participation of balancing or storage devices can be realized [28, 29]. When a fully connectivity is achieved, a more efficient grid can be implemented [30–32].

An unwanted side effect of such high connectivity in power systems is cybersecurity. This unprecedented data exchange creates issues on privacy of users and it may compromise safe operation of power systems. Therefore, research is needed to ensure this standardized communication is performed in a reliable way [33]. In new smartgrids many devices exchange data

selim.ustun@aist.go.jp

DOI: 10.1201/9781003663348-15

with each other as well as central control and management entities. This can be intercepted, stolen, manipulated and breached [34].

As recent events have shown, a small breach in an obscure part of the network can be effectively manipulated to reach different parts of the entire network. Considering the different layers of a power system, such as user information, communication layer, transaction layer, device layer etc, this can have very significant ramifications [35–38].

All of these considerations made it clear to professionals, researchers and policy-makers that future power systems require strong cybersecurity measures in place [39–42].

Continual research is being conducted to create encryption protocols, access control mechanisms, and intrusion detection systems specifically designed for the requirements of smart grids [43, 44]. Progress in machine learning, artificial intelligence, and blockchain tech have potential to enhance cybersecurity and privacy in these systems [45–47].

Ongoing research is striving to establish robust frameworks and standards to decrease privacy and cybersecurity risks in smart grid implementations by promoting collaboration among power systems engineers, cybersecurity experts, and policymakers [48–50]. Continual research and innovation are necessary to guarantee the resilience, security, and privacy of future energy systems as smart grids develop.

The rest of the paper is organized as follows; next section gives an overview of grid forming inverter operation. Section III discusses what kind of communication is required and related vulnerabilities. Examples of safeguard mechanisms to mitigate these are also presented in Section IV. Section V draws the conclusions of the paper.

Operation of Grid-Forming Inverters

Grid-forming inverters are crucial in contemporary power systems, particularly in microgrid and distributed energy resource contexts. These inverters have the ability to function separately in order to create and uphold grid stability.

They can control and manipulate frequency as well as voltage. They can use different controllers such as proportional integral (PI) or proportional integral derivative (PID) controllers. Other emerging controller topologies are also being tested and used.

By doing so, grid forming inverters try to establish a reference value that can be followed by other resources and or generators present in the system. Conventional controllers with long history of validation can be utilized. Alternatively, novel topologies can be used to increase efficiency and effectiveness.

These inverters can also serve under different grid conditions and loading situations. Different control algorithms are being investigated such as adaptive control algorithms or model predictive control (MPC) approaches. The benefit of these systems is that they can anticipate how a system will behave and preemptively take steps for increased stability.

They can also take part in droop control, a smart way of sharing the load utilized by large synchronous generators in traditional power systems. It is very robust and effective in distributing the load present in a system to different generators serving that system.

Droop curves can be designed in such a way that larger generators can take a bigger portion of the burden while smaller ones can contribute in proportion to their capacity. Advanced droop control approaches take into other events and parameters in the power system, instead of merely considering frequency change.

Communication Requirements and Related Vulnerabilities

As grid forming inverters become more popular and proliferate in future power systems, the need to increase their ability to operate and communicate with each other as well as other power system equipment becomes more apparent.

In any communication network, such needs are serviced by communication protocols so that different players can speak a common language and successfully exchange information. Modbud, DNP3 and recent rising start of this domain IEC 61850 can be mentioned as examples.

They enable different devices, most probably manufactured by different companies, to be able to use same variable names, exchange messages that are legible to everyone. There are different working groups and standardization efforts to include different subsections and devices present in a power system in this connected structure. For example, IEEE 1547 includes a set of recommendations so that distributed energy resources can be connected to each other in a power system.

Such recommendations ensure that devices such as grid forming inverters fulfill requirements outlined in grid codes. They are only connected to physical systems after ensuring that compatibility with these requirements and recommendations.

Traditional power systems were designed considering unidirectional power flow form bulk generation sites to bulk consumption sites. They did not consider dynamic distribution systems or bilateral power flow [51].

Therefore, dynamic devices such as smart inverters or grid forming inverters do not align well with traditional power system design and operation. In order to ensure mass deployment of such devices, it is required to tweak the system operation accordingly.

One main operational approach that needs to be changed it the control settings. Traditionally they are set as deterministic values that need to be followed. This did not pose a big problem as power systems rarely every changed. However, active distribution systems with high amount of dynamics require these fixed control parameters to be dynamic as well.

This can be achieved by constantly observing the system, monitoring the changes and system parameters and intervene with the system operation as required. It is obvious from this explanation that is requires a high volume of measurements to be performed and important intervention messages to be sent to relevant parties.

This background explains the motivation behind increasing connectivity in power systems and among power system devices. It also explains why a common standard approach is required when power system operators want to establish internet-of-things in their grids.

When this is achieved, it is easy to achieve high observability in power systems and control all devices effectively. However, it is not only easy for the grid operators who do these interventions in good faith. It is also possible for malicious parties to monitor the system and intervene.

Especially when well-known data modelling and communication standards such as IEC 61850 are used in power systems, these weaknesses are amplified. All the data modelling and the protocols used to read, write or manipulated the are publicly available. Therefore, additional safeguard mechanisms are required to stop bad players such as hackers.

Additionally, these inverters can prove to be useful in islanding conditions as they can help operate small portions on their own. Deliberately changing frequency settings can interfere with the islanding process, impeding successful restoration efforts and prolonging the blackout period.

Safeguarding Mechanisms to Increase Resilience Against

Figure 15.1 shows different safeguarding mechanisms and how they can be used to mitigate certain cybersecurity breaches. These are explained in detail below.

Cyberattacks

To avoid unauthorized entry, it is crucial to establish secure access techniques. One successful method is utilizing authorized access methods with digital certificates.

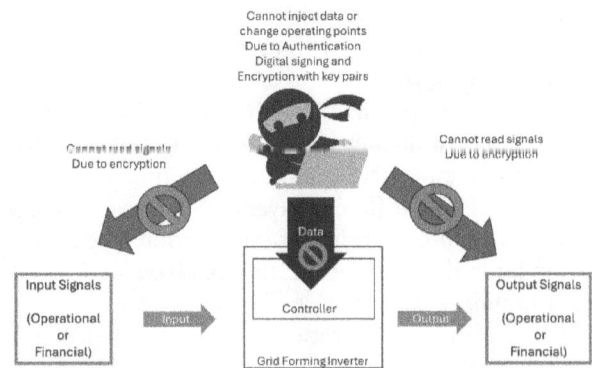

Figure 15.1 Safeguarding mechanisms for grid forming inverters
Source: Author

These guarantee that the system can only be accessed and modified by users who are deemed trustworthy. Another important measure is digital signing, which involves methods such as hashing. Hashing is the process of generating a distinct code, called a hash, from data. If an individual attempts to alter the information, the hash value will also be altered.

This simplifies the detection of any tampering. If data in a packet has been modified, the system can promptly reject it. This assists in preventing false data injection and masquerading attacks, in which an individual impersonates another to acquire entry.

End-to-end encryption is crucial for scenarios requiring complete privacy, such as financial transactions. This implies that data is encrypted during transmission. Only the person it was meant for can decode and understand the message.

If someone captures the data during transmission, they will not be able to view what is inside. This ensures that confidential data, like energy billing information, is protected from unauthorized viewing. Additionally, it is essential to utilize public and private key management. i.e. key pairs, to safeguard access to systems.

Conclusions

Grid-forming inverters play a crucial role in modern power systems as they support grid operation in addition to providing energy. They do this by actively intervening with the system parameters and this may be abused by hackers. They may be utilized to trigger false events or manipulate these parameters in wrong ways.

Therefore, it is important to study the cybersecurity aspect of these devices, as well as their other performance parameters. Grid-forming inverter technology is advancing through research and development, making these devices crucial for future power systems. They will assist in the progression towards grids that are more intelligent, robust, and effective.

References

[1] IEA (2024). Renewables 2023. Paris: IEA. https://www.iea.org/reports/renewables-2023.

[2] Yadav, A. K., Malik, H., Hussain, S. S., & Ustun, T. S. (2021). Case study of grid-connected photovoltaic power system installed at monthly optimum tilt angles for different climatic zones in India. *IEEE Access, 9*, 60077–60088.

[3] IEA (2024). CO2 Emissions in 2023. Paris: IEA. https://www.iea.org/reports/co2-emissions-in-2023.

[4] Das, A. (2022). A risk curtailment strategy for solar PV-battery integrated competitive power system. *Electronics, 11*, 1251.

[5] Al-Shetwi, A. Q., Issa, W. K., Aqeil, R. F., Ustun, T. S., Al-Masri, H. M., Alzaareer, K., et al. (2022). Active power control to mitigate frequency deviations in large-scale grid-connected PV system using grid-forming single-stage inverters. *Energies, 15*, 2035.

[6] Singh, N. K., Koley, C., Gope, S., Dawn, S., & Ustun, T. S. (2021). An economic risk analysis in wind and pumped hydro energy storage integrated power system using meta-heuristic algorithm. *Sustainability, 13*, 13542.

[7] Dey, B., Roy, B., & Datta, S. (2023). Forecasting ethanol demand in India to meet future blending targets: a comparison of ARIMA and various regression models. *Energy Reports, 9*, 411–418.

[8] Das, A., Dawn, S., & Gope, S. (2022). A strategy for system risk mitigation using FACTS devices in a wind incorporated competitive power system. *Sustainability, 14*, 8069.

[9] Dawn, S., Gope, S., Das, S. S., & Ustun, T. S. (2021). Social welfare maximization of competitive congested power market considering wind farm and pumped hydroelectric storage system. *Electronics, 10*, 2611.

[10] Kumar, K. K. P., Soren, N., Latif, A., Das, D. C., Hussain, S. S., Al-Durra, A., et. al. (2022). Day-ahead DSM-integrated hybrid-power-management-incorporated CEED of solar Thermal/Wind/Wave/BESS system using HFPSO. *Sustainability, 14*, 1169.

[11] GM Abdolrasol, M., Hannan, M. A., Hussain, S. S., Ustun, T. S., Sarker, M. R., & Ker, P. J. (2021). Energy management scheduling for microgrids in the virtual power plant system using artificial neural networks. *Energies, 14*, 6507.

[12] Farooq, Z. (2022). Power generation control of renewable energy based hybrid deregulated power system. *Energies, 15*, 5171.

[13] Ranjan, S., Das, D. C., Sinha, N., Latif, A., Hussain, S. S., & Ustun, T. S. (2021). Voltage stability assessment of isolated hybrid dish-stirling solar thermal-diesel microgrid with STATCOM using mine blast algorithm. *Electric Power Systems Research, 196*, 107239.

[14] Ustun, T. S., Ozansoy, C., & Zayegh, A. (2011). Implementation of Dijkstra's algorithm in a dynamic microgrid for relay hierarchy detection. In *2011 IEEE International Conference on Smart Grid Communications (SmartGridComm)*, Brussels, Belgium, (pp. 481–486).

[15] Barik, A. K., Das, D. C., Latif, A., Hussain, S. S., & Ustun, T. S. (2021). Optimal voltage–frequency regulation in distributed sustainable energy-based hybrid microgrids with integrated resource planning. *Energies, 14*, 27352.

[16] Ulutas, A., Altas, I. H., Onen, A., & Ustun, T. S. (2020). Neuro-fuzzy-based model predictive energy management for grid connected microgrids. *Electronics, 9*, 900.

[17] Latif, A. (2020). Price based demand response for optimal frequency stabilization in ORC solar thermal based isolated hybrid microgrid under salp swarm technique. *Electronics, 9*, 2209.

[18] Chauhan, A. (2021). Performance investigation of a solar photovoltaic/diesel generator based hybrid system with cycle charging strategy using BBO algorithm. *Sustainability, 13*, 8048.

[19] Dey, P. P., & Das, D. C. (2020). Active power management of virtual power plant under penetration of central receiver solar thermal-wind using butterfly optimization technique. *Sustainability, 12*, 6979.

[20] Chakraborty, M. R., Dawn, S., Saha, P. K., Basu, J. B., & Ustun, T. S. (2022). A comparative review on energy storage systems and their application in deregulated systems. *Batteries, 8*, 124.

[21] Ustun, T. S., Aoto, Y., & Hashimoto, J. (2020). Optimal PV-INV capacity ratio for residential smart inverters operating under different control modes. *IEEE Access, 8*, 116078–116089.

[22] Cheema, K. M. (2020). A comprehensive review of virtual synchronous generator. *International Journal of Electrical Power & Energy Systems, 120*, 106006.

[23] Hashimoto, J., Ustun, T. S., Orihara, D., Kikusato, H., Otani, K., Suekane, K., et al. (2023). Development of df/dt function in inverters for synthetic inertia. *Energy Reports, 9*(1), 363–371.

[24] Ustun, T. S., Hussain, S. M. S., Orihara, D., & Iioka, D. (2022). IEC61850 modeling of an AGC dispatching scheme for mitigation of short-term power flow variations. *Energy Reports, 8*(1), 381–391.

[25] Ustun, T. S. & Hussain, S. M. S. (2020). IEC 61850 modeling of UPFC and XMPP communication for power management in microgrids. *IEEE Access, 8*, 141696–141704.

[26] Ustun, T. S. & Hussain, S. M. S. (2019). Extending IEC 61850 communication standard to achieve internet-of-things in smartgrids. In *2019 International Conference on Power Electronics, Control and Automation (ICPECA)*, New Delhi, India, (pp. 1–6).

[27] Ustun, T. S. & Hussain, S. M. S. (2019). Implementation and performance evaluation of IEC 61850 based home energy management system. In *IEEE 8th Global Conference on Consumer Electronics (GCCE)*, Osaka, Japan, (pp. 24–25).

[28] Hussain, S. M. S., Aftab, M. A., Ali, I., & Ustun, T. S. (2020). IEC 61850 based energy management system using plug-in electric vehicles and distributed generators during emergencies. *International Journal of Electrical Power and Energy Systems*, 119, 105873.

[29] Hussain, S. M. S., & Aftab, M. A. (2020). Performance analysis of IEC 61850 messages in LTE communication for reactive power management in microgrids. *Energies*, 13, 6011.

[30] Ustun, T. S., Hussain, S. M. S., Syed, M. H., & Dambrauskas, P. (2021). IEC-61850-based communication for integrated ev management in power systems with renewable penetration. *Energies*, 14, 2493.

[31] Aftab, M. A., Hussain, S. S., Ali, I., & Ustun, T. S. (2020). IEC 61850-based communication layer modeling for electric vehicles: electric vehicle charging and discharging processes based on the international electrotechnical commission 61850 standard and its extensions. *IEEE Industrial Electronics Magazine*, 14(2), 4–14.

[32] Ustun, T. S., Ozansoy, C., & Zayegh, A. (2011). Extending IEC 61850-7-420 for distributed generators with fault current limiters. In *2011 IEEE PES Innovative Smart Grid Technologies*, Perth, Australia, (pp. 1–8).

[33] Roomi, M. M., Hussain, S. S., Mashima, D., Chang, E. C., & Ustun, T. S. (2023). Analysis of false data injection attacks against automated control for parallel generators in IEC 61850-based smart grid systems. *IEEE Systems Journal*, 17(3), 4603–4614.

[34] Hussain, S. M. S., Aftab, M. A., Farooq, S. M., Ali, I., Ustun, T. S., & Konstantinou, C. (2023). An effective security scheme for attacks on sample value messages in IEC 61850 automated substations. *IEEE Open Access Journal of Power and Energy*, 10, 304–315.

[35] Mbitiru, R. (2017). Using input-output correlations and a modified slide attack to compromise IEC 62055-41. In *2017 IEEE International Autumn Meeting on Power, Electronics and Computing (ROPEC)*, Ixtapa, Mexico, (pp. 1–6).

[36] Hussain, S. M. S. (2020). Smart inverter communication model and impact of cybersecurity attack. In *IEEE International Conference on Power Electronics, Drives and Energy Systems*, Jaipur, India, (pp. 1–5).

[37] Ustun, T. S. (2019). Cybersecurity vulnerabilities of smart inverters and their impacts on power system operation. In *2019 International Conference on Power Electronics, Control and Automation (ICPECA)*, New Delhi, India, (pp. 1–4S).

[38] Hussain, S. M. S., & Farooq, S. M. (2020). A method for achieving confidentiality and integrity in IEC 61850 GOOSE messages. *IEEE Transactions on Power Delivery*, 35(5), 2565–2567.

[39] Hussain, S. M.S. & Farooq S.M. (2022). A security mechanism for IEEE C37.118.2 PMU communication. *IEEE Transactions on Industrial Electronics*, 69(1), 1053–1061.

[40] Farooq, S. M., Hussain, S. S., Kiran, S., & Ustun, T. S. (2019). Certificate based security mechanisms in vehicular ad-hoc networks based on IEC 61850 and IEEE WAVE standards. *Electronics*, 8, 96.

[41] Farooq, S. M., Hussain, S. S., Ustun, T. S., & Iqbal, A. (2020). Using ID-based authentication and key agreement mechanism for securing communication in advanced metering infrastructure. *IEEE Access*, 8, 210503–210512.

[42] Farooq, S. M., Hussain, S. S., Kiran, S., & Ustun, T. S. (2018). Certificate based authentication mechanism for PMU communication networks based on IEC 61850-90-5. *Electronics*, 7, 370.

[43] Farooq, S. M., Hussain, S. S., & Ustun, T. S. (2019). S-GoSV: framework for generating secure IEC 61850 GOOSE and sample value messages. *Energies*, 12, 2536.

[44] Ustun, T. S., Farooq, S. M., & Hussain, S. M. S. (2020). Implementing secure routable GOOSE and SV messages based on IEC 61850-90-5. *IEEE Access*, 8, 26162–26171.

[45] Ustun, T. S., Hussain, S. M. S., Yavuz, L., & Onen, A. (2021). Artificial intelligence based intrusion detection system for IEC 61850 sampled values under symmetric and asymmetric faults. *IEEE Access*, 9, 56486–56495.

[46] Hussain, S. M. S., Farooq, S. M., & Ustun, T. S. (2019). Implementation of blockchain technology for energy trading with smart meters. In *2019 Innovations in Power and Advanced Computing Technologies (i-PACT)*, Vellore, India, (pp. 1–5).

[47] Ustun, T. S., Hussain, S. S., Ulutas, A., Onen, A., Roomi, M. M., & Mashima, D. (2021). Machine learning-based intrusion detection for achieving cybersecurity in smart grids using IEC 61850 GOOSE messages. *Symmetry*, 13, 826.

[48] Jha, A. V., Appasani, B., & Gupta, D. K. (2022). Analytical design of synchrophasor communication networks with resiliency analysis framework for smart grid. *Sustainability*, 14, 15450.

[49] Ustun, T. S., & Hussain, S. M. S. (2020). IEC 62351-4 security implementations for IEC 61850 MMS messages. *IEEE Access*, 8, 123979–123985.

[50] Ustun, T. S., & Hussain, S. S. (2020). An improved security scheme for IEC 61850 MMS messages in intelligent substation communication networks. *Journal of Modern Power Systems and Clean Energy*, 8(3), 591–595.

[51] Ustun, T. S., Ozansoy, C., & Zayegh, A. (2012). Simulation of communication infrastructure of a centralized microgrid protection system based on IEC 61850-7-420. In *2012 IEEE Third International Conference on Smart Grid Communications (SmartGridComm)*, Tainan, Taiwan, (pp. 492–497).

16 Techno-economic analysis of off-grid PV systems and batteries vs. diesel generators: life span and cost-effectiveness

Taha Selim Ustun[1,a] and Maher. G. M. Abdolrasol[2,b]

[1]Senior Researcher, Fukushima Renewable Energy Institute, AIST, Japan

[2]Post-doctoral Fellow, Institute of Sustainable Energy, Universiti Tenaga Nasional, 43000 Kajang, Selangor, Malaysia

Abstract

"This paper examines the life span and cost-effectiveness of off-grid photovoltaic (PV) systems compared to diesel generators. Although the initial investment for PV systems is higher than for diesel generators, the long-term cumulative costs reveal significant savings: the cost of PV systems is substantially lower than that of" diesel generators over a 25-year period, making diesel generators much more expensive. PV systems incur only a small percentage of their costs in operational and replacement expenses, while diesel generators account for a significantly larger portion. Additionally, PV systems provide clean energy, reducing greenhouse gas emissions. Off-grid solar systems can be a financially viable solution for the long run in Malaysia, thanks to progress in battery technology and high solar potential.

Keywords: Cost-effectiveness, diesel generators, photovoltaic systems, techno-economic analysis

Introduction

Renewable energy research and the push for replacing fossils fuels are very popular around the world [1–5]. Malaysia, another similar countries, need to phase out diesel gensets for financial and environmental motivations. Renewable energy based resources can be a good solution [5–10]. They can remove fuel costs associated with gensets and contribute to sustainable development goals. This double benefitted nature motivates researchers to look into these resources [11–15]. "In contrast, off-grid photovoltaic (PV) systems present a promising solution for generating clean energy, particularly in regions blessed with abundant sunlight. There are changing paradigms in power systems with novel solutions to accommodate more renewables [16–20]. In Malaysia the average is approximately 4.8 kWh/m²/day. The potential of off-grid PV systems in Malaysia" and around the world is vast, especially as the country seeks to improve access to electricity in rural and underserved communities. Researchers have been investigating novel systems and components to generate and deliver energy in unorthodox ways [21–29]. However, despite the growing recognition of these systems, comprehensive evaluations of their economic viability and lifespan compared to diesel generators and their alternative fuels are still limited [30, 31]. This paper aims to conduct a techno-economic analysis of the life span and cost-effectiveness of off-grid PV systems and batteries in comparison to traditional diesel generators.

By examining the durability and maintenance of PV systems, as well as their long-term economic benefits, this study seeks to illuminate the true cost of energy generation in Malaysia. The findings will highlight not only the potential savings in financial terms but also the environmental advantages of adopting solar technology over fossil fuels. Through this analysis, we aim to demonstrate how off-grid solar systems can provide a sustainable and economically feasible energy solution for Malaysia's energy needs, ultimately contributing to the country's goals for cleaner and more reliable energy access [32]. The increasing demand for electricity, the depletion of oil resources, and climate change goals highlight the need for better solutions. Malaysia, being a tropical country, receives sufficient sunlight [33]. This makes PV systems an excellent option for providing electricity to remote areas across country. "While the potential for off-grid PV systems has been acknowledged, detailed assessments of their actual capacity are still lacking. Some estimates suggest that the technical off-grid PV potential could reach around 900 MW, based on the number of households without electricity. Off-grid PV systems and batteries have emerged as sustainable alternatives to traditional diesel generators, especially in regions with ample solar resources like Malaysia [34]. The life span of PV

[a]selim.ustun@aist.go.jp, [b]maher@utm.edu.my

DOI: 10.1201/9781003663348-16

systems, including batteries and inverters, plays a crucial role in determining the overall cost-effectiveness compared to diesel generators, which have historically been relied upon for energy in remote areas This analysis examines the durability of off-grid PV systems, their components, and their economic feasibility in Malaysian Ringgit (MYR)," compared to the ongoing operational costs of diesel generators.

Life Span of Photovoltaic Systems

Over a 25-year period, a solar systems lifetime, panels are expected to lose approximately 12% to 25% of their capacity [33]. However, with proper maintenance and installation, they can continue to produce energy beyond their estimated life span, albeit at reduced efficiency. The photovoltaic modules used in this system are KD-M144, which are designed for strong sunlight regions. These panels feature N-type cells, half-cut cells, a 12-busbar design, and a power output of 415W, with eight sets included in the system. The Table 16.1 provides detailed specifications of the PV panel model [35, 36]. While manufacturers claim a life span of over 30 years for photovoltaic panels, most service providers base their calculations on a 20-year period. Performance typically decreases by 0.5% annually. For example, the ENF Solar MONO PERC 405W-415W 144 HALF-CELLS (158 mm) model offers 90.7% output power after 10 years and 80.2% after 25 years.

The choice of battery technology significantly impacts the life span of an off-grid PV system. Common battery types used in off-grid systems include lead-acid and lithium-ion (Li-ion) batteries.

Lead-acid batteries: These typically last between 3 to 7 years. They are a more affordable option upfront but require more frequent replacement due to their shorter life span and lower depth of discharge.

Lithium-ion batteries: Li-ion batteries offer a longer life span, generally between 10 to 15 years, with higher efficiency and greater depth of discharge compared to lead-acid batteries. While more expensive initially, they can offer better long-term savings due to their durability and lower maintenance needs [37]. Inverters typically last around 10 to 15 years, requiring replacement once or twice during the life span of the solar panels.

Cost-Effectiveness in Malaysian Ringgit

Off-grid PV systems

For this study a small 5 kW off-grid PV system is considered. The initial cost for such a system in Malaysia is between 40k and 50k MYR. In addition, average cost of annual maintenance ranges between 500 and 1000 MYR [38]. Life cycle costs: Given the life span of solar panels (25–30 years), the total life cycle cost of an off-grid system is heavily influenced by the need for battery and inverter replacements. Lithium-ion battery replacement every 10–15 years adds MYR 15,000 to MYR 20,000, while inverters cost around MYR 5,000 to MYR 10,000 every 10–15 years. Table 16.2 shows the initial costs and operational costs of off-grid pv systems vs. diesel generators.

Diesel generator costs

Initial costs: The cost of a 5 kW diesel generator ranges from MYR 5,000 to MYR 10,000, which is significantly lower than the initial cost of a PV system. Operational costs Diesel generators have high fuel and maintenance costs. Diesel fuel prices in Malaysia fluctuate, but on average, the cost of running a 5 kW generator for 8 hours per day can amount to MYR 50–70 per day (MYR 18,250–25,550 per year).

Table 16.1 Pre-crisis summary statistics

Module details	Rating
Rated maximum power (Pmax)	415 W
Module efficiency (%)	20.63%
Nominal operating cell temp (NOCT)	41°C ~ ±3°C
Power tolerance (W)	0 to +5W
Rated voltage (Vmpp)	40.9V
Rated current (Impp)	10.15 A
Open-circuit voltage (Voc)	19.6V
Short-circuit current (Isc)	10.66A
Maximum system voltage	1500 V
Maximum series fuse	15 A
Power temp. coef.	−0.34% / °C
Voltage temp. coef.	−0.26% / °C
Current temp. coef.	0.04%/ °C

Source: Author [39]

Table 16.2 Initial costs and operational costs of off-grid PV systems vs. diesel generators (in MYR)

Cost category	Off-grid PV System	Diesel generator
Initial cost	40,000 - 50,000	5,000 - 10,000
Annual Operational cost	500 - 1,000	18,250 - 25,550
Maintenance costs	500 - 1,000 per year	1,500 - 3,000 per year

Source: Author [40]

Maintenance costs for diesel generators, including oil changes and parts replacement, can range from MYR 1,500 to MYR 3,000 annually. Life cycle costs: Diesel generators typically last around 5,000 to 10,000 operational hours, translating to a life span of 2 to 5 years, depending on usage. Over 25 years, the need for multiple replacements, along with ongoing fuel and maintenance costs, makes diesel generators significantly more expensive in the long term.

Techno-Economic Cost Analysis

Over a 25-year period, the cost of an off-grid PV system is estimated to range from MYR 70,000 to MYR 90,000, taking into account the initial investment, battery replacements, and maintenance costs. A genset needs MYR 50000 for the same period. To help visualize the long-term cost comparison between an off-grid photovoltaic system and a diesel generator, Table 16.3 represents comparison of cumulative costs of an off-grid photovoltaic system and a diesel generator over a 25-year period, based on the assumptions from in the MATLAB environment. Over time, the diesel generator accumulates substantial operational costs due to fuel consumption and maintenance, totalling MYR 598,000 after 25 years. In contrast, the PV system's cumulative cost is much lower at MYR 119,000. Both systems require component replacements batteries and inverters for the PV system and the entire generator for the diesel system but the PV system's replacements are less frequent and less expensive in the long run. Despite the higher upfront cost, the off-grid PV system proves to be far more cost-effective over 25 years, making it the better financial choice, especially for long-term, sustainable energy needs as shown in Figure 16.1. Table 16.3 presents the summary of cost-effectiveness over 25 years.

Environmental and Practical Issues

Additionally, diesel generators are often associated with noise pollution, which can be a significant concern in residential or sensitive environments. The operation of these generators can lead to disturbances that negatively impact the quality of life for nearby residents and wildlife. On the other hand, PV systems operate silently, making them a more suitable choice for remote areas where tranquillity is essential.

Off grid solar systems are more resilient, once installed. They are not reliant to fuels imported into the community. This is especially important for isolated communities that can be disrupted by outside conflicts or price changes. Looking from sustainability perspective, off grid PV systems have many merits such as being clean, more resilient, and sustainable.

Table 16.3 Comparative cost analysis for 25 years (MYR)

Year	Off-grid PV system	Diesel generator
0	45,000	8,000
1	46,000	30,000
2	47,000	52,000
3	48,000	74,000
4	49,000	96,000
5	57,000 (*Battery replacement*)	126,000 (*Generator replacement*)
6	58,000	148,000
7	59,000	170,000
8	60,000	192,000
9	61,000	214,000
10	76,000 (*Battery replacement*)	244,000
11	77,000	266,000
12	78,000	288,000
13	79,000	310,000
14	80,000	332,000
15	88,000 (*Inverter replacement*)	362,000 (*Generator replacement*)
16	89,000	384,000
17	90,000	406,000
18	91,000	428,000
19	92,000	450,000
20	107,000 (*Battery replacement*)	480,000
21	108,000	502,000
22	109,000	524,000
23	110,000	546,000
24	111,000	568,000
25	119,000 (*Inverter replacement*)	598,000 (*Generator replacement*)

Source: Author

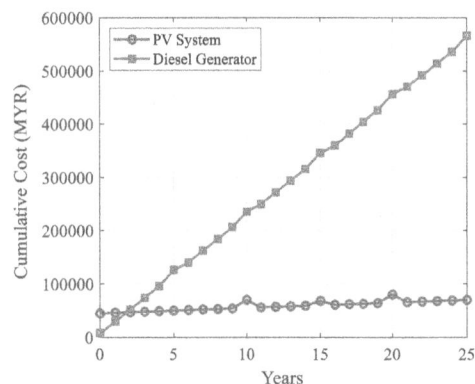

Figure 16.1 Cost comparison between PV system and diesel generator for 25 years
Source: Author

Conclusions

This research looks at the full life cycles costs of comparable PV systems and diesel genset systems used to energize an off grid community. Although initial costs are highly stacked against PV systems, MYR 45k versus 8k, the fuel costs rack up to a significant amount pretty quickly. Overall analysis shows that diesel genset system is 503 % more expensive with an overall cost of MYR 598000. PV system costs only reach MYR 119000 in 25 years. The operational and maintenance costs represent 87% and 21 % of these systems' overall cost, respectively. As battery technology continues to improve, the cost-effectiveness of PV systems will likely enhance further, reinforcing their position as the superior choice for off-grid power generation in regions with ample sunlight.

References

[1] Chakraborty, M. R., Dawn, S., Saha, P. K., Basu, J. B., & Ustun, T. S. (2022). A comparative review on energy storage systems and their application in deregulated systems. *Batteries*, 8, 124.

[2] Chauhan, A., Upadhyay, S., & Khan, M. T. (2021). Performance investigation of a solar photovoltaic/diesel generator based hybrid system with cycle charging strategy using BBO algorithm. *Sustainability*, 13, 8048.

[3] Ustun, T. S., Ozansoy, C., & Zayegh, A. (2011). Distributed energy resources (DER) object modeling with IEC 61850–7–420. In AUPEC 2011, Brisbane, QLD, Australia, (pp. 1–6).

[4] Ustun, T. S., Ozansoy, C., & Zayegh, A. (2011). Implementation of Dijkstra's algorithm in a dynamic microgrid for relay hierarchy detection. In 2011 IEEE International Conference on Smart Grid Communications (SmartGridComm), Brussels, Belgium, (pp. 481–486).

[5] Sahoo, A., Hota, P. K., & Sahu, P. R. (2023). Optimal congestion management with FACTS devices for optimal power dispatch in the deregulated electricity market. *Axioms*, 12, 614.

[6] Ranjan, S., Jaiswal, S., Latif, A., Das, D. C., Sinha, N., Hussain, S. S., et. al. (2021). Isolated and interconnected multi-area hybrid power systems: a review on control strategies. *Energies*, 14, 8276.

[7] Hussain, I. (2020). Performance assessment of an islanded hybrid power system with different storage combinations using an FPA-tuned two-degree-of-freedom (2DOF) controller. *Energies*, 13, 5610.

[8] Nayak, S. R. (2023). Participation of renewable energy sources in the frequency regulation issues of a five-area hybrid power system utilizing a sine cosine-adopted African vulture optimization algorithm. *Energies*, 16, 926.

[9] Ustun, T. S., & Aoto, Y. (2018). Impact of power conditioning systems with advanced inverter capabilities on the distribution network. In 2018 53rd International Universities Power Engineering Conference (UPEC), Glasgow, UK, (pp. 1–6).

[10] Dey, P. P., & Das, D. C. (2020). Active power management of virtual power plant under penetration of central receiver solar thermal-wind using butterfly optimization technique. *Sustainability*, 12, 6979.

[11] Barik, A. K., & Das, D. C. (2021). Latif., optimal voltage–frequency regulation in distributed sustainable energy-based hybrid microgrids with integrated resource planning. *Energies*, 14, 2735.

[12] Latif, A., Paul, M., Das, D. C., Hussain, S. S., & Ustun, T. S. (2020). Price based demand response for optimal frequency stabilization in ORC solar thermal based isolated hybrid microgrid under salp swarm technique. *Electronics*, 9, 2209.

[13] Hussain, S. M. S., Farooq, S. M., & Ustun, T. S. (2019). Implementation of blockchain technology for energy trading with smart meters. In 2019 Innovations in Power and Advanced Computing Technologies (i-PACT), Vellore, India, (pp. 1–5).

[14] Basu, J. B., Dawn, S., Saha P. K., & Chakraborty M. R. (2022). Economic enhancement of wind–thermal–hydro system considering imbalance cost in deregulated power market. *Sustainability*, 14, 15604.

[15] Hussain, S. M. S., Aftab, M. A., Nadeem, F., & Ali, I. (2020). Optimal energy routing in microgrids with IEC 61850 based energy routers. *IEEE Transactions on Industrial Electronics*, 67(6), 5161–5169.

[16] Sahu, P. R., Lenka, R. K., Khadanga, R. K., Hota, P. K., Panda, S., & Ustun, T. S. (2022). Power system stability improvement of FACTS controller and PSS design: a time-delay approach. *Sustainability*, 14, 14649.

[17] Pattnaik, S., Kumar, M. R., Mishra, S. K., Gautam, S. P., & Appasani, B. (2022). DC bus voltage stabilization and SOC management using optimal tuning of controllers for supercapacitor based PV hybrid energy storage system. *Batteries*, 8, 186.

[18] Rohmingtluanga, C., Datta, S., & Sinha, N. (2023). SCADA based intake monitoring for improving energy management plan: Case study. *Energy Reports*, 9, 402–410.

[19] Latif, A., Hussain, S. M. S., & Das, D. C. (2021). Optimization of two-stage IPD-(1+I) controllers for frequency regulation of sustainable energy based hybrid microgrid network. *Electronics*, 10, 919.

[20] Ustun, T. S., Ozansoy, C., & Zayegh, A. (2012). Simulation of communication infrastructure of a centralized microgrid protection system based on IEC 61850-7-420. In 2012 IEEE Third International Conference on Smart Grid Communications, Tainan, Taiwan, (pp. 492–497).

[21] Hussain, S. M. S., Aftab, M. A., Ali, I., & Ustun, T. S. (2020). IEC 61850 based energy management system using plug-in electric vehicles and distributed generators during emergencies. *International Journal of Electrical Power and Energy Systems*, 119, 105873.

[22] GM Abdolrasol, M., Hannan, M. A., Hussain, S. S., Ustun, T. S., Sarker, M. R., & Ker, P. J. (2021). Energy management scheduling for microgrids in the virtual

power plant system using artificial neural networks. *Energies*, 14, 6507.

[23] Mahafzah, K. A., Obeidat, M. A., Mansour, A. M., & Al-Shetwi, A. Q. (2022). Artificial-intelligence-based open-circuit fault diagnosis in VSI-Fed PMSMs and a novel fault recovery method. *Sustainability*, 14, 16504.

[24] Mahto, T., Kumar, R., Malik, H., Hussain, S. M. S. (2021). Fractional order fuzzy based virtual inertia controller design for frequency stability in isolated hybrid power systems. *Energies*, 14, 1634.

[25] Ulutas, A., Altas, I., Onen, A., & Ustun, T. S. (2020). Neuro-fuzzy-based model predictive energy management for grid connected microgrids. *Electronics*, 9, 900.

[26] Pati, A., Adhikary, N., Mishra, S. K., & Appasani, B. (2022). Fuzzy logic based energy management for grid connected hybrid PV system. *Energy Reports*, 8, 751–758.

[27] GM Abdolrasol, M., Hannan, M. A., Hussain, S. S., & Ustun, T. S. (2022). Optimal PI controller based PSO optimization for PV inverter using SPWM techniques. *Energy Reports*, 8, 1003–1011.

[28] Farooq, Z., Rahman, A., Hussain, S. M. S., & Ustun, T. S. (2022). Power generation control of renewable energy based hybrid deregulated power system. *Energies*, 15, 517.

[29] Singh, N. K., Koley, C., Gope, S., Dawn, S., & Ustun, T. S. (2021). An economic risk analysis in wind and pumped hydro energy storage integrated power system using meta-heuristic algorithm. *Sustainability*, 13, 13542.

[30] Szulczyk, K. R., Yap, C. S., & Ho, P. (2021). The economic feasibility and environmental ramifications of biodiesel, bioelectricity, and bioethanol in Malaysia. *Energy for Sustainable Development*, 61, 206–216.

[31] Dey, B., Roy, B., Datta, S., & Ustun, T. S. (2023). Forecasting ethanol demand in India to meet future blending targets: a comparison of ARIMA and various regression models. *Energy Reports*, 9, 411–418.

[32] Veldhuis, A. J., & Reinders, A. H. M. E. (2015). Reviewing the potential and cost-effectiveness of off-grid PV systems in Indonesia on a provincial level. *Renewable and Sustainable Energy Reviews*, 52, 757–769.

[33] Rodríguez-Gallegos, C. D., Gandhi, O., Bieri, M., Reindl, T., & Panda, S. K. (2018). A diesel replacement strategy for off-grid systems based on progressive introduction of PV and batteries: An Indonesian case study. *Applied Energy*, 229, 1218–1232.

[34] Yahoo, M., Mohd Salleh, N. H., Chatri, F., & Huixin, L. (2024). Economic and environmental analysis of Malaysia's 2025 renewable and sustainable energy targets in the generation mix. *Heliyon*, 10(9), e30157.

[35] Abdolrasol, M. G. M., Ayob, A., & Mutlag, A. H. (2023). Optimal fuzzy logic controller based PSO for photovoltaic system. *Energy Reports*, 9, 427–434.

[36] GM Abdolrasol, M., Jern Ker, P., Hannan, M. A., Tiong, S. K., Ayob, A., & Almadani, J. F. S. (2023). Optimized PV-battery systems using backtracking search algorithm for sustainable energy solutions. In *2023 IEEE International Conference on Energy Technologies for Future Grids*, (pp. 1–6).

[37] Lipu, M. S. H., Mamun, A. A., Ansari, S., Miah, M. S., Hasan, K., Meraj, S. T., et al. (2022). Battery management, key technologies, methods, issues, and future trends of electric vehicles: a pathway toward achieving sustainable development goals. *Batteries*, 8(9), 119.

[38] Aeggegn, D. B., Agajie, T. F., Workie, Y. G., Khan, B., & Fopah-Lele, A. (2023). Feasibility and techno-economic analysis of PV-battery priority grid tie system with diesel resilience: a case study. *Heliyon*, 9(9), e19387.

[39] Sunpro Power | MONO PERC 405W-415W 144 HALF-CELLS (158mm) | Solar Panel Datasheet | ENF Panel Directory." [Online]. Available: https://www.enfsolar.com/pv/panel-datasheet/crystalline/43356. [Accessed: 19-Mar-2025].

[40] "Off-Grid Solar System Sizing Calculator - malaysia-calculator.com." [Online]. Available: https://malaysia-calculator.com/off-grid-solar-system-sizing-calculator/. [Accessed: 19-Mar-2025].

17 Transforming waste into energy through a prototype model for sustainable and future-oriented solutions

Suparna Maity[1,a], Bapita Roy[1,b], Abhoy Ghosh[2,c], Debangi Chakraborty[2,d], Santana Das[1,e], Saikat Majumder[3,f], Sanghamitra Layek[4,g] and Bikas Mondal[4,h]

[1]Assistant Professor, Department of Electronics and Computer Science, Guru Nanak Institute of Technology, New Delhi, India

[2]Department of Electronics and Computer Science, Guru Nanak Institute of Technology, New Delhi, India

[3]Associate Professor, Department of Electronics and Computer Science, Techno Main Saltlake, West Bengal, India

[4]Assistant Professor, Department of Electronics and Computer Science, Narula Institute of Technology, West Bengal, India

Abstract

The waste-to-energy (WTE) process generates electricity from municipal solid waste (MSW). This method is increasingly being recognized as a sustainable and environmentally friendly energy recovery solution. MSW to Electricity this is a technology which takes Municipal solid waste as input, and convert power from it so that the energy in such form will be recovered. WTE is becoming an increasingly attractive alternative to traditional waste disposal methods, such as land filling, amid a greater awareness of the environmental implications. This project revolves around a fourth tier in the waste management hierarchy which refers to generating power from our wastes, and is based on 4R (Recycling — Recovery- Reduction – Reusing) concept. In this project we developed a prototype of the WTE composed by linking between solar cell and heat sensor, as resulting in system that is compact for energy generation. The heat sensor(s) from the solar cell to detect thermal energy release during MSW combustion. This energy is subsequently changed into electrical power, capable enough to charge any 4V rechargeable battery. The prototype employs the high calorific value of MSW and minimizes waste by product, thereby improving energy conversion efficiency. Initial performance tests in the prototype look very promising. The apparatus indeed makes power, indicating that it may be a viable solution for WTE electrical generation on a grand scale. The project will show how WTE can complement modern sustainable waste management strategies, decrease the environmental impact of MSW and contribute to renewable energy technology.

Keywords: 4R (Recycling — recovery-reduction – reusing), heat sensor, MSW, power plantation, solar panel, WTE prototype

Introduction

In this project, we are trying to convert the municipal solid waste (MSW) into electrical energy and store it in a 4V battery. This procedure is to create electricity by incinerating waste [1–3], include pieces such as heating panels". In steam plant that is powered by waste, the trash combustion produces heat and hence converted into electricity (the same can be measured using a multi-meter). Instead of a waste that nature cannot adequately process on its own and which ultimately causes considerable damage to the environment, it is an energy source. It combines to create an operational system that is 100% waste driven and requires no additional fuel. This not only reduces waste, but also creates sustainable power. The project underscores the environmental advantages of using waste-to-energy (WTE) technologies, which are less polluting than traditional power generation plants. Additionally, the initiative can be expanded to include any pollution that is generated while converting energy. Advanced emission control technologies can mitigate pollution in a WTE system by controlling harmful gases and particulate matter through scrubs, electrostatic precipitators and selective catalytic reduction.

Photo: AP/Francois Mori effective management of ash, includes measures like recycling bottom ash and clinkers from slag using brick manufacturing unit and treating fly ash to control environmental pollution so that generation of toxic waste is reduced. Other monitoring systems measure emissions to ensure they

[a]suparna.maity@gnit.ac.in, [b]bapita.roy@gnit.ac.in, [c]avoyghosh2022@gmail.com, [d]debangi.chakraborty2004@gmail.com, [e]santana.das@gnit.ac.in, [f]msaikat2004@gmail.com, [g]sanghamitra.layel@nit.ac.in, [h]bikas.mondal@nit.ac.in

DOI: 10.1201/9781003663348-17

stay within regulated levels. Furthermore, efficiency in combustion can be enhanced as well for lower emissions and less harmful compounds before the waste is being incinerated. With such steps taken, and the awareness to segregate waste imparted among people, WTE becomes a cleaner energy solution. The matters of emission control can be addressed in the potential future advancements as to make WTE a cleaner approach. This project demonstrates the critical role of WTE in sustainable waste management and power generation forms a pure energy source out of garbage.

Literature Review

The MSW management has rapidly evolved to become one of the more critical environmental issues of this century, with ever-increasing waste generation and related adverse impacts on the environment. Growing requirements for sustainable waste management solutions have resulted in a drive toward the development of technologies able to convert waste into useful resources of these technologies; WTE has become one of the most promising. The WTE technologies are now recovering energy and materials from over 300 million tons of municipal solid wastes worldwide [12]. This literature review evaluates the current status of WTE technology in waste reduction and the production of sustainable energy, together with associated environmental issues. WTE technology is applied in the conversion of wastes into electrical energy with several processes involved but most important being incineration. The greatest share in a WTE system is incineration where MSW is burnt to produce heat. The produced heat is further utilized for raising steam, which drives turbines to generate electricity. Several studies have documented that the incineration of MSW reduces the volume of waste by up to 90% [9, 10]. This eases a substantial load off landfills while producing a good deal of energy in the process. The energy has a potential of being captured for storage in batteries and used later, making WTE quite a genuine alternative to generating power using traditional fossil fuels. The already existing works available on this area of technology identifies the dual benefits associated with WTE technology: it not only helps solve environmental problems associated with wastes but also provides renewable energy. But converting waste that would have otherwise caused pollution and overcrowding landfills into a resource, present WTE systems play an important role in sustainable waste management [11]. The conversion of MSW into electrical energy through WTE technology involves a number of key components. These devices include heating panels, boosting coils, diodes, LEDs, capacitors, resistors, batteries, and PCB boards. All these

things inside work together, interacting with each other in providing a self-sustaining system that runs purely on waste and that needs no extra fuel input. The recent developments have focused on enhancing the efficiency of these components by using boosting coils and capacitors, which may actually help to improve the energy conversion efficiency of the WTE systems [4–7]. This will help in increasing the output energy from the same amount of wastes to make the process economically more viable. Furthermore, the inclusion of such high-tech elements in the present work ensures that the extracted energy will be stored and can be applied for practical purposes [8, 9]. For example, storing the generated electricity in a 4V battery, as is done in the current project, gives space to assert that WTE technology could not only contribute to immediate power generation but also be involved in creating a reliable energy reserve.

Methodology

Among the different methods of waste to energy conversion, in this work incineration method have been used. The energy produced depends on waste composition. Therefore, it is crucial to analyze waste in relation to energy generation. In this work, the waste is completely burnt to produce heat energy. A heat chamber is designed to convert the heat energy into electrical energy. A solar panel and a thermoelectric module used for energy conversion. The thermoelectric module produces energy using Peltier effect. The heat generated from burning the waste, used to increase the temperature at hot side of the module, accordingly electricity generates. Next a charging circuit is designed to store the produced electricity. The part of the circuit also enables the user to check the input and output status of the charging. The total unit is designed to charge the 12V battery, where 3nits of 4V batteries are used. The 12V DC is then converted to 230 V AC using a converter circuit.

Analysis of Waste

The type of waste and its composition plays an important role in waste to energy generation. Knowledge about the composition of waste is required to understand the characteristics of treated waste. MSW is used as source of waste in this project. MSW refers to the mixed waste that is collected from residential, commercial, and institutional sources. It typically consists of a wide variety of materials, some of which can be used for energy production, whereas RDF is produced from MSW by a combination of separation, size reduction, and possibly drying operations. Noncombustible fractions, for example metals, glass, and some organics are rejected, but all the combustibles in a residual waste, e.g. plastics, paper, and textiles go to RDF. This RDF can be used as a fuel in an energetic process. RDF, derived from MSW, is a key fuel in waste to energy projects. It allows for efficient energy recovery from waste, turning a disposal problem into a source of energy. To perform this project in large scale so that segregation of waste should be done properly and need to have the knowledge about MSW of nearby area of plant station and the content of RDF in it having the sufficient amount of calorific value to run the energy plant properly [10]. The chemical composition of the waste is also crucial since it may be used to determine how the waste will burn in the incinerator and what kind of emissions will likely arise from the proximate analysis and the ultimate analysis (elemental). During past 10 years the life and consumption style has been modified remarkably. Due to which quantity of organic waste has been decreased while packaging related waste has increased like plastic paper card-board which has high energy contents that progressively increase the overall heating value of MSW. Due to the lack of information regarding the MSW under study, the composition was calculated using sources from the literature [10]. Figures 17.1 and 17.2 show the physical and chemical analysis of MSW waste.

Population Range (million)	Paper %	Rubber, Leather & Synthetic %	Glass and Metals %	Combustibles %	Inerts %
0.1 – 0.5	2.91	0.78	0.89	44.57	43.59
0.5 – 1.0	2.95	0.73	0.67	40.04	48.38
1.0 – 2.0	4.71	0.71	0.95	38.95	44.73
2.0 – 5.0	3.18	0.48	1.08	56.67	49.07
> 5.0	6.43	0.28	1.74	30.84	53.90

Figure 17.1 Physical characteristics of MSW in Indian cities [10]
Source: Author

Population Range (million)	Moisture %	Organic matter %	Total N_2 %	P_2O_5 %	K_2O %	C/N Ratio %	CV (kcal / kg) %
0.1 – 0.5	25.8	37.1	0.7	0.6	0.8	30.9	1010
0.5 – 1.0	19.5	25.1	0.7	0.6	0.7	21.1	901
1.0 – 2.0	27.0	26.9	0.6	0.8	0.7	23.7	980
2.0 – 5.0	21.0	25.6	0.6	0.7	0.8	22.4	907
> 5.0	38.7	39.1	0.6	0.5	0.5	30.1	801

Figure 17.2 Chemical characteristics of MSW in Indian cities [10]
Source: Author

Energy Conversion

The detail block diagram for the energy conversion process of waste to energy is shown in Figure 17.3. A prototype model was prepared for this project, and Figure 17.5 shows the circuit schematic. Following section discusses the conversion procedure.

Heat chamber

The heat chamber is one of the crucial block of this project. It consists of a solar panel and a thermoelectric module. The solar panel is used to produce an output voltage of approximately 6 volts when operating under standard conditions of burning waste by harnessing the power of flame. This solar panel is able to convert light into electricity. On the other hand, the TEC1-12707 thermoelectric module is a specialized Peltier cooler that consists of a hot side and a cold side. This module can be utilized when a temperature gradient is present, for generating a small amount of electrical power. Due to the temperature gradient current flows and voltage is generated. Burning of waste generates heat which is fed to this module.

Charging circuit

The charging circuit, as shown in Figure 17.4, provides the essential information about the status of the charging process of the battery through the power generated from burning waste in heat chamber. In this charging circuit a common type of LED indicator has been used, which utilizes LED to visually represent the charging status of our project. Along which resistors and transistors and diode are used to make that circuit compatible with the electric power generated from heat chamber. In this charging circuit the green LEDs are used to indicate that the power is generating from heat chamber and passing through the circuit. The yellow LED specifies the active status of battery charging.

Rechargeable battery

In this project 4-volt rechargeable battery has been used which is typically classified as a small, low

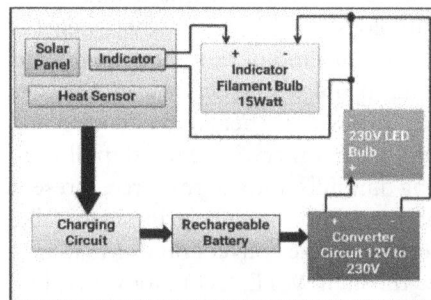

Figure 17.3 Block diagram
Source: Author

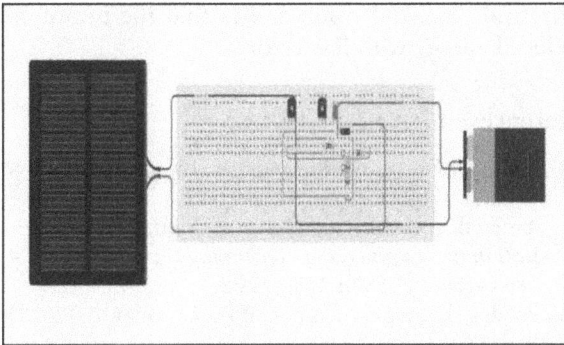

Figure 17.4 Charging circuit
Source: Author

Table 17.1 Experimental data

Sl. No.	Waste mass (gm)	Experimental charging time (minutes)	Energy produced (Ah)
1	70	2.50	0.28
2	120	4.29	0.49
3	220	7.86	1.06
4	320	11.44	1.50
5	420	15.02	2.03
6	520	18.59	2.44

Source: Author

voltage battery and is commonly used in electronic devices like toys, cameras, and small appliances. But in this project NiMH 4 volt rechargeable batteries are used with the advantages of more cost effectiveness but with the disadvantages of a lower energy density compared to Li-ion batteries. The capacity of the battery, measured in ampere-hours (Ah), indicates how much energy the battery can store.

Converter circuit
In this project the maximum 12volts DC is supplied by burning the waste in heat chamber which should be converted into 230volt AC supply to make it usable for home appliances. So, for converting 12 volts DC to 230 volts AC requires the use of an inverter circuit. This circuit, shown in Figure 17.3, is composed of various components that work together to transform direct current into alternating current.

Experimental Result

The energy converted from the waste is stored in rechargeable battery. In this work, the combustion of plastic material is considered. The battery capacity is 1.5 Ah, as provided in datasheet. But the capacity is examined which shows the capacity as 1.45 Ah. To conduct the experiment, different mass of the waste is burnt and the time is recorded for charging. For large amount of mass, multiple batteries are used.

Table 17.1 represents experimentally observed data.

Empirical Result

The electrical energy produced from the waste can be calculated using the formula as mentioned:

$$E = C * M * \eta \qquad (1)$$

Where E implies electrical energy, C is calorific value, M is the mass of waste and, η represents efficiency. Given, calorific value (plastic) = 35MJ/Kg;

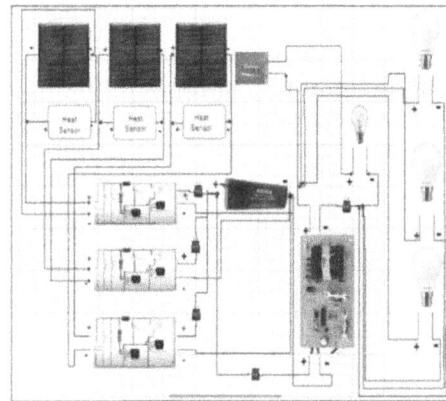

Figure 17.5 Circuit diagram of the prototype
Source: Author

mass of waste = 100 gm; burning time of waste = 3.10 minutes; efficiency = 100%; The energy released from burning waste,

E released = mass of waste x calorific value;

E released = 0.1 Kg x 35 J/Kg = 3.5J Putting all these into the formula we get, Power= (35000 J/gm x 100 gm) / (3.10/60 hours) = 180833 J/h = 181 kJ/h = 50 W Voltage to current: Assuming the output of each pair of heat sensors and solar panel is 6 V. So, the charging time for a battery to be charged fully from the burning waste can be calculated using the formula given below:

$$T = C_{Ah} * I \qquad (2)$$

Where T refers charging time (hours),C_{Ah} capacity (Ah), and I represents charging current (A) Given Voltage = 6V; Power = 50 W; Battery capacity = 1.5Ah; Charging Current = Power / Voltage Putting all these into the formula we get, 1.5 Ah / 8.33 A = 0.18 hours = 10 minutes Hence, for supplying 320gm of waste sequentially then it will take 10 minutes to charge the full battery of the prototype (12V). Table 17.2 shows the theoretical data for the prototype.

Result Analysis

Result analysis Figures 17.6 and 17.7 show the comparison between theoretical and experimental result. It is observed from the graph that, the experimental result provides very less error (Root mean square error (RMSE) = 0.0022).

Conclusion

In this work, a prototype for waste to energy has been developed. Initially, a study was conducted to analyze the MSW waste's composition. For this prototype development plastic is chosen as waste. By the incineration procedure combustion has been done. In order to convert heat energy to electrical energy, a heat chamber and required circuitry has been developed. The generated energy is stored in rechargeable batteries. The experimental data is then compared to the theoretical data and result shows that the prototype produced energy with less error.

Table 17.2 Theoretical data

Sl. No.	Waste mass (gm)	Theoretical charging time (minutes)	Calculated theoretical energy (Ah)
1	70	2.10	0.33
2	120	3.75	0.56
3	220	6.80	1.03
4	320	10.00	1.50
5	420	13.13	1.96
6	520	16.25	2.44

Source: Author

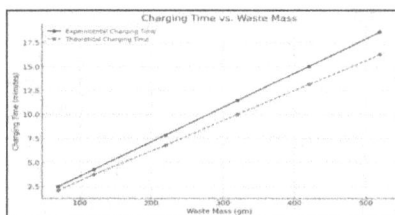

Figure 17.6 Theoretical vs experimental charging time for a given mass
Source: Author

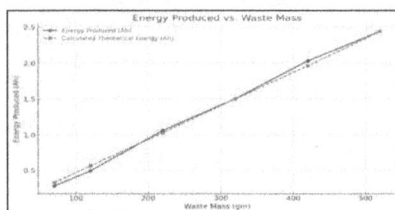

Figure 17.7 Theoretical vs experimental energy produced for a given mass
Source: Author

References

[1] Patil, R., Ghate, R., Karande, V., Bhingardeve, A., & Patil, B. (2022). Generate electricity by using waste material. *International Research Journal of Modernization in Engineering Technology and Science*, 4, 1861–1863. e-ISSN: 2582-5208.

[2] Pandey, A., Singh, A., Singh, P. D., & Yadav, A. (2021). Waste to energy: generation of electricity using waste materials. *International Journal of Creative Research Thoughts*, 9, 94–99. ISSN: 2320-2882.

[3] Sharanya, D., Babu, A. V., Sunay, K., Charan, B. K. C., & Abhilash, K. (2022). Electricity generated by waste material. International Research Journal of Engineering and Technology (IRJET), 9 (1651-1654). p-ISSN: 2395–007.

[4] Patil, N. B., Patil, O. V., Patil, S. A., Shaikh, A. A., & Mullani, M. B. (2022). International Journal of Research Publication and Reviews. To generate electricity from waste material and to reduce air pollution. 3(5), 3438–3441.

[5] Shukla, M., & Miyan, M. (2020). Waste to energy: a review on generating electricity in India. *SAMRIDDHI: A Journal of Physical Sciences, Engineering and Technology*, 12, 37–40. doi: 10.18090/samriddhi.v12i01.8.

[6] Pandal, A., Das, M., Dewangan, N., & Jaiswal, G. (2023). Generate electricity by waste materials. *International Research Journal of Modernization in Engineering Technology and Science*, 5, 2713–2716. e-ISSN: 2582–5208.

[7] Edinger, R., & Kaul, S. (2000). Renewable Resources for Electric Power: Prospects and Challenges. Quorum Books.

[8] Hawkes, A. D., & Leach, M. A. (2007). Cost-effective operating strategy for residential micro-combined heat and power energy. *Energy*, 32(5), 711–723.

[9] Traven, L. (2023). Sustainable energy generation from municipal solid waste: a brief overview of existing technologies. *Case Studies in Chemical and Environmental Engineering*, 8, 100491. ISSN 2666-0164, doi: 10.1016/j.cscee.2023.100491.

[10] Rao, T. S., Manoh, M. D., & Krishnaiah, B. (2023). Hyderabad MSW Energy Solutions Limited (2023). Detailed project report for expansion of existing waste to energy(WtE) plant from 19.8 MW to 48 MW at Jawa-harnagar. Survey No.173.

[11] Vuppaladadiyam, S. S. V., Vuppaladadiyam, A. K., Sahoo, A., Urgunde, A., Murugavelh, S., Šrámek, V., et al. (2024). Waste to energy: trending key challenges and current technologies in waste plastic management. *Science of the Total Environment*, 913, 169436. ISSN-0048-9697, https://doi.org/10.1016/j.scitotenv.2023.169436.

[12] Tian, Y., Dai, S., & Wang, J. (2023). Environmental standards and beneficial uses of waste-to-energy (WTE) residues in civil engineering applications. *Waste Disposal and Sustainable Energy*, 4, 323–350. 10.1007/s42768-023-00140-8.

18 Photoplethysmograph based biomedical signal analysis using machine learning and vital sign monitoring using a web application

Soumyadip Jana[1,a] and Partha Sarathi Pal[2,b]

[1]Assistant Professor, Electrical Engineering Department, Techno Main Salt Lake, Kolkata, West Bengal, India

[2]Professor, Electrical Engineering Department, Netaji Subhash Engineering College, Garia, West Bengal, India

Abstract

This paper presents a novel approach, for the multiclass classification of distinctive hypertension and cholesterol levels prediction in addition to the binary classification of cardiovascular diseases (CVD) prediction using different biomarkers after thoroughly processing, cleaning, and filtering the signal data. The models have been trained and validated with the real and noise-attenuated Photoplethysmograph (PPG) based CVD dataset from the UCI machine learning repository, Kaggle platform and also tested with the real-time data collected from the pathological laboratory. A comparative study between different algorithms has also been shown to get the most accurate results. The confusion matrix has been utilized to analyze the model's performance and calculate the model accuracy. Using the Streamlit Python programming library, a web application has also been created to predict diverse physiological conditions from real-time clinical data. The experimental results illustrate that the different proposed models can be a quick and automatic alternative for the conclusion of hypertension/blood pressure level, cholesterol level, and CVD prediction. To be more particular, the proposed ML technique has an accuracy of 99.8% in recognizing hypertension levels, 81.7% in cholesterol level prediction, and 83.1% in identifying CVDs. In addition, given the truth that an examination may be accomplished in only a few seconds, speedy response time is the primary advantage offered by the suggested method.

Keywords: Biomarkers, cardio vascular diseases, machine learning, photoplethysmography, web application

Introduction

Cardiovascular diseases (CVDs) are the most prominent cause of human death globally as per World Health Organization (WHO), taking an estimation of 17.9 million lives every year. Coronary artery diseases, cerebrovascular diseases, rheumatic heart diseases etc, which are basically different, types of disorders of heart and blood vessels, are coming under the category of CVDs. In India more than four out of five CVD deaths are caused due to heart attacks and one-third of these deaths occur rashly in individuals under 70 years of age. Subsequently, it is secure to assume that diagnosing heart diseases in their earliest stages is exceptionally essential.

Hypertension is one of the predominant risk components that further causes various types of CVDs. The treatment of chronic patients, suffering from CVD results in increased deaths from CVD around the world. Regular blood pressure (BP) monitoring is significant to anticipate CVDs as it can be controlled and analyzed through consistent monitoring. Physically diagnosing hypertension and CVD utilizing the ECG data is time-consuming and error-prone. In the modern day, the increasing rise of Photoplethysmograph (PPG) based signals is easing health monitoring and will become the go-to technology of modern-day healthcare. Analysis of data retrieved from PPG signals can predict different health parameters such as blood pressure, glucose level, cholesterol, cardiovascular disease, etc. With superior training, different machine learning (ML) techniques could be a better alternative for quick and automatic classification.

Therefore, this paper investigates best-suited machine learning techniques for the multiclass classification of hypertension levels, cholesterol levels, and binary classification for the presence of CVDs from real-time clinical data. Comparative analysis has been shown, and a web application, using Streamlit Python programming, which can predict diverse physiological conditions from real-time clinical data has also been developed.

The rest of the paper is structured as follows. Section 2 reviews literature. Section 3 describes the data and variables. Section 4 explains the methodology and model specifications of the proposed work. Section 5

[a]soumyadipjana@gmail.com, [b]pspal2k@gmail.com

DOI: 10.1201/9781003663348-18

discusses the key findings and comparative analysis. Section 6 summarizes the paper.

Literature Review

Photoplethysmography, as an extremely sensible diagnostic tool, has been used by Li et al. [1] for the determination of heart rate and respiratory rate based on a finite Gaussian basis. Woo et al. [2] in his paper has proposed a personal healthcare device based on a protocol conversion IoT system. Mobile health monitoring has been designed and implemented by Wannenburg and Malekian [3] by bio signal extraction method. Wijshoff et al. [4] proposed an algorithm for the removal of periodic motion artifacts, recovering artifact-reduced PPG signals for beat-to-beat analysis and monitoring in activities of daily living (ADL), cardiopulmonary exercise testing (CPX), and cardiopulmonary resuscitation (CPR) which hampers measurement of inter-beat-intervals (IBIs) and oxygen saturation (SpO2). PPG-based envelope-filtered method of enhancement for non-invasive Hemoglobin (Hb) monitoring has been described by Yuan et al. [5]. Zhang et al. [6] have proposed a framework, TROIKA which is capable of monitoring heart rate for fitness based on PPG signals. The application of various deep learning algorithms for the analysis of medical data has been depicted by Faust et al. [7]. Rahmani et al. [8] have successfully implemented IoT-based Smart e-Health Gateway called UT-GATE ubiquitous health monitoring systems, especially in clinical environments. Islam et al. [9] have addressed the application of IoT and eHealth policies in the context of healthcare and proposed some avenues for future research on IoT-based healthcare. Farahani et al. [10] have demonstrated multi-layer architecture based on a fog-driven IoT eHealth ecosystem for the transition from clinic-centric treatment to patient- centric healthcare. Iqbal et al. [11] have classified CAD patients using deep learning PPG (DL-PPG) waveforms and the study has shown that DL-PPG has performed better than the serum biomarkers to separate CAD patients from healthy control with an accuracy of 83.8%. CVDs involve factors related to blood pressure, cholesterol levels, glucose levels and lifestyle that can be controlled through medication and certain preventive measures. Many factors need to be considered because medication does not change age, ethnicity, and family history of CVDs. Therefore, to accurately predict CVD many factors need to be considered. In the domain of medical sciences artificial intelligence (AI) and machine learning (ML), play a pivotal role for effectively identifying numerous diseases in patients. ML is a cutting-edge tool that enables systems to automatically learn and improve based on their experience with the target system. ML techniques are reliable and efficient over naïve ML and regression methods in predicting CVDs which have been discussed by Alkadya et al. [12], Sardar et al. [13], and Chaouch et al. [14]. During the last decade, numerous ML algorithms have been proposed for prediction of CVDs by taking different attributes, datasets, and approaches. Nikam et al. [15] suggested ML approaches using the body mass index (BMI) for the prediction of CVDs. Different ML models have been utilized for the diagnosis of CVDs and among all the models the best accuracy at 84.25% has been achieved by the ANNs discussed by Meshref et al. [16].

From these studies, it can be concluded that continuous monitoring of different physical parameters of a patient is a highly emerging field in medical diagnosis. Hence accurate classification with best-fitted ML techniques in addition to implementation of real-time monitoring and prediction using Streamlit Python programming-based web application has also been demonstrated in this study with higher accuracy and in an effective manner.

Data and Variables

Study participants

The data set has been consolidated with the information taken from three primary sources namely the UCI machine learning repository-heart disease dataset, Kaggle- heart disease dataset by Yasser H, and real-time patients' clinical data from pathological laboratory which have been anonymized to ensure privacy.

In total 68,205 data have been collected from the first two said sources and 3000 data have been collected from pathology.

The proposed study has been carried out on both male and female subjects with an age group between 39 to 65 years. Out of 68,205 candidates, 23778 numbers of male subjects and 44427 numbers of female subjects have been present for this study. Figure 18.1 shows the distribution of different age groups of male and female candidates that were available in the dataset.

Out of 3000 real-time clinical data, 1690 numbers of female subjects and 1310 numbers of male subjects have been collected from the pathological laboratory. The distribution of male and female subjects has been shown in Figure 18.2.

Biomarkers and clinical measurements

Biological markers in short biomarkers are the measurable indicators of different biological conditions. Different features serving as biomarkers, provide quantifiable indicators of different physiological conditions. The choice of good biomarkers is very helpful for the determination of better health prediction and

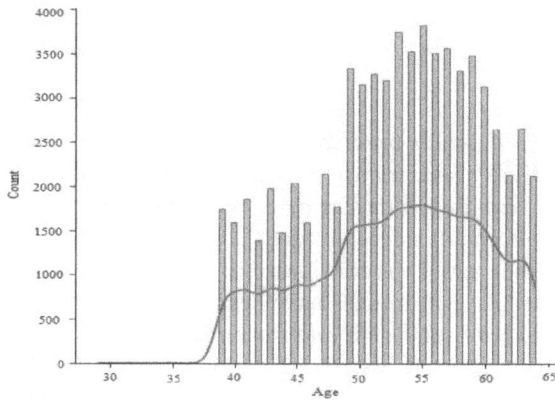

Figure 18.1 Distribution of age groups for male and female candidates
Source: Author

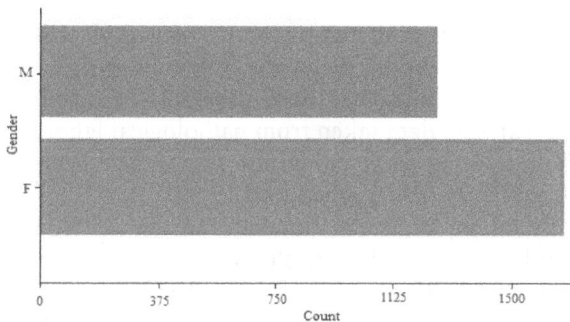

Figure 18.2 Distribution of male and female candidates for real-time clinical data
Source: Author

Table 18.1 Summary of biomarkers selection

Sl No	Type of Features	Name of the parameters	No of parameters
1	Objective	Age, Height, Weight, Sex	4
2	Examination	Systolic Blood Pressure, Diastolic Blood Pressure, Cholesterol, Glucose	4
3	Subjective	Smoking status, Alcohol intake, Physical Activity	3
		Total	11

Source: Author's compilation, attributes are considered from different datasets and real time clinical data

elevates the accuracy and precision of the proposed analysis.

In this study, biomarkers have been categorized into three features: objective features, examination features, and subjective features. Different features and their corresponding parameters that are taken into consideration have been shown in Table 18.1.

Variables in the dataset
The parameters, coming under objective features are age, height, weight, and sex. Age which influences disease susceptibility, treatment response, and overall health status, is the fundamental biomarker. Height and weight has been used for the derivation of body mass index (BMI) of a particular subject. The sex/gender of a subject has been taken as a categorical variable.

Examination features comprise measurable parameters that have been assessed during clinical evaluations. These features provide diagnostic clues, monitor disease progression, and evaluate diagnosis response. Systolic and diastolic blood pressure provides a comprehensive assessment of blood pressure

regulation, guiding hypertension management strategies to prevent cardiovascular events. Cholesterol, a lipid molecule, associated with cardiovascular risks is a categorical variable. Glucose level associated with diabetes diagnosis is also considered to be another categorical variable. Subjective features such as smoking status, alcohol intake, and physical activity have been considered as binary variables in this study.

The raw data set obtained has been filtered as per the requirement.

Methodology and Model Specifications

The workflow integrates data collection, biomarker selection, filtering of data, preparation of training and testing dataset, training the machine learning algorithm, prediction, and evaluation. The proposed methodology is shown in Figure 18.3.

Data collection refers to the process of gathering raw information from various sources such as databases, surveys, or other repositories for the analysis purposes.

In our context it has been gathered from UCI machine learning repository and Kaggle as mentioned above. These datasets typically contain structured or unstructured data which include combinations of text, numerical values, images or other types of information. Hence raw data have been filtered and

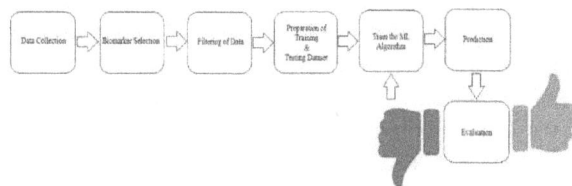

Figure 18.3 Flowchart of the proposed method
Source: Author

restructured in such a way so that it can be fed to different machine learning algorithm. From the filtered data the correlation matrix has been evaluated which represents the relationship between the variables. It is a square matrix showing the correlation coefficients between two variables. Different conclusions have been made from this correlation matrix such as CVD is highly dependent on systolic blood pressure, moderately dependent on cholesterol and age. Glucose and cholesterol are also highly dependent on each other.

Model specification

Model building in machine learning refers to the method of creating a mathematical representation of a real-world process or phenomenon using algorithms and data. It further encompasses various types of steps including data pre-processing, feature selection, algorithm selection, model training, evaluation, and optimization. The basic goal is to develop a model that generalizes to unseen data very well and effectively solves problems in a faster and more effective way.

Five different machine learning models such as logistic regression, support vector machine (SVM), K-nearest neighbor (KNN), decision tree, and random forest have been used in this study. The accuracies of each applied model have been compared to find out the effectiveness of the best-applied model.

Blood pressure level

This study used different ML techniques to find out best output for multiclass classification of blood pressure (BP) level. Blood pressure levels are classified as 1, 2, 3 and 4. These numeric values have been assigned for normal blood pressure as 1, elevated blood pressure as 2, hypertension stage 1 as 3, and hypertension stage 2 as 4 respectively. All the ML models have been trained with 70% of the total data and the remaining 30% of the data has been kept for testing purposes. Out of 68,205 data 47,743 data with 11 attributes have been fed for training. Therefore, the order of the training matrix is [47743 × 11], and the remaining 20462 data with 11 attributes have been used for testing purposes. Hence the order of the test matrix is [20462 × 11].

Cholesterol level

Cholesterol levels have been classified into three categories. 1,2,3 signifies normal, above normal, and well above normal respectively in this category of multiclass classification. For the classification of cholesterol level prediction, all the machine learning (ML) models have been trained with 90% of the total data and the remaining 10% of data has been kept for testing purposes. Out of 68205 data 61384 data with 6 attributes have been fed for training. Therefore, the

order of the training matrix is [61384 × 6], and the remaining [6821 × 6] data have been taken for testing purposes.

Cardiovascular prediction

The CVD diseases are present or not have also been classified. Five different machine learning approaches have been considered to get the best output for CVD and the example is considered to be a binary class ML model. All ML models have been trained with a similar 90% of total data and the remaining 10% of data have been used for testing purposes. Out of 68205 data 61384 data with 7 attributes have been fed for training and the order of the matrix is [61384 × 7]. The testing data have been taken as remaining 6821 numbers and the order of the matrix is [6821 × 7].

Health monitoring and predictive web application

A web application, using the Streamlit Python programming library has been developed. Considering the real-time data taken from pathological laboratory cholesterol levels, hypertension stages have been predicted by the said application.

By taking adjustable attributes in the web application like age, weight, systolic BP, diastolic BP, glucose level, and CVD present or absent, the cholesterol level of a person has been predicted. Figure 18.4 shows the cholesterol prediction interface which has been developed using Streamlit Python programming library.

Similarly, by adjusting these attributes like age, height, weight, gender, systolic BP, diastolic BP, physical activity, alcohol intake, glucose level, cholesterol level, CVD present or absent, different hypertension stages of a person have been predicted using the web application. Figure 18.5 depicts the hypertension prediction interface which has been developed using the same web application.

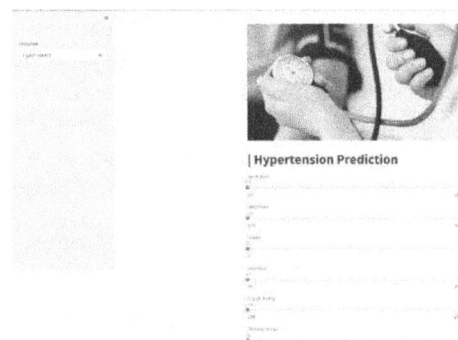

Figure 18.4 Cholesterol level prediction interface using streamlit python programming library
Source: Author

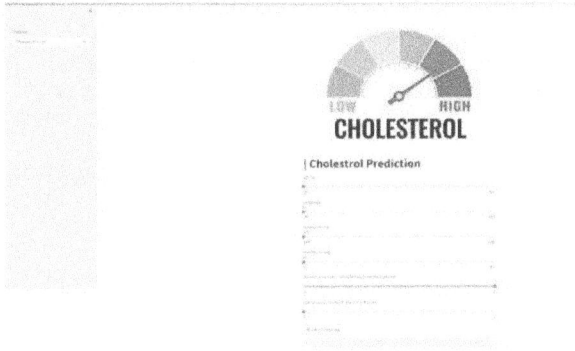

Figure 18.5 Hypertension prediction interface using streamlit python programming library

Source: Author

Empirical Results

Blood pressure level prediction results

For the multiclass blood pressure level prediction logistic regression, support vector machine, K-nearest neighbor, decision tree, and random forest ML techniques have been used. Among all the ML techniques random forest ML algorithm has been given 99.8% accuracy with 20436 correct output and 26 wrong output and it has been considered as the best ML algorithm for BP-level prediction. Different Hypertension Classification summary statistics, model accuracy, and comparative analysis between different ML models have been given in Table 18.2 and Figure 18.6.

Cholesterol level prediction results

On the other hand, for the multi-class cholesterol level prediction, the SV) ML technique is found to be more effective with an accuracy of 81.7% with 5577 correct results out of 6821 test data and considered the best-fitted model for cholesterol level prediction. Cholesterol Classification summary statistics, model accuracy, and comparative analysis between different ML models have been given in Table 18.3 and Figure 18.7.

Cardiovascular diseases prediction results

The CVD prediction is an example of binary classification. The logistic regression ML technique has been found the best-fitted model with 83.1% accuracy with 5668 correct results out of 6821 test data. CVD classification summary statistics, model accuracy, and comparative analysis between different ML models have been given in Table 18.4 and Figure 18.8.

Real-time cholesterol and blood pressure level prediction output

Real-time pathological known data have been taken to validate the proposed application for the said predictions. Out of 3000 real-time clinical data, 2990 data have been predicted correctly for cholesterol

level prediction using the web application achieving 99.6% accuracy and 2430 data have been predicted correctly for hypertension level prediction using the same achieving 81% accuracy by using the random forest ML technique that has been implemented in the web application.

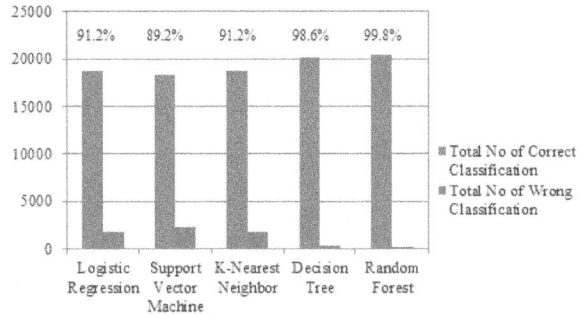

Figure 18.6 Hypertension classification summary statistics & model accuracy

Source: Author

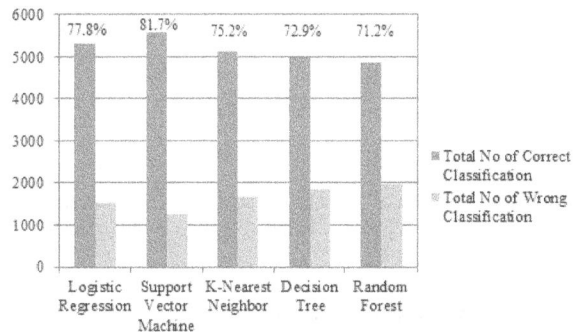

Figure 18.7 Cholesterol classification summary statistics & model accuracy

Source: Author

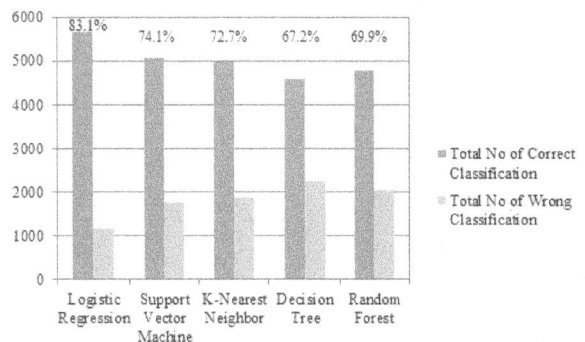

Figure 18.8 CVD classification summary statistics & model accuracy

Source: Author

Table 18.2 Hypertension classification summary statistics and model accuracy

Quality of state actual state	Predicted state Normal elevated Stage 1 Stage 2				Total number of correct classification	Classifier name	Classifier accuracy (%)
Normal	2805	125	125	0	2805		
Elevated	12	703	90	31	703	Logistic Regression	91.2%
Stage 1	34	107	11069	546	11069		
Stage 2	0	0	734	4081	4081		
Normal	2769	159	57	0	2769		
Elevated	0	0	18	79	0	Support Vector Machine (SVM)	89.2%
Stage 1	82	776	11424	523	11424		
Stage 2	0	0	519	4056	4056		
Normal	2805	125	125	0	2805		
Elevated	12	703	90	31	703	K- Nearest	91.2%
Stage 1	34	107	11069	546	11069	Neighbor (KNN)	
Stage 2	0	0	734	4081	4081		
Normal	2851	0	0	0	2851		
Elevated	0	830	83	22	830	Decision Tree	98.6%
Stage 1	0	37	11954	27	11954		
Stage 2	0	22	84	4552	4552		
Normal	2851	0	0	0	2851		
Elevated	0	933	2	1	933	Random Forest	99.8%
Stage 1	0	2	12016	21	12016		
Stage 2	0	0	0	4636	4636		

Source: Author's compilation, results achieved by different ML algorithms from confusion matrices

Table 18.3 Cholesterol classification summary statistics and model accuracy

Quality of state actual state	Predicted State Normal above normal well above normal			Total Number of correct classification	Classifier name	Classifier accuracy (%)
Normal	5019	785	496	5019		
Above normal	0	0	0	0	Logistic	77.8%
Well above normal	153	76	292	292	Regression	
Normal	5001	741	462	5001		
Above normal	0	157	0	157		81.7%
Well above normal	41	0	419	419	Support Vector Machine (SVM)	
Normal	4934	788	589	4934		
Above normal	152	50	47	50	K- Nearest	75.2%
Well above normal	86	23	152	152	Neighbor (KNN)	
Normal	4571	656	456	4571		
Above normal	332	141	68	141	Random Forest	72.9%
Well above normal	269	64	264	264		
Normal	4454	626	435	4454		
Above normal	412	144	95	144	Decision Tree	71.2%
Well above normal	306	91	258	258		

Source: Author's compilation, results achieved by different ML algorithms from confusion matrices

Table 18.4 Cardio vascular disease classification summary statistics and model accuracy

Quality of state actual state	Predicted state CVD_{Absent}	$CVD_{Present}$	Total number of correct classification	Classifier nName	Classifier accuracy (%)
CVD_{Absent}	3131	802	3131	Logistic Regression	83.1%
$CVD_{Present}$	351	2537	2537		
CVD_{Absent}	2812	1121	2812	Support Vector	74.1%
$CVD_{Present}$	643	2245	2245	Machine (SVM)	
CVD_{Absent}	2791	1142	2791	K- Nearest	72.7%
$CVD_{Present}$	717	2171	2171	Neighbor (KNN)	
CVD_{Absent}	2602	1331	2602	Decision Tree	67.2%
$CVD_{Present}$	906	1982	1982		
CVD_{Absent}	2693	1240	2693	Random Forest	69.9%
$CVD_{Present}$	815	2073	2073		

Source: Author's compilation, results achieved by different ML algorithms from confusion matrices

Conclusion

This study shows that PPG has transformative potential in modern healthcare by leveraging machine learning algorithms to predict various health parameters. The comprehensive approach involved meticulous data processing, cleaning, and filtering to ensure the reliability of the 50,000+ data points that have been utilized for training the machine learning models. Through rigorous testing and comparison of various ML algorithms, identifying the most accurate methods for predicting health conditions using binary and multi- class classification has been achieved. This comparative study highlights the importance of algorithm selection in achieving precise health predictions. Furthermore, the best-fitted ML model has been implemented for the development of a Streamlit Python programming library-based web application.

Overall, this study suggests that the different proposed ML techniques improve the idea of binary and multi- class classification methods in addition to the development of a web application which leads to lower diagnosis time.

References

[1] Li, D., Zhao, H., & Dou, S. (2015). A new signal decomposition to estimate breathing rate and heart rate from photoplethysmography signal. *Biomedical Signal Processing and Control*, 19, 89–95.

[2] Woo, M. W., Lee, J., & Park, K. (2018). A reliable IoT system for personal healthcare devices. *Future Generation Computer Systems*, 78, 626–640.

[3] Wannenburg, J., & Malekian, R. (2015). Body sensor network for mobile health monitoring, a diagnosis and anticipating system. *IEEE Sensors Journal*, 15, 6839–6852.

[4] Wijshoff, R., Mischi, M., & Aarts, R. (2017). Reduction of periodic motion artifacts in photoplethysmography. *IEEE Transactions on Biomedical Engineering*, 64, 196–207.

[5] Yuan, H., Memon, S. F., Newe, T., Lewis, E., & Leen, G. (2018). Motion artifact minimization from photoplethysmography based non-invasive hemoglobin sensor based on an envelope filtering algorithm. *Measurement*, 115, 288–298.

[6] Zhang, Z., Pi, Z., & Liu, B. (2015). TROIKA: a general framework for heart rate monitoring using wrist-type photoplethysmographic signals during intensive physical exercise. *IEEE Transactions on Biomedical Engineering*, 62(2), 522–531.

[7] Faust, O., Hagiwara, Y., Hong, T. J., Lih, O. S., & Acharya, U. R. (2018). Deep learning for healthcare applications based on physiological signals: a review. *Computer Methods and Programs in Biomedicine*, 161, 1–13.

[8] Rahmani, A. M., Thanigaivelan, N. K., Gia, T. N., Granados, J., Negash, B., Liljeberg, P., et al. (2015). Smart e-health gateway: bringing intelligence to internet-of-things based ubiquitous healthcare systems. In 2015 12th Annual IEEE Consumer Communications and Networking Conference (CCNC), (pp. 826–834).

[9] Islam, S. M. R., Kwak, D., Kabir, M. H., Hossain, M., & Kwak, K. S. (2015). The internet of things for health care: a comprehensive survey. *IEEE Access*, 3, 678–708.

[10] Farahani, B., Firouzi, F., Chang, V., Badaroglu, M., Constant, N., & Mankodiya, K. (2018). Towards fog-

driven IoT eHealth: promises and challenges of IoT in medicine and healthcare. *Future Generation Computer Systems*, 78, 659–676.

[11] Iqbal, S., Agarwal, S., Purcell, I., Murray, A., Bacardit, J., & Allen, J. (2023). Deep learning identification of coronary artery disease from bilateral finger photoplethysmography sensing: a proof of concept study. *Biomedical Signal Processing and Control*, 86(2023), 104993.

[12] Alkadya, W., ElBahnasy, K., Leiva, V., & Gad, W. (2022). Classifying COVID-19 based on amino acids encoding with machine learning algorithms. *Chemometrics and Intelligent Laboratory Systems*, 224, 104535.

[13] Sardar, I., Akbar, M. A., Leiva, V., Alsanad, A., & Mishra, P. (2023). Machine learning and automatic ARIMA/Prophet models-based forecasting of COVID-19:

Methodology, evaluation and case study in SAARC countries. *Stochastic Environmental Research and Risk Assessment*, 37(1), 345–359.

[14] Chaouch, H., Charfeddine, S., Aoun, S. B., Jerbi, H., & Leiva, V, (2022). Multiscale monitoring using machine learning methods: new methodology and an industrial application to a photovoltaic system. *Mathematics*, 10, 890.

[15] Nikam, A., Bhandari, S., Mhaske, A., & Mantri, S. (2020). Cardiovascular disease prediction using machine learning models. In Proceedings of the 2020 IEEE Pune Section International Conference, Pune, India, 16–18 December, (pp. 22–27).

[16] Meshref, H. (2019). Cardiovascular disease diagnosis: a machine learning interpretation approach. *International Journal of Advanced Computer Science and Applications*, 10, 258–269.

19 Performance analysis of nine level- inverter under different level shift modulation technique with variable frequency

Sagnik Lodh[1,a] and Jayanta Digar[2,b]

[1]B.Tech Student, Department of Electrical Engineering, Techno Main Saltlake, West Bengal, India

[2]Assistant Professor, Academy of Technology, Department of Electrical & Electronics Engineering, West Bengal, India

Abstract

Power quality issues are a major problem, especially the harmonics that occur in today's power systems. This research looks closely at how well multi-carrier pulse width modulation (PWM) method work to reduce harmonics produced from nine-level cascaded H-bridge inverters. The study examines three main modulation strategies: phase disposition (PD), alternative phase opposition disposition (APOD), phase opposition disposition (POD PWM. A detailed comparative analysis takes place over different carrier frequencies, ranging from 1 kHz to 10 kHz, without any filter. This study aims to identify the optimal modulation technique to minimize total harmonic distortion (THD) while keeping, output voltage waveform quality high.

The model of PWM-Inverter is constructed in MATLAB SIMULINK by adjusting the frequency. The results of this research help improve power quality in inverter-based systems.

Keywords: Alternative phase opposition disposition, multi-level inverters, phase disposition, phase opposition disposition, pulse width modulation

Introduction

Multi-level inverters (MLIs) are so much prevalent in both industrial applications and renewable energy applications. This fame is due to their extraordinary execution. When compared to conventional two-level inverters, substantially superior power quality is exhibited by MLIs and additionally, reduced total harmonic distortion (THD) is observed by them.

As flexible power electronic converters, the task of transforming DC power into AC power is vital for MLIs. Several sectors are engaged on by them, including the merging of green energy sources, the powering of industrial machinery, and the providing of residential appliances. Various power rangers are efficiently covered by them. [1, 2]. Cascaded H-bridge inverters, a type of multilevel inverter, offer a high-efficiency variation ratio, reduced electromagnetic interference, and lower power fall in energy variation processes compared with traditional inverters [3, 4]. They provide smoother voltage waveforms with more levels than traditional inverters, allowing for smaller filters and a reduced system space requirement. Additionally, they decrease the THD at the output voltage, improving the system's overall power quality [2, 5].

Numerous modulation methods are used to set voltage levels in inverters. In multicarrier PWM methods, these are divided into two types- level shift and phase shift. Under level shift approaches there are three common PWM methods - phase disposition (PD), alternative phase opposition disposition, phase opposition disposition (POD) [6, 7]. In this study, the nine-level cascaded H-bridge (CHB) inverter is examined, and the effects are analysed across three multicarrier PWM strategies, under switching frequencies values ranging between 1 kHz to 10 kHz.

Cascaded H-Bridge Inverter (CHB)

A cascaded H-bridge multilevel inverter (CHBMI) is a type of inverter which has several DC sources. Among both classic and current multilevel inverter topologies, the CHBMI is notably recognized for its segmented architecture, great mobility, and inherent fault-tolerant characteristics. Also, in situations where DC sources are usually split up, such as in photovoltaic (PV) systems and vehicle propulsion, It's best to have the CHBMI because it makes it easy to add to or change the layout of the PV plant or battery system. even under fault situations [8, 3]. The output of the cascaded multilevel inverter (CMLI) is the cumulative result of the H-bridge.

Four inverters linked in series. As more H-bridge inverters are added, the number of levels in the output waveform grows, making the waveform shape more (a) (b) closely resemble a sinusoid [7]. Figure 19.1 shows the nine-level cascaded H-bridge multilevel inverter and it's output waveform. The inverter is fed

[a]sagniklodh@gmail.com, [b]digarjayanta@gmail.com

DOI: 10.1201/9781003663348-19

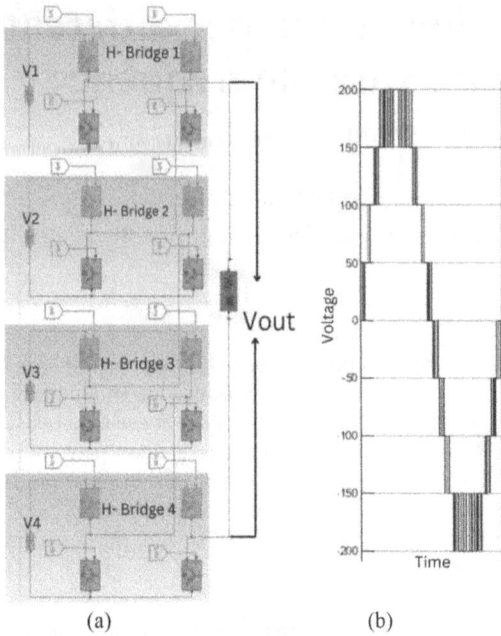

Figure 19.1 (a) Circuit diagram of 9-level inverter (b) output voltage waveform
Source: Author

by four separate DC sources each giving 50V and produces nine-levels output.

$$V_{out} = V_1 + V_2 + V_3 + V_4 \qquad (1)$$

To generate a multilayer waveform, the AC outputs from each H-bridge cell are coupled in series. The resulting voltage waveform is the cumulative sum of the outputs from all the inverters. The total number of voltage levels in the output phase of a cascaded inverter can be computed using the formula [7, 9]:

$$M = 2X + 1 \qquad (2)$$

Where X represents the number of DC sources. Here, each complete bridge inverter can produce three levels -Vdc, 0 and +Vdc in output.

Modulation Techniques

Flow switching frequency the carrier ratio is maintained within 10. Considering the necessity of eliminating high frequency voltage harmonics, it is desirable to enhance the switching frequency. However, it's vital to remember that larger inverter switching losses are incurred as a result. Therefore, it is advised to maintain the switching frequency as low as 6 kHz or 20 kHz, coinciding with the frequencies that the human ear can tolerate[10, 11]. In this study switching frequency is considered from 1 KHz to 10 KHz.

Multicarrier PWM
Cascaded inverters are fed from multi-carrier PWM approaches, which require modulating numerous triangular or sawtooth carrier signals. Popular approaches under this category are PD, alternative phase opposition disposition (APOD) and POD PWM schemes [2, 12].

PD-PWM: PD-PWM modulation involves utilizing multiple carrier signals in conjunction with a solitary reference waveform. These carrier signals are identical in phase and strategically positioned to produce a series of synchronized voltage levels. This method essentially compares a sinusoidal reference signal with a set of vertically offset carrier waveforms to generate the PWM switching patterns [2, 13]. Figure 19.4, displays how PDPWM is set up in a cascaded inverter to switch between nine voltage levels. The carrier signals required for each switch of the nine-level inverter are detailed in Table 19.1.

APOD-PWM: APOD PWM utilizes a series of vertically stacked triangular carrier signals with180-degree phase difference between adjacent ones. These carrier waveforms are then compared to a single sinusoidal reference signal to determine the switching states of the cascaded H-bridge multilevel inverter [12, 13]. In a cascaded inverter, Figure 19.6 shows how APOD-PWM is set up to switch between nine voltage levels. The same carrier signal arrangement is used for each switch in APOD-PWM, given in the Table 19.1.

POD-PWM: In POD, triangular carrier waveforms are stacked, with those above the zero-point shifted by 180 degrees relative to those below [12]. These carrier signals are superimposed on a single reference signal to determine the switching instants for the MLI's power switches [13]. The POD-PWM switching pattern for

Table 19.1 Carrier signal arrangement

Signals	Positive	Negative
Carrier 1	S1	S2
Carrier 2	S3	S4
Carrier 3	S5	S6
Carrier 4	S7	S8
Carrier 5	S9	S10
Carrier 6	S11	S12
Carrier 7	S13	S14
Carrier 8	S15	S16

Source: Common switching techniques for cascaded H-bridge multilevel inverter.(for verification purpose reference [1] & [2] can be considered).

a nine-level inverter is shown in Figure 19.8 with the same carrier arrangement used for all switches, given in the Table 19.1.

Result and Simulation

Our model a 9-level inverter has been generated in MATLAB Simulink, comparable to a 5-level inverter. The fundamental variation is in the number of carrier signals. Here, eight carrier signals are used. Four different carrier signals are used during the positive half cycle of modulating signal. The other four operate during the negative half cycle. From these carrier signals, sixteen PWM signals are produced and fed to IGBT switches.

The pulses are generated using different level-shifting techniques like PDPWM, PODPWM, and APODPWM. The system, which features multicarrier PWM control for the regulated inverter, is tested with an 'RL' load. This setup is illustrated in Figure 19.2. The DC input voltage for each source is set at 50 V. Furthermore, the system parameters for this designed model are outlined in Table 19.2.

In this study, the carrier frequencies have been set to 8 kHz. Nine voltage levels are produced on the load side for PD-PWM method, depending on switching states. The other methods are incapable of producing nine level voltage. Figure 19.3, illustrates the voltage waveforms of the output overtime FOR different PWM techniques. In this case study, voltage waveform is captured for 10 periods.

Figures 19.4, 19.6 and 19.8 demonstrates the carrier arrangement of PD-PWM, APOD-PWM, POD-PWM respectively and Figure 19.5, 19.7 and 19.9 demonstrates the THD performance of the evaluated multi-carrier PWM schemes for output voltages respectively. This method is tested at 8 kHz, and no output filter is applied. APOD-PWM and POD-PWM demonstrate THD performances of 17.51% and 12.86%,

Table 19.2 System parameters

Parameters	Value
Input voltage	50V DC
Transistor	IGBT
Load	Resistor - 100 Ω, Inductor - 2mH
Reference signal frequency	$\sin(100\pi t)$
Carrier signal frequency	1 KHz – 10 KHz
Repeating sequence time value	0 – 0.2 – 0.4
Modulation index	1

Source: Author

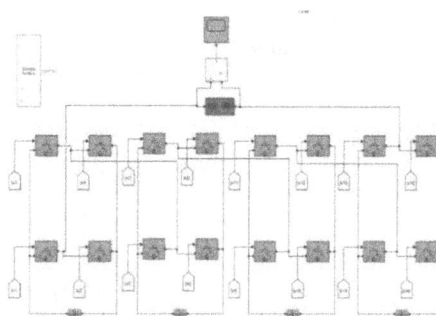

(c)

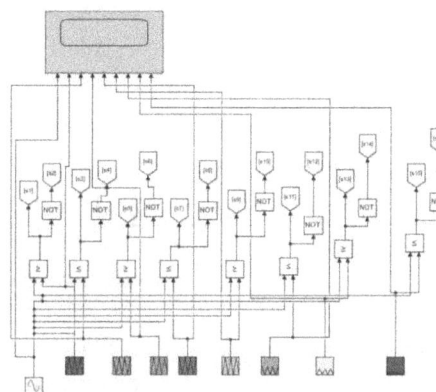

(d)

Figure 19.2 (c) Inverter circuit (d) Control circuit
Source: Author

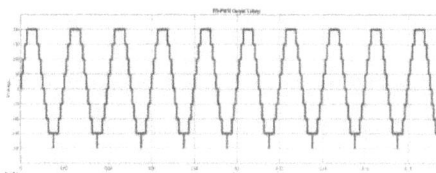

(e)

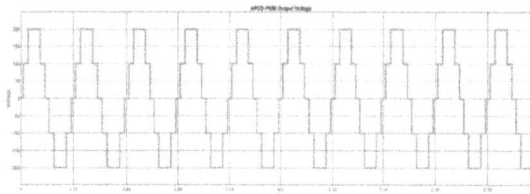

(f)

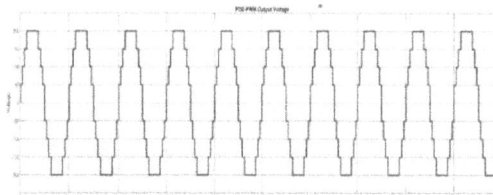

(g)

Figure 19.3 Output voltage over time (e) for PDPWM (f)for APODPWM (g) for PODPWM
Source: Author

Table 19.3 THD values for multi-carrier PWM methods for different carrier frequency values

| Sl, No | THD Values | | | |
	Frequency	PD-PWM	APOD-PWM	POD-PWM
1	1 KHz	14.51	14.44	14.41
2	2 KHz	16.29	16.43	15.83
3	3KHz	13.44	13.56	13.46
4	4 KHz	28.10	21.36	21.51
5	5 KHz	14.51	14.32	14.01
6	6 KHz	12.84	12.72	13.01
7	7 KHz	13.70	13.76	13.62
8	8 KHz	10.66	17.51	12.86
9	9 KHz	13.70	13.71	13.74
10	10 KHz	16.29	16.43	15.83

Source: Matlab simulation result

respectively. In a remarkable breakthrough, PD-PWM obtains a superior THD value of 10.66%, producing a greater value in performance. The multi-carrier PWM algorithms are examined with varied carrier frequencies. In the performance analysis, the carrier frequency is varied from 1 kHz to 10 kHz. All data are listed in Table 19.3.

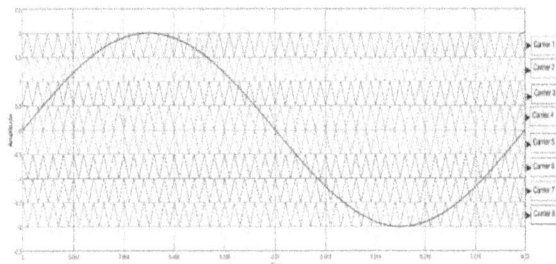

Figure 19.4 Carrier Arrangement of PD-PWM
Source: Author

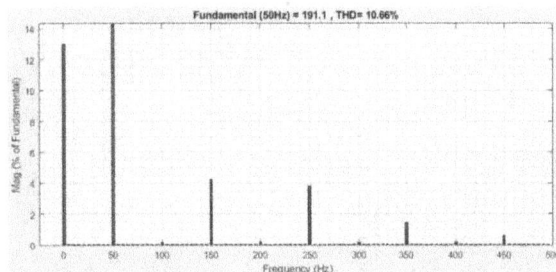

Figure 19.5 THD of PD-PWM at 8 KHz
Source: Author

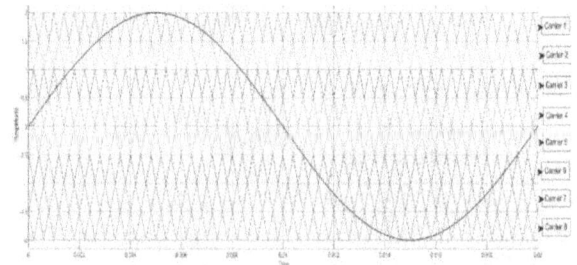

Figure 19.6 Carrier arrangement of APOD-PWM
Source: Author

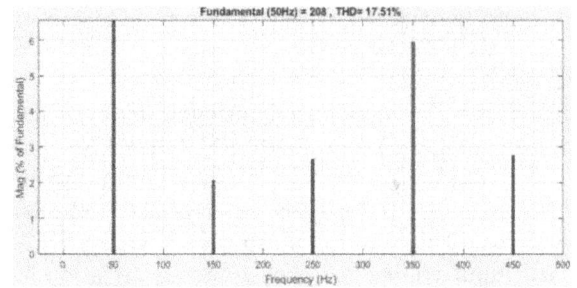

Figure 19.7 THD of APOD-PWM at 8 KHz
Source: Author

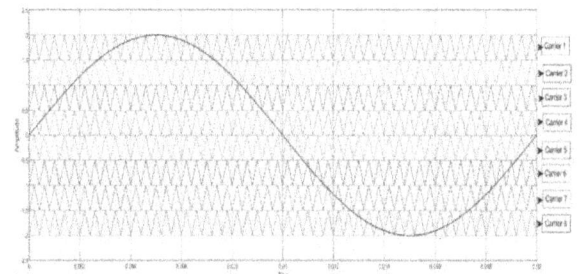

Figure 19.8 Carrier arrangement of POD-PWM
Source: Author

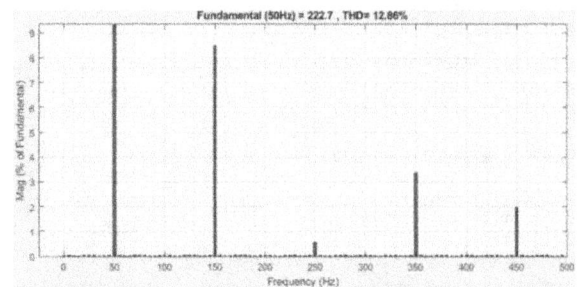

Figure 19.9 THD of POD-PWM at 8 KHz
Source: Author

Future Scope

As this study has evaluated in MATLAB, later it can be implemented in Hardware setup. And other modulation strategies such as PS-PWM, SHE-PWM, SVPWM etc can also be used.

System Configuration

Software-MATLAB 2018b

PC Configuration- Windows 10, 8gb ram, 512gb hard disk, AMD Athlon 200GE 2cores 3.2GHz processor.

Conclusion

This research offers an examination into the comparative performance of several multi-carrier PWM algorithms in generating high voltage levels. Techniques such as PD-PWM, POD-PWM, and APOD-PWM, each employing various carrier signals, are studied for their usefulness in providing high voltage outputs. That's why these approaches are simulated in a nine-level cascaded H-bridge inverter and the performance of inverter is evaluated. The modulation schemes are applied on 'RL' loads and assessed throughout a frequency range extending from 1 kHz to 10 kHz. It is discovered that the APOD PWM methodology gives somewhat better results (12.72%) than other techniques at specified frequencies, such as 6 kHz. However, based on the data obtained, the PD-PWM approach displays greater performance at 8 kHz compare to APOD-PWM and POD-PWM. The result of THD value is 10.66%, which is substantially lower than those attained by other approaches. Also it is noticeable that when switching frequency is raised, the THD values are lowered. However, at some point, the THD values grow, thus a correct switching frequency is essential to produce better THD values. The THD values corresponding to different carrier frequency changes have been provided.

So, the implications of this research are to enhance the power quality or to improve the efficiency and reliability and also to reduce the equipment wear and maintenance cost.

References

[1] Al-Badrani, H., & Salman, L. (2022). Study the effect of switching frequency on THD of multilevel inverter. *Journal of Modern Computing and Engineering Research (JMCER)*, 2022, 74–83.

[2] İnci, M. (2019). Performance evaluation of multi-carrier PWM techniques: PD, POD and APOD. *International Journal of Applied Mathematics Electronics and Computers*, 7, 38–43. 10.18100/ijamec.569660.

[3] Vijaybabu, S., Naveen Kumar, A., & Rama Krishna, A. (2013). Reducing switching losses in cascaded multilevel inverters using hybrid modulation techniques. *International Journal of Engineering Science Invention*, 2(4), 26–36.

[4] Kannadhasan, S., Saravanapandi, M., & Gurunathan, C. (2017). Switching strategies based cascaded multilevel inverters using modulation techniques. *IJIRMPS*. 5(1). DOI 10.17605/OSF.IO/4569E

[5] Amamra, S.-A., Meghriche, K., Cherifi, A., & Francois, B. (2017). Multilevel Inverter Topology for Renewable Energy Grid Integration. *IEEE Transactions on Industrial Electronics*, 64 (11), pp.8855–8866. 10.1109/TIE.2016.2645887 . hal-01717607

[6] Rodriguez, J., Franquelo, L. G., Kouro, S., Leon, J. I., Portillo, R. C., Prats, M. A. M., et al. (2009). Multilevel converters: an enabling technology for high-power applications. *Proceedings of the IEEE*, 97(11), 1786–1817. doi: 10.1109/JPROC.2009.2030235.

[7] Babkrani, Y., Naddami, A., Hayani, S., Hilal, M., & Fahli, A. (2017). Simulation of cascaded H - bridge multilevel inverter with several multicarrier waveforms and implemented with PD, POD and APOD techniques. In 2017 International Renewable and Sustainable Energy Conference (IRSEC), Tangier, Morocco, (pp. 1–6). doi: 10.1109/IRSEC.2017.8477370.

[8] Busacca, A., Di Tommaso, A. O., Miceli, R., Nevoloso, C., Schettino, G., Scaglione, G., et al. (2022). Switching frequency effects on the efficiency and harmonic distortion in a three-phase five-level CHBMI prototype with multicarrier PWM schemes: experimental analysis. *Energies*, 15, 586. https://doi.org/10.3390/en15020586.

[9] Barman, P., & Choubey, A. (2016). An overview of different multi-level inverters. *International Journal of Recent Research in Electrical and Electronics Engineering*, 3(3), 18–22.

[10] Wang, Z., Hao, S., Han, D., Jin, X., & Gu, X. (2021). Low switching frequency operation control of line voltage cascade triple converter. *Electronics*, 10(24), 3059. https://doi.org/10.3390/electronics10243059.

[11] Asker, M. E., & Kiliç, H. (2017). Modulation index and switching frequency effect on symmetric regular sampled SPWM. *European Journal of Technic*. 7, 102–109. 10.23884/ejt.2017.7.2.04.

[12] Baimel, D., Tapuchi, S., & Baimel, N. (2017). A review of carrier based PWM techniques for multilevel inverters' control. *WSEAS Transactions on Power Systems*, 12, 165–170.

[13] Meena, P., Kansal, K., Khan, A. H., & Sharma, A. K. (2020). Comparative investigation of 7-level cascaded multilevel inverter. *International Research Journal of Engineering and Technology (IRJET)* e-ISSN: 2395-0056 7(7) | July 2020 www.irjet.net p-ISSN: 2395-0072 © 2020, IRJET | Impact Factor value: 7.529 | ISO 9001:2008 Certified Journal | Page 3586.

20 Optoelectronic characteristics of exotic GaN/AlGaN based avalanche photodetector

Debraj Modak[1,a], Karabi Ganguly[2,b], Indranath Sarkar[2,c] and Moumita Mukherjee[3,d]

[1]Assistant Professor, AIEM, Mogra, West Bengal, India

[2]Professor, JISCE, Kalyani, West Bengal, India

[3]Professor, Adamas University, Kolkata, West Bengal, India

Abstract

In this research article, the authors have introduced an exotic type $Al_xGa_{1-x}N/GaN/Al_xGa_{1-x}N$ photodetector with SiC substrate material. This device shows significant improvement in photoresponsivity and quantum efficiency. The study is done through TCAD Silvaco simulator tool. This proposed structure incorporates the AlN buffer layer grown on SiC substrate. The new class of high electron mobility transistor (HEMT) device offers high mobility, better electrical conductivity, photocurrent and lower dark current at terahertz frequency. The quantum corrected drift diffusion (QCDD) simulator provides high responsivity (91%), quantum efficiency (80%), high photocurrent of 1.1×10^8 A. The simulation observation is verified with the experimental data. After the fluctuation of Al mole fraction, the author observed a good response of $Al_xGa_{1-x}N/GaN/Al_xGa_{1-x}N$ devices as an application of SPAD in UV region.

Keywords: HEMT, QCDD, silvaco, SPAD, substrate

Introduction

The avalanche photodiode (APD) is a highly sensitive, normal p-n junction semiconductor diode. It is operated in reverse bias voltage. It converts light into an electrical current. Due to the impact ionization, APD can detect the individual photon in low light conditions, generating electron-hole pairs. Under reverse bias voltage, when the electric field increases above the threshold voltage, the photon achieves sufficient energy, hits the other crystal atom and creates new electron-hole pairs. This is known as Avalanche Multiplication.

In the UV region, AlGaN/GaN-based APDs are highly recommended. Due to large bandgap, the UV region's optical characteristics determination is challenging. We resolve these issues by quantum-corrected drift diffusion (Q-CDD) model, minimizing the leakage current and enhancing the device performance [1]. We achieve high photo responsivity, high quantum efficiency, improved carrier mobility and lower dark current at lower operating voltage. The breakdown voltage of the heterostructure APDs model may be higher than in Si-based devices due to the large bandgap. The authors optimized the device structure to achieve lower breakdown voltages with increasing avalanche gain. The heterostructure model increases the avalanche gain due to the band alignment and the separation of absorption and multiplication regions. Due to the large bandgap, lattice mismatch maintenance on high-quality interface, material quality optimization and high noise could occur in the APD device. Calibrated and precise doping concertation and Monte-Carlo simulation can minimize the defect with lattice mismatch and noise in heterostructure devices.

In vivid research, silicon (Si) photodiodes are considered for detecting the optical signalling from UV to Infrared regions. Most researchers and the semiconductor industry use the silicon substrate for its high responsivity, low cost, ease to fabrication and availability. In the present era, researchers choose new improved compound materials such as SiC, 4H-SiC, and GaN for better response and characteristics like high functionality, better mobility, compact size, fabrication design. In this research work, the SiC substrate is preferred due to high-temperature coefficient, high mobility, and higher bandgap energy of the semiconductor material. In 2019 Su et al. described the recent advancement in SiC UV single-photon-counting avalanche photodiodes (APDs) with high multiplication gain significant efficiency and lower dark current [2].

In 2023, Wang et al. designed and simulated a comprehensive analysis of heterojunction bipolar transistor characteristics of AlGaN/GaN. It is observed

[a]debrajmodak.aiem@jisgroup.org, [b]karabiganguly73@gmail.com, [c]indranath.sarkar@jiscollege.ac.in, [d]drmmukherjee07@gmail.com

DOI: 10.1201/9781003663348-20

that, a non-linear relationship with increasing Al mole fraction and current gain of 73 at breakdown voltage 1270V. In Munoz et al. [4] described (Al, In, Ga) N-based III-nitrides photodetector. Here, overviews the optoelectronic materials properties in the UV region. In Khan et al. [5] reported the 0.2 um gated super-lattice GaN/AlGaN photodetector with 3000 A/W for 200-365 nm wavelength and 0.2 ms response time. This photodetector was incorporated with the He-Cd laser and significant output was observed in 325nm. Kumar et al. [6] demonstrated the HEMT-based InAlN/GaN-on-Si UV detector with photocurrent to dark current ratio. Ti/Au has been used as an ohmic contact with 20 nm depth to pinch-off the 2D electron gas between Source-drain pads. The author observes the spectral responsivity of 34 A/W at 5v incorporated with a very high photocurrent to dark current ratio at 367 nm wavelength. In Jiang et al. [7] designed and developed Visible-Blind MSM photodetectors on undoped HEMT AlGaN/GaN heterostructure devices with dark current 1.8×10^8 A/cm2 at 10 v bias voltage, peak responsivity 114mA/W and quantum efficiency 40% at 350 nm.

So, the authors have attempted to develop a new class of superlattice AlGaN/GaN/AlGaN photodetector to achieve high photo responsivity, high quantum efficiency, high photocurrent and low dark current. The characteristics of different materials are listed in Figure 20.1.

This research article shows the GaN/AlxGa1-xN heterostructure on the new substrate SiC materials for improving electro-optical characteristics. It observed

that a 40%-50% Al mole fraction can be applied for good conductivity and better output [8]. In this research article, the author performed 45% of Al mole fraction for better output. The Conventional models do not include all the characteristics of the semiconductor devices, such as the hot-carrier effect, ion mobility, photon absorption and other aspects of carrier transport phenomenon. Due to inadequate attributes of the conventional model, Author proposed the QCDD model for achieving the output after verified by experimental data. After validation, the authors demonstrated the p+ n- n n+ output of opto-electronics effect by TCAD simulator. This result exhibited the proposed highly sensitive, high-speed superlattice devices for UV region.

Device Modelling

The authors have designed and simulated $Al_{0.45}Ga_{0.55}$N/GaN/ $Al_{0.45}Ga_{0.55}$N APD heterostructure photodetector using MATLAB and TCAD simulator in the UV region. In this architecture, 200 um SiC substrate is used. On the top of the substrate, 50nm AlN buffer layer, 20 nm $Al_{0.45}Ga_{0.55}$N layer, 30 nm GaN and 20 nm $Al_{0.45}Ga_{0.55}$N layer is considered. The active region of the device has been vertically doped. The photons are struck on the top of ARC interface. The ARC coating has been utilized to prevent the reflection of absorbed photons. Ti/Au (20nm/50nm) ohmic contact has been used for better conductivity. The Authors have considered AlN nucleation layer above the SiC substrate. It improves the strain of the AlGaN and GaN interface. Under the reverse bias voltage, the asymmetrical doping concentration is applied to the p+ n-n n+ APD device in the UV region. The doping concentrations are 5×10^{18} cm, low doped 3×10^{17}cm, relatively high doped 1×10^{18} cm, and high doped 5×10^{18} cm at p+ n- n n+ respectively. A schematic APD photodiode structure is shown in Figure 20.2.

Material Parameter	Si	SiC	GaN	AlGaN	AlN
Band gap (ev)	1.1	3.26	3.42	3.91	6.12
Effective Conduction Band density of state (10^{19} cm−3)	3	1.5	2.24	2.7	4.42
Relative permittivity	11.7	9.6	9.5	9.7	8.5
Electron Mobility (cm2/Vs)	1450	700	1500	1000	300
Breakdown Electrical Field (10^6)	0.3	3.0	3.3	5	1
Carrier lifetimes	10^{-6}	$1-2 \times 10^{-6}$	10^{-9}	10^{-10}	35×10^{-3}
Electron saturation velocity (10^6) cm/sec	10	22	25	20	14
Electron affinity	4.05	3.1	4.1	4.02	1.45

Figure 20.1 Different material properties of compound semiconductor
Source: Author

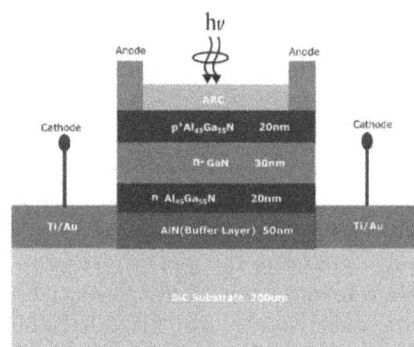

Figure 20.2 2D Structure of $Al_{0.45}Ga_{0.55}$N/GaN/$Al_{0.45}Ga_{0.55}$N heterostructure device
Source: Author

Result and Discussion

The Authors compare and analyze the performance of $Al_{0.45}Ga_{0.55}N/GaN/$ $Al_{0.45}Ga_{0.55}N$ and $Al_{0.45}Ga_{0.55}N/$ GaN devices to be grown in MOCVD technique by QCDD model coupled with the TCAD Silvaco simulator. The QCDD model incorporates carrier transport mechanism is expressed as [9],

$$\frac{\partial^2}{\partial x^2}V_{ex}(x,t) = -\frac{q}{\varepsilon}[N_{dc}(x,t) - N_{ac}(x,t) + C_h(x,t) - C_e(x,t)] \quad (1)$$

$$\frac{\partial}{\partial x}C_h(x,t) = -\left(\frac{1}{q}\right)\frac{\partial}{\partial x}J_{ih}(x,t) + G_{Ah}(x,t) + G_{Th}(x,t) - R_{ih} \quad (2)$$

$$\frac{\partial}{\partial x}C_e(x,t) = \left(\frac{1}{q}\right)\frac{\partial}{\partial x}J_{ie}(x,t) + G_{Ae}(x,t) + G_{Te}(x,t) - R_{ie} \quad (3)$$

Where, $V_{ex}(x, t)$ known as electric-field potential, $R_{ih,ie}$ (x, t), $J_{ih}(x, t)$, $G_{Ah,Ae}(x, t)$, $\mu_{h,e}$, $N_{ac}(x, t)$, and represents the recombination rate of electrons and holes, carriers current density, avalanche carrier generation rates, mobility of electrons and holes, acceptor and donor concentration, and carrier's permittivity respectively.

The Author describes and analyses the optoelectronics characteristics such as Photocurrent, Dark current, Photo responsivity and Quantum efficiency of the $Al_{0.45}Ga_{0.55}N/GaN/$ $Al_{0.45}Ga_{0.55}N$ heterostructure device in the UV region. Before analyzing the optoelectronic characteristics of APD sensors, the QCDD model was experimentally verified. This photocurrent of the devices is expressed in equation 4,

$$I_p = P_{in}\left(q\lambda/M_sM_{sc}\right)(1 - R)\left[1 - e^{(-\alpha x)}\right] \quad (4)$$

Here, M_s is the refractive index of the medium, M_{sc} is the refractive index of the material, P_{in} is the incoming photon flux, and α is the absorption coefficient of the material.

The photoresponsivity and quantum efficiency described in equation 5 and equation 6,

$$R = \left(q\lambda/hc\right)\eta \quad (5)$$

$$\eta = \left[\frac{\left(I_p/q\right)}{\left(P_{in}/hv\right)}\right] \quad (6)$$

where R and η represent the photo responsivity and quantum efficiency respectively.

Figure 20.3 illustrates the photo responsivity output of experimental work with a conventional AlGaN profile counterpart. Figure 20.4 compares various mole fraction values for $Al_xGa_{1-x}N/GaN/Al_xGa_{1-x}N$ profile. The Authors determine the photo responsivity of various mole fractions such as x = 0.40, 0.42, 0.45, 0.49. As we observe the $Al_{0.45}Ga_{0.55}N/GaN/$ $Al_{0.45}Ga_{0.55}N$ gives the better result of this heterostructure device. Figure 20.5. illustrates the photocurrent of Al0.45Ga0.55N/GaN/ Al0.45Ga0.55N and Al0.45Ga0.55N/GaN APD. The author obtains the

maximum photocurrent of 0.061A and 0.046A at 1v respectively and $Al_{0.45}Ga_{0.55}N/GaN/$ $Al_{0.45}Ga_{0.55}N$ gives better output under reverse bias conditions. Figure 20.6. Shows the Dark Current of $Al_{0.45}Ga_{0.55}N/$ GaN/ $Al_{0.45}Ga_{0.55}N$ photosensor devices by QCDD simulator and obtains better output of 14% than the $Al_{0.45}Ga_{0.55}N/GaN$ at -1volt. Figures 20.7 and 20.8. depicts the photoresponsivity and quantum efficiency of the superlattice APD photosensor devices. The author observes the photo responsivity of $Al_{0.45}Ga_{0.55}N/$ GaN/ $Al_{0.45}Ga_{0.55}N$ and $Al_{0.45}Ga_{0.55}N/GaN$ are 0.088A and 0.072A at 272nm respectively and maximum quantum efficiency are 84% and 75% respectively within 0-500 nm wavelength. Due to better carrier mobility of AlGaN/GaN/AlGaN device gives a significant output of photoresponsivity and quantum efficiency than AlGaN/GaN. From the above discussion, we reveals that $Al_{0.45}Ga_{0.55}N/GaN/Al_{0.45}Ga_{0.55}N$ photosensor device shows a significant outcome of optoelectronics characteristics in the UV wavelength region.

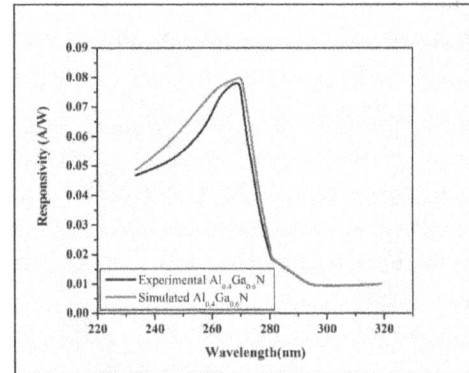

Figure 20.3 Experimental and simulation data of photo responsivity of AlGaN counterpart

Source: Author

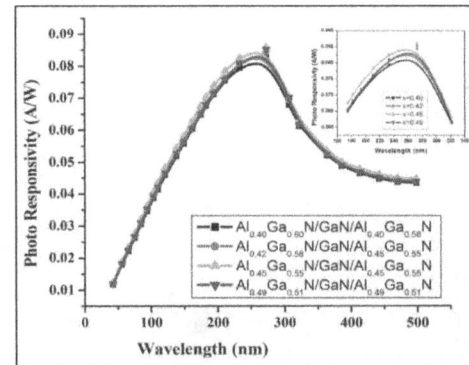

Figure 20.4 Comparison various mole fraction value of AlxGa1-xN/GaN/AlxGa1-xN profile

Source: Author

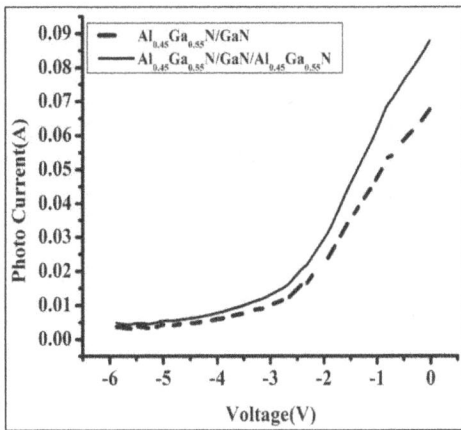

Figure 20.5 Photo current analysis of AlGaN/GaN/ Al-GaN, AlGaN/GaN APD (QCDD model)

Source: Author

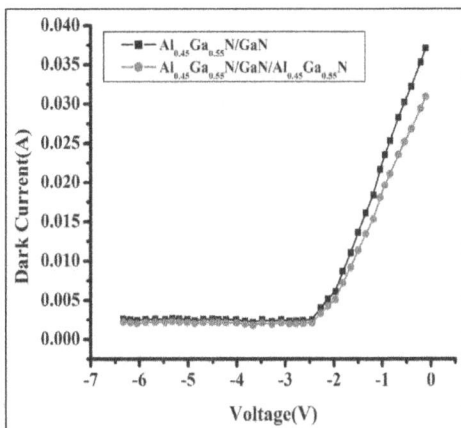

Figure 20.6 Dark current analysis of AlGaN/GaN/ Al-GaN, AlGaN/GaN APD (QCDD model)

Source: Author

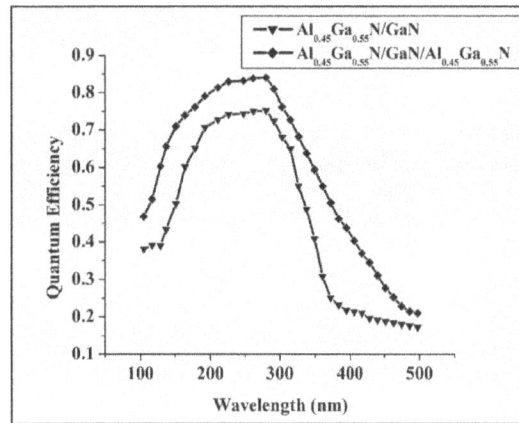

Figure 20.7 Comparison of photo responsivity for Al-GaN/GaN/AlGaN, AlGaN/GaN APD (QCDD model)

Source: Author

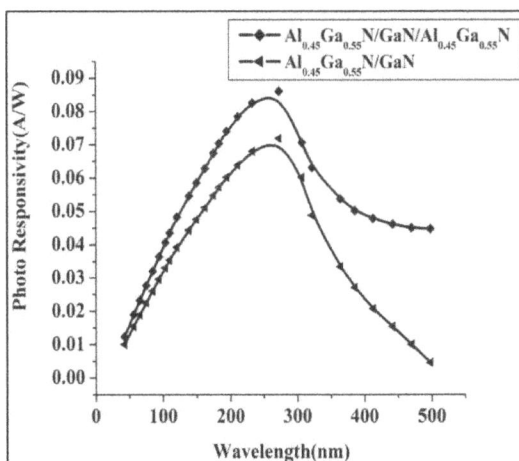

Figure 20.8 Comparison of quantum efficiency for Al-GaN/GaN/AlGaN, AlGaN/GaN APD (QCDD model)

Source: Author

Conclusion

The authors have designed and developed a new class of $Al_{0.45}Ga_{0.55}N/GaN/Al_{0.45}Ga_{0.55}N$ heterostructure high-performance APD device in the UV detection zone. The authors compare the AlGaN/GaN/AlGaN structure with AlGaN/GaN device. It reveals that AlGaN/GaN/AlGaN gives better output and leads to a significant carrier transport phenomenon. The high optoelectronics characteristics are shown in the result for AlGaN/GaN/AlGaN profile and generate at least 23% better photocurrent, 14% less dark current, 18% and 11% better output of Photo-responsivity and Quantum Efficiency respectively. According to the authors' information, it is first report of AlGaN/GaN/AlGaN over the SiC substrate with asymmetric doping concentration. This report can be used as an application for civil, defence and commercial sectors.

References

[1] Modak, D., Nair, S. D., Kundu, A., Sarkar, I,. Ganguly, K., & Mukherjee, M. (2024). Low noise GaN/AlxGa1-xN on SiC vertical p-n diodes for improved optoelectronic performance. In IEEE Space, Aerospace and Defence Conference (SPACE).

[2] Su, L., Zhou, D., Hai, L., Zhang, R., & Zheng, Y. (2019). Recent progress of SiC UV single photon counting avalanche photodiodes. *Journal of Semiconductors*, 40(12), 1–11.

[3] Wang, X., Zhang, L., Jiaheng He, J., Cheng, Z., Liu, Z., & Zhang, Y. (2023) Simulation and Comprehensive Analysis of AlGaN/GaN HBT for High Voltage and High Current. Electronics, 12(17), 3590.

[4] Muñoz, E. (2007). (Al,In,Ga) N-based photodetectors. some materials issues. *Physica Status Solidi B*, 2859, 244.

[5] Khan, M. A., Shur, M. S., Chen, Q., Kuznia, J. N., & Sun, C. J. (1995). Gated photodetector based on GaN/AlGaN heterostructure field effect transistor, *Electronics Letters*, 31(5), 398–400.

[6] Kumar, S., Pratiyush, A. S., Surani, D. B., Tripathi, S., Muralidharan, R., & Nath, D. N. (2017). UV detector based on InAlN/GaN-on-Si HEMT stack with photo-to-dark current ratio > 107. *Applied Physics Letters*, 111(25), 1–12.

[7] Jiang, H., Egawa, T., Ishikawa, H., Shao, C., & Jimbo, T. (2004). Visible-blind metal-semiconductor-metal photodetectors based on undoped AlGaN/GaN high electron mobility transistor structure. *Japanese Journal of Applied Physics*, 43(5B), 683.

[8] Boeykens, S., Leys, M., Germain, M., Poortmans, J., Daele, B. V., Tendeloo, S. V., et al. (2005). Growth, processing and characterization of GaN/AlGaN/SiC vertical n-p diodes. *MRS Online Proceedings Library (OPL)*, 892, 0892–FF13.

[9] Modak, D., Kundu, A., & Mukherjee, M. (2020). Multiple-graphene layer based p++n- n- - n++ device on Si/3C-SiC (100) substrate: a high sensitive visible photo-sensor. *Semiconductor Science and Technology*, 35, 1–14.

21 Realization of an innovative high efficiency n-type TOPCon solar cell using a lower band gap dielectric oxide

Shiladitya Acharyya[1,2,a], Dibyendu Kumar Ghosh[1,b], Sukanta Bose[2,c], Partha Banerjee[2,d] and Santanu Maity[1,e]

[1]SAMGESS, Indian Institute of Engineering Science and Technology, Shibpur, Howrah, West Bengal, India

[2]Academy of Technology, Adisaptagram, West Bengal, India

Abstract

This paper presents a high-efficiency n-type tunnel oxide passivated contact (TOPCon) photovoltaic cell employing titanium dioxide (TiO_2) as the dielectric layer in place of the conventional silicon dioxide (SiO_2). Our simulation results using AFORS- HET software demonstrate that TiO_2 not only provides excellent passivation of silicon surfaces but also significantly improves the electrical characteristics of the solar cell at higher tunneling thickness of TiO2. The innovative integration of TiO_2 in the TOPCon structure achieved notable efficiency at higher tunneling thicknesses compared to traditional SiO_2-based TOPCon designs. It also showed promising results with lower wafer thicknesses which reduce the use of the amount of silicon material in photovoltaic cells. This breakthrough undermines the potentiality of alternative dielectric materials in advanced silicon PV technologies, offering a innovative pathway toward more efficient and cost-effective solar energy providence.

Keywords: Band-offset, passivation, TOPCon, tunneling

Introduction

The realization of high-efficiency n-type solar cells exhibiting TOPCon structure represents a significant breakthrough in wafer-based silicon photovoltaic technology. The continuous quest for greater efficiency and cost-effectiveness in Silicon PV technologies has driven significant and remarkable advancements in design of silicon PV cells and use of various innovative materials. Among the leading innovations, the TOPCon cell has proven as an important technology because of its excellent passivation of silicon surfaces and superior electrical properties. Traditional TOPCon solar cells typically utilize ultra-thin silicon dioxide (SiO_2) as the dielectric layer, which serves as an essential component for passivating the silicon surface and reducing recombination losses as well as allowing the tunneling of charge carriers. However, some recent research articles has studied the potentiality of replacing SiO_2 with lower band gap dielectric oxides in enhancing PV performance parameters of solar cells [1–3].

TiO_2 offers several advantages over SiO_2, including a lower band gap, lower tunneling mass, which can facilitate better electron transport and reduce optical losses. Implementation of TiO_2 in the TOPCon structure improves passivation quality and enhanced electrical properties, contributing to higher overall efficiency. By leveraging TiO_2, researchers aim to push the boundaries of PV cell parameters, making renewable energy market more competitive and sustainable. The shift to using TiO_2 could pave the path for further developments in solar energy, driving progress toward more efficient and cost-effective solar power solutions. Sugiura et al studied tunneling properties of TiO_2 but the results were slightly anomalous at thicknesses exceeding 2nm because of their consideration of negative barrier height with silicon [3]. Our study is different as we simulated a slightly different model of TiO_2 with a smaller positive barrier height that allows higher tunneling thicknesses, upto 3nm. Study was also done on thinner wafers. Through some simulations and optimization, we demonstrate the potentiality of TiO_2 to not only match but surpass the maximum dielectric thickness of SiO2 in conventional SiO_2-based TOPCon solar cells, since it is already known how difficult it is to deposit ultra-thin layers with good uniformity, even with advanced PECVD and ALD systems. The study emphasizes the vital weightiness of material selection in determining photovoltaic performance and lays the groundwork for innovations in silicon photovoltaics. It also investigates the potential of thinner silicon wafers, offering a means to significantly decrease silicon consumption in wafer-based silicon photovoltaic systems.

[a]shiladitya.acharyya@aot.edu.in, [b]dibyendu.kumarghosh@aot.edu.in, [c]sukanta.bose@aot.edu.in, [d]partha.banerjee@aot.edu.in, [e]smaity@cegess.iiests.ac.

DOI: 10.1201/9781003663348-21

Physics

The insulator quantum-mechanical tunneling theory, also referred to as oxide tunneling theory [4], explains the tunneling current through the following equation:

$$J_{tunnel} = \frac{qn_{io}^2 v_{th}}{N_{eff}} e^{-\frac{2t_{ox}}{\hbar}\sqrt{2qm_{eff,ox}\Phi_{ox}}} = \frac{qn_{io}^2 v_{th}}{N_{eff}} P_{tunnel} \quad (1)$$

here, Jtunnel represents the current density due to tunneling phenomena, q is the elementary charge of 1 electron, nio denotes the intrinsic silicon charge carrier density, Neff is the effective density of charge carriers, v_{th} refers to the velocity of the thermally excited charge carriers due to lattice vibrations Additionally, meff,ox is the effective mass of electrons/holes in the oxide, Φox signifies the oxide barrier height (determined by the valence band offset for holes or offset at the conduction band for electrons), tox is the thickness of the oxide/dielectric, Ptunnel is the Probability of tunneling.

In Equation (1), all parameters related to crystalline silicon and other materials are treated as constants, except for some properties of the dielectric, which are oxide thickness (t_{ox}), effective tunneling hole/electron mass ($m_{eff,ox}$), and oxide barrier (Φ_{ox}). The terms "dielectric" and "oxide" are used interchangeably in this vein. From equation (1), it is evident that reducing the insulator thickness improves the likelihood/probability of quantum tunneling and, consequently, the tunneling current. Similarly, a lower tunneling mass and a smaller oxide barrier height increase the likelihood/probability of quantum-mechanical tunneling. According to the equation, the current decay exhibits exponential behavior.

$$J_{tunnel} = K_1 e^{-\frac{2t_{ox}}{\hbar}\sqrt{2qm_{eff,ox}\Phi_{ox}}}$$

$$t_{ox} \propto \frac{K}{\sqrt{m_{eff,ox}\Phi_{ox}}} \quad (2)$$

, where K and K$_1$ are constants

Equation (2) indicates that a lower tunneling mass of charge carriers and a reduced oxide/Si interface barrier height allow for a greater maximum permissible oxide thickness to enable carrier tunneling. However, since the insulator or dielectric layer is only a few nanometers thick, fabricating such an ultra-thin oxide layer is highly challenging. Therefore, materials with a smaller oxide barrier offset and lower carrier tunneling mass are preferred. From equation (2),

$$t_{ox} \propto \frac{1}{\sqrt{m_{eff,ox}}}; \text{ for one specific value of } \Phi_{ox} \quad (3)$$

$$t_{ox} \propto \frac{1}{\sqrt{\Phi_{ox}}}; \text{ for one specific value of } m_{eff,ox} \quad (4)$$

Tunneling becomes possible at greater thicknesses when the carrier tunneling effective mass and oxide barrier-offset are reduced. Insulators typically have higher barrier heights because of their broader bandgap relative to silicon. However, these barrier heights can be lowered using effective band-offset engineering, especially with asymmetric materials (i.e. materials with different band gap, work function and electron-affinity). The objective is to optimize the oxide thickness in order to facilitate quantum-mechanical tunneling, overcoming the difficulties of producing ultra-thin insulating layers. Furthermore, the performance parameters of solar cells could be analyzed by adjusting the silicon wafer thickness. Figures 21.1 and 21.2 represent the schematic of the proposed device architecture and the band diagram of the rear side of the cell structure.

Simulation Parameters and Results

Simulation is carried out using Automat FOR Simulation of HETerostructures i.e. AFORS-HET.

Key solar cell parameters, including open-circuit voltage (Voc), short-circuit current density (Jsc), fill factor (FF), and efficiency (Eff), have been analyzed for solar cells featuring an n-type TOPCon structure involving TiO2 as the tunneling layer. The n+-poly-Si doping variation, TiO2 thickness and Silicon wafer thickness curves are shown in in Figures 21.3–21.5 respectively with tables in Tables 21.3–21.5 respectively.

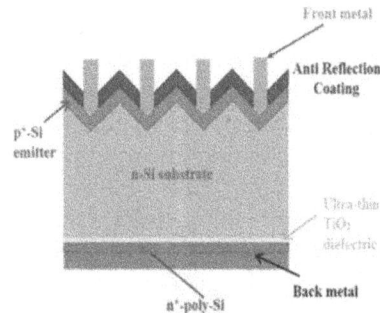

Figure 21.1 n-type TOPCon with TiO2 as tunneling dielectric

Source: Author

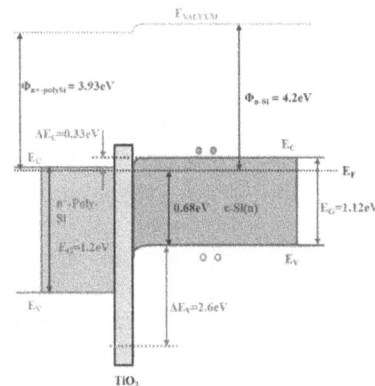

Figure 21.2 Band diagram of the BSF side of a n-TOPCon PV cell when TiO2 instead of SiO2

Source: Author

Table 21.1 PV Cell general parameters

Variable	Value
Silicon Wafer lifetime (n-type)	1ms
Front ARC SiNx thickness	75nm
Silicon band-gap	1.12eV
Poly-Si band-gap	1.2eV
Emitter thickness	0.3μm
Emitter doping	2×10^{19} cm^{-3}
Substrate/Wafer thickness	Variable (finally fixed at 180 μm)
Substrate/Wafer doping (n-type)	1×10^{16} cm^{-3}
BSF layer thickness	30nm
BSF layer n$^+$-polySi doping	Variable (finally fixed at 5×10^{19} cm^{-3})

Source: Author

Table 21.1 and 21.2 demonstrate the used parameters for conducting the simulation. Table 21.3 represent the impact of n+ poly-Si carrier concentration on the device output parameters whereas 21.4 shows the influence of TiO2 thickness on the device performance. The impact of wafer thickness on the device performance has been shown in Table 21.5. All the tables have been formed by the authors, the required data have been cited within the tables itself.

Table 21.2 Selected tunneling properties for TiO2 in comparison with SiO2

Parameters	SiO2	TiO2
Electron tunneling mass (me-ox)	0.41m$_0$ [5]	0.31m$_0$[7]
Hole tunneling mass (mh-ox)	0.32m$_0$ [5]	0.31m$_0$[7]
Bandgap (Eg)	9.00eV [5]	3.20eV[6]
Electron Affinity (χ)	0.90eV [5]	4.01eV[6]

Source: Author

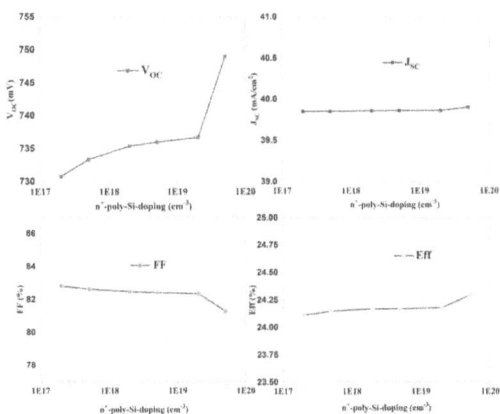

Figure 21.3 The variation of solar cell parameters, including Voc, Jsc, fill factor, and efficiency, on n$^+$-polySi doping has been studied
Source: Author

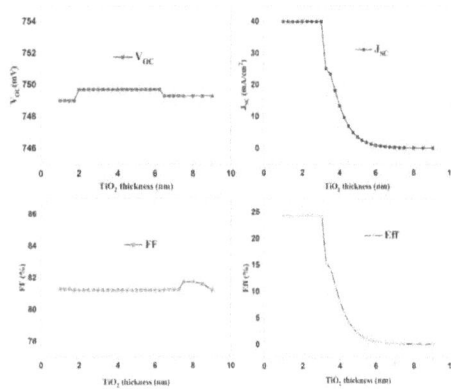

Figure 21.4 The variation of solar cell parameters, including Voc, Jsc, fill factor, and efficiency, on TiO$_2$ layer thickness has been studied
Source: Author

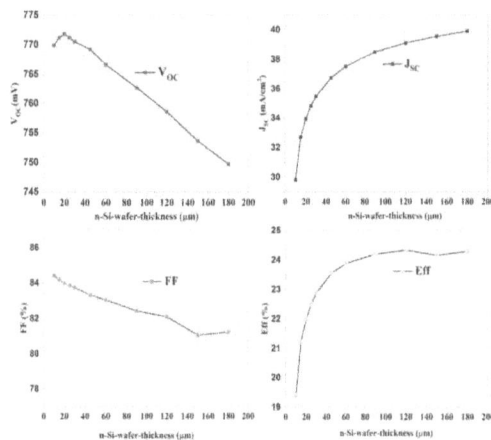

Figure 21.5 The variation of solar cell parameters, including Voc, Jsc, fill factor, and efficiency, on n-Si wafer thickness has been studied
Source: Author

Table 21.3 Variation with n+-poly-Si doping

n$^+$-poly-Si doping (cm^{-3})	V$_{OC}$ (mV)	J$_{SC}$ (mA/cm^2)	Fill factor (%)	Efficiency (%)
2.00E+17	730.8	39.85	82.81	24.11
5.00E+17	733.4	39.85	82.63	24.15
2.00E+18	735.4	39.86	82.47	24.17
5.00E+18	736.0	39.86	82.41	24.17
2.00E+19	736.7	39.86	82.34	24.18
5.00E+19	749.0	39.90	81.27	24.29

Source: Author

Results and Conclusion

This simulation study using AFORS-HET has successfully demonstrated the realization of high-efficiency n-type TOPCon PV cells using titanium dioxide (TiO₂)

Table 21.4 Variation with TiO2 thickness

TiO$_2$ thickness (nm)	V$_{OC}$ (mV)	J$_{SC}$ (mA/cm^2)	Fill factor (%)	Efficiency (%)
1	749.0	39.90	81.27	24.29
2	749.7	39.90	81.21	24.29
3	**749.7**	**39.90**	**81.21**	**24.29**
4	749.7	13.36	81.21	8.135
5	749.7	3.545	81.21	2.158
6	749.7	0.9372	81.21	0.5706
7	749.3	0.2525	81.24	0.1537
8	749.3	0.06881	81.73	0.04214
9	749.3	0.01886	81.22	0.01148

Source: Author

Table 21.5 Variation with n-Si wafer thickness

n-Si wafer thickness (μm)	V$_{OC}$ (mV)	J$_{SC}$ (mA/cm^2)	Fill factor (%)	Efficiency (%)
180	749.7	39.90	81.21	24.29
120	758.6	39.10	82.08	24.33
60	766.6	37.52	83.02	23.88
30	770.5	35.47	83.73	22.88
20	771.8	33.95	83.95	21.99
10	769.9	29.82	84.41	19.38

Source: Author

as the dielectric layer, replacing the conventional silicon dioxide (SiO$_2$). First n$^+$- poly-Si doping concentration has been varied and it was observed that the Voc increases with the increase in doping concentration, which in turn improves the Efficiency. While increasing the TiO2 tunneling layer thickness, it has been found that beyond 3nm thickness the short current density Jsc decay is exponential, which exactly follows the exponential equation for the decay of tunneling current with oxide thickness. Hence the decay in Efficiency Eff follows the same pattern. TiO$_2$ allows tunneling up to 10 nm whereas SiO$_2$ provides steady efficiency up to 1.5nm with tunneling terminating completely at 1.9nm as in our previous literature [1]. In our next study the Si wafer thickness was varied and it was found that the Voc improves due to lesser recombination in silicon bulk and the Jsc decreases due to less generation of charge carriers throughout the Si wafer which is our main absorber material. Jsc curve follows a nice exponential nature. The best results are obtained with 120μm Si wafer thickness. Thus, lesser amount of Si material used can also be

promoted, which not only reduce the cost but also opens the path for bendable (flexible) solar cell structures. Thus, our simulation results using AFORS-HET software confirmed that TiO$_2$ improves the electrical characteristics of the solar cell, particularly at tunneling thicknesses beyond 2nm, whereas in case of SiO$_2$ the value ranges from 1 to 1.5nm. This advancement addresses the fabrication challenges of ultra-thin oxide layers and demonstrates that TiO$_2$ can achieve high efficiency even with slightly thicker dielectric layers. Moreover, TiO$_2$ showed promising results with lower wafer thicknesses, which can decrease the quantity of silicon material needed in photovoltaic cells, thereby lowering costs and resource usage.

Our findings highlight the potentiality of alternative dielectric materials in advancing photovoltaics that provide a enlightened pathway toward more efficient and cost-effective solar energy. Future research will focus on further optimization and the practical industrial compatibility of this technology, aiming to push the boundaries of photovoltaic efficiency and broadly adopt them for sustainable energy.

References

[1] Acharyya, S., Sadhukhan, S., Panda, T., Ghosh, D. K., Mandal, N. C., Nandi, A., et al. (2022). Performance analysis of different dielectrics for solar cells with TOPCon structure. *Journal of Computational Electronics*, 21, 471–490.

[2] Acharyya, S., Ghosh, D. K., Banerjee, D., & Maity, S. (2024). Analyzing the operational versatility of advanced IBC solar cells at different temperatures and also with variation in minority carrier lifetimes. *Journal of Computational Electronics*, 23(6), 1170–1194. https://doi.org/10.1007/s10825-024-02232-y.

[3] Sugiura, T., Matsumoto, S., & Nakano, N. (2021). Numerical analysis of tunnel oxide passivated contact solar cell performances for dielectric thin film materials and bulk properties. *Solar Energy*, 214, 205–213. ISSN 0038-092X, https://doi.org/10.1016/j.solener.2020.11.032.

[4] Van der Vossen, R. (2017). Master of Science Thesis at the Delft University of Technology, May 17, 2017.

[5] Yeo, Y. C., Lu, Q., Lee, W. C., King, T., Hu, C., Wang, X., et al. (2000). Direct tunneling gate leakage current in transistors with ultrathin silicon nitride gate dielectric. *IEEE Electron Device Letters*, 21, 540–542.

[6] Dette, C., Pérez-Osorio, M. A., Kley, C. S., Punke, P., Patrick, C. E., Jacobson, P., et al. (2014). TiO2 anatase with a bandgap in the visible region. *Nano Letters*, 14(11), 6533–6538. DOI: 10.1021/nl503131s.

[7] Yu, H., Schaekers, M., Barla, K., Horiguchi, N., Collaert, N., Thean, A. V. Y., et al. (2016). Contact resistivities of metal-insulator-semiconductor contacts and metal-semiconductor contacts. *Applied Physics Letters*, 108(17), 171602.

22 Investigating the photovoltaic performance and thermal stability of AZO/Si hetero-junction solar cells with an analytical approach

Dibyendu Kumar Ghosh[1,2,a], Senjuti Das[3,b], Priyanjana Santra[3,c], Shiladitya Acharyya[3,d] and Sukanta Bose[3,e]

[1]School of Advanced Materials, Green Energy and Sensor Systems (SAMGESS), Indian Institute of Engineering Science and Technology (IIEST), Shibpur, Howrah, West Bengal, India

[2]Department of Electrical Engineering, Academy of Technology (AOT), Adisaptagram, Hooghly, West Bengal, India

[3]Department of Electronics and Communication Engineering, Academy of Technology (AOT), Adisaptagram, Hooghly, West Bengal, India

Abstract

In this study, the automat FOR simulation of HETero-structures (AFORS HET) v 2.5 simulation software was used to assess the potential efficiency and thermal stability of aluminum-doped zinc oxide (AZO)/c-Si(p) solar cells. Specifically, the study examined how the thickness (d) and carrier concentration (n) of the AZO layer affect the efficiency potential of the device. The results showed that with a 75 nm thickness and a carrier concentration of 9×10^{19} cm^{-3}, the conversion efficiency could reach up to 8.893%. Additionally, the study looked at how the carrier concentration of the AZO layer impacts the thermal stability of the AZO/c-Si(p) cell architectures.

Keywords: Automat FOR simulation of HETero-structures, Al-doped ZnO, heterostructure

Introduction

Solar photovoltaics (SPV) are widely regarded as one of the most important renewable energy sources due to their significant potential for sustainable and cost-effective electricity generation. Currently, among various solar cell architectures, advanced c-Si-based solar cell modules are extensively used worldwide. The abundance of raw materials, high efficiency, long-term stability, and scalability are the key driving forces behind advanced c-Si-based solar cells. However, the use of high-temperature processes during their fabrication increases production costs, energy consumption, and thermal budget. Therefore, careful selection of suitable solar cell structures for electricity generation is crucial. One effective approach may involve the fabrication of low-temperature processed metal oxide (doped or undoped)/c-Si heterostructure solar cells, as this type of cell architecture offers several advantages, including material flexibility, precise layer control, inherent surface passivation, and low-temperature processing.

Among various metal oxides, ZnO and its derivatives (AZO, BZO, GZO, IZO, etc.) are considered as promising TCO (transparent conducting oxide) materials due to their inherent advantages such as abundance, low processing costs, environmental friendliness, and compatibility with various deposition techniques [1–4]. ZnO can also serve as an emitter layer in ZnO/c-Si hetero-junction solar cells [1–4]. Furthermore, its potential to function as an anti-reflecting coating in ZnO/c-Si hetero-junction has generated significant interest in the solar photovoltaic research community. Aluminum doped zinc oxide (AZO) is a well-known n-type TCO material, and its doping concentration and conductivity can be adjusted by varying the aluminum concentration percentage in the zinc oxide matrix [5]. Moreover, this layer can be deposited using different physical (sputtering, ALD, thermal evaporation, etc.) and chemical methods (sol-gel, hydrothermal, spray pyrolysis, etc.) [5]. The refractive index of the AZO layer is suitable for providing anti-reflecting properties, thereby eliminating the need for a separate anti-reflecting coating (ARC).

Simulation Set Up

In this research, AFORS-HET v2.5 simulation software was utilized to analyze the photovoltaic efficiency and thermal stability of AZO/c-Si(p) hetero-junction solar cells [6]. First, the impact of AZO layer thickness on the device performance was evaluated. Subsequently,

[a]dibyendu.kumarghosh@aot.edu.in, [b]senjuti.das.21@aot.edu.in, [c]priyanjana.santra.22@aot.edu.in, [d]shiladitya.acharyya@aot.edu.in, [e]sukanta.bose@aot.edu.in

DOI: 10.1201/9781003663348-22

the influence of the photovoltaic performance and the cell structure's thermal stability on the carrier concentration of the AZO layer was investigated in order to determine the optimal conditions for future practical device fabrication. Figures 22.1 and 22.2 illustrate the cell architecture and the band diagram of the proposed device.

Results and Discussion

Variation of AZO layer thickness

The emitter layer thickness is undoubtedly considered as a pivotal role in determining the photovoltaic performance of solar cell structures. Consequently, this study explored how changing the thickness of the AZO layer affected the output parameters of the device. By adjusting the AZO layer thickness from 1 nm to 1000 nm, the optimal performance was observed at a thickness of 1 nm, resulting in the following output PV parameters: V_{oc} = 336.3 mV, J_{sc} = 34.01 mA/cm^2, FF = 73.75%, and efficiency = 8.436%. However, creating such a thin AZO layer while maintaining stoichiometry proved to be complicated. It was observed that the device performance deteriorated upon increasing the thickness of the AZO layer. For instance, the conversion efficiency decreased to 7.331% when the

AZO layer thickness was 10 nm and dropped even further to 6.32% when the thickness was 1000 nm. This decrease in efficiency was attributed to less light reaching the junction when the AZO layer thickness was increased, leading to more carrier recombination at the junction and within the bulk emitter. However, as no additional anti-reflecting layer was introduced, the optimized AZO layer thickness was chosen as 75 nm to provide the anti-reflecting attribute of the AZO layer. This decision was made based on a specific formula [7]:

$$t = \lambda/4n \qquad (1)$$

Where t is the required thickness of the AZO layer, λ is the wavelength of the incident light, and n is the refractive index (assumed to be 2 at a wavelength of 600 nm). In this instance, the photovoltaic efficiency was determined to be 7.108%. Therefore, the thickness of the AZO layer will be assumed as 75 nm for the remaining part of the study.

Impact of carrier concentration of AZO layer

On the photovoltaic performance

Apart from the emitter layer's thickness, another factor that influences performance is the conductivity of the emitter layer. Accordingly, the carrier concentration (n) of the 75 nm AZO layer was altered from 1×10^{17} to 9×10^{19} cm^{-3} to analyze its effect on the device's output. A notable improvement in device efficiency, especially in terms of V_{oc} and FF, was perceived upon increasing in carrier concentration of the AZO layer. An increment in V_{oc} was realized from 138.8 mV to 354.8 mV when the carrier concentration was raised from 1×10^{17} to 9×10^{19} cm^{-3}. Correspondingly, the FF value rose from 52.96% to 74.39% with this increase in carrier concentration. Figure 22.3 illustrates the

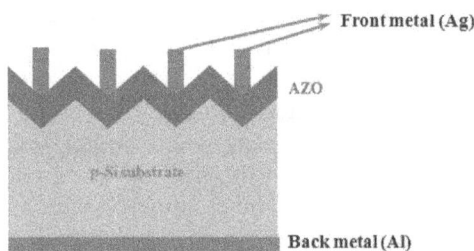

Figure 22.1 Schematic of AZO/Si hetero-junction solar cell

Source: Author

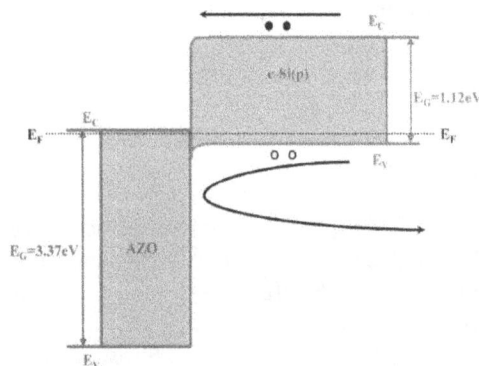

Figure 22.2 Energy band structure of the AZO/Si solar cell

Source: Author

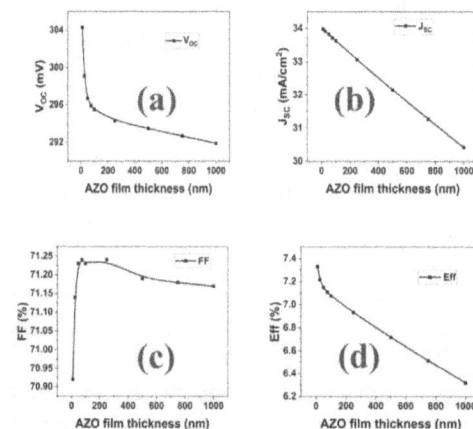

Figure 22.3 Impact of AZO layer thickness on (a) V_{oc}, (b) J_{sc}, (c) FF and (d) efficiency

Source: Author

effect of AZO layer carrier concentration on device output parameters. The carrier concentration of the AZO layer was not increased further to avoid the possibility of (i) higher Auger recombination rate and (ii) reduced optical transparency in the NIR region of the incident light due to increased free carrier absorption loss at higher carrier concentrations [8]. Figure 22.4 represents the change of solar cell output parameters with respect to AZO carrier concentration

On the thermal stability of the device

The performance of a solar cell is influenced by several output parameters such as temperature, humidity, sunhours, possibility of dust accumulation, and size and shape of the dust particles. Among these parameters, temperature is particularly crucial. In this study, we examined the photovoltaic performance of the AZO/Si solar cell at different operating temperatures ranging from 250K to 375K with intervals of 25K. The study also examined varying carrier concentrations in the AZO layer to investigate its correlation with the device's thermal stability.

The changes in the PV output parameters due to the variation in the device's operating temperature and different AZO carrier concentrations are illustrated in Figure 22.5. We observed a decline in V_{oc}, FF, and consequently the conversion efficiency, as the operating temperature increased, irrespective of AZO layer's carrier concentration. This is likely due to the increase in recombination current density (J_0) with the operating temperature [9]. However, we found that increment in the concentration of the carriers of the AZO layer led to better thermal stability, i.e., less deterioration in terms of V_{oc} and FF. This demonstrates the importance of enhancing the thermal stability by increasing the conductivity of the AZO layer, to achieve both higher power output and thermal stability.

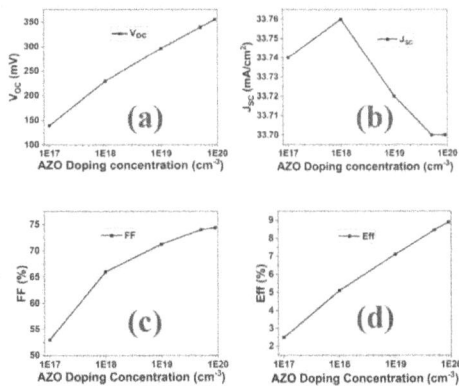

Figure 22.5 Variation in(a) V_{oc},(b) J_{sc},(c) FF and (d) efficiency due to change in the working temperature
Source: Author

Challenges and future scope of work

Apart from optimizing the emitter layer, the quality of the interface is another vital factor in deciding the power output of any type of heterojunction solar cell. Therefore, it is essential to conduct a comprehensive study on interface engineering in AZO/c-Si(p) solar cells to understand the impact of interface quality on device output. It is crucial to find strategies to enhance interface quality in order to significantly improve conversion efficiency. Additionally, fabricating practical solar cells while maintaining theoretical efficiency is a challenging task. Another aspect to consider is reducing base material consumption, such as wafers, in order to lower production costs. Taking all these factors into consideration, it can be concluded that improving the cost/watt-peak of AZO/Si heterojunction solar cells may be the future direction of research. The below Table 22.1 shows a comparison of the present work with some other earlier published one.

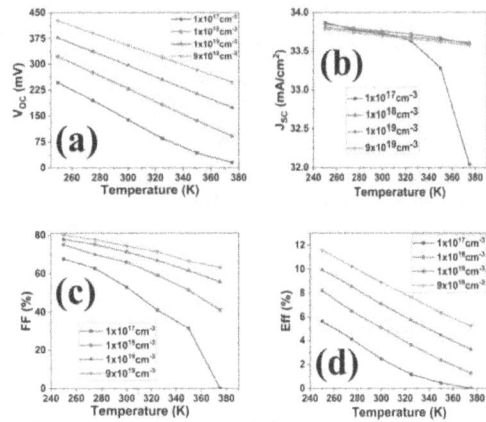

Figure 22.4 Impact of AZO layer carrier concentration on (a) V_{oc}, (b) J_{sc}, (c) FF and (d) efficiency
Source: Author

Table 22.1 Comparison with the practical devices fabricated in different labs worldwide

Cell Structure	V_{oc} (mV)	J_{sc} (mA/cm²)	FF (%)	Efficiency (%)	Ref
ZnO/Si	375	25	72	6.79	[10]
AZO/ ZnO:Mg (nanorod)/Si	496	28.3	65	9.1	[11]
AZO/buffer layer/bulk-Si	360	28.2	58.2	5.91	[12]
AZO/Si	354.8	33.70	74.39	8.893	This work

Source: Author

Conclusion

The potential of the photovoltaic performance of AZO/c-Si(p) solar cells was explored using AFORS HET v 2.5 simulation software. Initially, we examined the impact of the AZO layer's thickness on device performance, followed by studying the influence of the concentration of the carriers of the AZO layer. Our study revealed that by keeping the thickness and the carrier concentration of the AZO layer at 75 nm and 9×10^{19} cm^{-3}, respectively, the power conversion efficiency could reach 8.893%. Additionally, we investigated how the carrier concentration of the AZO layer affects the thermal stability of the cell architecture between 250 and 375K, at 25K intervals, our findings suggested that increasing the same could enhance the layer's thermal stability. Finally, we addressed the challenges and discussed the potential future work in our study.

References

[1] Roy, A., & Benhaliliba, M. (2023). Investigation of ZnO/p-Si heterojunction solar cell: showcasing experimental and simulation study. *Optik*, 274, 170557. doi:10.1016/j.ijleo.2023.170557.

[2] Chen, L., Chen, X. L., Liu, Y. M., Zhao, Y., and Zhang, X. D. (2017). Research on ZnO/Si heterojunction solar cells. *Journal of Semiconductors*, 38(5), 054005. doi: 10.1088/1674-4926/38/5/054005.

[3] Naim, H., Shah, D. K., & Bouadi, A. (2022). An in-depth optimization of thickness of base and emitter of ZnO/Si heterojunction-based crystalline silicon solar cell: a simulation method. *Journal of Electronic Materials*, 51, 586–593. doi:10.1007/s11664-021-09341-5.

[4] Hussain, B., Aslam, A., Khan, T. M., Creighton, M., & Zohuri, B. (2019). Electron affinity and bandgap optimization of zinc oxide for improved performance of ZnO/Si heterojunction solar cell using PC1D simulations. *Electronics*, 8, 238. doi: 10.3390/electronics8020238.

[5] Sarkar, J. (2014). Chapter 6 - Sputtering Targetsand Thin Films for Flat Panel Displays and Photovol-taics. Sputtering Materials for VLSI and Thin Film Devices (pp. 417–499). William Andrew Publishing. doi:10.1016/B978-0-8155-1593-7.00006-0.

[6] Varache, R., Leendertz, C., Gueunier-Farret, M. E., Haschke, J., Muñoz, D., & Korte, L. (2015). Investigation of selective junctions using a newly developed tunnel current model for solar cell applications. *Solar Energy Materials and Solar Cells*, 141, 14–23. doi:10.1016/j.solmat.2015.05.014.

[7] PV Education. "Anti-Reflection Coatings of Solar Cells." *PVCDROM – Design of Silicon Cells*, https://www.pveducation.org/pvcdrom/design-of-silicon-cells/anti-reflection-coatings. Accessed 4 August 2024.

[8] Bose, S., Arokiyadoss, R., Bhargav, P. B., Ahmad, G., Mandal, S., Barua, A. K., et al. (2018). Modification of surface morphology of sputtered AZO films with the variation of the oxygen. *Materials Science in Semiconductor Processing*, 79, 135–143. doi:10.1016/j.mssp.2018.01.027.

[9] Ghosh, D.K., Acharyya, S., Bose, S. *et al.* A Detailed Theoretical Analysis of TOPCon/TOPCore Solar Cells Based on p-type Wafers and Prognosticating the Device Performance on Thinner Wafers and Different Working Temperatures. *Silicon* 15, 7593–7607 (2023). https://doi.org/10.1007/s12633-023-02606-0

[10] Ismail, R. A., Al-Jawad, S. M. H., & Hussein, N. (2014). Preparation of n-ZnO/p-Si solar cells by oxidation of zinc nanoparticles: effect of oxidation temperature on the photovoltaic properties. *Applied Physics A*, 117, 1977–1984. doi: 10.1007/s00339-014-8605-y.

[11] Pietruszka, R., Witkowski, B. S., Ozga, M., Gwozdz, K., Placzek-Popko, E., & Godlewski, M. (2021). 9.1% efficient zinc oxide/silicon solar cells on a 50 μm thick Si absorber. *Beilstein Journal of Nanotechnology*, 12, 766–774. doi: 10.3762/bjnano.12.60.

[12] Kozarsky, E., Yun, J., Tong, C., Hao, X., Wang, J., & Anderson, W. A. (2012). Thin film ZnO/Si heterojunction solar cells: Design and implementation. In 2012 38th IEEE Photovoltaic Specialists Conference, Austin, TX, USA, (pp. 001217–001219). doi: 10.1109/PVSC.2012.6317821.

23 Safe terrestrial route recommendation system using geographical information system

Shuvodeep Debnath[1,a] and Yumnam Momojit Singh[2,b]

[1]Guru Nanak Institute of Technology, New Delhi, India
[2]Medhavi Skills University, New Delhi, India

Abstract

This paper proposes an efficient and safe terrestrial route recommendation system using geographical information system (GIS). Due to the increase in vehicular densities all over the world, a significant number of deaths have been registered due to road accidents. A probable solution that comes into our mind is to evolve our current transportation system into a sustainable transport system that must manage to provide mobility and reliability to all the users in a safe and circumstances (environment) friendly mode of transport. But this is a bit complex task and less practical as well because demands of people are different from one another and in a sense conflicting too. In Order to promote roads safety, we need to review the various measures that are been taken to reduce the conflicts arose due to increase in the road traffic and the conditions of road, road environment and several other important aspects. A particular interest is taken in understanding the impact of road safety issues on the wellbeing of all sorts of users and analysis is done for the various causes of road accidents.

Keywords: Evolution, Fiona, geographical information system, OpenMap, safest route recommendation

Introduction

Road safety is considered to be a significant issue and one amongst the foremost causes with the number of injuries and casualties across the world, leading more than one million demise as reported within the global statistics report on road safety index in previous years. Approximately more than 90% of those casualties happened in developing countries and 15% alone are encountered only in India [9]. Road traffic fatalities accounts for 16.6% of all deaths, making this the sixth leading reason behind death in India. A total of 4,80,652 road accidents happened in the year 2016 which caused injuries to 4,94,624 persons and claimed 1,50,785 lives [2]. We believe that the major reason behind road accidents is that the user is not cautious enough while driving. So, we will try to make the user cautious at the beginning of his/her journey. Users will enter the coordinates of the source and destination, and our model will show the possible routes and will try to rank those possible routes. The ranking will be done on the basis of the data we have analyzed in the paper. A more detailed structure of our model has been described in the paper.

More than 75% of road accidents fatalities occur in low- and middle-income countries such as India. Nearly half of the accidents involve pedestrians, cyclists and two power wheelers and due to this they are collectively termed as vulnerable road users. We need to know the cause of accidents to reduce the number of accidents. For that we need to consider the factors that are happening, the road mishaps and based on those factors we will carry out our analysis. Within the framework, we will be considering roads as one of the main components of this analysis, since it is the roads that determine the driver's behavior and as a consequence, is a direct participant in road traffic accidents [1]. We will be understanding the causes of accidents by classifying them into different categories. "It is obvious from Reference[10] that" instead of "Table 1 shows that".

Accidents classified by type of neighborhood

Table 1 shows that the traffic congestion in residential, institutional and market areas will be high, and these areas are more likely to involve a higher number of accidents but the data received shows a higher percentage of accidents in open areas [10]. The reason behind this can be over speeding or lower enforcement presence.

Accidents classified by road features

It may be directly noted that straight roads involve a maximum number of accidents; the reason may be the over speeding or any other traffic rule violations [10]. The number of accidents in form of deaths and injuries in previous years include different structures of the road like bridges, tunnel, conduit, ongoing construction of roads or roads under maintenance are increasing day by day and need attention for the engineers

[a]shuvodeep.debnath@gnit.ac.in, [b]yumnam.s@msu.edu.in

DOI: 10.1201/9781003663348-23

who are being assigned for maintaining these roads on a daily basis with the pedestrian who are using this type of roads in their daily life.

Accidents classified by road junction type

Road junctions are traffic merging points and hence are prone to accidents [10]. Though the statistical data presented by higher authority claimed that approximately more than 25% of the accidents in last 5 years took place at various types of connection point like "T shaped junction", "V shaped junction" or "Y shaped junction" with the majority of the mishaps happening in the "others" road feature in the complex network. In between all types of junction's options, T junction accounts for the largest share of accidents, persons injured and killed junction is a three-way node junction, a special options like street connection with different unequal sides, as one arm can be a pedestrian road connecting a highway or busy roads. Y junction is also a three-way junction, another option like road connection structure where three sides are equal, or more than two arms are equal.

Literature Survey

In this paper, we propose the safest route proposal system to generate optimal geometric safest route for the user in a given road network. The proposed work is relevant to a personalized safest route recommendation system for vehicles and pedestrians in a particular city [8]. In this section, they tried to illustrate similar work with real-time accident data. Also, another research is based on the safest route recommendation in terms of fewer crime avoidance roads. All previous work is based on the statistical analysis of data. The personalized route recommendation system based on real-time data is highly beneficial for the government, traffic control department as well as different commercial applications like Ola, Uber, etc.

Ministry of Road Transport and Highways (MoRTH), Govt. of India inaugurated the integrated road accident detection project with accident database (iRAD) project [10] on 15th February 2021 which was also funded by World Bank, with the objective to improve road safety in the country. IRAD was developed by IIT MADRAS and implemented by the National Informatics Centre [10]. Based on real-time traffic changes [1], enhanced ambulance dispatching by using a time-dependent shortest path method. Created dynamic vehicle routing for emergency services that takes unpredictable traffic patterns into account [2]. The main obstacles are traffic congestion, obstructions, and erratic traffic patterns. Conventional routing algorithms do not take sustainability, energy efficiency, or network resiliency into consideration [7].

Emphasized the requirement for intelligent transportation systems that incorporate adaptive route modifications and real-time data [3].

Emphasized how crucial resilient and sustainable routing systems are becoming in crowded metropolitan settings [4].

Methodology

The special GIS Software applications such as ArcGIS, QGIS, PostGIS etc. offer an interface to do evaluation for use of Python scripting. We'll use libraries which might be GIS specific as well as popular libraries in python in order to perform data pre-processing, feature extraction as well as model implementation.

GeoPandas

GeoPandas is an undertaking to feature a guide for geographic records to pandas gadgets. Kinds of Geoseries and Geodata frames are implemented by Geopandas. Geoseries and Geodata frames are subclasses of pandas Series and pandasData Frame respectively. GeoPandas gadgets can act shapely geometry gadgets and carry out geometric operations.

GeoPandas geometry operations are Cartesian in nature. The coordinate reference system (crs) may be saved as a characteristic on an object, and is routinely set while loading from a report. Objects can be converted to new coordinate structures with the to_crs() method. There is presently no enforcement of like coordinates for operations, however that could alternate with inside the future. In our project geopandas library is used to retard the shape(.shp) file and create the data frame from that shape file.

Shapely

Shapely is a popular python package that deals with set-theory based operation and manipulation of planar capabilities by using features from the widely known and broadly deployed GEOS library. The first premise of Shapely is that Python programmers need to be capable of carrying out PostGIS kind geometry operations out of doors of an RDBMS. Not all geographic records originate or are living in a RDBMS or are fine processed the use of SQL We have used the shapely library in python to do the spatial operations like "Touches", "Join", "Intersects", "Overlaps" that is to calculate numbers of different junction types present in a road.

We can extract the data from the open street map API that requires place names as the parameter. It can also take regions or residential areas as parameters. From this place, we can create two different data frames one for nodes and another for edges. The node data frame contains the geometry as a point and number of streets attached with it; it helps to find the junction types present in the road network. The edge data frame contains the geometry line string with the road name and id of the road. Besides this, the edge data frame also contains the road features like whether it is a bridge or not and whether the road is under construction or not. This data tells the road name with road features. Apart from that, the road data frame also contains the road type defining whether it is primary or secondary or it passes through any market region or residential region.

In the first part of implementation there are different types of road networks, road junctions, weather conditions and the corresponding data for accidents which are statistically shown in the statistical data [5]. All this data will be stored on the server. Here we will perform geo information analysis on spatial geo information data which is collected as input in the geographical information system. At first, we label the data in different categories based on the previous year statistical data model. Based on the statistical data (No of accidents, person killed, person injured) we can label it (HIGH accident prone, LOW accident prone). And this labelled data is also stored on the server. So, when the user enters the source point and destination point it shows the trajectory road with junctions or road where there is high prone denoted by red section or low prone zone denoted by green section and mid accident-prone zones are denoted by orange section [6]. In the following graphs we can see graphically about the accident with whether the person is injured or died in that road junction. Now all things should be displayed to the user to give in a graph to know the percentile change in data. Also, which section belongs to red (high zone) or yellow or green. In the second part of this model we will take the data of some roads as a shape file and read the file in python. For this part we have used different libraries and modules like shapely, osmnx etc. to store the shape file and perform spatial queries and operation (touches, intersects, crosses) with this data. So, we are particularly considering line string data for this implementation. We are finding different types of junction using these line string data. For this implementation we have used the geopandas and shapely module. Using geopandas library we read road's shape file and store it in a data frame. Then we extract the "geometry" part from that data frame and store it in a line string object. Now we calculate the different types of junction in the road using spatial operations like "touches", "cross", "intersects", "overlaps". For example, to find four arm junction means one road intersect another road. If such a intersect point is found then we append the intersection point. We can also plot the intersection point into the map using plot() function. Now we can

label this point which we calculated in the first step. Now there are some road features like "bridge", "tunnel", "straight road", "curved road", "potholes" we can directly detect by the line string features [5]. Now we use the "osmnx" module, this module will directly load the open street map api and convert into data frame. From this data frame we get the road network containing nodes and the edges. Edges are denoted as roads and from there we get the features like whether it is a bridge or not or whether it is a tunnel or not along with the name of roads. We label this road feature "bridge", "pothole", "Road under construction" as Red zone, Yellow zone or green zone based on the count of mishaps, count of casualties due to road mishaps and count of injured persons in the mishap. Now the third part in this problem describes how users have to give the source nodes and the final nodes. We must find all possible roads that are connected from the source point and to the final nodes. In Figure 23.1 we have described the proposed methodology.

Now we calculate the weight of the roads including the junction of two connecting roads. This weight is calculated based on the roads or junction in which zone it is falling under e.g. red zone has higher weights green zone has lower weights. Our task is to find the route which has lower weights like Dijkstra's Algorithm [6]. First we start with the source point. And find the roads connected to this source point and then push it into the stack and pop the nodes one by one from the stack and then find all the road's connecting with this point. Here we will use the tolerance that is used to modify the points value as each point may be disconnected by some value. This optimum route is recommended to the user for his journey. A architecture of model is as shown in Figure 23.2.

Now this model has other functionality like- maintaining the integrity of the database i.e. The accident data can only be entered by the government authorities and in some cases users can also enter the data provided that the data should be verified by Government authorities. Users can also know the accidents that happened in the previous year through this application

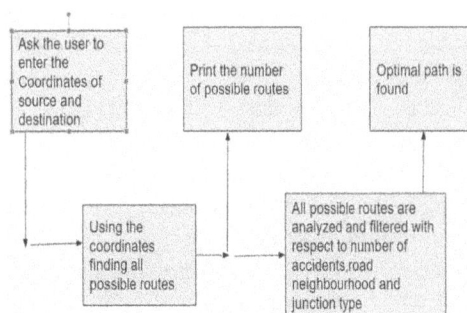

Figure 23.1 Flowchart
Source: Author

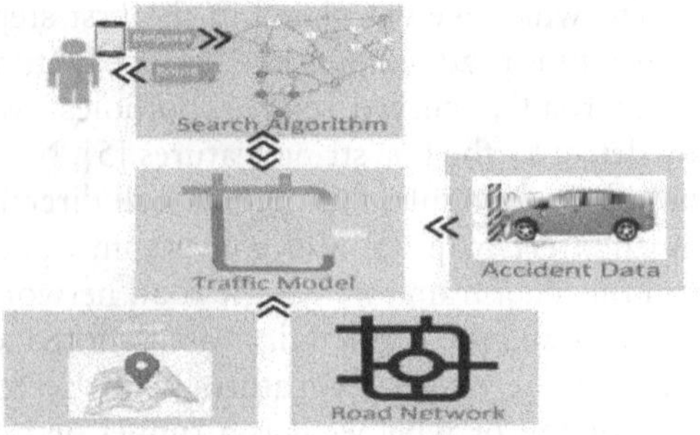

Figure 23.2 Model architecture
Source: Author

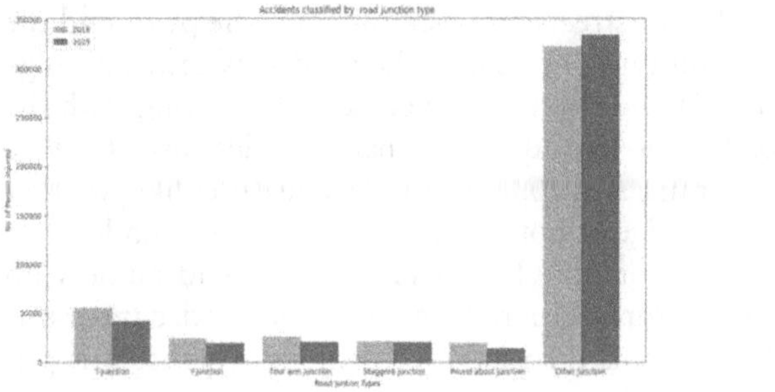

Figure 23.3 Accident data analysis with no of person injured
Source: Author

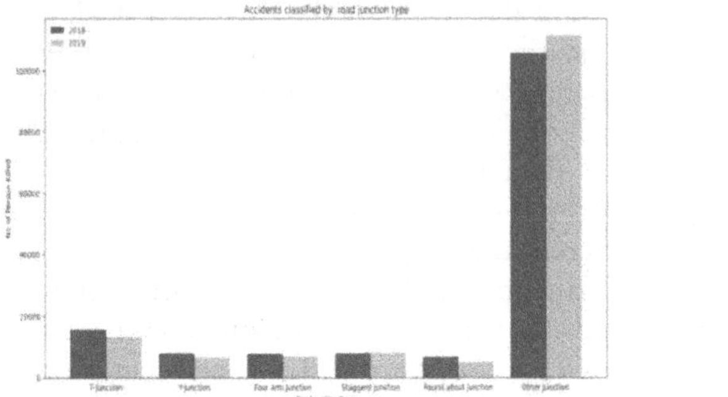

Figure 23.4 Accident data analysis with no of person killed
Source: Author

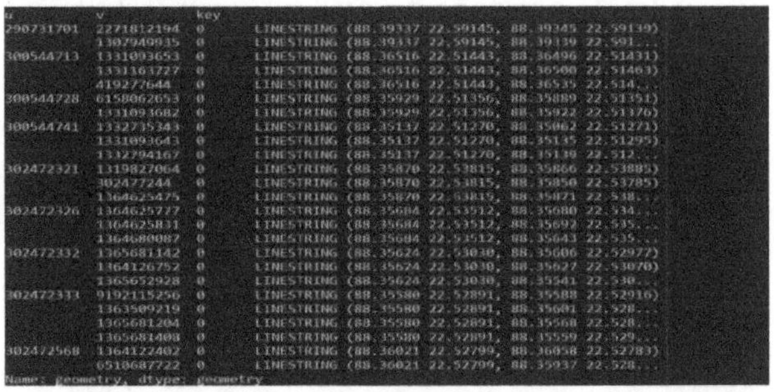

graphically .At last we calculate the weather condition of the current day and based on the weather condition we label the day as suitable for a journey or not (for example if weather is foggy then it falls under the red section as no accidents increases with foggy weather.

Algorithm used for the model is as follows :

//Input: Road network G consists of linestring,name of the road etc,label data with accident information,weather information and source and destination in terms of longitude and latitude.

//Output: safest route in terms of name of the roads

Algorithm Safest-Route-Recommendation(G,L,W,S,D)
$pt1 \leftarrow Find(G,S)$ //find the points nearest to source that is any of the end point of a road

$pt2 \leftarrow Find(G,D)$ //find the points nearest to destination that is any of the end point of a road

while $p1 \neq p2 + \delta$ // Delta is the tolerance
    $sol \leftarrow FindAllPossibleRoads(p1,p2)$
$wt \leftarrow FindWeight(sol,L)$ // get the weight age of every route with label information

$opt \leftarrow MinWeight(wt)$ //Find minimum weighted route
return opt

Results

Figures 23.3 and 23.4 will help users to graphically differentiate between the count of deaths due to mishap and count of people wounded year wise at different junctions of road.

From Figure 23.5 we can see that if we extract the "geometry" from the features it will give us a line string data frame. This line string contains the longitude-latitude value of each road in "Kolkata". In Figure 23.6 we find the names of the bridges in Kolkata. In similar fashion we can calculate the other roads features, types of neighborhood , junction types etc.

This (Figure 23.7) is the optimum safest route when the user gives a longitude and latitude

If we plot the retrieved points on google maps the optimum path will be shown like this (Figure 23.8). This optimal path may not be the shortest route from source to destination, but it guarantees the safest

Figure 23.5 Node Data frame from "osmnx" library.
Source: Author

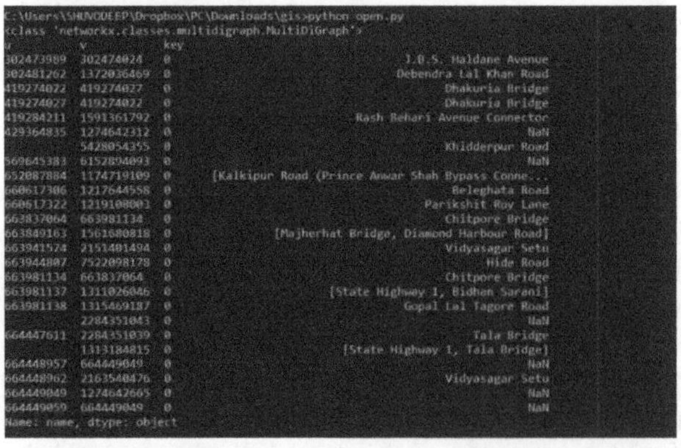

Figure 23.6 Edge data frame from "osmnx" library
Source: Author

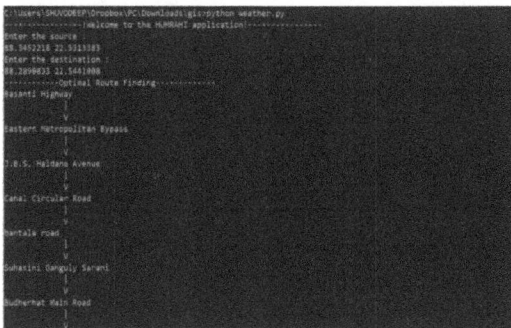

Figure 23.7 Optimal route after entering source and destination
Source: Author

Figure 23.8 Optimal route in graphical user interface
Source: Author

route from source to destination. In this model, we can alternatively use the Dijkstra or A* algorithm just by assigning each road junction a weight. That weight should be calculated from the accident label data. Then the Dijkstra algorithm finds the shortest path based on the lower weight.

Conclusion

The proposed model suggests the user a safe path between the specified starting and ending nodes based on the road features and the road junction type mainly. Apart from that there are other causes like weather conditions or types of neighborhood also causes the road accident. Also there is no user interference in this proposed model so user has to write the source, destination as a longitude latitude value or x-y coordinate value. Also, this model needs to be secure that no one can change any data without proper permission. Here we take the accident data as a yearly report but the results will be more correct if we analyze the monthly data. From the pedestrian and vehicle point of view the application. Now this application would be beneficial for both pedestrians and vehicles. By evaluating the complex network parameters like node degree, path length, betweenness centrality, clustering coefficient we can analyze the road network from the source to destination. They will get enough idea about previous year's accidents and also get to know the places in where more accidents are occurring in

the map. This app also be beneficial for finding the weather data according to the accidents. So this application will definitely reduce the number of accidents from the previous year accidents. Apart from that this application also helps the government of India and the traffic police to track the accident data and routes may vary based on the accidents. This application can also be attached with different emergency based transport system like ambulance, firefighting services to detect the sustainable route between one point to another point. This safest route recommendation interface is applicable for different cab services and essential product delivery services.

References

[1] Novikov, , Borovskoy, , Gorbun, , Terentyev, , & Pletney, (2021). Geographic information systems to improve road safety. In 2021 Systems of Signals Generating and Processing in the Field of on Board Communications, (pp. 1–6). doi: 10.1109/IEEE-CONF51389.2021.9416096.

[2] Kaur, A., Bala, P., & Santhi, A. (2018). GIS based road safety and traffic accident data analysis. *International Journal of Advanced Research in Science, Engineering and Technology (IJARSET)*, 5(4), 1–11. ISSN: 2350-0328.

[3] Budzynski, M., Kustra, W., Okraszewska, R., Jamroz, K., & Pyrchla, J. (2018). The use of GIS tools for road infrastructure safety management. In E3S Web of Conferences, (Vol. 26, p. 00009). 10.1051/e3s-conf/20182600009.

[4] Abd ALrhman, , & Jabbar, (2019). Decision support system in GIS to find best route. In 2019 First International Conference of Computer and Applied Sciences (CAS), (pp. 73–78). doi: 10.1109/CAS47993.2019.9075719.

[5] Automating GIS-processes (n.d.). Course information Retrieved October 30,2021. https://automating-gis-process-github.io/CSC18/course-info/course-info.htm.

[6] Wang, H., Li, G., Hu, H., Chen, S., Shen, B., Wu, H., et al. (2014). R3: a real-time route recommendation system. *Proceedings of the VLDB Endowment*, 7(13), 1549–1552. https://doi.org/10.14778/2733004.2733027.

[7] Zhou, X., Su, M., Liu, Z., Hu, Y., Sun, B., & Feng, G. (2020). Smart tour route, planning algorithm based on naïve bayes interest data mining machine learning. *ISPRS International Journal of Geo-Information*, 9(2), 112.

[8] Liu, L., Xu, J., Liao, S., & Chen, H. (2014). A real-time personalized route recommendation system for self-drive tourists based on vehicle to vehicle communication. *Expert Systems with Applications*, 41, 3409–3417. 34110.1016/j.eswa.2013.11.035.

[9] Ministry of Road Transport and Highways(MoRTH). 2019. Road acidents of India. https://morth.nic.in/road-accident in india.

[10] Ministry of Road Transport and Highways (MoRTH). (n.d.). iRAD. https://irad.parivahan.gov.in/

24 Electromagnetic imaging system for monitoring hip stem micromovements

Saurav Kumar Roy[1,a], Tirtharaj Banerjee[2,b], Debasis Mitra[1,c], Rik Chattopadhyay[1,d] and Souptick Chanda[2,e]

[1]Department of Electronics & Telecommunication Engineering, IIEST, Shibpur, Howrah, India

[2]Department of Biosciences and Bio Engineering, IIT Guwahati, North Guwahati, India

Abstract

In this study, our objective is to work on a proof-of-concept design and development of a non-ionizing, non-invasive, portable electromagnetic imaging system for continuous monitoring of micromovements of hip stem, necessary for ensuring the stability of the stem post operation of a hip replacement surgery. The imaging system consists of six-element antenna followed by a microwave imaging which processes the datasets from the data collection unit. A fast frequency-based imaging algorithm is applied to get reconstructed images. The images reconstructed from the imaging algorithm show the sensitivity of the imaging system to get different images according to the displacement of the stem.

Keywords: Antenna, electromagnetic imaging, hip replacement surgery, implant, micromovement

Introduction

Hip fracture and osteoarthritis is a serious concern particularly among the elderly people. The incidences of hip fracture in Indian population for the people aged 50 and above is roughly about 20 fractures per 1,00,000 persons [5]. India has an annual occurrence of 6,00,000 hip fractures due to osteoarthritis [23]. Urgent medical intervention is required in order to prevent potentially fatal complications. Clinicians and orthopedic surgeons often prescribe total hip replacement (THR) surgery for hip joint-related complications that cannot be addressed using non-invasive treatments such as physical therapies or medications [7]. In THR surgery, the damaged part of the femur is replaced with a cemented or cementless artificial prosthetic stem implant. The femoral head is removed and replaced with a stem that fits with the medullary cavity [22], such that it mimics the function of the normal hip joint. Doctors are required to continuously monitor the patients for their safety postoperatively. The success of the cementless total hip replacement surgeries depends on the primary stability of hip stem [18] ensuring minimal stem-bone interfacial micromotions. The relative micromovement values exhibit clinical significance since they influence the long-term secondary stability of the stem through osseointegration [14]. Literature study has shown the range of the micromovement of the stem metal implant to be about 50–150 μm which helps in developing osseointegration [4, 8, 17, 18]. Detection and monitoring of the relative interfacial micromotion (in μm) post-operatively becomes clinically significant immediately after the THR procedure since the primary stability of the stem influences its longer-term integration with the host bone and thus, the success of the surgery.

Earlier, there has been reported research in the measurement of micromotion of hip stem using in vitro measurement [3]. But the study done is invasive in nature and done in preclinical phase only. The stem micromotion monitoring is being done mainly using X-ray and CT-scan images. A study on the dynamic monitoring of micromovement of metal implant is performed by measuring the change in amplitude comparison function using a hybrid microwave-photonic sensor [11]. But the above system does not have the adequate resolution and precision. In the current market, the trending techniques for monitoring of hip related fractures and hip stem micromovement are X-ray and CT-scan imaging. However, these methods are not feasible for continuous monitoring due to several reasons, e.g., high ionizing radiation effect and non-portability [13, 15]. These techniques use ionizing radiation and so these cannot be used for continuous monitoring in patients as it may damage the healthy tissues. Also, a portable, non-ionizing, non-invasive system for the primary detection of diseases

---

[a]roysaurav.ju@gmail.com, [b]tirtharaj.workmail@gmail.com, [c]debasis.mitra.telecom@faculty.iiests.ac.in, [d]chatterjee.rik.84@gmail.com, [e]csouptick@iitg.ac.in

DOI: 10.1201/9781003663348-24

is currently in demand in the biomedical market. A study has shown that microwave-based sensors have been readily used in different fields of biomedical research applications like glucose monitoring, wireless telemetry of pacemakers, breast cancer detection, detecting anomalies in the lung, patient heart rate monitoring, etc [16]. Also, another study [19] examined that microwave technologies may be used as an imaging system that uses non-ionizing electromagnetic (EM) signals. From these studies it appears that EM imaging (EMI) is emerging as an alternative imaging technique in the biomedical field compared to the traditional medical imaging techniques. As it uses non-ionizing, low power EM signals, using microwave-based sensors for imaging is a low health-risk option. These microwave-based imaging systems are being readily used to monitor *in vivo* abnormalities in the healthcare domain now-a-days. Another study has been performed [10] on microwave-based imaging techniques that have been used in healthcare applications to visualize and study the internal human body organs and monitor bone fracture involving the use of EM signals transmitted into human body phantom and receiving the backscattered signals. The reflected signal power is used to monitor the healing of the fracture as the reflected power is different for normal and fractured bones. Bone fracture healing with time is also reported in a study [1] that uses transmission characteristics with two antenna systems. EMI has been reported [20] for onsite detection of knee injuries using array of eight antennas, vectored network analyzer (VNA) and microwave imaging unit (PC) which does the imaging algorithm. The processing of the data is done using imaging algorithms mainly in the PC or laptop.

In this work, our objective is to work on the design and development of a portable, non-invasive, non-ionizing microwave imaging system that will be used for the continuous monitoring of hip stem micromotion, necessary for ensuring the primary stability of the stem post operation of a hip replacement surgery. First, this study gives a brief summary of the hip fracture problem and its post-surgical monitoring problem which cannot be done using traditional imaging systems. Second, this study signifies the importance of EM based systems that have the potential to become an alternative to the X-rays and CT-scan systems. Third, use of the EMI systems in the bio-medical field in detection of diseases and anomalies is discussed and studied. The whole paper has been structured as follows. The methodology of the EMI system is described in Section 2 of the paper. Section 3 describes the results and discussions on the images reconstructed from the imaging system. Section 4 summarizes the paper.

Methodology

In this research work, our study is to do a continuous monitoring of the micromovement of the hip stem implant relative to the bone which is a key factor for ensuring the stability of the metal stem post hip replacement surgery operation using an EMI system. The proposed EMI system for the dynamic monitoring of micromovements (in µm) of hip stem consists of six antenna element that is positioned to encircle the hip of a human body as a wearable antenna is shown in the Figure 24.1.

The detection of micromovement (in µm) demands high resolution imaging, which requires wideband feature of an antenna so that we can reconstruct images with good resolution [2]. A wideband slot antenna operating from 0.5 to 1.788 GHz has been selected as our antenna element [21]. Lower microwave frequency range is selected in this imaging study as EM signals have a high penetration depth inside body tissues. Our study has been done on multilayer tissue models of a human body model. The six antennas are placed over the cylindrical human thigh model in CST microwave studio software. Our human thigh model consists of average skin, fat and muscle (SFM) layer, an average bone layer, assigning them with their respective permittivity and loss tangent value [12] in which a CAD model of stem is implanted inside the bone layer. The stem is made up of Ti64 alloy.

The antenna system of six elements is placed around the cylindrical body model of human thigh as shown

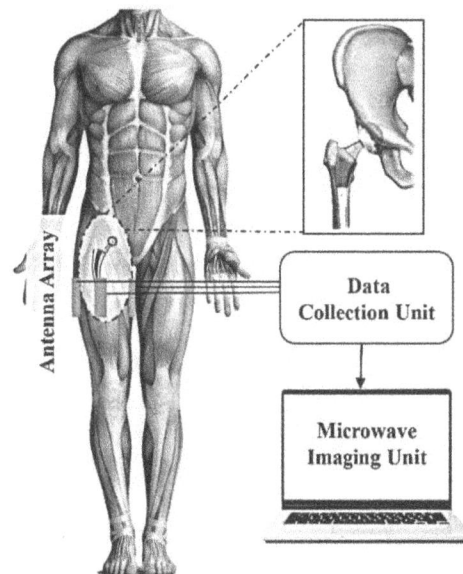

Figure 24.1 Overview of the proposed EMI system for the continuous monitoring of micromovements of stem post hip replacement surgery
Source: Author

in the Figure 24.2. The antenna system is excited and the scattering parameters (S_{11}, S_{12}, ..., S_{66}) are collected in the data collection unit. Furthermore, a fast frequency-based imaging algorithm is used to reconstruct the 2D images of the imaging plane. The variation in dielectric properties of human body tissues is utilized to monitor micromovements of stem implant. There is a change in scattering parameters of the six antenna elements with the micromotion of the metal implant as the dielectric properties of the thigh region will change due to the conducting nature of the metal implant. Tracking the micromovements of hip stem metal implant is done by further sending the scattering parameter data to the microwave imaging unit which applies the fast frequency-based algorithm to the data sets. The scattering parameters data utilization for the six-element antenna system for imaging is done on the PC which is done in following steps. The hip stem which is implanted in the bone is displaced in z direction and the scattering parameters are collected from simulation. The typical range of the micromovement of the stem metal implant is about 5-150 μm according to the literature. Contrarily, some authors have mentioned that the range of stem micromovements can be about 500–750 μm [9]. A stem movement of 100 micrometers (μm) causes very little shift in the resonant frequency in the reflection coefficient plot in the simulations. Using this information, we have performed the simulation for the stem displacement in z direction only in cases of Δz = 0 micrometers, 0.1 micrometers, 0.2 micrometers, 0.3 micrometers, 0.4 micrometers and 0.5 micrometers, where Δz is micromovement of the stem or the small displacement given to the stem in positive z direction. So, the scattering parameters (S_{11}, S_{12}, ..., S_{66}) are collected in the data collection unit for the displacement of stem for Δz = 0 to 500 μm with a step of 100 μm and then further this data is sent to the imaging unit where the fast frequency-based imaging algorithm is applied to reconstruct the 2D images in the imaging plane for the different amount of micromovements of the stem in z direction.

Results and Discussions

Six element antenna system simulation for imaging system

The imaging system consists of a six-element antenna system which is placed around the cylindrical model of human thigh bone in a conformal manner. The simulation of the antenna system is done for different Δz and the scattering parameters (S_{11}, S_{12}, ..., S_{66}) are collected in the data collection unit for the displacement of stem for Δz = 0 μm to 500 μm with a step of 100 μm. The simulation for stem displacement of Δz = 0 μm (condition for no movement) is taken as reference point which is the distance of the ports of the antenna system from the top of the cylindrical model. Then, the simulations for the rest of the displacements of stem Δz = 100 μm to 500 μm with a step size of 100 μm are carried out. Simulation for the Δz = 0 μm subsidence of stem is given in the Figure 24.3. The reflection coefficient parameters of the six antenna (S_{11}, S_{22}, S_{33}, S_{44}, S_{55} and S_{66}) element are shown in the figure for no subsidence of the stem (Δz = 0 μm).

Validation of EMI system

Six different cases of micromovements (Δz = 0 μm to 500 μm with a step of 100 μm) are studied in the simulation environment where Δz = 0 μm is taken as the reference image. We have applied the fast frequency-based imaging algorithm to reconstruct the 2D images in the x-y plane for the six different cases of micromovements taking the stem movement of Δz = 0 μm as the reference. The total numbers of scattered parameters from the six-antenna element system with six ports obtained are 16 × 16 = 256, and each signal is

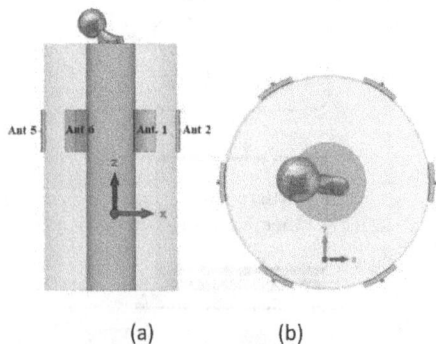

(a)   (b)

Figure 24.2 Overview of six antenna elements placed around the cylindrical model for human thigh with stem implanted into the bone layer. (a) Side view of placement of the six-antenna element; (b) Top view of the placement of six antenna element

Source: Author

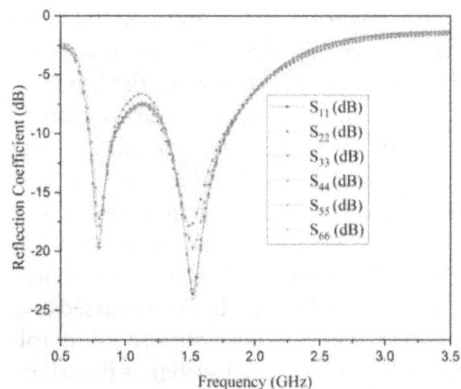

Figure 24.3 Reflection Coefficients of six antenna element for reference condition (in case of no stem movement)

Source: Author

sampled at 1001 frequency points. Figure 24.4 depicts the reconstructed 2D images of the different cases of micromovements of the stem. Case 1-6 gives us the six different levels of micromovements of stem (Δz = 0 to 500 μm with a step of 100 μm). The images are constructed in a 20 cm × 20 cm square area in x-y plane at the level of the location of the ports.

To assess the reconstructed images, two factors are used for differentiating the obtained images from simulations. Mean and normalized histogram plot of the six cases is calculated and plotted for the images respectively. The mean (μ) calculates the average value of all the pixels in an image. The normalized histogram $p(r_k)$ is obtained by dividing all elements of $h(r_k)$ by the total number of pixels in the images [6].

$$p(r_k) = \frac{h(r_k)}{n} \qquad (1)$$

where, r_k is the kth intensity level in the interval [0, G] and n is total the number of pixels in the image and G refers to the maximum possible pixel intensity value for an image. Table 24.1 shows the mean values of the images for the different cases of the stem movement. The reconstructed images of normalized histogram plot for different cases of displacement of stem Δz in z direction are shown in the Figure 24.5.

It can be observed that from Table 24.1 that the mean values are increasing in order of the increase of displacement of stem from the reference position. The histogram plot shows that the relative frequency of the pixels for intensities closer to 150 is decreasing

with the increase of displacement of stem from the reference position. So, these different images of the six cases of displacement can be mapped according to their respective stem displacement from reference position and we can get the measure the micromovements of the stem with the microwave imaging unit.

Conclusion

Hip stem micromovement becomes a crucial parameter in the aftermath of hip replacement surgery which plays a key factor for the stability of the stem. In this research work, we have done continuous monitoring of the micromovements of the hip stem implant relative to the bone with a microwave imaging system. Six different cases of stem motion are studied with the simulations done on a multi-layered tissue model of a hip. The images are reconstructed using a fast frequency-based imaging algorithm. Furthermore, the comparison of mean values and the normalized histogram plots of the reconstructed images gives information about the estimation of the stem displacement for different values of the micromovements. The future study will be to work on improving the imaging algorithm

Table 24.1 Comparison between different cases of stem displacement Δz (in μm)

Case	Stem displacement Δz (in μm)	Mean (μ)
1	0	121.5278
2	100	122.9989
3	200	126.3974
4	300	137.9008
5	400	153.0034
6	500	158.2085

Source: This table presents a comparison of the mean values of reconstructed images corresponding to six different types of stem displacement through simulation done in CST Microwave Studio software.

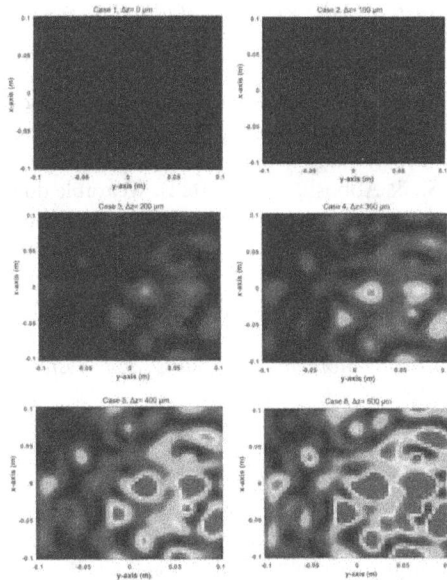

Figure 24.4 Reconstructed images of different cases of displacement of stem Δz in z direction. The horizontal axis (in m) and the vertical axis (in m) denotes the xy-plane in which the images are reconstructed at the level of the locations of the ports

Source: Author

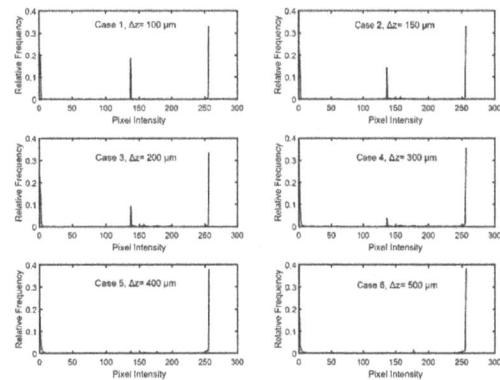

Figure 24.5 Reconstructed images of normalized histogram plot for different cases of displacement of stem Δz in z direction

Source: Author

so that we can measure the very small displacements of the hip implant.

Acknowledgement

This research work acknowledges "Development of the Microwave-Photonic Hybrid Wearable Sensor for in vivo Monitoring of Hip Stem Micromovements", a project by DST-SERB, GOI (Sanction Order No. SCP/2022/000232).

References

[1] Akdoğan, V., Özkaner, V., Alkurt, F. Ö., & Karaaslan, M. (2022). Theoretical and experimental sensing of bone healing by microwave approach. *International Journal of Imaging Systems and Technology*, 32(6), 2255–2261.

[2] Alqadami, A. S. M., Stancombe, A. E., Bialkowski, K. S., & Abbosh, A. (2020). Flexible meander-line antenna array for wearable electromagnetic head imaging. *IEEE Transactions on Antennas and Propagation*, 69(7), 4206–4211.

[3] Baleani, M., Cristofolini, L., & Toni, A. (2000). Initial stability of a new hybrid fixation hip stem: experimental measurement of implant–bone micromotion under torsional load in comparison with cemented and cementless stems. *Journal of Biomedical Materials Research*, 50(4), 605–615.

[4] Engh, C. A., O'connor, D., Jasty, M., McGovern, T. F., Bobyn, J. D., & Harris, W. H. (1992). Quantification of implant micromotion, strain shielding, and bone resorption with porous-coated anatomic medullary locking femoral prostheses. *Clinical Orthopaedics and Related Research (1976-2007)*, 285, 13–29.

[5] George, J., Sharma, V., Farooque, K., Mittal, S., Trikha, V., & Malhotra, R. (2021). Injury mechanisms of hip fractures in India. *Hip and Pelvis*, 33(2), 62–70.

[6] Gonzalez, R. C. (2009). Digital Image Processing. Pearson Education India.

[7] Mayo Clinic Staff, 2022, "Hip Replacement" [Online]. Available: https://www.mayoclinic.org/tests-procedures/hip-replacement/about/pac-20385042. [Accessed: 19-Sep-2024].

[8] Jasty, M., Bragdon, C., Burke, D., O'connor, D., Lowenstein, J., & Harris, W. H. (1997). In vivo skeletal responses to porous-surfaced implants subjected to small induced motions. *Journal of Bone and Joint Surgery (JBJS)*, 79(5), 707–714. https://journals.lww.com/jbjsjournal/fulltext/1997/05000/in_vivo_skeletal_responses_to_porous_surfaced.10.aspx.

[9] Kohli, N., Stoddart, J. C., & van Arkel, R. J. (2021). The limit of tolerable micromotion for implant osseointegration: a systematic review. *Scientific Reports*, 11(1), 10797.

[10] Lin, X., Chen, Y., Gong, Z., Seet, B.-C., Huang, L., & Lu, Y. (2020). Ultrawideband textile antenna for wearable microwave medical imaging applications. *IEEE Transactions on Antennas and Propagation*, 68(6), 4238–4249.

[11] Mitra, S., Mitra, D., Chanda, S., Chattopadhyay, R., Mandal, B., & Augustine, R. (2023). Hip Implant Micromotion monitoring using microwave-photonic hybrid device. In 2023 17th European Conference on Antennas and Propagation (EuCAP), (pp. 1 4).

[12] Nappi, S., Gargale, L., Naccarata, F., Valentini, P. P., & Marrocco, G. (2021). A fractal-RFID based sensing tattoo for the early detection of cracks in implanted metal prostheses. *IEEE Journal of Electromagnetics, RF and Microwaves in Medicine and Biology*, 6(1), 29–40.

[13] Ou, X., Chen, X., Xu, X., Xie, L., Chen, X., Hong, Z., et al. (2021). Recent development in x-ray imaging technology: future and challenges. *Research*. 1–18. https://doi.org/10.34133/2021/9892152.

[14] Schaer, M. O., Finsterwald, M., Holweg, I., Dimitriou, D., Antoniadis, A., & Helmy, N. (2019). Migration analysis of a metaphyseal-anchored short femoral stem in cementless THA and factors affecting the stem subsidence. *BMC Musculoskeletal Disorders*, 20, 1–9.

[15] Shbeer, A. (2024). Radiation in the intensive care units: a review of staff knowledge, practices, and radiation exposure. *Journal of Radiation Research and Applied Sciences*, 17(2), 100849. https://doi.org/10.1016/j.jrras.2024.100849.

[16] Singh, A., Mitra, D., Mandal, B., Basuchowdhuri, P., & Augustine, R. (2023). A review of electromagnetic sensing for healthcare applications. *AEU-International Journal of Electronics and Communications*, 171, 154873.

[17] Søballe, K., Hansen, E. S., B.-Rasmussen, H., Jørgensen, P. H., & Bünger, C. (1992). Tissue ingrowth into titanium and hydroxyapatite-coated implants during stable and unstable mechanical conditions. *Journal of Orthopaedic Research*, 10(2), 285–299.

[18] Soballe, K., Toksvig-Larsen, S., Gelineck, J., Fruensgaard, S., Hansen, E. S., Ryd, L., et al. (1993). Migration of hydroxyapatite coated femoral prostheses. a roentgen stereophotogrammetric study. *The Journal of Bone and Joint Surgery British*, 75(5), 681–687.

[19] Sultan, K. S., & Abbosh, A. M. (2022). Wearable dual polarized electromagnetic knee imaging system. *IEEE Transactions on Biomedical Circuits and Systems*, 16(2), 296–311.

[20] Sultan, K. S., Mahmoud, A., & Abbosh, A. M. (2021). Textile electromagnetic brace for knee imaging. *IEEE Transactions on Biomedical Circuits and Systems*, 15(3), 522–536.

[21] Sultan, K. S., Mohammed, B., Manoufali, M., & Abbosh, A. M. (2021). Portable electromagnetic knee imaging system. *IEEE Transactions on Antennas and Propagation*, 69(10), 6824–6837.

[22] P., Sheth Neil, R.H., F. J., 2024, "Total Hip Replacement - OrthoInfo - AAOS" [Online]. Available: https://orthoinfo.aaos.org/en/treatment/total-hip-replacement/. [Accessed: 19-Sep-2024].

[23] Yadav, L., Tewari, A., Jain, A., Essue, B., Peiris, D., Woodward, M., et al. (2016). Protocol-based management of older adults with hip fractures in Delhi, India: a feasibility study. *Pilot and Feasibility Studies*, 2, 1–6.

25 A comparative analysis of liquid level transmitter using force resistive sensor and regression technique of machine learning

Anindya Ghosh[1,a], Brajesh Kumar[2,b], Suman Kumar[2,c], Rajan Sarkar[3,d] and Shib Sankar Saha[4,e]

[1]AssistantProfessor, Swami Vivekananda Institute of Science and Technology, Kolkata, West Bengal, India

[2]Assistant Professor, Government Engineering College Jamui, Bihar, India

[3]Associate Professor, Gaya College of Engineering, Bihar, India

[4]Professor, Kalyani Government Engineering College, Kalyani, West Bengal, India

Abstract

In this manuscript a comparative analysis presented for elevation of liquid level measurement with force resistive sensor and regression technique of machine learning. Here is the elevation of liquid level measurement by a sensor using force resistive sensor. It works primarily as a sensing element and response elevation of liquid level in terms of resistance. The sensor response is decreasing resistance value exponentially accordingly of increasing of elevation of liquid level. This exponential response change in linear resistance values due to elevation of liquid level here applied ridge regression technique of machine learning. A converter that can change resistance to voltage has been used, resistance before applied machine learning and after applied machine learning. The voltage signals which get resistance value applied before machine learning is nonlinear in nature with respect to elevation of liquid level. To achieve linear response used regression technique of machine learning. The mathematical and experimental results are reported here. The machine learning response has linear in nature with very small percentage deviation, and good repeatability.

Keywords: Force resistive, linear, liquid level, machine learning, percentage deviation, sensor

Introduction

In the process industries different kinds of liquid play an important role in running properly these industries. This liquid stored in a storage tank and maintaining proper height of liquid level is essential for maintaining the quality of process. To maintain quality of process it is vital to precise measurement of elevation of liquid level of a storage tank. There are two ways contact and non-contact type categorized all liquid level measuring devices. In contact type capacitive type, capacitive probe type, displacer type, differential pressure type and temperature-based methods are used in measurement of height of liquid level. In non-contact type liquid level measurement system, radiation, absorption, ultrasonic, and hydrostatic pressure etc. [1–4].

In the capacitance type system, the liquid level is measured in terms of change in capacitance. These changes in capacitance pass through corresponding bridges circuit to get elevation of level of liquid information in respect of voltages and current signal[15].

In the pressure type measurement system of elevation of liquid level, measured pressure of a storage tank in terms of height of liquid level. The difference time between received and transmitted signal from surface of liquid produce information of elevation of liquid level in ultrasonic type system. Radiation passes through a liquid column of a storage tank, reduction of intensity of radiation produces information of elevation of liquid level. All methods of height of liquid level measurement system are used as per requirement of process industries. There are many works that has been reported in last two decades in the field of elevation of liquid level measurement systems. Reverter et al. [5] have been used as a remote grounded capacitive senor for precise height of liquid level. Khan et al. [6] have been developed a transducer based on non-contact capacitance for measurement of height of liquid level and liquid identification. Canbolat [7] has been used three parallel plates capacitive structures for level, reference, and air. In this designed minimized liquid and air moisture for précised measurement of elevation of liquid level. Terzic et al. [8]

[a]anindya.ciem@gmail.com, [b]brajesh.nitrkl@gmail.com, [c]sumankumar86028@gmail.com, [d]rajansarkar77@gmail.com, [e]sahashib@hotmail.com

DOI: 10.1201/9781003663348-25

have developed a liquid level measurement system in movable situation based on capacitive sensor. Singh et al. [9] have suggest a new method for continuous measurement of elevation of level and volume of liquid using transparent cylindrical based on non–intrusive optical strategy. Nakagawa et al. [10] have suggested a method for measurement of the height of liquid in opaque containers based on millimeter wave absorption in liquid. A Doppler sensor is used for generating millimeter waves. The elevation of liquid level monitoring inside a tanks based on time domain reflectometry has been proposed by Cataldo et al. [11]. Marick et al. [12] have proposed a transducer for precise of height of liquid level float type. Here float type transducer developed of modification of inductance. In this method local level measurement can be done by a float and a level switching and adjusted differential inductance type electromechanical transducer that is used for transferring the measured value of liquid. A modified capacitive based sensor for measurement of height of liquid level measurement has been presented by Bera et al. [13]. This sensor measured the height of liquid level as well as eliminated the effect of self-inductance of metallic rod. Chakraborty et al. [14] have proposed a contact less liquid level sensor using infrared light and a float. In this sensor infrared type light reflects from defusing area of a float. The reflected infrared light intensity changes accordingly height of liquid level. This sensor further modified for conducting liquid as well as minimized the effect of fringe and parasitic capacitance by Chakraborty et al. [16]. A remote liquid level measurement as well as control of height of liquid level has been developed by Roy and Das [17]. In this system a capacitive gauge class capacitive transmitter has been used for precise height of liquid level. Kumar and Mandal [18] have proposed a liquid level transmitting system using Mach-Zehnder interferometer based on electro optic material. The output of this level system is optical signal. The primary sensing element is a float. Lata et al. [19] design and developed a force resistive sensor for measurement of elevation of liquid level based on hydrostatic pressure. The resistive output throughout elevation of level of liquid is not linear. After some adjustment the system response has linear in nature. Its hysteresis and nonlinearity error minimize by Lata and Mandal [20] using artificial neural network (ANN) [21].

In the mentioned manuscript, we have developed a very simple design structure and cost-effective liquid level transmitter. It can measure several types of liquids using a force resistive sensor and ridge regression technique for machine learning. In the above system, at the bottom of tank a force resistive sensor is placed. The force resistive sensor action is in terms

of decreasing resistance is maintained by increasing in elevation of liquid level. The change of resistance according to the liquid level is converted into voltage signal with the help of resistance to voltage converter circuit. The nature of response both force resistive and resistance to voltage converter are not in linear nature with respect to height of liquid level. To achieve linear response of force resistive sensor and resistance to voltage converter response, we have applied a ridge regression technique of machine learning. The machine learning responses of both cases are stated in this paper.

Course of Action

Let us consider the pressure at the bottom of the tank P at elevation of liquid level h, then,

$$P = \rho g h \tag{1}$$

$$F = P \times A = A\rho g h = k_1 h \tag{2}$$

where liquid density is ρ, the cross-sectional area of the guide tube is A, the gravitation force is g and is constant.

From the above equation 2 it is clear that force is directly proportional to height of liquid level. The block diagram and experimental setup shown in Figures 25.1 and 25.2 respectively.

Where, X^T is transpose matrix of X response decreasing with height of liquid level. It has a negative slope, that's why we used ridge regression technique of machine learning to achieve linear response with increasing mode. In this technique we assumed resistance (X) values as input and level (Y) value as output in ($nx1$) matrix form. Where n are number of samples of taken data. Then we calculate the following parameters.

$$W = (X^TX + \lambda I)^{-1}X^TY \tag{3}$$

I is identical matrix

λ is Regularization parameters.

$$b = \frac{\sum_{i=1}^{n} Y - m \times \sum_{i=1}^{n} X}{n} \tag{4}$$

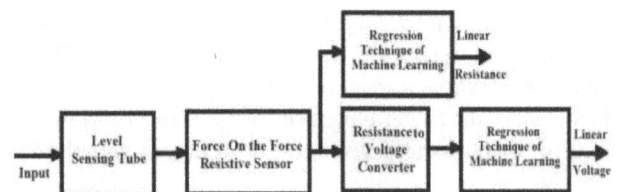

Figure 25.1 Block diagram of proposed model
Source: Author

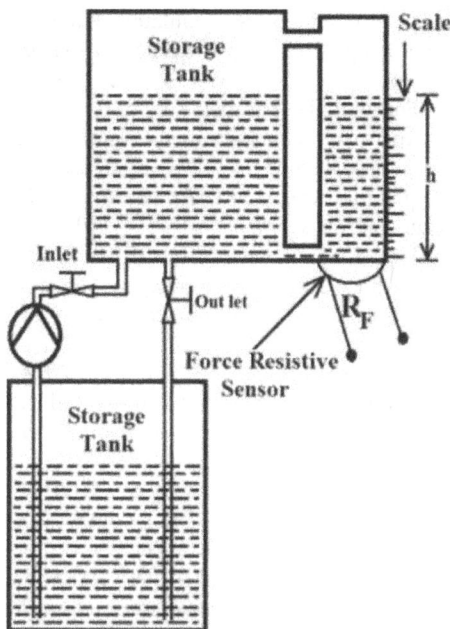

Figure 25.2 Proposed level setup experimentally we found that force resistive

Source: Author

The above equations (3) and (4) have been used internally for ridge regression technique. Where W is co-efficient of function and b is intercept value. The predicted equation of ridge regression technique is

$$\hat{y}=y_{Pred} = XW+b \tag{5}$$

First measured force resistive response in terms of elevation of liquid level and corresponding voltages which contain information of elevation of liquid level. The force resistive output and voltage converter output are not in linear nature with respect to elevation of liquid level. So, in this case the regression technique of machine learning for linear output is applied. Linear regression is a supervised machine learning technique for finding trend lines of input data. In this technique, we renamed level and resistance as X & Y, then we calculate the following parameters X^2, and XY then calculate coefficient of function and intercept value of regression function.

$$= \frac{n\times\sum_{i=1}^{n}XY-\sum_{i=1}^{n}X\sum_{i=1}^{n}Y}{n\times\sum_{i=1}^{n}(X)^{2}-\left(\sum_{i=1}^{n}X\right)^{2}}\ m \tag{6}$$

$$b=\frac{\sum_{i=1}^{n}Y-m\times\sum_{i=1}^{n}X}{n} \tag{7}$$

With the help of equation (3) and (4) we get regression function:

$$y=y_{Pred} =m\times X+b \tag{8}$$

Results and Discussion

The experiment has been performed in two steps first step collect force resistive response in k- ohm range corresponding elevation of liquid level which is not linear with elevation of liquid level shown in blue dot in Figure 25.3. It has negative slope with respect to elevation of liquid level. To achieve an increasing slope with linear we applied here ridge regression technique. This data applied in machine learning then got linear with respect to height of liquid level shown in Figure 25.3.

In the second step force resistive output pass through resistance to voltage converter circuit which give the information of elevation of level of liquid in terms of voltage. It is also not linear in nature so applied linear regression technique of machine learning. The machine learning response is linear in nature with respect to elevation of level of liquid. The resistance to voltage converter's response before applied machine learning and after is shown in Figure 25.4.

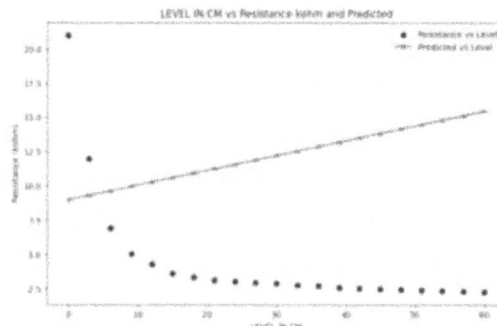

Figure 25.3 Force resistive output vs elevation of liquid level

Source: Author

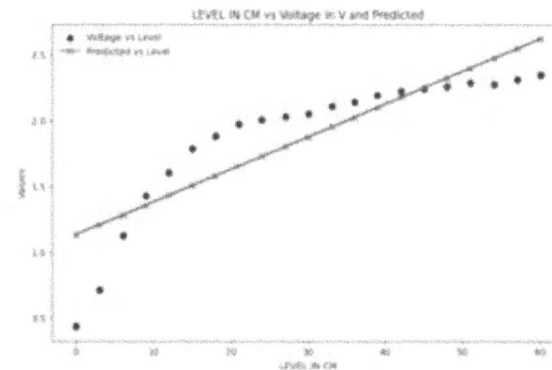

Figure 25.4 Elevation of level of liquid in voltage

Source: Author

The response after machine learning is linear and there is no need to do special adjustment. It is very easy and simple to use. The ridge regression technique has been used internal mathematical equations (3),(4) and (5) to give linear response. The linear regression technique applied for voltage signal used equations internally (6), (7) and (8). The response of machine learning is linear which showed in Figures 25.3 and 25.4. The machine learning performs with the help of sk-learn, numpy, pandas and matplotlib python libraries, help of python programming language.

References

[1]   Doeblin, E. O. (1990). Measurement System and Application and Design. 4th edn. New York, NY, USA: McGraw- Hill.

[2]   Liptak, B. G. (2003). Process Measurement and Analysis. 4th edn. Oxford, U.K.: Butterworth Heinman Ltd.

[3]   Patranabis, D. (2010). Principle of Industrial Instrumentation. 3rd edn. New Delhi: Tata McGraw Hill.

[4]   Golding, E. W., & Widdis, F. C. (1963). Electrical Measurement and Measuring Instruments. 5th edn. London, U.K.: Pitman.

[5]   Reverter, F., Li, X., & Meijer, G. C. M. (2007). Liquid-level measurement system based on a remote grounded capacitive sensor. *Sensors and Actuators A: Physical*, 138(1), 1–8.

[6]   Khan, S., Htike, K. K., Ali, M., Alam, A. H. M. Z., Islam, M. R., Khalifa, O. O., et al. (2008). A non-contact type level transducer for liquid characterization. In IEEE Proceedings of Int. Conference on Computer and Communication Engineering, Kuala Lumpur, Malaysia, 13-15 May, 2008.

[7]   Canbolat, H. (2009). A novel level measurement technique using three capacitive sensors for liquids. *IEEE Transactions on Instrumentation and Measurement*, 58(10), 3762–3768.

[8]   Terzic, E., Nagarajah, C. R., & Alamgir, M. (2010). Capacitive sensor-based fluid level measurement in dynamic environment using neural network. *Artificial Intelligence*, 23(4), 614–619.

[9]   Singh, H. K., Chakroborty, S. K., Talukdar, H., Singh, N. M., & Bezboruah, T. (2011). A new non-intrusive optical technique measure transparent liquid level and volume. *IEEE Sensors Journal*, 11(2), 391–398.

[10] Nakagawa, T., Hyodo, A., Kogo, K., Kurata, H., Osada, K., & Oho, S. (2013). Contactless liquid-level measurement with frequency- modulated millimeter wave through opaque container. *IEEE Sensor Journal*, 13(3), 926–933.

[11] Cataldo, A., Piuzzi, E., De Benedetto, E., & Cannazza, G. (2014). Experimental characterization and performance evaluation of flexible two-wire probes for TDR monitoring of liquid level. *IEEE Transaction on Instrumentation and Measurement*, 63(12), 2779–2788.

[12] Marick, S., Bera, S. K., & Bera, S. C. (2014). A float type liquid level measuring system using a modified inductive transducer. *Sensors and Transducers*, 182(11), 111–118.

[13] Bera, S. C., Mandal, H., Saha, S., & Dutta, A. (2014). Study of a modified capacitive-type level transducer for any type of liquid. *IEEE Transaction on Instrumentation and Measurement*, 63(3), 641–649.

[14] Chakraborty, S., Mandal, N., & Bera, S. C. (2014). Study of and IR defusing surface as non-contact level sensor. *Sensors and Transducers*, 183(12), 53–59.

[15] Kumar, B., Rajita, G., & Mandal, N. (2014). A review on capacitive-type sensor for measurement of height of liquid level. *Measurement and Control*, 47(7), 219–224.

[16] Chakraborty, S., Bera, S. K., Mandal, N., & Bera, S. C. (2015). Study on further modification of non-contact capacitance type-level transducer for a conducting liquid. *IEEE Sensor Journal*, 15(11), 6678–6688.

[17] Roy, J. K., & Das, S. (2015). Low cost non contact capacitive gauge glass level transmitter suitable for remote measurement & control. In 9th IEEE Proceedings of International Conference on Sensing Technology, 8-10 December 2015, (pp. 570–574).

[18] Kumar, B., & Mandal, N. (2016). Study of an electro- optic technique of level transmitter using mach-zehnder interferometer and float as primary sensing elements. *IEEE Sensor Journal*, 16(11), 4211–4218.

[19] Lata, A., Kumar, B., & Mandal, N. (2017). Design and development of a level transmitter using force resistive sensor as a primary sensing element. *IET Science, Measurement and Technology*, 12(1), 118–125.

[20] Lata, A., & Mandal, N. (2021). ANN-based liquid level transmitter using force resistive sensor for minimisation of hysteresis and non-linearity error. *IET Science, Measurement and Technology*, 14(10), 923–930.

[21] Ghosh, A., Choudhary, R. K., Kumar, B., Lata, A., Saha, S. S., & Sarkar, R. (2023). A PC-based real-time level measurement and transmitting technique using optical channel. *IEEE Sensors Journal*, 23(20), 24721–24728.

26 Prediction of external quantum efficiency for organic tin based perovskites

Saurabh Basak[1,a], Tanmay Sinha Roy[2,b] and Bansari Deb Majumder[3,c]

[1]Assistant Professor, Electronics and Communication Engineering, Budge Budge Institute of Technology, Kolkata, India.

[2]Associate Professor, Electrical and Electronics Engineering, Institute of Engineering and Management (IEM), Kolkata, India.

[3]Associate Professor, Electrical Engineering, Narula Institute of Technology, Kolkata, India.

Abstract

Today organic tin based perovskite solar cell is a new emerging technology. It has been proven by different experimental research that the best suited candidate of this group of perovskite is $FASnI_3$. The experimental data of external quantum efficiency, integrated short circuit current density, photo responsivity is easily available in different research papers. This X site anion i.e iodine can be replaced by bromine (Br), chlorine (Cl) also but no such experimental data of the above said parameters of $FASnX_3$, $MASnX_3$ (X = Cl, Br) is available in literature. In this research we predicted the nature of the external quantum efficiency curve (EQE) with respect to frequency of light by simulation. In this research MA, FA means MethylAmmonium and FormAmidinium respectively.

Key words: External quantum efficiency, simulation, solar cell, tin based organic perovskite

Introduction

Perovskite based solar cells are a new emerging technology for solar cell. Among them tin-based organic perovskite solar cells are very suitable candidate for research due to easy synthesis process and reasonably high photo conversion efficiency which is increasing day by day. Kojima et al first invented this perovskite based solar cell in 2009 having photo conversion efficiency 3.8% [1–5]. Since then, photo conversion efficiency has been increased to more than 25% [9–20] by today. But this perovskites are lead (Pb) based which causes several environmental hazards So, tin based organic perovskite based solar cells came to picture to replace the former perovskite based solar cell. After huge research on tin-based organic perovskite solar cell the photo conversion efficiency becomes 14% for $FASnI_3$ [2, 6–8]. Since there are no such research papers showing the variation of external quantum efficiency (EQE) with respect to frequency of incident light for $MASnX_3$, $FASnX_3$ we predict the above said variation by simulation.

Paper Organization

The full research paper is structured as mentioned. Section 3 provides the related work and literature study of different types of organic tin-based perovskites used by other researchers. Section 4 details about the methodology adapted in the proposed analysis. Section 5 describes the results and discussion. Eventually, Section 6 summarizes the conclusions.

Related Works

The variation of EQE with respect to input optical wavelength for $FASnI_3$ was done by many researchers like Meng et al. [7], Zhu et al. [2], Nasti et al. [6] etc. but no such works on other organic tin-based perovskite was done before. In this work we have done the variation of EQE vs input optical frequency by simulation for different organic tin based perovskites whose general formula is $MASnX_3$ and $FASnX_3$, where X = Cl, Br, I i.e. halogen elements.

Methodology

It is known to us that for any semiconductor light absorption takes place only when the light energy,, where $E \geq E_g$ is the band gap of the semiconductor. Otherwise, it can be stated that if the frequency of the incident light $f \geq f_c$, where f_c is the cut off frequency of light absorption of that semiconductor then only absorption occurs otherwise this semiconductor will be acting as window layer for that semiconductor. So, if $f < f_c$ then EQE = η = 0. According to Fang Hong et al [1] it is suggested that the best suitable candidate of tin based organic perovskite is $FASnI_3$. So, extensive research is going on that particular perovskite. In this

[a]saurabhzen435@gmail.com, [b]tanmoysinha.roy@gmail.com, [c]bansarideb.research@gmail.com

DOI: 10.1201/9781003663348-26

research we first make an algorithm based on the basic optoelectronic property of FASnI$_3$ for light absorption to determine the variation of EQE with respect to incident light frequency. After that we will tally that simulated result with data obtained from literature. If the result is close, then we accept this simulation result and generalize it for other tin based organic perovskites of different cut off frequencies due to different band gaps.

Cut off frequency $f < f_c$ can be written as $f_c = \dfrac{E_g}{h}$, where h = Planck's constant. For FASnI$_3$ the E_g = 1.41 ev, f_c = 340.47 THz & maximum value of EQE = η = 76% at f = 800 THz [2]. If 375 < f <500 THz, then EQE = η = 65%. The variation of EQE is from 60% - 76% within the frequency range of 375 – 800 THz. EQE decreases sharply after 800 THz due to surface absorption.

The algorithm of finding EQE of FASnI$_3$ is given below.

- Step 1: Start.
- Step 2: Read the input frequency (f) in THz.
- Step 3: Check if $f < f_c$, then η = 0, else if f <375, η = 0.2 then else if f < 428.57, η = 0.2, else if f < 800, η = 0.8, else if f < 860, η = –0.012 * f + 10.4, else η = 0.
- Step 4: Stop.
- We consider that for other tin based perovskites if $f > f_c$, then η = 0.8,0.7,0.5 for X = I, Br, Cl respectively.

Result and Discussion

The EQE vs frequency plot of FASnI$_3$ is given below. In this plot the actual as well as simulated result is shown here.

Boundary conditions of the algorithm

- From the EQE of FASnI$_3$ it is clear that after f >857 THz, the wavelength become so small that it will be absorbed totally within the surface.

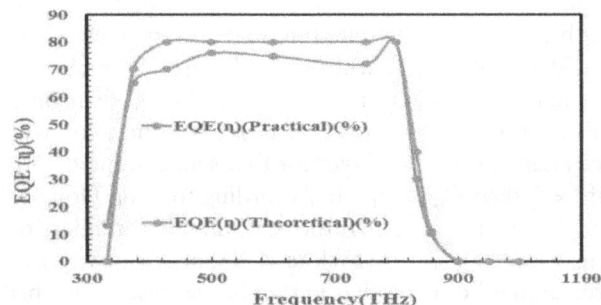

Figure 26.1 Frequency vs EQE plot of FASnI$_3$
Source: Author

- So, for higher value of f, the EQE will be sharply decreased.
- As the electronegativity of Cl, Br is higher than I so for ASnCl$_3$, ASnBr$_3$, η must be lesser than FASnI$_3$. Where A = MA, FA i.e., MethylAmmonium and FormAmidinium respectively.
- If f > 800 THz, then the wavelength of the incident light is very small. For that reason EQE after 800 THz decreases sharply due to heavy surface absorption. We take this frequency value same for other tin based organic perovskites.
- For FASnI$_3$ at f = 800 THz, η = 0.8 & at f = 857 THz, η = 0.1. We connect these two points through straight line whose equation is η = –0.012 * f + 10.4 Putting f = 833 THz, we get η = 0.4 which is very close to actual value η = 0.3 So, we accept this straight-line approximation.

From the Figure 26.1 it can be seen that there is a excellent match between the simulated (theoretical) and observed (practical) data. So, we accept this simulation model for other tin based organic perovskites EQE variation with respect to frequency. In Figure 26.2 the EQE variation with respect to incident light frequency is plotted for MASnX$_3$ perovskites. In Figure 26.3 the EQE variation with respect to incident light frequency is plotted for FASnX$_3$ perovskites. In Figure 26.1 it can be also seen that in experimental data the EQE increment is observed before 340.47 THz i.e cut off frequency (f_c) of FASnI$_3$ due to defect state absorption but in simulated data we consider the optical absorption takes place only when $f \geq f_c$. After 800 THz due to extreme surface absorption EQE falls sharply for increasing incident optical frequency. This happens because beyond 800 THz the optical wavelenght is very small. The cut off frequencies and band gaps of different tin-based perovskites are given below in Tables 26.1 and 26.2.

Since the band gap of FASnCl$_3$ is unavailable in literature so it is not given here.

From Figure 26.2 it is evident that due to the highest electronegativity value of Cl with respect to other halogen elements Br and I the EQE value is lowest and due to lowest electronegativity value of I the EQE yield is highest. We consider the EQE values as 0.8, 0.7 and 0.5 for X = I, Br, Cl respectively but it may be much less due to many other factors like defect states, ion migration etc. Since the cut off frequency of the MASnCl$_3$ is the highest among all so the EQE will decrease very sharply similar to other organic tin-based perovskites.

Frequency vs EQE plot of FASnX$_3$ is given in the Figure 26.3.

Table 26.1 List of cut-off frequencies for MASnX3 [19]

Name of perovskites	Band gap (ev)	Cut off frequency (f_c) (THz)
MASnCl$_3$	3.69	891.03
MASnBr$_3$	2.15	519.16
MASnI$_3$	1.25	301.84

Source: Das, T., Liberto, G. D. , & Pacchioni, G. (2022) Density functional theory estimate of halide perovskite band gap : when spin orbit coupling helps. The Journal of Physical Chemistry C, 126, 2184-2198.

Table 26.2 List of cut-off frequencies for FASnX3. [19]

Name of perovskites	Band gap (ev)	Cut off frequency (f_c) (THz)
FASnBr$_3$	2.55	615.75
FASnI$_3$	1.41	340.47

Source: Author

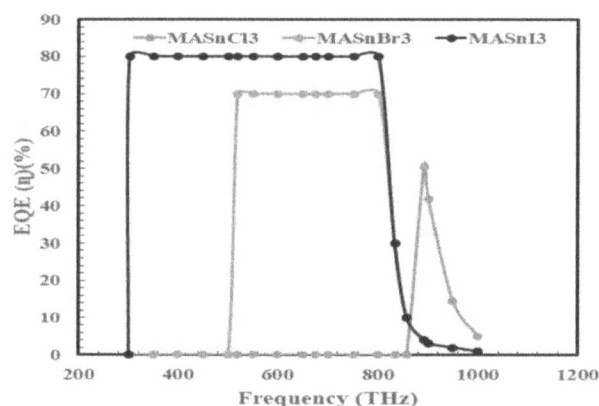

Figure 26.2 Frequency vs EQE plot of MASnX$_3$
Source: Author

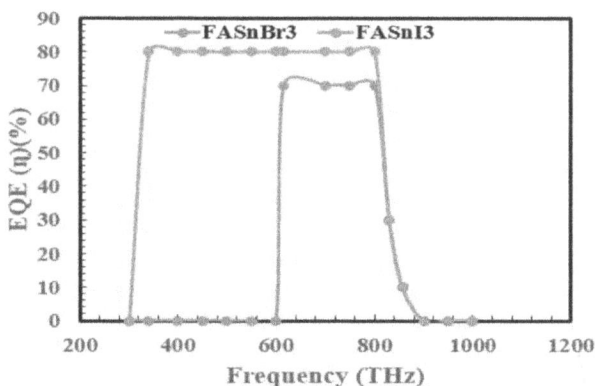

Figure 26.3 Frequency vs EQE plot of FASnX$_3$
Source: Author

From Figure 26.3 it can be seen that the nature of the curve is similar to the previous case.

Conclusion

The predicted external quantum efficiency (EQE) curve of MASnX$_3$, FASnX$_3$ has been plotted. Here it will help us for further study related to tin based organic perovskite solar cell. The authordeclaringdeclare that there are no conflict of interest among them and others.

Acknowledgement

The authors gratefully acknowledge the overwhelming support of their colleagues of respective department.

References

[1] Fang, H. H., Adjokatse, S., Shao, S., Even, J., Loi, & M. A. (2018). Long - lived hot - carrier light emission and large blue shift in formamidinium tin triiodide perovskites. *Nature Communications*, 9(243), 243.

[2] Zhu, Z., Jiang, X., Yu, D., Yu, N., Ning, Z., & Mi, Q. (2022). Smooth and compact FASnI$_3$ films for lead – free perovskite solar cells with over 14% efficiency. *ACS Energy Letters*, 7, 2079–2083.

[3] Baig, F., Khattak, Y. H., Mari, B., Beg, S., Ahmed, A., & Khan, K. (2018). Efficiency enhancement of CH$_3$NH$_3$SnI$_3$ solar cells by device modelling. *Journal of Electronic Materials*, 47(9), 5275–5282.

[4] Farooq, W., Tu, S., Iqbal, K., Khan, , Rehman, , Khan, A. D., et al. (2020). An efficient non -toxic and non – corrosive perovskite solar cell. *IEEE Access*, 8, 210617–210625.

[5] Moiz, S. A., Alahmadi, A. N. M., & Aljohani, A. J. (2021). Design of a novel lead – free perovskite solar cell for 17.83% efficiency. *IEEE Access*, 8, 54254–54263.

[6] Nasti, G., Aldamasy, M. H., Flatken, M. A., Musto, P., Matczak, P., Dallmann, A., et al. (2022). Pyridine controlled tin perovskite crystallization. *ACS Energy Letters*, 7, 3197–3203.

[7] Meng, X., Wu, T., Liu, X., He, X., Noda, T., Wang, Y., et al. (2020). Highly reproducible and efficient FASnI$_3$ solar cells fabricated with volatilizable reducing solvent. *The Journal of Physical Chemistry Letters*, 11, 2965–2971.

[8] Nakamura, T., Handa, T., Murdey, R., Kanemitsu, Y., & Wakamiya, A. (2020). Materials chemistry approach for efficient lead – free tin halide perovskite solar cells. *ACS Applied Electronic Materials*, 2, 3794–3804.

[9] Wang, M., Wang, W., Ma, B., Shen, W., Liu, L., Cao, K., et al. (2021). Lead-free perovskite materials for solar cells. *Nano-Micro Letters*, 13, 62.

[10] Li, M., Li, F., Gong, J., Zhang, T., Gao, F., Zhang, W. H., et al. (2021). Advances in Tin(II)-based perovskite

solar cells: from material physics to device performance. *Small Structures*, 2100102, 1–30.

[11] Abd Mutalib, M., Ahmad Ludin, N., Nik Ruzalman, N. A. A., Barrioz, V., Sepeai, S., Mat Teridi, M. A., et al. (2018). Progress towards highly stable and lead-free perovskite solar cells. *Materials for Renewable and Sustainable Energy*, 7(7), 1–13.

[12] Boix, P. P., Nonomura, K., Mathews, N., & Mhaisalkar, S. G. (2014). Current progress and future perspectives for organic/inorganic perovskite solar cells. *Materials Today*, 17, 16–23.

[13] Green, M. A., Ho-Baillie, A., & Snaith, H. J. (2014). The emergence of perovskite solar cells. *Nature Photonics*, 8, 506–514.

[14] Baranwal, A. K., & Hayase, S. (2022). Recent advancements in tin halide perovskite-based solar cells and thermoelectric devices. *Nanomaterials,* 12, 4055.

[15] Li, C., Lu, X., Ding, W., Feng, L., Gao, Y., & Guo, Z. (2008). Formability of ABX_3 (X = F, Cl, Br, I) halide perovskites. *Acta Crystallographica B*, 64, 702–707.

[16] Wang, L., Ou, T., Wang, K., Xiao, G., Gao, C., & Zou, B. (2017). Pressure-induced structural evolution opti-

cal and electronic transitions of nontoxic organometal halide perovskite-based methylammonium tin chloride. *Applied Physics Letters*, 111, 233901.

[17] LU, C. H., McGee , G.B. , Liu, Y. , Kang, Z., & Lin Z. (2020) Doping and ion substitution in colloidal metal halide perovskite nanocrystals. *Chemical Society Reviews*, 14, 1–130.

[18] Das, T., Liberto, G. D., & Pacchioni, G. (2022). Density functional theory estimate of halide perovskite band gap: when spin orbit coupling helps. *The Journal of Physical Chemistry C*, 126, 2184–2198.

[19] Zhou, F., Qin, F., Yi, Z., Yao, W., Liu, Z., Wu, X., et al. (2021). Ultra-wideband and wide-angle perfect solar energy absorber based on Ti nanorings surface plasman resonance. *Physical Chemistry Chemical Physics*, 23, 17041.

[20] Saliba, M., Matsui, T., Domanski, K., Seo, , Ummadisingu, A., Zakeeruddin, S. M., et al. (2016). Incorporation of rubidium cations into perovskite solar cells improves photovoltaic performance. *Science*, 354, 206–209.

27 Beyond sight: ML techniques for image classification

Jayashree C. Nidagundi[a], Muskansaba M. Shahiwale[b], Preeti V. Savanur[c], * and Sneha S. Yaligar[d]

Department of Electronics & Communication Engineering, SDM College of Engineering & Technology, Dharwad, Karnataka, India

Abstract

This paper aims to present the possibility of using machine learning algorithms for purposes of classification of images and specifically, it proposes to use support vector machines and convolutional neural networks. The authors put forward the idea of image classification in the domain of SVMs where the feature extraction and classification of images have been discussed. The factors used to measure the performance of the proposed technique are accuracy, precision, recall as well as F1-score. These results prove the efficiency of the proposed approach in the image classification problem. As a whole, this work advances the establishment of effective image classification approaches employing machine learning theories.

Keywords: Convolutional neural network, hyperparameter tuning, image classification, machine learning, support vector machine

Introduction

Machine learning is a subfield grown from the roots of pattern recognition and allows algorithms to make predictions in some manner by learning from data. Of which particular application is image recognition, which makes a categorical assignment based on image characteristics. A great deal of statistical learning support independently has established the algorithm support vector machine (SVM) in this area. Deep learning enforces the image recognition further to a higher level and AI-based methods have also been adapted for the completion of video analysis along with classification using SVM and artificial neural networks (NN) [1].

We started with traditional machine learning techniques, putting in place a wide and general understanding of image classification, and then moved to deep learning methods. We started the implementation with SVM because of its solid theoretical background, even though NN had higher accuracy. This turned out to be a good starting point for SVM, but NN might be the better advanced choice [2]. Traditional machine learning algorithms, such as SVM, tend to saturate in performance with large datasets. Therefore, deep learning methods are at the forefront, with performance close to nearing optimal with complicated data and in feature learning [3]. As we have seen along our journey in the SVM world, its strengths and weaknesses were exposed; consequently, it was found to be of great importance to further dissect this topic

under three key areas for an in-depth understanding: Visualization, tuning parameters, and implementation and analysis [4]. Basically, image classification tries to bring the ability of computers closer to human vision by training computers to make sense of visual information [5]. Artificial intelligence research conducted over the last couple of decades has been on enabling machines to perceive and interpret visual information contained in images and videos [6].

Rest of the paper is structured as follows. Section 2 reviews the extant literature. Section 3 describes the sample and variables. Section 4 explains the research methodology. Section 5 discusses the empirical findings. Section 6 summarises the paper.

Literature Review

In another study, Hassan et al. studied about image classification [7]. Image classification is the method of partitioning pixel into distinct groups by comparing the values in the data set. To classify a pixel to any particular class then that pixel should meet certain conditions to be able to be placed in that certain class. The classes may be known if the user was able to select the number of classes of the data depending on the training data, or unknown. Particularly, the image classification process entails the extraction of the features of the image and subsequent classification of the features. Hence, what kind of features need to be extracted from images and how those extracted

[a]jayaprajwal8@gmail.com, [b]shahiwalemuskan@gmail.com, [c]savanurpreeti3@gmail.com, [d]snehayaligar03@gmail.com

DOI: 10.1201/9781003663348-27

features must be further analyzed forms the major framework of image classification.

Traditional methods use first order architectural discrete attributes primarily at low or mid gray level density; and color, texture-space arrangement, shape, and position. Mid-level features are in general down sampled using bag of visual words (BoVW) approach; this was earlier used in image classification. After the feature extraction process, one arrives at the object label predications via classifiers that include SVM or else random forest.

Another study studied about SVM[8]. They are a subgroup of supervised learning methods of ML that are used for classification and regression analyses. The basic assumption of SVMs is that one has to look for a hyperplane within the high-dimensional space that would efficiently distinguish between the given classes. This hyperplane provides the largest amount of distance between the classes keeping the white space, which is the minimum distance between the support vectors and the hyperplane to the maximum. SVMs are most useful for problems with clean boundaries and can take linear as well as non-linear data by mapping it into a higher dimension through the use of a kernel. Inclusive, although it is a powerful model, the SVMs may select time and have some sensitive parameters and kernel functions that may need optimal tuning. As shown in Figure 27.1, there are three distinct lines in the model. The marginal line, also called the margin of separation, is defined as w.x-b=0.The lines on the either side of the line of margin are w.x - b = 1 and w.x – b = -1. The three lines together create the hyperplane that separates the given patterns. The pattern that lies on the edges of the hyper plane is called Support Vectors. The perpendicular distance between the line of margin and the edge of hyper plane is called margin.

Pedregosa et al. [9] helped us to gain knowledge about Scikit-learn. Scikit-learn offers the best available implementations for many machine learning algorithms with a very clean and easily accessible interface deeply integrated with Python. This addresses the need for users with no programming background in the software ad web industries as well as, biology, physics, and other fields to analyze statistical data.

Scikit-learn differs from other machine learning toolboxes in Python for various reasons: i) it is distributed under the BSD license ii) it incorporates compiled code for efficiency unlike MDP (Zito et al., 2008) and pybrain (Schaul et al., 2010) iii) it depends only on NumPy and SciPy to facilitate easy distribution unlike pymvpa (Hanke et al., 2009) which has optional dependencies like R and shogun and iv) it works more on The package is best written in Python, but the C++ libraries LibSVM by Chang and Lin (Chang and Lin, 2001), and LibLinear by Fan et al. (Fan et al., 2008) which offer the reference implementation of SVMs and generalized linear models respectively that are compatible with this license. Hello world Binary packages exist on a large number of platforms such as windows and POSIX.

In the another article [10] there are listed a number of factors that make python to be regarded as one of the most popular languages. According to Harrington (2012) the Python programming language is described as simple and elegant language with perfectly transparent grammar having very easy text comprehension and handling capabilities and a dedicated community of developers resulting in availability of adequate documentation.

Python also offers an interactive shell through which python can be interactively used by programmers to edit parts of a program and observe other parts in comparison at the same time. Lists, queues, tuples, etc., It enables programmers to freely write and test code and execute the line of code at the same time. It is employed to try out small bits of code, to debug the code and to learn more about the capabilities of Python in an improvised mode. Lists, Queues, and Tuples used in the programming language are easily viewable and changeable within the shell by the programmer. The shell is also capable of editing some portions of a program and at once showing the changes made. This makes it very helpful especially for use in teaching and in the creation of proofs of concept [11].

Methodology

The methodology is as shown in Figure 27.2 which depicts a flowchart.

a) Data collection: We have collected three datasets of images: Fluffy Red Panda, Car Images and Pretty Sunflower. Each dataset contains 60

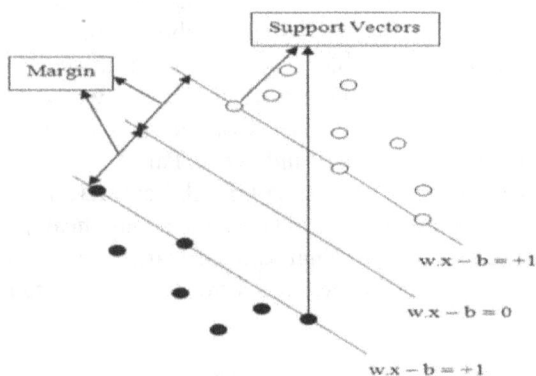

Figure 27.1 SVM model [8]
Source: Author

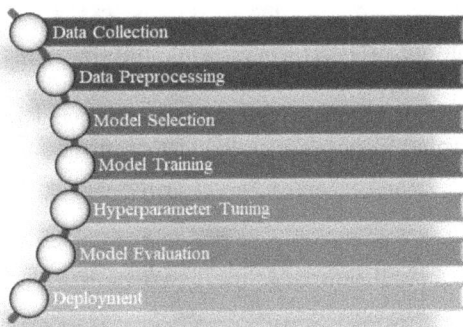

Figure 27.2 Flowchart of proposed methodology
Source: Author

images. We took the help of 'Bing Image downloader'. The Bing Image Downloader is a Python tool that speeds up the process of searching and downloading images from Bing based on specified keywords. It's useful for quickly building image datasets for machine learning and research purposes [12].

b) Data preprocessing: This process involved image resizing and image flattening. The images were resized into the dimensions of 150 × 150 pixels with three color channels (RGB). The 'resize' function is typically used for standardizing image sizes for providing them as an input to the model. The 'flatten ()' method converts the image matrix into a single dimensional array. This approach is common in tasks like feature extraction for machine learning or data preprocessing in image classification.

c) Model selection: This process involves choosing the best algorithm or model that fits the data. This involves evaluating different models using techniques like cross validation and metrics comparison. Tools like 'GridSearchCV' and 'RandomizedSearchCV' help optimize model parameters to improve performance [13].

d) Model training: The dataset was divided into training and testing sets. Modules from scikit-learn library were used to carry out this task. The split allocated 70% of the data for training and 30% for testing. The 'random_state=109' parameter is used to control the randomness of data splitting. It ensures that every time the code is run, the data is split in the same way.

e) Hyperparameter tuning: In machine learning, it involves optimizing the parameters that control the behavior of a model. We used 'GridSearchCV' to perform hyperparameter tuning for an SVM classifier from 'sklearn' library. The 'svc' model is passed to 'GrisdSearchCV' and trains the model using training data, selecting the optimal hyperparameters for better performance

f) Model evaluation: This process assesses the performance of a trained machine learning model using metrics and techniques that measure how well the model predicts or classifies new data. Common evaluation metrics include accuracy, precision, recall, F1 score [14].

g) Deployment: Deploying a model using Streamlit and ngrok allowed us to create a quick and shareable web application. First the machine learning model is included within a Streamlit app for interactive user input and predictions. Ngrok is used to create a public URL that connects the local machine to web. It handles networking efficiently allowing the user to share their app with others.

Results and Discussion

Before we start developing our Streamlit application, we need to set up an integrated development environment (IDE) or a suitable code editor. Streamlit is an open-source framework for building web apps in Python, which is especially useful for data science and machine learning projects. It works well with popular Python libraries like scikit-learn, NumPy, and pandas. After setting up the IDE, we need to install Python and the package installer (PIP) on our system to facilitate development.

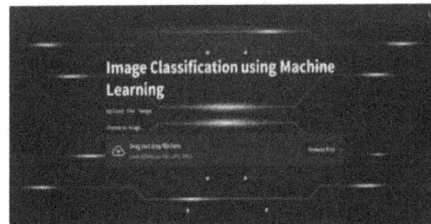

Figure 27.3 Streamlit App
Source: Author

In Figure 27.3, The car images are uploaded in Streamlit application using Bing Image Downloader it is most often that the images are in JPEG (. jpg) format. There is a limit of 200MB to upload file since large images sometimes takes time to process in the application.

The following Figure 27.4, shows the predicted result after uploading of a car image as presented in Figure 27.5. After taking the transformation, the images are aligned in a vertical manner in order to compare them. It should be noted that pictures in the cropped folder are used to train a model. Training our classifier is the raw image as well as the transformed wavelet image. The training files contain 70% of the images, the testing files make up for the rest 30% of the images that are used to test the reliability of the model.

Figure 27.4 Uploaded image
Source: Author

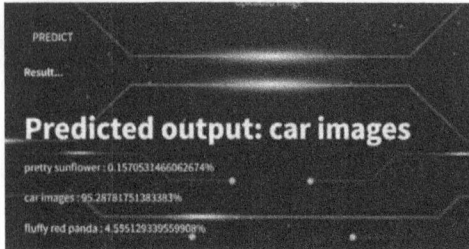

Figure 27.5 Predicted output
Source: Author

In Figure 27.5, The predicted output as follows:

- Pretty Sunflower:0.157%
- Car Image: 95.28%
- Fluffy red panda: 4.55%

The highest predicted output was 'Car Image'.

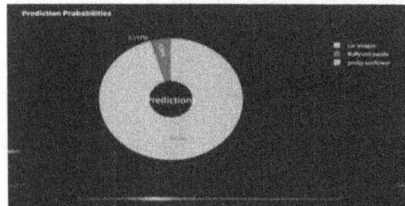

Figure 27.6 Prediction probabilities in pi-chat
Source: Author

In Figure 27.6 the predicted probabilities are presented using a pie chart.

2.Now we are uploading the mixed images (like Sunflower and Car) in the Streamlit application using Bing Image Downloader it is most often that the image is in JPEG (.jpg) format.

Figure 27.7 Uploaded image
Source: Author

The following figure shows the predicted result after uploading of a car image, and sunflower image as present in Figure 27.7. After undergoing the transformation, coordinates of images are arranged in a vertical manner in order to predict one output.

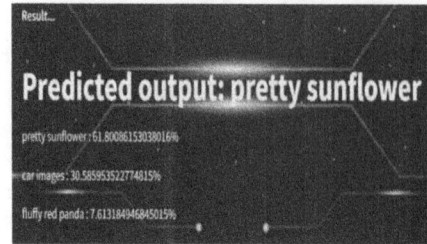

Figure 27.8 Predicted output
Source: Author

In the Figure 27.8, a mixed image of a car and a sunflower was processed, resulting in a predicted output of 'Pretty Sunflower.' The predictions are as follows:

- Pretty Sunflower: 61.80%
- Car Image: 30.58%
- Fluffy Red Panda: 7.61%

The highest predicted output was 'Pretty Sunflower'.

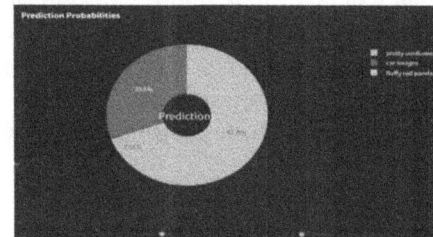

Figure 27.9 Prediction probabilities in pi-chart.
Source: Author

In Figure 27.9 the predicted probabilities are presented using a pie chart.

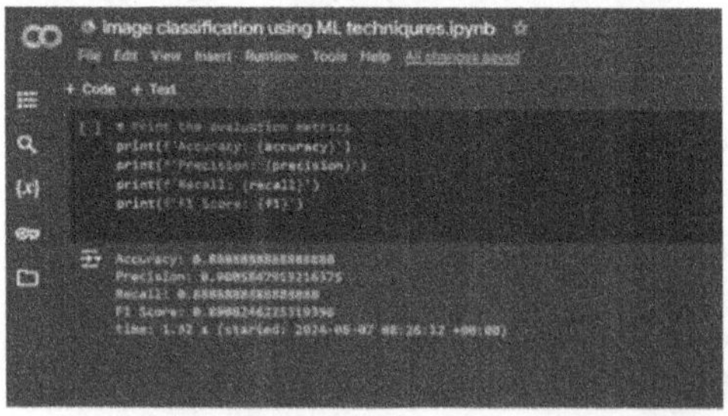

Figure 27.10 Overall performance of the model
Source: Author

In the Figure 27.10, the image classification process using machine learning techniques, the evaluated output are presented using Python code as represented here,

- Accuracy: is computed as the number of instances that has been classified correctly divided by the total number of instances in the dataset.
- Precision: is the closeness of the agreement between two or more measurements or observations taken on the same magnitude under unchanged condition of measurement.
- Recall: reflects the performance of a model or system in terms of its capacity to find all positive samples in data set, in terms of the percentage of actually positive data points successfully identified.
- F1-score: Which is the harmonic mean of precision and recall that gives a single measure that takes into account both false positive and false negatives for the entire model.

Overall performance of the model: -

- Accuracy :0.888
- Precision: 0.900
- Recall: 0.888
- F1-score: 0.890.

Conclusion

Nowadays, image processing plays an increasingly important role in our lives, with image recognition serving as a crucial method for efficiently extracting needed images from a large set. However, low-quality images are challenging to recognize due to their poor quality [15]. Despite efforts to avoid them, low-quality images are often unavoidable due to various operational and environmental factors. As a result, effectively identifying low-quality images has emerged as a significant new challenge. In simple word Image recognition is divided in different steps i.e. information acquisition, pre-processing, feature extraction and selection (classification), classifier design etc.An effective image detection technique is presented in this paper utilizing SVM methods. Features are extracted from the image Classification Task. After preprocessing and feature extraction, SVM classifier is used for the forgery image classification. The performance of the proposed technique is evaluated is measured in accuracy 0.888, precision 0.900, recall 0.888, F1-score: 0.890.

Acknowledgment

We would like to thank all our colleagues and peers for their suggestions and assistance throughout the project. The authors would like to express their gratitude for the support granted by SDM College of Engineering and Technology, Dharwad for the conduct of this research. This paper aims at contributing towards the development in the area of image classification.

References

[1] Mishra, D. M., Mishra, S., Jena, S., & Salkuti, S. R. (2023). Image classification using machine learning. *Indonesian Journal of Electrical Engineering and Computer Science*, 31(3), 1551–1558. ISSN:2502-4752, DOI:10.11591/ijeecs. v31.i3.

[2] Chaganti, S. Y., Prudhvith, T. G., Nanda, I., Kumar, N., & Pandi, K. R. (2020) Image classifiction using SVM and CNN. *IEEE Journal.*

[3] Patil, A. N. Image recognition using machine learning SSRN 3835625, 2021.

[4] Mangasarian, O. L., & Musicant, D. R. Active support vector machine classification, Advances in neural information processing systems, 2000.

[5] Pavlidis, P., Wapinski, I., & Noble, W. S. (2004). Support vector machine classification on the web. *Bioinformatics*, 20(4), 586–587. DOI: 10.1093/bioinformatics/btg461.

[6] Kecman, V. (2005). Support vector machines – an introduction. In Support Vector Machines: Theory and Applications. Berlin, Heidelberg: Springer Berlin Heidelberg, (pp. 1–47). ResearchGate, DOI: 10.1007/10984697_1.

[7] Hassan, A., Refaat, M., & Hemeida, A. M. (2022). Image classification based deep learning: a review. *Aswan University Journal of Science and Technology*, 2(1), 11–35.

[8] Pradhan, A. (2012). Support vector machine -a survey. *International Journal of Emerging Technology and Advanced Engineering*, 2(8), 82–85. ISSN 2250-2459.

[9] Pedregosa, F., Varoquaux, G., Gramfort, A., Michel, V., Thirion, B., Grisel, O., et al. (2011). Scikit-learn: machine learning in Python. *Journal of Machine Learning Research*, 12, 2825–2830.

[10] Sodhia, P., Awasthi, N., & Sharmac, V. (2019). introduction to machine learning and its basic application in Python. In Proceedings of 10th International Conference on Digital Strategies for Organizational Success.

[11] Abu, M. A., Indra, N. H., Rahman, A. H. A., Sapiee, N. A. & Ahmad, I. (2019). A study on image classification based on deep learning and tensorflow. *International Journal of Engineering Research and Technology*, 12(4), 563–569. ISSN 0974-3154.

[12] Krishna, M. M., Harshali, M., & Rao, M. V. G. (2018). Image classification using deep learning. *International Journal of Engineering and Technlogy*, 7(2.7), 6147–617.

[13] He, T. (2019). Image quality recognition technology based on deep learning. *Journal of Visual Communication and Image Representation*, 65, 102654.

[14] Deshmukh, S. C. (2023). Study of image recognition using machine learning. *LJRASET Journal Research in Applied Science and Engineering Technology*, 11(6), 3229–3231.

[15] Garg, M. K. (2022). Image forgery detection and classification using support vector machine. *Journal of Emerging Technologies and Innovative Research (JETIR)*, 9(7), 129–131.

28 Analysis of inductor topologies experimentally: a modification of triangular model

Soumyendu Bhattacharjee[1,a], Reshmi Chandra[1,b], Jinia Datta[2,c] and Goutam Kumar Das[1,d]

[1]Assistant Professor, Abacus Institute of Engineering and Management, West Bengal, India

[2]Professor, Abacus Institute of Engineering and Management, West Bengal, India

Abstract

This paper reveals a typical geometrical analysis of three-dimensional (3D) inductor which is able to work with the high 'Q'-values for a wide band of frequencies. Here two different types of design have been proposed for inductor. The design is a slight modification of triangular 3D inductor model of inductor on a same layer. This design shows the phase transition characteristics. As a circuit switch, a thin film of vanadium dioxide bar is used to make a full spiral coil of inductor. The experimental result is verified through the simulation by an authentic software tool. The experimental result shows a 29% variation in the value of inductance. This paper also shows the analysis of Q factor of the entire 3D inductor. This analysis found it better with respect to the triangular model for a smaller number of turns and may be beneficial for the designing low power microwave circuit.

Keywords: 3-Dimensional inductor, band width, Q-Value, vanadium dioxide

Introduction

The efficiency of the tunable inductor has improved a lot. There is no doubt that the designs of tunable inductor are flexible in design and also stable when it is used in RF frequency range. The adjustable inductors are designed as versatile and to work consistently in RF and microwave frequencies. There are currently four common methods for sketching a tunable inductor: (a) using ON/OFF to segregate or make group of inductor segments; (b) controlling materials capable of being magnetized, in the tunable inductors to tune the magnetic field in the inductors; (c) controlling the magnetic coupling coefficient to tune the mutual-inductance between the wound coil which is primary & secondary coil; and (d) Placing a metallic structure on an inductor coil can significantly influence the magnetic flux and the inductance of the coil. This technique is often used in electromagnetic engineering to tune the inductor. Each of these methods is utilized frequently to produce high-performing adjustable inductors. In this research work, only interest is about the geometry of the 3D inductor where the main intension is to improve the Q factor over a wide range of frequency.

Literature Review

Assadsangabi et al. [1] shows that tunable inductors as essential parts for circuits requiring re-configurability, channel selectable, RF filters and circuitry to adaptive matching networks in power systems. In this paper, it is shown that a brand-new continuously tunable inductor works by physically stretching the inductor traces. A pneumatic bubble actuator is surrounded by flexible conductors which are made of liquid metal, that enable the inductor to be compressed or expand under pressure. With each technology node, integrated circuit feature size gets smaller, but connection delay has a bigger impact on overall delay. Thus, one of the most crucial jobs is to minimize the wire-length for the physical design of high performance circuits. Banerjee et al. [2] demonstrate a 3-D design pipeline for vertically integrated circuits in this study. The floor-planning and placement results indicate up to 50% less of total wire-length and longest net length. As a result, we show that vertical integration can significantly reduce connection delay. Chow et al. [3] shows three-dimensional (3-D) architectures and innovative IC, electrical through wafer connected internally is ETWI, which connects devices on both sides of a substrate. These are essential parts for micro-electromechanical systems (MEMS) and integrated circuits (IC). In addition to being incompatible with currently available sensors, previously ETWI are shown which are particularly application-specific hence these are difficult to combine with conventional semiconductor production techniques and it is not allowed for substantial processing on both sides of the wafer

[a]s.microwave@gmail.com, [b]reshmichandra.2024@gmail.com [c]dattajinia@gmail.com, [d]goutamkrdas.aiem@jisgroup.org

DOI: 10.1201/9781003663348-28

while in a research article Choi et al. [4] conducted a studies involving the void-free Cu filling of a TSV (10-20 mm in diameter with an aspect ratio of 5–7) by adjusting the plating DC current density and the additive SPS concentration in an effort to address the challenges associated with the fabrication of high-aspect-ratio TSVs. The change in the plating DC current density is used to estimate the copper deposit development mode in and around the trench and the TSV. Copper was used to electroplate at different rates at the upper and lower portion of the trench depending on the change in plating current density. Chuang et al. [5] introduces a double-layer atom chip which offers users additional flexibility in the design of the magnetic field and also increases variety in the design of the wire patterns. For use in atomic physics experiments, this is more practical. An insulating layer made of SU-8 negative photo resist was utilized in between the top and bottom copper wires. As per the electrical measurement results, cables with a width of 100 m on top and bottom can support a 6 A current without burning out. This work also focuses on the double-layer atom chips that have been anodically bonded to a Pyrex glass cell and integrated with the silicon via (TSV) approach, making it a desirable vacuum chamber of atomic physics experiments. Domann et al. [6] developed and optimized the electroplating method for filling high aspect ratio through-silicon-vias (TSVs) in packed environments. The experimental power source was pulsed power, and different additive concentrations were prepared for the electroplating solution. A control variable experiment which is desired were carried out to find the best approach. The association between numerous experimental factors, such as flow density (0.25-2 A/dm2), additive concentration (0.5-2 mL/L), and various forms of TSVs (circle, oral, and square), was extensively examined in the control variable studies. TSV technology has been proposed by Gambino et al. [7] for use in a long time. However, the introduction of this technique into high scale manufacturing is quite new. The details of TSV fabrication processes including metallization, are summarized totally in this publication. Along with the difficulties that come with using TSVs for backside processing, assembly, metrology, design, packaging, dependability, testing, and yield. The integration of silicon ICs in a heterogeneous fashion with high performance cheaply, is made possible by a 3D Circuit technology platform. Gutmann et al. [8] describes a three-step thinning procedure, copper damascene patterning, and dielectric adhesive bonding. These are used to create inter-wafer interconnects between fully processed, wafer-to-wafer ICs which are aligned. The viability of the process flow is indicated by daisy-chain inter-wafer through test structures and compatibility of the

production steps with 130 nm CMOS semiconductor devices and circuits. The development and presentation of fabrication technology for building parts for the use in switched-mode power supply is based on MEMS procedures. We design, construct, and characterize a sacrificial multilayer electroplated capacitors, magnetic cores, and inductors [9]. A study had been conducted by the researcher Huang [10] that developed and optimized the electroplating method for filling high aspect ratio through-silicon-vias (TSVs) in packed environments. Different additive concentrations were prepared for the electroplating solution and the experimental power source was pulsed power. To find the best approach, designed control variable experiments were carried out. A multi-step current density method to simultaneously fill TSVs with various aspect ratios (20 m 120 m, 30 m 130 m, 40 m 140 m, and 50 m 150 m) without voids was examined by Wang et.al. [11, 12].

Methodology and Model Specifications

In this article a new method has been proposed as a designing a 3D inductor which is nothing but a slight modification of the triangular Model, where the upper mask layer geometric structure of a coil with three turns is shown in Figure 28.1.

Here, it is hypothesized that ownership concentration and institutional ownership affect the stock return of listed companies. Based on this hypothesis, the following empirical research models are developed. within the non-parallel category. But this paper shows a design of how the initial top wires and other bottom segments are arranged geometrically. The total inductance of the entire design is expressed as follows.

$$M_{Horizontal\ (Non\ Prallel)} = \sum_{j=i+1}^{N} \sum_{i=1}^{i=N-1} k. M_{Non-parallel}(required\ variable) \quad (1)$$

where, "$M_{Non-parallel}(i, j)$" is the mutual inductance of any two non-parallel wires that are on separate levels as shown by the equation (1). The top wires are represented by the odd 'i' and the bottom wires are represented by the even 'i'. Now the angle 'ε' for our proposed design is given by the following equation.

$$\varepsilon = 2 \times tan^{-1}\frac{W}{2L} \quad (2)$$

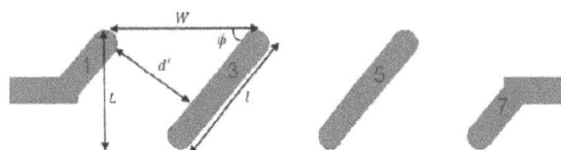

Figure 28.1 Geometrical configuration of parallel wire on the same layer
Source: Author

Let the height of each wire to be '*l*' for this design and other relevant parameters like '*μ*' & '*v*' can be expressed by the following equation.

$$\mu = v = \frac{(j-i-1)}{2} \times l \qquad (3)$$

According to the above construction, the design formulas are given below.

$$M_{Total(Trianglar)} = M_{Vertical} + M_{Horizontal(Paralleled)} +$$
$$M_{Horizontal(Non\ Paralleled)} \qquad (4)$$

The squared form with the parallel wires and cases with straight rectangular wires are also investigated in this article which is shown in Figures 28.3 and 28.4. In the squared model, there are no wires which are not parallel. The equations which already exist in the zigzag model, has been used to compute the self and mutual-inductance between them.

Below an abstractual configuration of the proposed design is given in the following figure.

Depending on the position of the wires, three cases for the paralleled wires are formed. As shown in Figure 28.4, the first scenario involves all wires being on a horizontal layer and in Y direction. The conclusion might be expressed as follows. In Figure 28.2, a special type of structure has been incorporated using triangular geometry which increases number of flux

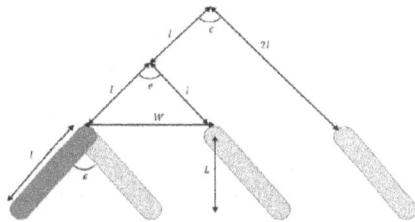

Figure 28.2 The geometrical configuration of the proposed design
Source: Author

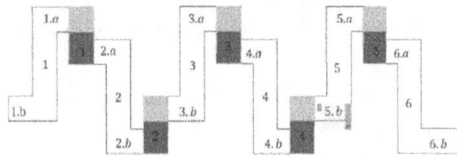

Figure 28.3 Pictorial view of the square model
Source: Author

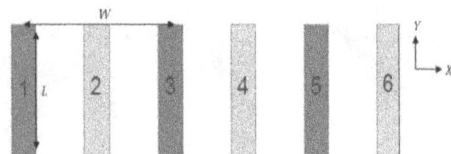

Figure 28.4 Geometric configuration of the Y direction wire
Source: Author

w.r.t. normal structure. Not only that the structure is having comparatively high q-factor.

$$M_{Horizontal}(Y) = \sum_{j=i+1}^{N} \sum_{i=1}^{i=N-1} (-1)^k M_{Horizontal(i,j)} \qquad (5)$$

$M_{Horizontal\ (i,\ j)}$ in the above equation the parameter '*k*' is equal to 1 if $(i+j)$ is odd. Otherwise '*k*' is equal to zero. The distance '*x*' for the above equation can be expressed as follows.

$$x = \sqrt{(k+h)^2 + (k^* + h)^2} \qquad (6)$$

In the above equation, $k^* = \frac{(j-i)}{2}$.

Result and Discussion

The MATLAB simulation for zigzag coils and simulation for our proposed design for 3D Inductor are given in this section. Due to the increased mutual inductance in the abstract structure, as given in Tables 28.1 and 28.2, the simulated data demonstrated that the

Table 28.1 Inductance of 3D inductor (Both Zigzag and proposed design)

Specification & Number of turns		$L_{Mutual}(nH)$	Wire length (μM)	$L_{-Self}(nH)$	$L_{-Total}(nH)$
Zigzag Model	1 Turns	.000356	271	0.23	0.23
	2 Turns	.0034	301	0.277	0.342
	3 Turns	.0673	206	0.398	0.331
Proposed Model	1 Turns	-.000356	251	0.410	0.23
	2 Turns	-.00034	123	0.341	0.322
	3 Turns	-.0122	321	0.351	0.05

Source: Author

Table 28.2 Inductance of 3D inductor (Both Zigzag and proposed design)

Specification & Number of turns		$L_{Mutual}(nH)$	Wire length (μM)	$L_{-Self}(nH)$	$L_{-Total}(nH)$
Zigzag Model	1 Turns	.00341	163	0.22	0.19
	2 Turns	.0703	229	0.30	0.16
	3 Turns	.069	197	0.311	0.303
Proposed Model	1 Turns	-.00998	161	0.388	0.321
	2 Turns	-.0022	101	0.4	0.423
	3 Turns	-.0122	98	0.412	0.016

Source: Author

abstract coils expand than the proposed design when higher turns are utilized in the sample. But in cases of a smaller number of turns, the proposed design is better than zigzag coil.

The Q factor and self-inductance of the zigzag coil are plotted in the above Figure 28.6 with the measured frequency ranging from 0.52GHz to 20.2 GHz. With more turns, the inductance rises from 0.37 nH to 0.41 nH, increasing the mutual and self-inductance. However, because resistance is rising more quickly, the Q factor also falls at the same instant.

Now we are going to investigate the experimental results of our proposed design. The Q factor and inductance of the newly proposed coil are plotted in the below Figure 28.18 with the output frequency ranging from 0.51 GHz to 20.2 GHz. With more turns, the inductance rises from 0.24nH to 0.36nH and hence increases the mutual and self-inductance. However, because resistance is rising more quickly, the Q factor also increases at the same period.

From the above Figure 28.8(a) and (b), it has been observed that, with the increment of frequency, the inductance falls in an almost linear fashion which is the another aspect of the proposed design.

In fig 28.5 a comparative study of mutual inductance has been depicted very clearly. It has been observed that at initial stage our design shows linearity which is necessary for the simple design.

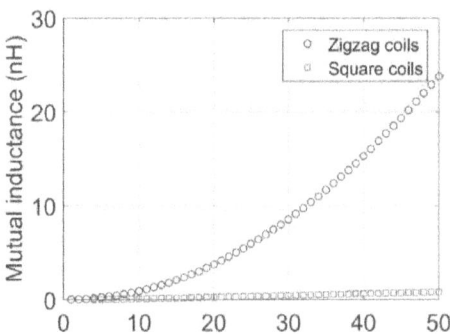

Figure 28.5 Mutual inductance of proposed model.
Source: Author

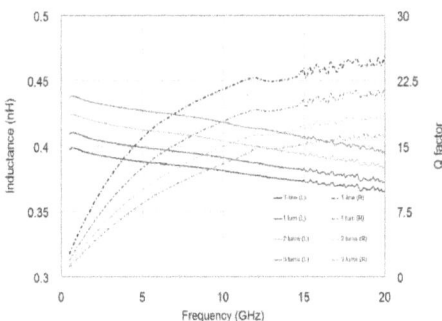

Figure 28.6 Inductance and Q factor Vs Frequency.
Source: Author

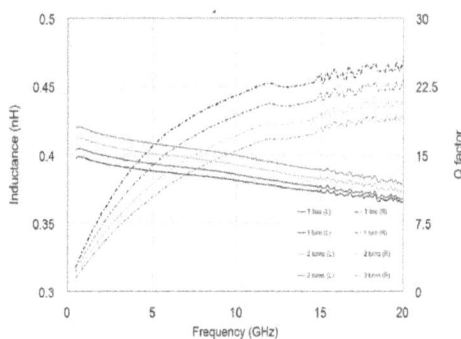

Figure 28.7 Inductance & Q factor Vs frequency for our proposed design
Source: Author

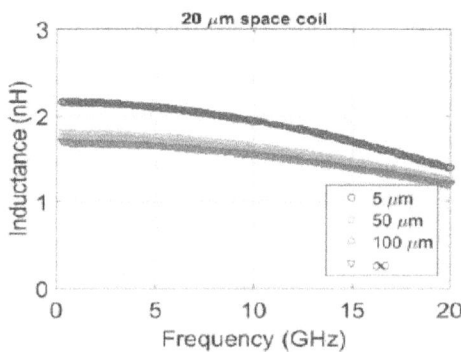

Figure 28.8 (a) Inductance Versus Frequency for different space between two turn. In Figure 28.7 simulation result is achieved
Source: Author

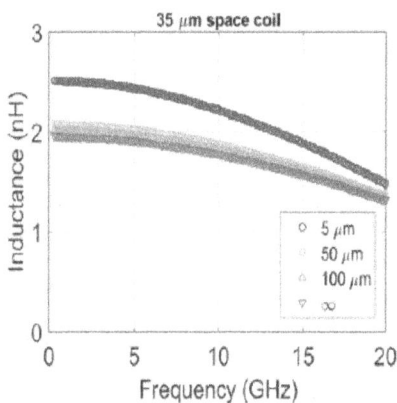

Figure 28.8 (b) Inductance versus frequency for different space between two turn.
Source: Author

Conclusion

In this research work, two distinct configurations of a 3D inductor are proposed and tested via simulation. Due to its distinctive non-parallel configuration, the abrupt structure always offers the highest mutual-inductance, and it also increases considerably more quickly than other arrangements. When greater than 55 abrupt turns are employed in the inductors, the inductance of zigzag structure can reach 20nH, with the exception of self-inductance. Only 3nH of additional inductance may be obtained for 55 turns inductors in the other configurations. On the other hand, due to the unexpected resistance the simulation result of the triangular 3D inductors demonstrates that higher turns result in higher inductance but worse Q-factor. As a result, both the zigzag coils and triangular coil shows lowest quality factor. In this article, two unique inductor types have been produced. Only a little modification has been made to the same-layer in the triangular 3D inductor model. As a result, the design demonstrates the phase transition properties. There are different techniques to make a 3D inductor, however in this proposed concept; a circuit switch made of a thin coating of vanadium dioxide on a bar is used before a full spiral inductor coil is made. The experimental result is validated through simulation, which makes use of a trustworthy software tool. The experiment caused only a 29% change in inductance value with high Q value.

References

[1] Assadsangabi, B., Ali, M. S. M., & Takahata, K. (2013). Planar variable inductor controlled by ferrofluid actuation. *IEEE Transactions on Magnetics*, 49(4), 1402–1406.

[2] Banerjee, K., Souri, S. J., Kapur, P., & Saraswat, K. C. (2001). 3-D ICs: a novel chip design for improving deep-submicrometer interconnect performance and systems-on-chip integration. *Proceedings of the IEEE*, 89, 602–633.

[3] Chow, E. M., Chandrasekaran, V., Partridge, A., Nishida, T., Sheplak, M., Quate, C. F., et al. (2002). Process compatible polysilicon-based electrical through-wafer interconnects in silicon substrates. *Journal of Microelectromechanical Systems*, 11, 631–640.

[4] Choi, E. H., Lee, Y. S., & Rha, S. K. (2012). Effects of current density and organic additives on via copper electroplating for 3D packaging. *Korean Journal of Materials Research*, 22, 374–378.

[5] Chuang, H. C., Li, H. F., Lin, Y. S., Lin, Y. H., & Huang, C. S. (2013). The development of an atom chip with through silicon vias for an ultra-high-vacuum cell. *Journal of Micromechanics and Microengineering*, 23, 085004.

[6] Domann, J. P., Chen, C., Sepulveda, A. E., Candler, R. N., & Carman, G. P. (2018). Micro-motors with deterministic single input control. arXiv:1802.09420.

[7] Gambino, J. P., Adderly, S. A., & Knickerbocker, J. U. (2015). An Overview of Through-Silicon-Via Technology and Manufacturing Challenges. Amsterdam, The Netherlands: Elsevier Science Ltd.

[8] Gutmann, R. J., Lu, J. Q., Devarajan, S., Zeng, A. Y., & Rose, K. (2004). Wafer-level three-dimensional monolithic integration for heterogeneous silicon ICs. In Proceedings of the Topical Meeting on Silicon Monolithic Integrated Circuits in RF Systems, Atlanta, GA,USA, 8–10 September, (pp. 45–48).

[9] Gallé, W. P. (2012). MEMS-Based Fabrication of Power Electronics Components for Advanced Power Converters. Atlanta, GA, USA: Georgia Institute of Technology.

[10] Huang, C. (2015). High performance through-silicon-via in 3D interconnection (TSV). Beijing, China: Tsinghua University. *Business Studies*, 39(7), 11491168.

[11] Salvia, J., Bain, J. A., & Yue, C. P. (2020). Tunable on-chip inductors up to 5 GHz using patterned permalloy laminations. In Proceedings of the IEEE International Electron Devices Meeting (IEDM), Washington, DC, USA, 5 December 2005; IEEE: Manhattan, NY, USA.

[12] González, J. L., Aragonés, X., Molina, M., Martineau, B., & Belot, D. (2020). A comparison between grounded and floating shield inductors for mmW VCOs. In Proceedings of the IEEE European Conference on Solid-State Circuits (ESSCIRC), Seville, Spain, 14–16 September 2010; IEEE: Manhattan, NY, USA.

29 Thermal behaviour of a 3D printing nozzle

Abhijit Dutta[a], Debangan Bhattacharjee[b] and Rajarshi Ghosh[c]

Department of Mechanical Engineering, MCKV Institute of Engineering, Liluah, Howrah, West Bengal, India

Abstract

This paper addresses the thermal response of a 3D printing nozzle used in a fused deposition modelling printer. The key findings include temperature distribution, entropy change and strain energy distribution of the nozzle assembly. The energy input has been considered from an electrical source. An electrical heater is imposed in the mounting block to melt the filament. The system is considered as steady state. Finite element software has been used to solve the energy equation along with the boundary condition. The results show that heaters and the mounting block are reflecting significant thermal change. The magnitude of the rate of change in entropy of the filament material at the nozzle inlet is 8.59×10^{-07} J/s. K. A significant 98% drop-in rate of change of entropy of the filament is seen at the nozzle tip. However, the strain energy at the nozzle inlet is around 6.194×10^{-04} J whereas at the outlet a significant rise in magnitude around 6.685×10^{-07} J is observed.

Keywords: Entropy, fused deposition modelling, nozzle, strain energy, thermal response

Introduction

Fused deposition modelling (FDM) is one of the methods of additive manufacturing where polymer based molten filaments are deposited layer by layer upon the bed to get desired object. The hot end nozzle is used to flow the molten metal on the bed. The temperature distribution of the nozzle exhibits significant impact on the printing performance of the nozzle. Most of the printing defects are caused by uneven temperature distribution. Hence, in this work authors are interested to observe temperature distribution of a 3D printer nozzle situated at the center of excellence of 3D printing and scanning in the MCKV Institute of Engineering, West Bengal, India.

A brief review of open literature is now discussing in this section. The influence of pressure, nozzle diameter, viscosity and particles on the printing process of molten TNT based explosive have been carried out by Zong et al. [11]. Zheng et al. [10] proposed a local radial basis function collocation method for electrochemical 3D printing process and established a good agreement between numerical and experimental results. The flow of filament materials like ABS in the nozzle and its velocity, temperature fields were addressed in a literature [2]. Raja et al. [7] reviewed different models of food printing technology. They have found that the food printing technology is a complex phenomenon where velocity, flow, pressure, and shear distribution play pivotal role. Mixing of two different color of filaments inside nozzle in FDM process may cause blockage in the intersection region. To avoid the blockage of filament the design of the nozzle

should be improved [3]. Wan Muhamad et al. [8] investigated flow characteristics of filament through the nozzle of a FDM machine. They concluded that the nozzle with a brass material and having 120° die angle, 0.4 mm outer diameter selected as the best optimum performer.

While thermal analysis on the 3D printing nozzle has been carried out as already discussed, to our knowledge comprehensive thermodynamic analysis including rate of change of entropy of the filament material and strain energy distribution is not covered in the open literature. Therefore, the aim of this work is to evaluate thermal responses of a 3D printer nozzle assembly considering in a realistic scenario.

Model Description

The model is an assembly component which is consists of four members that includes Heater Cartridge, Nozzle, Mounting part and Filament (see Figure 29.1). Here the model name of the nozzle which is fitted into the mounting part is E3D V6 Brass Nozzle. The nozzle is composed of brass material [6]. The Heater Block used in it is composed of steel alloy named AISI 304 [1]. The filament used here for melting in the nozzle is composed of Acrylonitrile Butadiene Styrene (ABS) [9]. Lastly the mounting part which holds all the components is composed of aluminum alloy named 1060-H12. The main function of the heater cartridge is to convert the electrical power into heat power to provide heat in order to melt the filament inside the nozzle. That's why the heater cartridge is composed of steel alloy because of its capacity of wear resistance

[a]abhijitdutta@mckvie.edu.in, [b]debangan80@gmail.com, [c]ghosh.rajarshi3009@gmail.com

DOI: 10.1201/9781003663348-29

and toughness and also the capability to endure high temperatures. The nozzle particularly helps in holding the filament and to melt the filament in order to print the required items. In that case the nozzle is composed of material brass because it possesses optimal thermal properties for printing. The nozzle also holds a crucial role in printing because the solid state of the filament changes to the molten state in it. The filament is actually the printing element which melts and flows out to take the desired shape as per the required CAD model. Finally, the mounting part also has its crucial role in order to hold each member in a proper manner. The aluminum alloy of the mounting part helps in the efficient heat transfer in mode of conduction where the heater cartridge generates the heat to provide the melting temperature in order to melt the filament inside the nozzle and to attain the printing temperature to print the molten filament on bed.

Methodology and Model Specifications

The model consists of four different components i.e., heater block, mounting block, nozzle and filament. A mounting part is a component where it holds the heater block and the nozzle. The filament is being fitted inside the nozzle where it will melt and get deposited on the bed. The important thermal properties include thermal conductivity (k), density () and specific heat) are depicted in Table 29.1 for the assembly components.

The thermodynamic behavior due to the heat transfer in a 3D printing nozzle assembly has been assessed. In the nozzle assembly the heat is generated in the heater block due to the Joule heating. A heater block where the heat is being generated due to heat power originated from electric source and then heat flows through the mounting block to melt the filament inside the nozzle. Thermodynamics is a subject which

Table 29.1 Thermo-physical properties of the components

Component	Material	Properties
Nozzle	Brass	W/(m.K) kg/m^3 J/(kg.K)
Heater Block	AISI 304 Steel Alloy	W/(m.K) 8000 kg/m^3 J/(kg.K)
Filament	ABS (Acrylonitrile Butadiene Styrene)	0.2256W/(m.K) kg/m^3 J/(kg.K)
Mounting	1060-H12 Aluminum Alloy	W/(m.K) 05 kg/m^3 J/(kg.K)

(Source: Material Impact Data is based on bulk raw material computed from MetalPrices.com, 2012)[5]

governs all natural and artificial phenomena. It gives a clear idea about the performance of a system (nozzle assembly). The thermodynamic entropy is a degree of randomness which is associated with the loss of useful work. Therefore, in this work entropy change has been studied over the filament length.

The 3D model of the assembly was created in a solid modelling package software as per the dimension available in manual of 3D printing machine present in center of excellence in 3D printing lab at MCKVIE. The sectional view of the model is depicted in Figure 29.2 (a). Then the model was imported into the simulation environment for numerical analysis.

Figure 29.2 (a) shows an assembly part of a nozzle, a heater block and a filament along with a mounting block where all the parts have been fitted. Along with Figure 29.2 (b) depicts the assembly part which was pre-processed with solid mesh type along with standard mesh which has been used. The mesh results

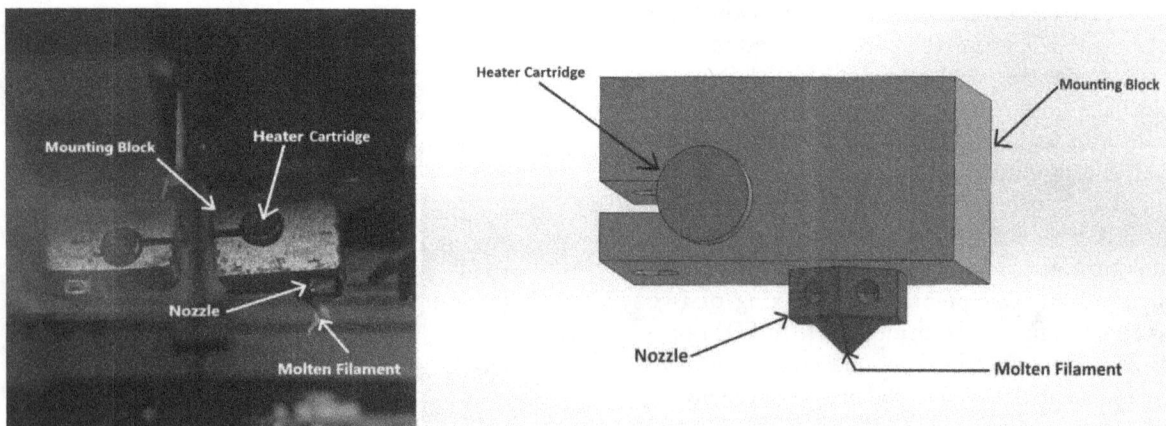

Figure 29.1 (a) Nozzle assembly (b) Corresponding 3D model
Source: Author

(a)

(b)

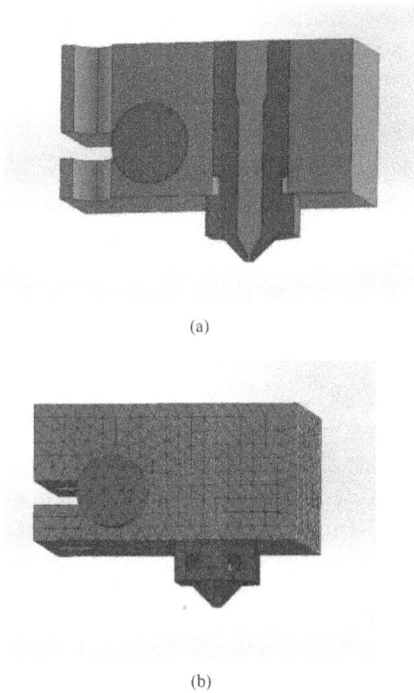

Figure 29.2 (a) Sectional view of nozzle assembly (b) corresponding mesh
Source: Author

contain a total of four Jacobian points for high quality mesh along with 33,758 number of total nodes, 22,679 number of total elements, 1.13598 mm of element size, 0.0567988 mm of Tolerance and Aspect Ratio of 12.258.

For a particular domain, the heat conduction equation has been solved with specified boundary conditions to observe the spread of temperature. The conduction equation of heat transfer is given in Eq. (1)

$$\nabla . q + Q_g = 0 \qquad (1)$$

In Eq. (1) $\nabla . q$ represents conductive heat transfer in W along the three dimensions. The heat generation in the domain (i.e., control volume) is given as Q_g also in W. This term is only applicable in the heater zone where electrical heat production (Q_g) is estimated from the test set as: $Q_g = 12\,W$.

The exposed portions of the assembly are subjected to heat exchange with the surrounding air through convection the corresponding equation is depicted as Eq. (2)

$$Q_{Conv} = h.A.(T - T_{Ambient}) \qquad (2)$$

Where, h is the convective heat transfer coefficient in W/m^2K, A presents area of convective heat transfer in m^2, T is the temperature of the surface in K and $T_{Ambient}$ represents ambient temperature of the assembly. However, in this work convective heat transfer

coefficient is assumed as 25 W/m^2K Kosky [4] and the ambient temperature is set to 27°C.

The strain energy of the nozzle assembly can be computed by solving Eq. (3)

$$U = \sum \left[(\sigma_x \in_x + \sigma_y \in_y + \sigma_z \in_z + \tau_{xy} \varnothing_{xy} + \tau_{xz} \varnothing_{xz} + \tau_{yz} \varnothing_{yz}) . \frac{V.W_{(i)}}{2} \right] \qquad (3)$$

Where, U is the total strain energy in J, σ represents normal stress, ∈ presents change in dimensions along the selected reference geometry, τ_{xy} stands for shear in Y direction on YZ plane, shear strain in the Y direction in the YZ-plane of the selected reference geometry is denoted by $\varnothing_{xy}$, V is the volume, $W_{(i)}$ is the weighted constant at integration point i.

Entropy change along the filament length is depends on its initial temperature T_0 is depicted in Eq. (4)

$$S_{Change}(T) = C_p \ln \frac{T}{T_0} \qquad (4)$$

In Eq. (4) T is the temperature at a fixed location on the filament within the nozzle.

Results and Discussion

Thermal response of different region of the assembly component as well as analysis of thermal effects on different parts is presented in this section. Temperature has been tracked at different location of inside the filament present in the nozzle and fixed value of these locations is recorded for analysis.

Thermal response

The spread of thermal field in steady state condition of the nozzle assembly is depicted in Figure 29.3. The areas of connecting part of heater block and mounting part undergo thermal changes due to generation of heat power of 12 W. A minor change in temperature is observed with the following location range between 1-6. The area is significant because it is the actual region where the filament is melting. A significant span of temperature difference is noted within the whole assembly which results in enhancement of the thermal stress in the nozzle and filament respectively.

Figure 29.3 portrays that the maximum temperature can be seen at the end of the heater cartridge around 248°C (red color) while the intermediate region of the cartridge maintaining around 1.5 % drop in temperature (green). Similarly, the mounting region above the cartridge is around 2.27 % less temperature as marked by blue colored. The temperature drops at the outlet of the nozzle, covered with purple region is about 2.6% compared to maximum temperature observed at the heater cartridge. Also, there is temperature drop change of about 0.20 % between the inlet and outlet region of the nozzle where the filament changes from solid state to its molten state. Due to the latent heat

(a)

(b)

(c)

(d)

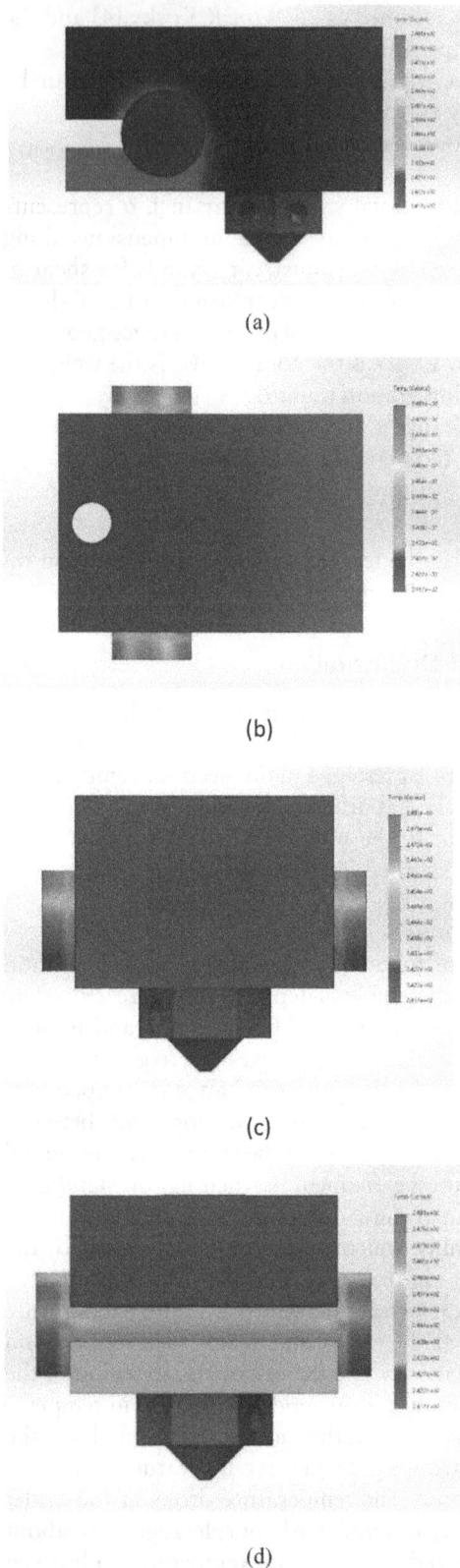

Figure 29.3 Thermal effect on the nozzle assembly (a) Elevation (b) Plan (c) Right Hand Side View (d) Left Hand Side View

Source: Author

transfer during phase change of the filament material the temperature remain constant. Therefore, from the top view no significant temperature difference of the filament is seen.

Rate of entropy change in the filament

Entropy change of the filament material in the nozzle is significant from the vantage of thermodynamics which plays an indication of losses involved in energy interaction and internal irreversibility. There are seven different locations of the filament inside the nozzle the rate of entropy change has been estimated. The entropy change of the abs material along the length of the nozzle are depicted in Figure 29.4.

In Figure 29.4, the distance at which entropy is estimated has stated at the nozzle inlet hence it is designated as 0. The length of the nozzle is 16.5 mm therefore; this is the 7th location where entropy of the filament has been computed (see inset). However, the rate of change of entropy decreases with the increase in length. The magnitude at the inlet is 8.59×10^{-07} J/s.K. After that it increases around 12.5% because of receiving more heat from the heater at this location and then it decreases. Between the distance 4.76 mm to 14.58 mm the rate of change of entropy steadily decreases as the length increases. A significant 98% drop-in rate of change of entropy of the filament is observed at the tip of the nozzle. It can be observed that the entropy change is maximum at the adjacent to the heater and gradually decreases away from it. Higher heat transfer from the heater to filament leads to enhancing kinetic energy of the filament material.

Strain energy distribution

Strain energy is a type of potential energy that stored in the body due to deformation within elastic limit. The strain energy distribution is portraying in Figure 29.5.

Figure 29.5 (a) depicts the strain energy spread over the entire assembly. At the nozzle inlet, the magnitude of the strain energy is observed around 6.194×10^{-04} J and at the outlet the value is around 6.685×10^{-04} J (ref. Figure 29.5 (b)). The strain energy signifies the energy which is stored inside the body due to deformation. In this case, the deformation is mainly due the thermal expansion. Figure 29.5 (a and b) shows that the energy stored at the filament region is quite higher than the energy stored at the outlet area of the nozzle. The exposed part of the heater and nozzle exhibits low strain energy.

Conclusions

Understanding the thermal behavior of a FDM nozzle assembly is connected to the quality of the printing

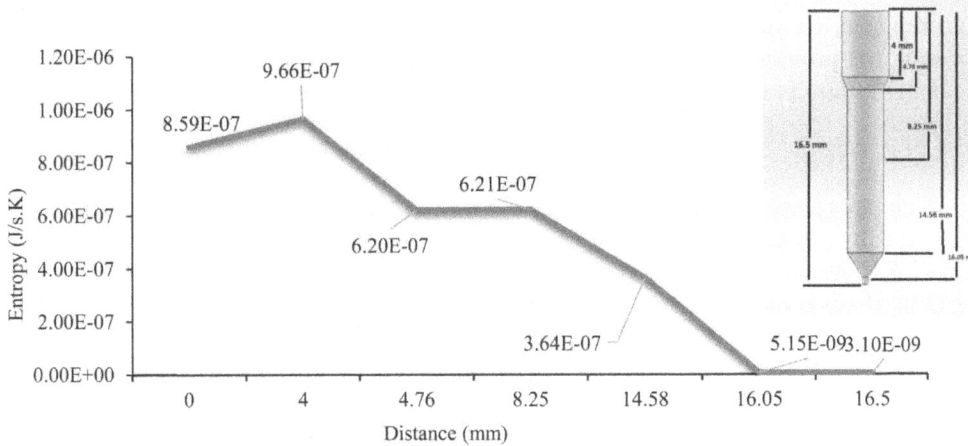

Figure 29.4 Rate of entropy change in the filament
Source: Author

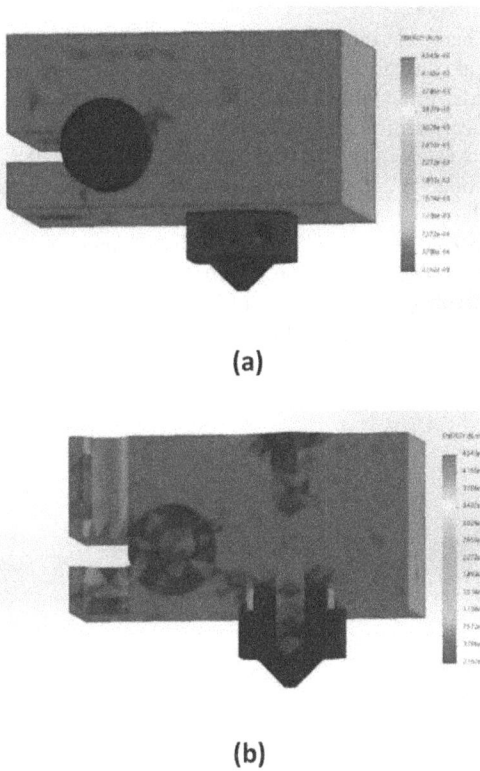

(a)

(b)

Figure 29.5 (a) Strain energy distribution of the nozzle assembly (b) Corresponding sectional view
Source: Author

process. Under this perspective, temperature distribution, rate of change in entropy and strain energy spread in a nozzle assembly are the focus of the current work. The results of this work claim that the 3D printer nozzle assembly exhibits significant thermal response which refers to the following conclusions.

The maximum temperature can be seen at the end of the heater cartridge around 248°C (red color) while the intermediate region of the cartridge maintaining around 1.5 % drop in temperature (green). Similarly, the mounting region above the cartridge is around 2.27 % less temperature as marked by blue colored. The temperature drops at the outlet of the nozzle, covered with purple region is about 2.6% compared to maximum temperature observed at the heater cartridge. Also, there is temperature drop change of about 0.20 % between the inlet and outlet region of the nozzle where the filament changes from solid state to its molten state.

The rate of change of entropy decreases with the increase in length. The magnitude at the inlet is 8.59 $\times 10^{-07}$ J/s.K. After that it increases around 12.5% because of receiving more heat from the heater at this location and then it decreases. Between the distance 4.76 mm to 14.58 mm the rate of change of entropy steadily decreases as the length increases. A significant 98% drop-in rate of change of entropy of the filament is observed at the tip of the nozzle.

The strain energy signifies the energy which is stored inside the body due to the deformation. At the nozzle inlet, the magnitude of the strain energy is observed around 6.194×10^{-04} J and at the outlet the value is around 6.685×10^{-07} J which is significantly higher.

Thermal analysis of a 3D printing nozzle is important to understand the performance of a nozzle as well as for the remedial measures of failure. FDM printing is time consuming, therefore, nozzle is under stressed conditions for prolonging time. Selection of nozzle material and filament material plays a vital role on the printing quality. The future scope of this work is to design upgraded nozzle geometry along with new material to minimize irreversibility during printing.

Acknowledgments

The authors express their gratitude to the Authority of MCKV Institute of Engineering for providing access to the Centre of Excellence in 3D Printing Laboratory.

References

[1] ISSN International Centre (2006) The ISSN register. https://cartridgeheaters. co.uk/technical-specifications/#:~:text=Materials,%E2%80%93%20a%20 ceramic%20insulator. Accessed 04 June 2024.

[2] Han, Z., Liu, Q., & Dong, Y. (2018). Numerical simulation of printer nozzle extrusion mechanism based on ANSYS. *International Journal of Trend in Research and Development*, 5(4), 290–294.

[3] Han, S., Xiao, Y., Qi, T., Li, Z., & Zeng, Q. (2017). Design and analysis of fused deposition modeling 3D printer nozzle for color mixing. *Advances in Materials Science and Engineering*, 1–12. https://doi. org/10.1155/2017/2095137.

[4] Kosky, P. (2021). Exploring Engineering ǁ Mechanical Engineering. An Introduction to Engineering and Design. 5th edn. (pp. 317–340). Academic Press. doi:10.1016/B978-0-12-815073-3.00014-4.

[5] The source for SOLIDWORKS® and SOLIDWORKS Simulation material properties in the "SOLIDWORKS Materials" library is the "Metals Handbook Desk Edition (2nd Edition)", ASM International.

[6] Melo, J. T., Santana, L., Idogava, H. T., Pais, A. I. L., & Alves, J. L. (2022). Effects of nozzle material and its lifespan on the quality of PLA parts manufactured by FFF 3D printing. *Engineering Manufacturing Letters*, 1(1), 20–27. 10.24840/2795-5168_001-001_0005.

[7] Raja, V., Nimbkar, S., Moses, J. A., Ramachandran Nair, S. V., & Anandharamakrishnan, C. (2023). Modeling and simulation of 3D food printing systems—scope, advances, and challenges. *Foods*, 12(18), 3412. https://doi.org/10.3390/foods12183412.

[8] Wan Muhamad, W. M., Saharudin, M. S., Abd Wahid, K. A., Saniman, M. N. F., & Reshid, M. N. (2020). Nozzle design for fused deposition modelling 3D printing of carbon fibre reinforced polymer composite component using simulation method. *PalArch's Journal of Archaeology of Egypt / Egyptology*, 17(9), 4192–4204.

[9] Xu, X., Qiu, W., Wan, D., Wu, J., Zhao, F., & Xiong, Y. (2024). Numerical modelling of the viscoelastic polymer melt flow in material extrusion additive manufacturing. *Virtual and Physical Prototyping*, 19(1), e2300666. https://doi.org/10.1080/17452759.2023.2 300666.

[10] Zheng, H., Lu, X., Jiang, P., & Yang, Y. (2022). Numerical simulation of 3D double-nozzles printing by considering a stabilized localized radial basis function collocation method. *Additive Manufacturing*, 58, 103040. https://doi.org/10.1016/j.addma.2022.103040.

[11] Zong, H., Cong, Q., Zhang, T., Hao, Y., Xiao, L., Hao, G., et al. (2022). Simulation of printer nozzle for 3D printing TNT/HMX based melt-cast explosive. *The International Journal of Advanced Manufacturing Technology*, 119, 3105–3117. https://doi. org/10.1007/s00170-021-08593-z.

30 Mechanical properties and rheological investigation of aluminum A356: dual approach through experimental and numerical simulations

Somnath Mitra[1,a], Joydip Paul[2,b], Bikash Banerjee[1,c] and Hiranmoy Samanta[2,d]

[1]Abacus Institute of Engineering and Management, Hooghly, West Bengal, India

[2]Gargi Memorial Institute of Technology, South 24 parganas, West Bengal, India

Abstract

This paper represents the casting of the aluminum in sand casting process and hence the numerical model has been developed to understand the transport phenomenon of solidification. The coarse and fine-grain sand has been used for the casting, and it was found that the fine grain has a better surface finish than the coarse grade. The semi-implicit finite volume method (FVM) has been used to solve the set of governing differential equations for mass, momentum, and energy. using a line-by-line Tri Diagonal Matrix method solver and based on the SIMPLER method. The experimental data has been used to validate numerical simulation. It has been discovered that the chance of remelting increases with cavity height, leading to a little rise in the solidification growth rate. It is clear from the melt fraction and isotherm plots that the aluminum solidified. It is thought to be between 707°C and 800°C. After doing the grid independence test, 142×82 uniform grids with a time-step of 0.1 were ultimately discovered. The most promising method for achieving improved mechanical properties with metal-matrix composites is to incorporate nano and micro-sized support particles into the lattice defined for better mechanical properties. The convergence of residuals for continuity and momentum equations is below 10^{-5}, and for energy equation 10^{-8}. Solidification is the primary concern to determine the properties of a cast. It involves heat and mass transfers, tracking of solid-liquid interface etc. Rejection of solute, micro- and macro-segregation etc. It involves complex transport phenomena: experimental determination is difficult and expensive. To examine the transport processes during solidification, a numerical analysis is taken into consideration.

Keywords: Casting, computation fluid dynamics, rheology, solidification

Introduction

Al-alloy, which is used as the lattice in MMCs with the Si in A356 aluminum amalgam, is inexpensive, easy to maintain, of excellent quality and ductility, and resistant to corrosion and climatic deterioration. In the composites, difficult particles like SiC and Al2O3 are frequently used as support phases. In the automobile and aircraft industries, the use of Al2O3 or SiC molecule enhanced aluminum combination lattice composites for cylinders, barrel heads, interface bars, etc., is gradually growing. Tribological properties of the materials are exceptionally critical sand casting, as the name implies. Al MMC, or aluminum metal matrix composite, was introduced by Natarajan et al. [1]. Grey cast iron and sliding 82 automotive friction material were used to compare the material's wear behavior. At state dry conditions, Cree et al. [2] assessed and demonstrated the sliding wear and friction characteristics of A356 aluminum alloy and a hybrid composite of A356 aluminum alloy. SiC reinforcements in heterogeneous nucleation sites for Si during solidification of Al–Si–SiC composites were demonstrated by Nagarajan et al. [3]. Heidary et al. [4] studied, and investigated experimentally as well as compared with theoretical predictions and it shows a good relationship between them. The impact of cooling conditions and mold size on aluminum solidification was investigated and numerically modeled by Egole et al. [5]. In order to construct a mold cavity and form the desired casting result, they will be prepared in the same manner as the tests for moisture content, clay content, refractoriness, strength, permeability, flow-ability, mold hardness, etc. [6–8]. Sand types and collection zones differ in their mechanical characteristics and chemical makeup. The grain size affects the casting properties in terms of heat generation, stress and strain and the mechanical properties and behavior Figure 30.1 shows the layout of the work. So, Metal products considering the knowledge f casting at he remedies of error and defects formation [9–11]. The numerical simulation is based on the experimental data obtained from the experiment. The solidification time and other parameters involved.

asomnathmitra.aiem@jisgroup.org, bjoydip.me_gmit@jisgroup.org, cbikashbanerjee25@gmail.com, dhiranmoy.me_gmit@jisgroup.org,

DOI: 10.1201/9781003663348-30

Research gap and research questions:

- The effect of varying temperatures on both mechanical properties and flow behavior is minimal.
- Quantitative relationships and simulation-based predictions remain sparse.
- Limited work exists using high-resolution techniques in tandem with simulations.

R1: To find the best suited GFN for aluminum casting from different GFN segregation

R2: Melting phenomenon of casting

Material and Methods

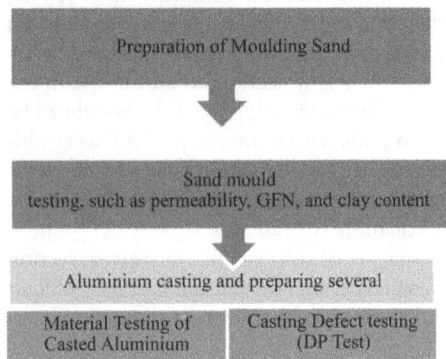

Figure 30.1 Flow chart of this work
Source: Author

Table 30.1 Chemical composition of aluminum A356

Composition	Percentage
Aluminum	92.05
Silicon	7
Magnesium	0.35
Iron	0.20
Copper	0.20
Manganese	0.10
Zinc	0.10

Source: Author [8, 9]

Figure 30.1 (a) molding box preparation. (b) pouring, and (c) solidification and casting
Source: Author

The pattern and sand in a gating system are seen in Figure 30.1(a). As illustrated in Figure 30.1(b), melt the metal into the mold cavity. The liquid aluminum alloy turns into a solid cast product, as seen in Figure 30.1(c).

Numerical simulation of aluminum casting in a rectangular cavity (Table 30.1–30.3).

Numerical modeling
The assumptions are made for the solidification modelling.

(i) The liquid is laminar and unsteady in nature
(ii) Newtonian characteristics and incompressibility
(iii) In the fluid Boussinesq approximation is satisfied,
(iv) No internal stresses in solid form
(v) Zero shrinkage during solidification

✓ Convergence

$$\frac{\emptyset - \emptyset_{max}}{\emptyset_{min}} < 10^{-5}$$

✓ *Calculation of liquid fraction*

Considered enthalpy update scheme

$$[\Delta H_p]^{n+1} = [\Delta H_p]^n + \frac{a_p}{a_p^0}[\{H_p\}^n - C_p T_M]$$

Liquid fraction, $f_l = \frac{\Delta H_p}{L_a}$

Figure 30.2. Solidification curve for alloy.

Solid fraction, $f_s = 1 - f_l$

Mathematical modeling

Conservation of mass

$$\frac{D\rho}{Dt} + \rho(\nabla \vec{V}) = 0$$

Conservation of Momentum

$$\rho \frac{D\vec{V}}{Dt} = -\nabla P + \nabla . (\mu\nabla)\vec{V} + \rho\vec{f}_b$$

$$+ A\vec{V} + \rho\vec{g}\beta_T(T - T_M)$$

where $A = -\frac{C(1-f_l)^2}{f_l^3 + b}$

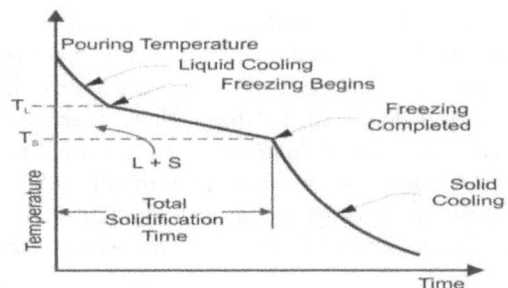

Figure 30.2 Solidification curve for alloy
Source: Author

Conservation of energy

$$\frac{D(\rho c_p T)}{Dt} = \nabla \cdot (k\nabla T) + S_h$$

Where,

$$S_h = -\left\{\frac{\partial}{\partial t}(\rho f_l \Delta H) + \nabla \cdot (\rho \vec{U} \Delta H)\right\}$$

$$S_h = -\left\{\frac{\partial}{\partial t}\right.$$

S_h={∂/∂t(⊢ρ f_l ΔH)+∇·(ρ U ⁻ΔH)⊣ }

Enthalpy update scheme

Conservation of species

$$\frac{\partial(\rho C_l)}{\partial t} + \nabla \cdot (\rho \vec{U} C_l) = -\frac{\partial(\rho_s f_s C_s)}{\partial t}$$

A356 alloy properties	Values
c_{pl}	963 J/kgK
c_{ps}	963 J/kgK
k_s	60 W/mK
k_l	160 W/mK
ρ_s	2685 kg/m^3
ρ_l	2685 kg/m^3
μ	1.13×10^{-3} kg/m s
D_l	1.0×10^{-9} m^2/s
L	397,700 J/kg
β_T	2.1×10^{-5} K^{-1}
β_S	0.025 K^{-1}
T_M	660 °C
T_E	578 °C
k_p	0.1

The rectangular cavity is shown in the Figure 30.3. height: 0.1; width 0.3. Radiation, convection and conduction mode are considered for the optimum calculation of space heat.

Left side: constant temperature.
Right: convection
Top face: convection and radiation
Bottom surface: insulated.

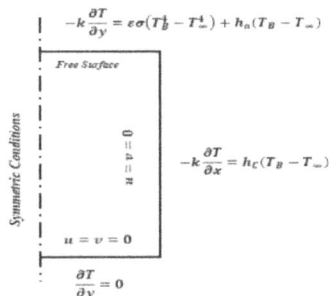

Figure 30.3 Numerical domain considered for calculation
Source: Author

Results and Discussion

From Figure 30.3 we understand the surface imperfections (porosity) of the cast goods from this experiment. We had tested the dye penetration on the specimen's three neighboring surfaces.

Through the course of this investigation, it has been found that the coarse grain sample's Brinell hardness value, which is 57.74 and in fine grain 69.16. According to the results of the impact test (Izod and Charpy), the fine grain sand product's toughness is higher (Izod: -8, Charpy: 162) than the coarse grain sand product's (Izod: -6, Charpy: 76).

However, fine grain has lower porosity, so permeability was lower in fine grain sand (175.84) than in coarse grain sand (296.35).

The validation curve shows a good agreement with the graph generated by Egole et. al. after successful grid independence test it was found that 142 × 82 grids is the best fitted grid.

Figure 30.4 Testing of the aluminum product
Source: Author

Table 30.2 Comparative analysis between the fine grain sand and coarse grain sand cast products

Sl. No	Name of the tests	Coarse grain sand product	Fine grain sand product
1	Brinell hardness number	57.74	69.16
2	Izod number	6	8
3	Charpy number	76	162

Source: Author

Table 30.3 Comparative analysis between the fine grain sand and coarse grain sand

Sl. No	Name of the tests	Coarse grain sand product	Fine grain sand product
1	Grain fineness number	51.65	122.79
2	Permeability	296.35	175.84
3	Clay content	14	17.25

Source: Author

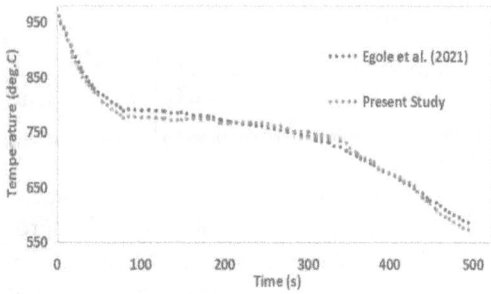

Figure 30.5 Variation of temperature with time of present study and Egole et al. [5] study
Source: Author

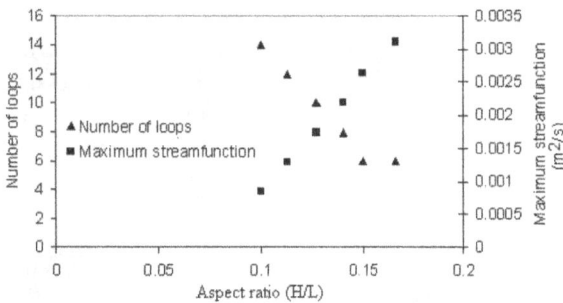

Figure 30.6 Changes in the maximum stream function and number of loops with aspect ratio
Source: Author

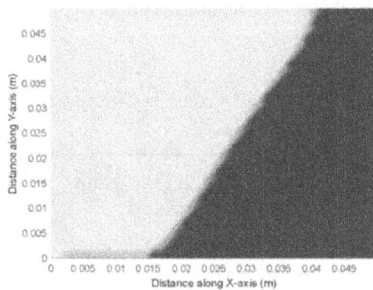

Figure 30.7 Solid- liquid path tracking during solidification of A356[7070C 200Sec]
Source: Author

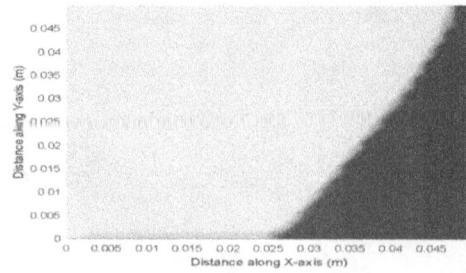

Figure 30.8 Solid- liquid path tracking during solidification of A356[800°C 200Sec]
Source: Author

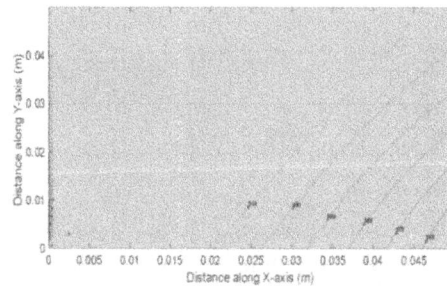

Figure 30.9 Isotherm plot at different temperature [800°C 200Sec]
Source: Author

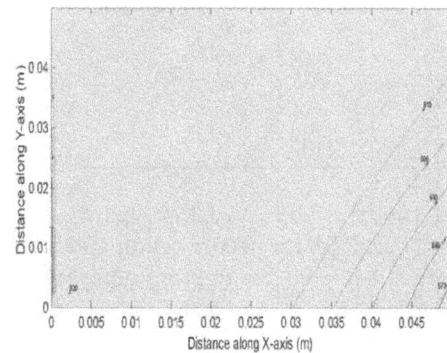

Figure 30.10 Isotherm plot at different temperature [707°C 200Sec]
Source: Author

The rectangular domain is heated with constant temperature at 900°C. The melting fraction shows the solidification pattern of the molten Aluminium.at 707°C and 800°C at different time of 200 Sec. Ambient temperature(T_∞) 26°C The time for solidification have been updated with the numerical simulation.

The isotherm plot shows the temperature distribution at different points of the rectangular cavity with respect to time. Left wall maintained at fixed temperature.

Conclusions

1. The fine grain sand (17.25%) has a higher clay concentration than coarse grain sand (14%). Hence fine grain sand has stronger mold binding qualities.
2. The fine grain sand product has a higher hardness and toughness (Izod number-8, Charpy number-162) than the coarse grain sand product whose Izod number is 6 and Charpy number-is 76.
3. This solidification study giving an idea of temperature distribution.

4. During solidification of A356 the solid-liquid path has been easily tracked.

References

[1] Natarajan, N., Vijayarangan, S., & Rajendran, I. (2006). Wear behaviour of A356/25SiCp aluminium matrix composites sliding against automobile friction material. *Wear*, 261, 812–822.

[2] Cree, D., & Pugh, M. (2011). Dry wear and friction properties of an A356/SiC foam interpenetrating phase composite. *Wear*, 272, 88–96.

[3] Nagarajan, S., Dutta, B., & Surappa, M. K. (1999). The effect of SiC particles on the size and morphology of eutectic silicon in cast A356/SiCp composites. *Composites Science and Technology*, 59, 897–902.

[4] Heidary, D. S. B., & Akhlaghi, F. (2011). Theoretical and experimental study on settling of SiC particles in composite slurries of aluminum A356/SiC. *Acta Materialia*, 59, 4556–4568.

[5] Egole, C. P., Mgbemere, H. E., Sobamowo, G. M., & Lawal, G. I. (2021). Numerical modelling of the effect of cooling conditions and mould size during solidification of Al-4.5%Cu alloy in static casting process. In IOP Conference Series: Materials Science and Engineering (Vol. 1107, p. 012019). DOI 10.1088/1757899X/1107/1/012019.

[6] Bhagavath, S., Cai, B., Atwood, R., Li, M., Ghaffari, B., Lee, P. D., et al. (2019). Combined deformation and solidification-driven porosity formation in aluminum alloys. *Metallurgical and Materials Transactions A*, 50, 4891–4899.

[7] Lewis, R. W., & Ransing, R. S. (1998). A correlation to describe interfacial heat transfer during solidification simulation and its use in the optimal feeding design of castings. *Metallurgical and Materials Transactions B*, 29(2), 437–448. doi:10.1007/s11663-998-0122-y.

[8] Chou, S. N., Huang, J., Lii, D., & Lu, H.-H. (2006). The mechanical properties of Al2O3/aluminum alloy A356 composite manufactured by squeeze casting. *Journal of Alloys and Compounds*, 419, 98–102.

[9] Sajjadi, S. A., Ezatpour, H. R., & Parizi, M. T. (2012). Comparison of microstructure and mechanical properties of A356 aluminum alloy/Al2O3 composites fabricated by stir and compo-casting processes. *Materials and Design*, 34, 106–11.

[10] Miskovic, Z., Bobic, I., Tripkovic, S., & Rac, A. (2006). The structure and mechanical properties of an aluminium A356 alloy base composite with Al2O3 particle additions. *Tribology in Industry*, 28, 23–27.

[11] Yang, C., Liu, Z., Zheng, Q., & Cao, Y. (2018). Ultrasound assisted in situ casting technique for synthesizing small-sized blocky Al3Ti particles reinforced A356 matrix composites with improved mechanical properties *Journal of Alloys and Compounds*, 747, 580–590.

[12] Bang, H.-S.; Kwon, H.-I.; Chung, S.-B.; Kim, D.-U.; Kim, M.-S. Experimental Investigation and Numerical Simulation of the Fluidity of A356 Aluminum Alloy. Metals 2022, 12, 1986. https://doi.org/10.3390/met12111986.

[13] Sun, C., Cao, Z., Jin, Y., Cui, H., Wang, C., Qiu, F., et al. (2024). Numerical simulation of lost-foam casting for key components of A356 aluminum alloy in new energy. *Materials*, 17(10), 2363.

[14] Guo-chao Gu, Li-xin Xiang, Rui-fen Li, Hong-liang Zheng, Yu-peng Lu, Raphaël Pesci, (2023). Microstructure, segregation and mechanical properties of A356 alloy components fabricated by rheo-HPDC combined with the swirled enthalpy equilibration device (SEED) process, Journal of Materials Research and Technology, 26, 7803–7815, ISSN 2238-7854, https://doi.org/10.1016/j.jmrt.2023.09.153.

[15] Dong, G., Li, S., Ma, S., Zhang, D., Bi, J., Wang, J., et al. (2023). Process optimization of A356 aluminum alloy wheel hub fabricated by low-pressure die casting with simulation and experimental coupling methods. *Journal of Materials Research and Technology*, 24, 3118–3132. http://dx.doi.org/10.2139/ssrn.4365913.

31 Numerical modeling of thermal phenomena and optimization of process parameters in friction stir welding of ZE42 magnesium plates

Sayon Dey[1,2,a] and Rayapati Subbarao[3,b]

[1]Department of Mechanical Engineering, Kolkata, Brainware University, West Bengal, India

[2]Department of Mechanical Engineering, Nadia, Maulana Abul Kalam Azad University of Technology, West Bengal, India

[3]Department of Mechanical Engineering, Kolkata, NITTTR, West Bengal, India

Abstract

Friction stir welding (FSW) can join materials that cannot be joined using traditional fusion welding methods. Detecting flaws such as flash, insufficient heat input, inadequate material flow and mixing, etc. is made easier by studying heat transfer in FSW. This work used modeling using finite elements for the magnesium alloy ZE42 to estimate the transient temperature distribution during FSW. The model assessed the thermal phenomenon and forecasted the peak temperature. A three-level factorial design was used to study the various process parameters like tool rotational speed, tool travel speed, tool shoulder diameter, and tool pin diameter affected the heat generation over the plates. This study aims to numerically model the thermal phenomena and optimize the process parameters of ZE42 magnesium plates in Friction Stir Welding (FSW) using Response Surface Methodology to regulate brittle intermetallic compounds (IMCs) the formation. The published experimental data were utilized to validate the model results. Peak temperature was discovered to be inversely related to tool travel speed and tool shoulder diameter. The plate's maximum temperature generation was directly proportional to pin diameter and tool rotation speed.

Keywords: Friction stir welding (FSW), peak temperature, response surface methodology (RSM), ZE42

Introduction

The automobile industry's need for weight reduction necessitates the precise selection and connecting of lightweight materials. Magnesium is the dominant option for various automobile applications. The traditional fusion welding procedure fails to create good quality welds in magnesium alloys because of the creation of brittle intermetallic compounds (IMCs) formation along the weld line. Being a solid-state welding method, FSW joins materials using heat produced by the tool's simultaneous rotating and sliding motion under axial pressure. Friction and plastic flow provide heat, producing defect-free, high-quality welding in ferrous and non-ferrous materials [1]. FSW is applicable for joining the majority of magnesium alloys that are challenging to fuse using traditional fusion methods. The possible applications of FSW include the joining of non-ferrous materials, including aluminum, copper, lead, titanium, zinc, and other alloys [5]. In the FSW technique, a uniquely designed rotating tool is developed to move over the adjacent surfaces of the joint. Frictional heat is produced by the tool and workpiece's relative motion. The translation of the

spinning tool along the joint line (i.e. forward) under axial stress generates plasticized material flow resulting in a solid phase joint behind the tool [7].

This research analyses numerical modeling of the thermal phenomenon on ZE42 magnesium alloy plates for friction stir welding. The optimization of the process parameters effect is studied using RSM. During the study, the heat transfer analysis is omitted along the edge borders of the plate [2]. A two-factor, three-level design was used to examine how welding parameters like tool rotational speed, tool traverse speed, tool shoulder diameter, and tool pin diameter affected temperature distribution.

Materials and Methodology

ZE42 magnesium alloy plates have been taken on both sides for FSW. The model dimensions were 400 mm × 200 mm × 6 mm [7]. As a tool material, the H13 steel has been selected and as per DoE suitable tool geometries are used for modeling. The temperature dependency on the yield strength of ZE42 alloy is shown in Figure 31.1.

[a]dey.sayon@gmail.com, [b]rsubbarao@nitttrkol.ac.in

DOI: 10.1201/9781003663348-31

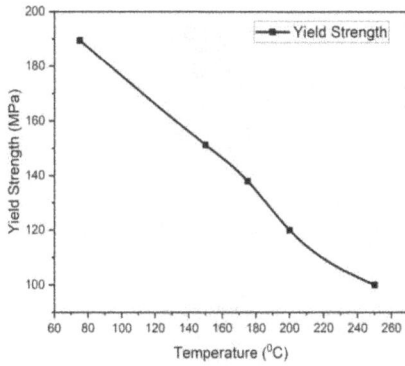

Figure 31.1 Changes in yield strength with temperature
Source: Author

Finite Element Modeling

COMSOL Multiphysics 6.1 software was used to create the FSW model used in this study. The model's dimensions were 400 mm × 200 mm × 6 mm, and its x-axis was surrounded by two infinite domains. Nonuniform boundary conditions were described as the temperature at the top side of the workpiece changes based on the modification in the process parameters and the tool design [4].

Response Surface Method

Response variable models where several independent features influence a dependent response are modeled and improved using the response surface method (RSM). It blends mathematical and statistical instruments. A central composite rotatable second-order design (CCD) matrix is used to optimize the process parameters [3]. The FSW experiment design is shown in Table 31.1. The Actual and predicted values of

maximum and minimum temperature distribution over the surface of the similar ZE42 magnesium plates are shown in Figure 31.3. Based on Tool rotational speed (TRS), the welding speed's (WS) effect on heat generation is shown in Figure 31.5. The entire study shows that the actual value is near the predicted value. The standard error in the proposed model design based on varying tool rotational speeds (TRS) and welding speeds (WS) is shown in the Figure 31.5(a). ANOVA was used to determine the relative impact of each factor and factor combination on the responses [6, 7].

Figure 31.2 illustrates the residuals based on the run number, which has been designed through the Design of Experiment software. The detailed table is shown in Table 31.1. Figure 31.4 shows responses based on different TRS.

Results and Discussion

Process parameters like tool rotational speed, welding or tool traverse speed, tool shoulder and tool pin diameter have been studied. In response, the temperature

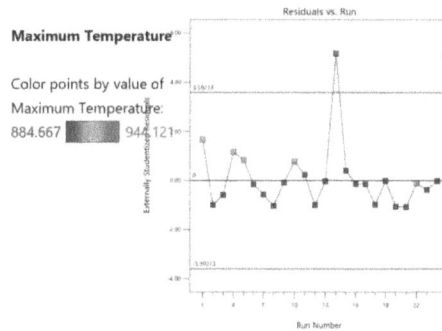

Figure 31.2 Residuals vs run number
Source: Author

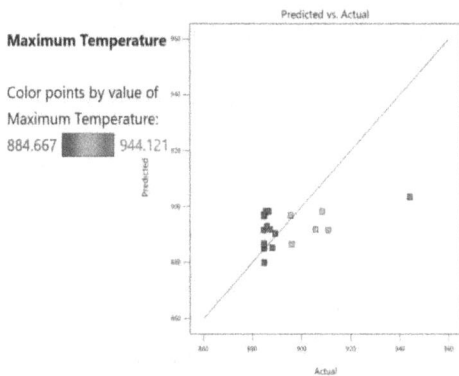

(a)

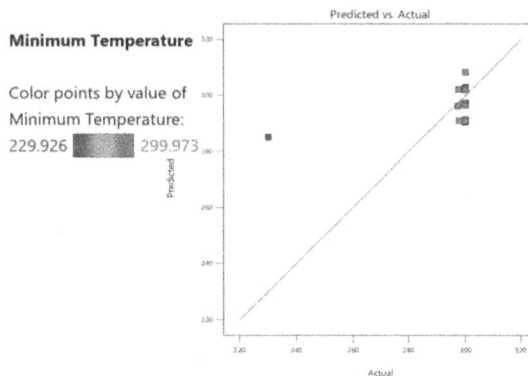

(b)

Figure 31.3 Actual vs predicted temperature: (a) For maximum temperature (b) For minimum temperature
Source: Author

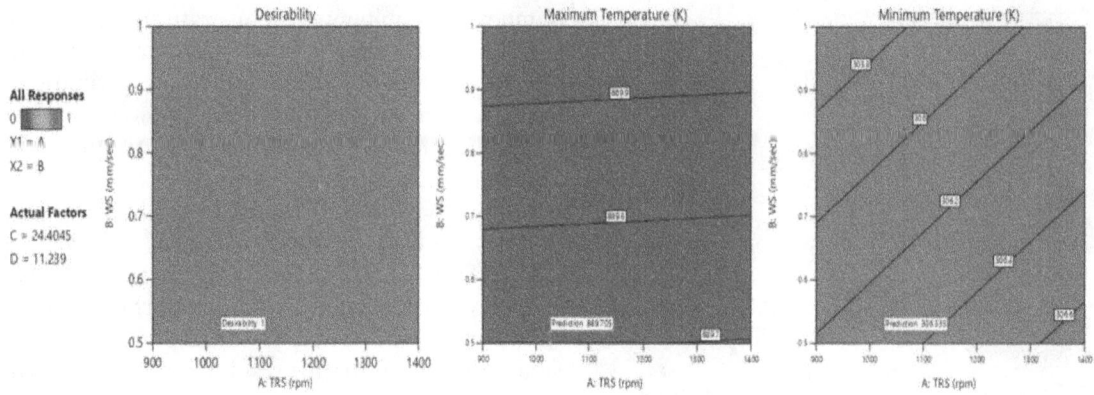

Figure 31.4 Responses based on TRS
Source: Author

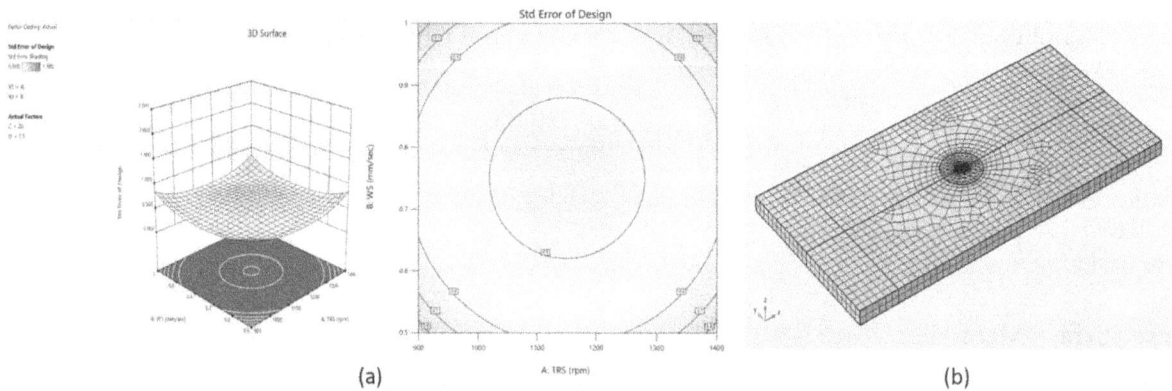

Figure 31.5 (a) For Std error of design (b) For mesh view of the proposed model
Source: Author

distribution on the plate surface is obtained through the simulation. Table 31.1 shows the obtained temperatures based on different parameters. It has been observed that there is a significant impact of the welding parameters to compute the maximum and minimum temperature. Figure 31.6 shows the effect of maximum or peak temperature on different process parameters used in this study. From Figure 31.6, it can be observed that there is a significant effect of the process parameters in the heat generation over the ZE42 magnesium plates.

The extremely fine mesh of the model has been developed through the numerical analysis on COMSOL Multiphysics software, shown in Figure 31.5(b). The highest temperature ranges from 884.68K (611.53°C) to 944.121 K (670.971°C) using 1150 rpm TRS and other parameters like Welding speed (1 mm/s, 0.75 mm/s), shoulder diameter (15 mm, 20 mm) and pin diameter (6 mm,).

Conclusions

To consider the maximum and lowest temperatures detected by the heat production in the FSW of ZE42 magnesium alloy plates, a thermal analysis along with optimization of various process parameters was carried out in this study. A central composite rotatable design technique was the experimental plan. For determining the relative impact, ANOVA was utilized in each factor and factor combination on the responses. Based on an analysis it was found that there was a significant impact of process parameters in heat generation on the plates. Maximum temperature of 944.121K (670.971°C) and minimum temperature of the plates of approximately 24°C have been obtained at 1150 and 900 rpm respectively. Other parameters like welding speed and tool geometry have also an impact on heat generation which regulates the formation of IMCs for the FSW of ZE42 magnesium

Table 31.1 Computed temperatures during FSW of ZE42 plates

Exp. no.	Tool rotational speed (rpm)	Welding speed (mm/s)	Shoulder diameter (mm)	Pin diameter (mm)	Maximum temperature (°C)
1	1400	0.5	20	9	637.735
2	1150	0.5	20	12	611.734
3	1150	0.75	15	6	612.798
4	1400	1	20	9	632.595
5	900	0.75	15	9	635.257
6	**1150**	**1**	**20**	**6**	**611.53**
7	900	0.5	20	9	611.541
8	1150	1	20	12	611.517
9	1150	0.75	25	12	616.19
10	900	0.75	20	6	633.853
11	1150	1	25	9	614.973
12	1400	0.75	20	12	611.742
13	900	0.75	25	9	611.657
14	**1150**	**0.75**	**15**	**12**	**670.971**
15	1150	0.75	25	6	611.741
16	1150	0.5	20	6	611.561
17	1400	0.75	20	6	611.597
18	1150	1	15	9	613.446
19	1150	0.5	25	9	611.735
20	1150	0.5	15	9	612.313
21	1400	0.75	15	9	612.284
22	900	0.75	20	12	622.435
23	900	1	20	9	614.1
24	1400	0.75	25	9	611.821

Source: The table is generated using the Design of Experiments (DoE). Tool rotational speed, welding speed, shoulder diameter, and pin diameter are considered as input factors. The maximum temperatures are recorded as output responses through simulations in COMSOL Multiphysics software.

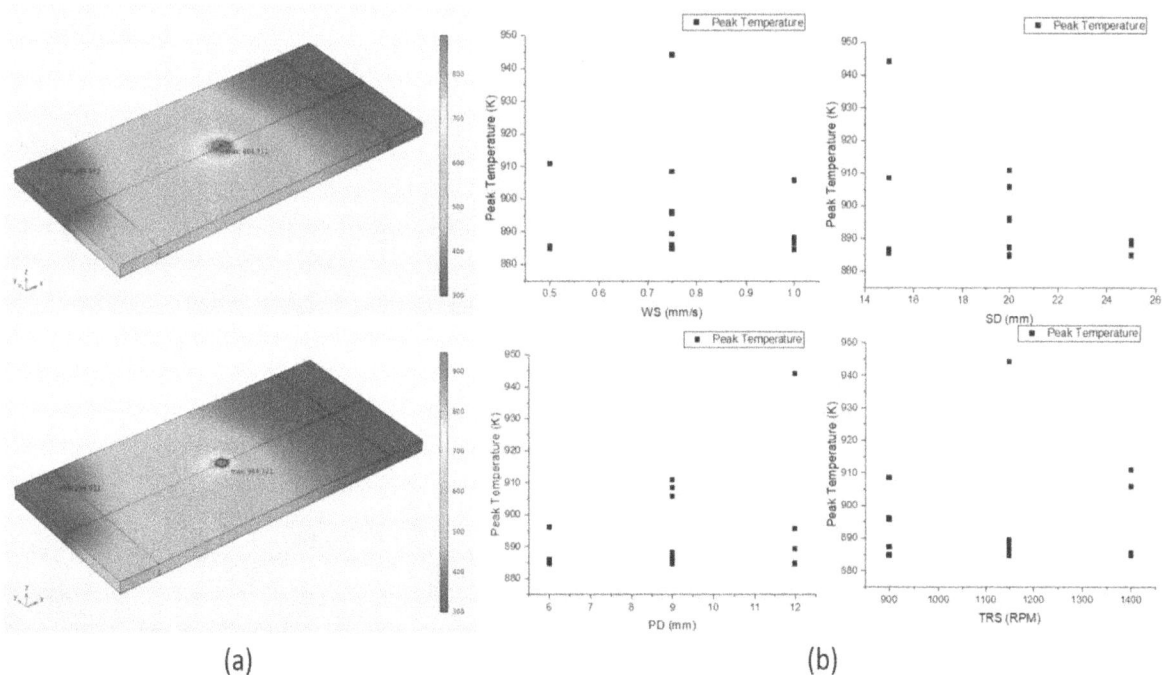

(a)   (b)

Figure 31.6 (a) Thermal analysis on plates; (b) the effect of maximum or peak temperature on different process
Source: Author

alloy. Peak temperature was discovered to be inversely related to welding speed and shoulder diameter. The tool pin diameter and rotation speed were directly correlated with the peak temperature. This study provides detailed modeling of thermal phenomena addressing a gap in the literature and supports the use of FSW in the automobile industry, optimizing processes for lightweight magnesium alloys.

References

[1] Akbari, M., Asadi, P., & Sadowski, T. (2023). A review on friction stir welding/processing: numerical modeling. *Materials*, 16(17), 5890. Multidisciplinary Digital Publishing Institute (MDPI). https://doi.org/10.3390/ma16175890.

[2] Akbari, M., & Rahimi Asiabaraki, H. (2023). Modeling and optimization of tool parameters in friction stir lap joining of aluminum using RSM and NSGA II. *Welding International*, 37(1), 21–33. https://doi.org/10.1080/09507116.2022.2164530.

[3] Elatharasan, G., & Kumar, V. S. S. (2012). Modelling and optimization of friction stir welding parameters for dissimilar aluminium alloys using RSM. *Procedia Engineering*, 38, 3477–3481. https://doi.org/10.1016/j.proeng.2012.06.401.

[4] Lemi, M. T., Gutema, E. M., & Gopal, M. (2022). Modeling and simulation of friction stir welding process for AA6061-T6 aluminum alloy using finite element method. *Engineering Solid Mechanics*, 10(2), 139–152. https://doi.org/10.5267/j.esm.2022.2.001.

[5] Nishant, Jha, S. K., & Prakash, P. (2023). Numerical analyses of underwater friction stir welding using computational fluid dynamics for dissimilar aluminum alloys. *Journal of Materials Engineering and Performance*, 33, 12620–12637. https://doi.org/10.1007/s11665-023-08824-2.

[6] Rajakumar, S., Razalrose, A., & Balasubramanian, V. (2013). Friction stir welding of AZ61A magnesium alloy: A parametric study. *International Journal of Advanced Manufacturing Technology*, 68(1–4), 277–292. https://doi.org/10.1007/s00170-013-4728-0.

[7] Vignesh, R. V., Padmanaban, R., Arivarasu, M., Thirumalini, S., Gokulachandran, J., & Ram, M. S. S. S. (2016). Numerical modelling of thermal phenomenon in friction stir welding of aluminum plates. In IOP Conference Series: Materials Science and Engineering, (Vol. 149, no. 1, p. 012208). https://doi.org/10.1088/1757-899X/149/1/012208.

32 Heat transfer and failure analysis in a Ti–6Al–4V gas turbine compressor stator blade

Nityanando Mahato[1, 2,a] and Rayapati Subbarao[3,b]

[1]Department of Mechanical Engineering, Brainware University, India, West Bengal, India

[2]Department of Mechanical Engineering, MAKAUT, West Bengal, India

[3]Department of Mechanical Engineering NITTTR, Kolkata, West Bengal, India

Abstract

The problem of high temperatures during gas turbine operations, which can cause material deterioration in compressors and as well turbine blades, here we address in the study of compressor stator blades. The goal is to examine and contrast the mechanical and thermal properties of titanium alloys Ti-8Al-1Mo-1V and Ti-6Al-4V under these circumstances. Heat transmission and thermal stress in a compressor stator blade were assessed using COMSOL Multiphysics software. COMSOL Multiphysics software is used to numerically investigate heat transmission and related to the thermal stresses of compressor stator blades made of titanium alloys. The effects are compared with Ti-8Al-1Mo-1V alloy. As a result of its smaller thermal gradient, Ti–6Al–4V has a more uniform temperature distribution than Ti-8Al-1Mo-1V alloy. Ti–6Al–4V overall displacement was slightly more than Ti-8Al-1Mo-1V alloy because of its lower thermal expansion. Moreover, the Von Mises stress emphasis is more concerned than that of displacement. Alloy Ti–6Al–4V exhibited a lower value than that of Ti-8Al-1Mo-1V with the same boundary conditions. The results shed light on material performance, which is crucial for enhancing the longevity of the compressor stator blades.

Keywords: Compressor blades, gas turbine, thermal stress, titanium alloys

Introduction

In the growing worldwide population power consumption is rising faster than ever in a number of energy and technological industries within the next 60 years, the average annual growth rate of consumption is predicted to quadruple [4]. According to this prediction, industrial and energy-consumable technologies will be essential to supplying the world's energy needs going forward. Although renewable energy sources have demonstrated significant although there has been progress with different appliances and positive feedback from ongoing research, gas and steam turbines are still needed to meet the final mass-energy requirement [2]. Recently, a number of active and passive strategies for cooling the turbine blade have been considered. Nevertheless, the majority of these procedures involve thermal cooling of the blades, which causes a significant drop in gas turbine efficiency [11]. The first-stage compressor rotor blades of gas turbines that are susceptible to the high-cycle fatigue mechanism (HCF) have been examined in a failure analysis of an Inconel 718 gas turbine rotor blade [5]. Since the invention of the gas and air jet turbine, GTD 111 DS, an alloy based on nickel, has been mainly utilized as the material for the blades [8]. After comparing four distinct Nimonic materials—Nimonic 105, 90, 80, and 263—both numerically and empirically, Subbarao and Mahato came to the conclusion that Nimonic 80A has a moderate result for thermal stress, distribution, and displacement in comparison to other materials [9]. Although a greater compressed gas temperature will boost the gas turbine cycle's efficiency, the stator blade's constant exposure to high temperatures is linked to a number of metallurgical issues. Hou et al. [12] Lee and Joo [13] Nimonic 80A is a superalloy based on nickel, chromium, titanium, and aluminum that finds extensive use in industrial applications. Nimonic 80A has been shown to have superior wear resistance, high hardness and relatively high toughness, sufficient thermal shock characteristics, excellent chemical stability, and good machinability [10]. In the present work gas turbine compressor stator blade materials are examined, and analysis of thermal stress and heat transfer were carried out in the materials of Ti-6Al-4V and Ti-8Al-1Mo-1V alloy.

[a]nityanandomahato@gmail.com, [b]rsubbarao@nitttrkol.ac.in

DOI: 10.1201/9781003663348-32

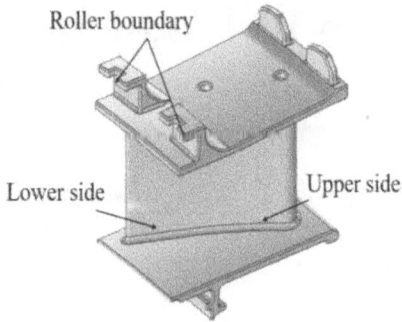

Figure 32.1 Gas turbine compressor stator blade geometry details with roller boundary
Source: Author

Methodology

In Figure 32.1 shows the applied geometry of the gas turbine compressor stator blade, which is an improved version of the design found in Power Turbine [6]. Some common escalating designs and cooling are included in the aforementioned field. At extremely high operating temperatures, metallurgical issues including sulfurization, oxidation, and corrosion are feasible. Therefore, cooling the blades is essential for avoiding these issues. The compressor's output air moves between the internal cools the blades using ducting Figure 32.2 depicts the cooling ducts within of blades that have been employed in this endeavor, but for ease of computation, the specifics of the rib geometry within the ducts are disregarded, and the association between the average Nusselt number and computed to determine the coefficient of heat transmission [1]. The atmosphere, as a cooling fluid, is taken into consideration at 30 bar of pressure and a temperature of 600 K.

Governing equation

Cooling duct

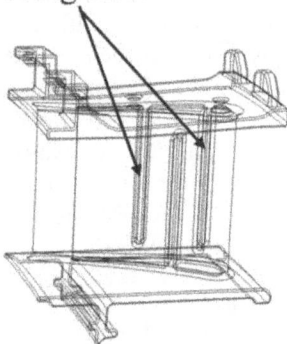

Figure 32.2 Gas turbine compressor stator blade with cooling duct
Source: Author

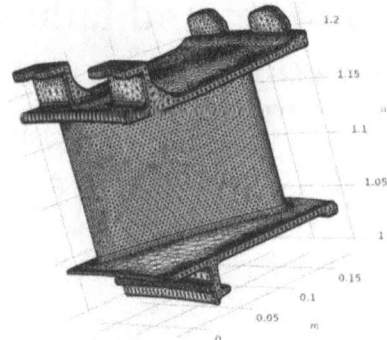

Figure 32.3 Applied unstructured tetrahedral mesh
Source: Author

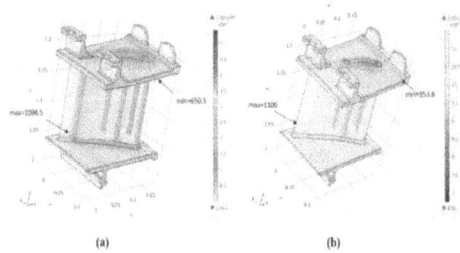

Figure 32.4 (a) and (b) Temperature distribution compressor stator blade made of Ti-4V-6Al and alloy Ti-8Al-1Mo-1V
Source: Author

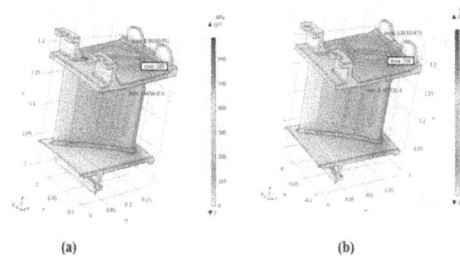

Figure 32.5 (a) and (b) von Misses stress distribution compressor stator blade made of Ti-4V-6Al and alloy Ti-8Al-1Mo-1V
Source: Author

By use of a convection and conduction mechanism, the hot air delivers heat to the blades material. The distribution of temperature throughout the blade produces thermal stress due to the permanent support of the blade. For this purpose, dual sets of governing formulas are required to handle the thermal and mechanical properties of the blades. The governing equations are linear elastic equations and heat diffusion equations. The heat diffusion equation is shown as Eq. (1)

$$\rho \, c \, \partial T / \partial t \; t = \Delta. \, (k \Delta T) \tag{1}$$

In this paper, the physical properties and thermal properties of the materials of the blade are as symbolized.

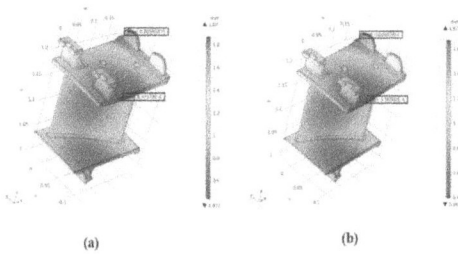

Figure 32.6 (a) and (b) Blade displacement contour of Ti-4V-6Al and alloy Ti-8Al-1Mo-1V
Source: Author

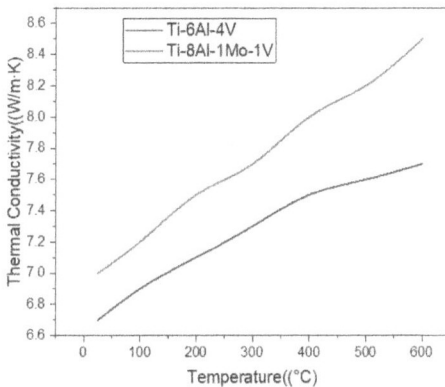

Figure 32.7 Thermal conductivity coefficient of blade Ti-4V-6Al and alloy Ti-8Al-1Mo-1V
Source: Author

Table 32.1 Displacement value of Ti-4V-6Al and Ti-8Al-1Mo-1V

Material	Maximum displacement (mm)	Minimum displacement (mm)
Ti-4V-6Al	0.00187475	4.37E-04
Ti-8Al-1Mo-1V	0.00167451	3.97E-04

Source: Author

Here heat transferred as follow the cooling law

$$q_0 = h\ (T_{ext} - T) \qquad (2)$$

In where h represents the co-efficient of convective heat transfer and q_0 represents the heat flux. Turbine components rupture due to their higher rotational velocity, Ti-4V-6Al is a titanium alloy that exhibits ductility properties and materials have a high strength-to-weight ratio, and high-temperature resistance that is greater than their equivalent tensile values.

Materials properties
The stator blades are not subject to centrifugal force since they are stationary. Therefore, the main causes of stress generated in the blades (stator) are the flow forces and the thermal. With tensile stresses resulting

from the centrifugal force being disregarded. The stator blades experience compressive thermal stresses, which are predominant. Because of this, titanium alloy is an excellent choice due to its high-temperature resistance. In this research, the stator blade material is, which has superior mechanical and thermal properties, which is also quantitatively examined. The regulating formulas are resolved using the COMSOL program. COMSOL Multiphasic requires the mechanical and thermal characteristics of Ti-4V-6Al and Ti-8Al-1Mo-1V alloy to address the governing equations [3, 7]. In contexts, findings a numerical evaluation, the blade and associated mounting composed of the elevated tensile Ti-4V-6A and Ti-8Al-1Mo-1V alloy are examined. There has been no attention to any coating.

Boundary conditions
In Figure 32.3, the mesh geometry is shown. Different sized unstructured tetrahedral elements were created in order to produce mesh-independent results, and mesh independence was attained by 147481 constituents. The convective heat transfer coefficient is used to calculate the heat flux on the blade surfaces. Regarding this, the suction and pressure sides resemble two flat plates. utilizing externally driven convection with the local heat transfer coefficient. The hot gas flows across the blade at 30 bars of pressure and a temperature of about 600K the surface. The velocity of the upper surface (suction side) of the blade, the platform, and the lower surface (pressure side) wall speeds are calculated as follows 300 m/s, 350 m/s. and 450m/s.

Results and Discussion

The analysis of heat transmission and thermal stress were achieved to find out practicability of the titanium alloy which is using gas turbine compressor stator blade. First, using identical geometry and boundary conditions. After that, both the alloys' simulations were carried out Figure 32.4 provides a better understanding of the cooling effect by displaying the temperature distribution in the cooling ducts and blades. It is evident that the cooling duct is where the minimum temperature of 650.5 K. However, the mounting components and the trailing edge are located at a height distance from the cooling that is why the highest temperature of 1098.5 K. It might be summarized, cooling is significantly influenced by internal cooling blade-processing method. The blades' highest and lowest temperatures comprised Figure 32.4 materials are displayed. The variations in heat conduction and retention in the compressor stator blades are depicted in this image. Improved resilience to thermal gradients is suggested by the more uniform distribution of Ti-6Al-4V, which may lessen localized stress concentration spots that fuel fatigue. It can be seen that the minimum temperature is lower in Ti-4V-6Al alloys as compared to alloy Ti-8Al-1Mo-1V. The temperature distribution is not uniform due to internal cooling. Thermal stress is primarily caused by variations in the distribution of temperatures. The von Mises Stress distribution is displayed in Figure 32.5. in stress dispersal Ti-4V-6Al alloy showed less stress as compared to the alloy Ti-8Al-1Mo-1V, the stress responses of the two alloys under operating conditions are depicted in this image. Greater fatigue potential is shown by the higher stress values in Ti-8Al-1Mo-1V, particularly in regions that are farther away from cooling ducts and are subjected to stronger thermal gradients. Perhaps as a result of inadequate cooling or the greater coefficient convection of the high-velocity burning gases and extremely high temperature is reached at the trailing edge that is about equivalent to the temperature of the gases produced during burning. The increased temperature differential could cause blade failure by causing cracks to form and spread close to the internal cooling ducts. Figure 32.6 shows the contours of displacement at several locations on the Ti-4V-6Al-made of compressor blade. This contour map illustrates how thermal expansion has affected the blade's displacement at various locations. Where structural reinforcements would be required to preserve blade integrity under high heat loads can be determined using contour data. The thermal insulation wall experiences a minimum displacement of 4.37E-04 mm, whereas the top wall has a maximum displacement of 0.001874751 mm. Table 32.1 shows the displacement of the blade which is lowered from the leading edge to the trailing edge for both materials. Because the leading edge is near the internal cooling duct, which creates a greater temperature difference, it has more displacement than the trailing edge as a result of this thermal gradient. In Figure 32.7 shows the thermal conductivity of both materials which shows the temperature distribution throughout the compressor stator blade, a lesser thermal conductivity lowers the heat generation alloy Ti-4V-6Al alloy has low heat generation as compared to the Ti-8Al-1Mo-1V.

Conclusions

The current work uses the finite element method to numerically examine heat transmission and thermal stress analysis of a gas turbine compressor stator blade, considering both steady and transient states. The material utilized for the blade body is titanium alloy with a notable high temperature strength and comparatively high heat conductivity. Additionally, the alloy Ti-8Al-1Mo-1V is compared with temperature distribution and associated displacements that are obtained. The results showed displacement is slightly higher in Ti-4V-6Al alloy but thermal stress is more concerned with blade failure, Ti-4V-6Al alloy has less thermal stress than the blade made of Ti-8Al-1Mo-1V alloys.

Future Work and Recommendations

Examine surface coatings for enhanced wear, thermal fatigue, and oxidation resistance, particularly for Ti-6Al-4V, which displayed higher stress concentrations.

References

[1] Bredberg, J. (2002). Turbulence Modelling for Internal Cooling of Gas-Turbine Blades. Göteborg, Sweden: Chalmers University of Technology.

[2] Gholizadeh, T., Vajdi, M., & Rostamzadeh, H. (2020). Exergoeconomic optimization of a new trigeneration system driven by biogas for power, cooling, and freshwater production. *Energy Conversion and Management*, 205, 112417. https://doi.org/10.1016/j.enconman.2019.112417.

[3] Kermanpur, A., Sepehri Amin, H., Ziaei-Rad, S., Nourbakhshnia, N., & Mosaddeghfar, M. (2008). Failure analysis of Ti6Al4V gas turbine compressor blades. *Engineering Failure Analysis*, 15(8), 1052–1064. https://doi.org/10.1016/j.engfailanal.2007.11.018.

[4] Liu, Y., Zhou, Y., & Wu, W. (2015). Assessing the impact of population, income and technology on energy consumption and industrial pollutant emissions in China. *Applied Energy*, 155, 904–917. https://doi.org/10.1016/j.apenergy.2015.06.051.

[5] Maktouf, W., & Saï, K. (2015). An investigation of premature fatigue failures of gas turbine blade. *Engineering Failure Analysis*, 47(PA), 89–101. https://doi.org/10.1016/j.engfailanal.2014.09.015.

[6] Power Turbine (n.d). Glenn Research Center. Available from: www.grc.nasa.gov/WWW/K-12/ airplane/powturb.html.

[7] Rao Muktinutalapati, N. (n.d.). Materials for Gas Turbines-an overview. Available from: www.intechopen.com.

[8] Scheibel, J. R., Aluru, R., & Van Esch, H. (2018). Mechanical properties in GTD-111 alloy in heavy frame gas turbines. In Turbo Expo: Power for Land, Sea, and Air (Vol. 51128, p. V006T24A009). American Society of Mechanical Engineers. http://www.asme.org/about-asme/terms-of-use.

[9] Subbarao, R., & Mahato, N. (2019). Computational analysis on the use of various nimonic alloys as gas turbine blade materials. *International Gas Turbine Institute*, 2. https://doi.org/10.1115/GTINDIA2019-2398.

[10] Xu, Y., Jin, Q., Xiao, X., Cao, X., Jia, G., Zhu, Y., et al. (2011). Strengthening mechanisms of carbon in modified nickel-based superalloy Nimonic 80A. *Materials Science and Engineering: A*, 528(13–14), 4600–4607. https://doi.org/10.1016/j.msea.2011.02.072.

[11] Zecchi, S., Arcangeli, L., Facchini, B., & Coutandin, D. (2004). Features of a cooling system simulation tool used in industrial preliminary design stage. In Turbo Expo: Power for Land, Sea, and Air, (Vol. 41685, pp. 493–501). http://www.asme.or.

[12] Hou, J., Wicks, B. J., & Antoniou, R. A. (2002). An investigation of fatigue failures of turbine blades in a gas turbine engine by mechanical analysis. *Engineering Failure Analysis*, 9(2), 201–211. https://doi.org/10.1016/S1350-6307(01)00005-X.

[13] Lee, S. W., & Joo, J. S. (2017). Heat/mass transfer over the cavity squealer tip equipped with a full coverage winglet in a turbine cascade: part 2 – data on the cavity floor. *International Journal of Heat and Mass Transfer*, 108, 1264–1272. https://doi.org/10.1016/j.ijheatmasstransfer.2016.12.026.

33 Optimizing fluid delivery technique for enhanced grinding of titanium grade 5: a multi-criteria decision-making approach using combined AHP-TOPSIS method

Sourav Adhikary[1,a], Santanu Das[2,b] and Sirsendu Mahata[3,c]

[1]M.Tech in Production Engineering Student, Mechanical Engineering Department, Kalyani Government Engineering College, Kalyani, West Bengal, India

[2]Professor, Mechanical Engineering Department, Kalyani Government Engineering College, Kalyani, West Bengal, India

[3]Associate Professor, Mechanical Engineering Department, Kalyani Government Engineering College, Kalyani, West Bengal, India

Abstract

Countless industries have strong demand for titanium and its alloys for possessing an outstanding strength-to-weight ratio. For fulfilling the demand of high-quality surface finish and accurate parts in industries, grinding is one of the most popular machining processes. Titanium grinding is challenging due to increased temperature, small plastic deformation zone, and high wheel wear rate, often requiring grinding fluids to reduce cutting temperature. The goal of this study was to select the optimal grinding fluid delivery technique by applying hybrid multi-criteria-decision-making (MCDM) tools. Grinding experiments were conducted on Titanium Grade 5 (Ti-6Al-4V) specimens under five environmental conditions: dry, drop-by-drop delivery, micro pump-based flood cooling, LCO_2 (Liquid Carbon dioxide) coolant delivery technique and cooling using micro pump based LCO_2 delivery technique. The AHP and TOPSIS, two popular MCDM techniques, were combined to optimize grindability of titanium alloy based on the five criteria, namely specific energy, surface roughness, volume of material removed, work hardening effect, and grinding ratio. Considering all the fundamental characteristics and ecological constraints, the combined AHP-TOPSIS technique indicates that LCO_2 with a micropump is the best selection out of five.

Keywords: AHP, grinding, MCDM, titanium grade 5, TOPSIS

Introduction

Grinding is a high speed precision machining process in which an abrasive bonded rotating wheel is used as the cutting tool [6]. Many grinding flaws are caused by the high temperature that is created at the grinding zone during the grinding process. Grinding fluid is typically used to compensate for this heat related flaws. When grinding high strength temperature resistant exotic materials, effectiveness of using a grinding fluid varies greatly depending on the type of workpiece material. This type of material includes titanium alloys because of their usual adhesive capabilities and mechanical properties. Titanium alloys, especially Titanium Grade 5, are employed in aeronautics and automobile sectors as well as medical applications for having enough strength, corrosion resistance and biocompatibility. As it presents a challenge to the machinist, determining the ideal conditions for optimal grinding performance becomes imperative [7]. Grinding fluid delivery technique plays a crucial role in grinding titanium grade 5. Different small quantity lubrication (SQL) techniques are used to provide cooling and lubrication during grinding. These techniques help reduce friction, heat generation, and wheel loading, leading to improved surface quality, lower specific energy consumption, and better chip formation. Proper selection and application of coolant helps in enhancing grindability, reducing tool wear, and achieving optimal grinding performance for Titanium Grade 5 alloy [13].

Integration of the AHP and TOPSIS in decision-making improves accuracy and efficiency by creating a decision matrix, normalizing it and calculating the weighted normalized matrix. MCDM methodologies have been utilized by numerous researchers to decide on the strategy of applying cutting fluid in machining. Shah et al. [20] suggested that LB2000 vegetable-based oil is the best-cutting fluid for use in machining titanium alloys. Chatterjee and Chakraborty [5] demonstrated that the choice of weighting method significantly influences the outcome of the optimization process, highlighting the importance of selecting an appropriate weighting

[a]adhikarysourav833@gmail.com, [b]sdas.me@gmail.com, [c]mahatasirsendu@gmail.com

DOI: 10.1201/9781003663348-33

method for accurate decision-making. Abhang and Hameedullah [1] used the combined AHP-TOPSIS technique and discovered that, 10% boric acid in SAE-40 base oil is the best option for a minimum quantity lubricant. Chaudhury et al. [7] demonstrated that 20 μm infeed compound nozzle fluid delivery is the ideal setting for grinding titanium grade 1. In the study of Dwivedi and Sharma [10], PAG cutting fluid was detected much effective compared with synthetic ester and vegetable-based canola oil. Dubey et al. [9] used response surface methodology for the design of experiment towards its optimization. Using nanoparticle concentration of 1.5% and favorable process parameters, optimal responses could be arrived at with a fixed weight factor with each MCDM technique. On the other hand, Van Thanh et al. [23] used entropy method to select weight of criteria and Multi-Attributive Ideal-Real Comparative Analysis method. Sofuoğlu [22] employed suitable cutting conditions which were determined using 4 different normalizing procedures and 15 different MCDM methods.

The Spearman correlation test was used by Sinha et al. [21] to compare the ranks obtained. They showed that Gaussian process regression performed better at properly anticipating the data than other methods. Patil et al. [18] evaluated the impact of using biodegradable coolants and hybrid nanoparticles and adjustment of Computer Aided Manufacturing strategies to achieve efficient 3D finish milling operation. The greatest connection between the MOORA and VIKOR techniques was discovered by Özakin [17]. Three multiple criteria decision-making strategies were employed by Jayant and Neeru [11] to help the decision-making process when choosing a green cutting fluid such as VIKOR, enhanced PROMETHEE, and ELECTRE III. They developed a model to maximize objectives to detect the corresponding cutting fluid. Adopting Taguchi approach, Wang et al. [24] improved cooling/lubrication conditions and grinding settings for obtaining desired grindability maintaining eco-friendliness.

Mahata et al. [14] noted that when grinding with SQL and liquid CO_2 setup, there was a noticeable decrease in grinding forces. Mukhopadhyay and Kundu [16] developed a new technology called limited quantity lubrication (LQL) after an experimental examination of the up-grinding of Ti-6Al-4V using an alumina wheel. Mandal et al. [15] determined air layer pressure at different wheel conditions and grinding speeds. According to Chakraborty et al. [4], both AHP and ANN could be utilized successfully in multicriteria decision-making to choose the best grinding environment and produce the required grinding performance. Robot belt grinding, according to Xiao et al. [25] could resolve the additive titanium alloy blade

processing issue. According to Li et al. [12], greater grinding temperature is associated with desirable surface roughness and high rates of material removal. Using multi-attribute decision-making (MADM) method, Azizi et al. [2] found out specific energy to be useful criterion for assessing process efficiency, while changes in G-ratio and surface roughness showed similar trends.

Researchers utilized MCDM methods to evaluate appropriate process parameters and cutting fluid for the respective titanium alloy, but most of the fluid was not sustainable.

Objective of this investigation has been to explore importance of process variables (specific energy, grinding ratio etc.) in determining the appropriate fluid delivery method for improved grinding of titanium alloy grade 5; to achieve this, a hybrid MCDM technique combining the AHP and TOPSIS is used and this is the novelty of this investigation.

Methodology for Applying MCDM Tools

In this study, two multi-attribute decision making tools namely the analytical hierarchy process (AHP), created by Saaty [19], and the Technique for Order of Preference by Similarity to Ideal Solution (TOPSIS), developed by Hwang and Yoon are employed for optimizing grindability of titanium alloy grade 5. Admissibility of these models is established with a consistency ratio below 10%. TOPSIS tries to derive a solution close to the ideal one, further away from negative ideal one. The following steps are taken for the implementation of TOPSIS and the AHP.

Step 1: The AHP is utilized to transform complex decision problems to a hierarchical structure to identify the goal and relevant evaluation criteria. Goals are arranged at the top of a tree-like hierarchy, followed by criteria, sub-criteria, and alternatives that present the optimum solution [1].

Step 2: A decision matrix based on all available information describing grinding fluid delivery attributes is used in this step. Each row and column relate to an alternative and a criterion respectively. The value of an element d_{ij} in a decision matrix D represents the jth attribute's original, on-normalized value for the ith alternative. An MxN decision matrix can be represented with N attributes and M alternatives [2]. It can be written as, $[D]_{MxN} = [D]_{5x5}$

Step 3: Forming the original decision matrix (D) and using vector normalization to normalize its elements in order to produce the normalized decision matrix (N) [5].

$$D=[d_{ij}]\times_n; \ N=[n_{ij}]\times n, \text{ where, } n_{ij}=\frac{x_{ij}}{\sqrt{\sum_{u=1}^{m} x_{uj}^2}} \qquad (1)$$

Step 4: The AHP prioritizes qualities based on their impact on the objective using pairwise comparisons and a preference scale (Table 1) of Saaty [19] considering reflexive property and intermediate values.

An element is compared with the elements of next higher level to create pairwise comparison matrices. The local priority weights are determined in part by this. Each element of a matrix is for a preference of an alternative over the other related to a criterion. Values of aij are selected from the ratio scale enlisted in Table 1. In the next step, consistency ratio (CR), which may be calculated using the formula, CR = (CI/RI), is then used to verify the consistency of the matrix [7, 19]. Consistency is said to be there if CR becomes 0.1 or less, ensuring that all criteria have a total priority relative to the alternatives, indicating acceptable decision-making. The priority vector created is used to rank each of the five alternatives [8].

Step 5: The weighted decision matrix is created by dividing the normalized performance ratings of each alternative by the weight of the corresponding criterion [5].

Step 6: To identify the PIS (v^+_j) and NIS (v^-_j) with respect to each criterion [5].

Step 7: To use PIS and NIS to calculate the Euclidean distance of each option [5].

$$PIS\ (S_I^+) = \sqrt{\sum_{j=1}^{n}(v_{ij} - v_j^+)^2} \quad \text{and,}$$

$$NIS\ (S_I^-) = \sqrt{\sum_{j=1}^{n}(v_{ij} - v_j^-)^2} \qquad (2)$$

Step 8: The proximity of each option to the PIS is calculated using the negative and positive separation metrics.

$$CC_i = \frac{s_i^-}{s_i^- + s_i^+} \qquad (3)$$

It is simple to observe that $0 \le CC_i \le 1$. The value of Si^+ must tend toward 0 when CC_i approach 1. Consequently, a greater CC_i value denotes a closer alternative to the PIS. As a result, the options are arranged according to decreasing CC_i, with the optimal option having the highest CC_i [5].

Results and Discussion

The study utilized the combined AHP-TOPSIS method to optimize fluid delivery for enhanced titanium alloy grinding, considering experimental data of Mahata [13]. Standard alumina grinding wheel was used to surface grind titanium grade 5 in up-grinding mode. The method ranked cutting fluid delivery techniques based on criteria namely, specific energy (C1), roughness (C2), grinding ratio (C3), material removal volume (C4) and work hardening (C5) with alternative methods like dry (A1), drop-by-drop (A2, A3), and micro pump-based (A4, A5) delivery techniques. These methods and their respective outcomes are discussed as follows:

Calculation of criteria weights using the AHP

Table 2 is pair-wise comparison matrix for criteria that was constructed using the details shown in Table 1. The element values are totally dependent on the decision makers. Table 3 shows the normalized values of Table 2. Normalization is done to conduct an easy evaluation process of the given problem.

Table 33.2 Pair-wise comparison matrix for criteria

	C1	C2	C3	C4	C5
C1	1	1/4	5	6	7
C2	4	1	6	7	9
C3	1/5	1/6	1	2	4
C4	1/6	1/7	1/2	1	2
C5	1/7	1/9	1/4	1/2	1

Source: Author

Table 33.3 Normalized pair-wise comparison matrix

	C1	C2	C3	C4	C5
C1	0.18	0.15	0.39	0.36	0.30
C2	0.73	0.60	0.47	0.42	0.39
C3	0.04	0.10	0.08	0.12	0.17
C4	0.03	0.09	0.04	0.06	0.09
C5	0.03	0.07	0.02	0.03	0.04

Source: Author

Table 33.1 Preference scale of Saaty [19]

Value	Importance Intensity	Interpretation
9	Absolute	Superior
7	Critical	Greater
5	Strong	Finer
3	Moderate	Improved
1	Equivalent	Unconcerned
2,4,6,8	Intermediate	Intermediate
	importance intensity	interpretations

Source: Saaty, T. L. [19]

Table 33.4 Calculated criteria weights following the AHP.

Criteria	Criteria weights
Specific Energy (J/ mm³)	0.28
Roughness (Ra)	0.52
Grinding ratio	0.10
Volume of material Removed (mm³)	0.06
Work hardening (HV)	0.04
Sum	1.00

Source: Author

The AHP uses a normalized matrix to calculate criteria weights with Roughness Average and Specific Energy being the most crucial, while Work Hardening effect is less significant.

Calculated Consistency Ratio is 0.069 which is less than 0.1 meaning the decision acceptable.

Selection of the best alternative using TOPSIS
Hwang and Yoon's TOPSIS method ranks alternatives based on proximity to the ideal solution, balancing beneficial and non-beneficial factors by considering the option closest to the positive solution [3, 5]. Table 5 considers the experimental data of Mahata [13] with the five alternatives and five parameters.

Decision matrix shown in Table 5 is now normalized using Equation {1}. Normalized values of decision matrix are indicated in Table 6.

With the help of pre-calculated criteria weights from the analytic hierarchy process (Table 4), a weighted normalized decision matrix (Table 7) is formulated from normalized squared decision matrix. Also, the V^+ and V^- values are calculated using the help of Equation {2}.

Table 8 shows respective Euclidean distance from positive and negative ideal solutions denoted by Si^+ and Si^- respectively. This helps compute closeness coefficient.

Taking help of Equation {3}, the closeness coefficient is calculated. A high closeness coefficient denotes the best alternative. The analytic hierarchy process and TOPSIS are calculated using Microsoft Excel Software, with closeness coefficient values indicating optimal parameter combinations.

Based on the results obtained (Table 9), out of the five alternative techniques, cooling with liquid CO_2 and a micropump is evaluated to be the most effective fluid delivery method for improved titanium grade 5 grinding. The system can be enhanced by adding additional

Table 33.5 Decision matrix

	C1	C2	C3	C4	C5
A1	73.31	0.86	0.55	122.4	325
A2	42.62	0.77	0.66	129.6	328.33
A3	35.25	0.53	0.82	136.8	348.66
A4	47.36	0.74	0.73	129.6	362.66
A5	44.19	0.41	1.04	144	384

Source: Author

Table 33.6 Normalized decision matrix

	C1	C2	C3	C4	C5
A1	0.652	0.564	0.316	0.413	0.415
A2	0.379	0.505	0.379	0.437	0.419
A3	0.314	0.347	0.471	0.461	0.445
A4	0.421	0.485	0.420	0.437	0.463
A5	0.393	0.269	0.598	0.485	0.490

Source: Author

Table 33.7 Weighted normalized decision matrix

	C1	C2	C3	C4	C5
A1	0.183	0.293	0.032	0.025	0.017
A2	0.106	0.262	0.038	0.026	0.017
A3	0.088	0.181	0.047	0.028	0.018
A4	0.118	0.252	0.042	0.026	0.019
A5	0.110	0.140	0.060	0.029	0.020
V+	0.088	0.140	0.060	0.029	0.020
V-	0.183	0.293	0.032	0.025	0.017

Source: Author

Table 33.8 Calculated Euclidean distance

Alternatives	Si^+	Si^-	Si^++Si^-	CC_i
A1	0.183	0.000	0.183	0.000
A2	0.126	0.083	0.209	0.396
A3	0.043	0.148	0.191	0.775
A4	0.118	0.077	0.195	0.396
A5	0.022	0.172	0.194	0.885

Source: Author

Table 33.9 Ranking on the basis of closeness coefficient

Alternatives	CC_i	Rank
A1	0.000	5
A2	0.396	3
A3	0.775	2
A4	0.396	4
A5	0.885	1

Source: Author

parameters, transforming it into a versatile decision-making tool applicable to various MCDM problems.

Conclusion

This research is vital for industry as it tries to address key challenges related to grindability, economy and sustainability. This way, it helps manufacturers improve their operations and adapt to the evolving demands of advanced materials and technologies.

The AHP-TOPSIS method enhances titanium grade 5 grinding by optimizing machining parameters and solving the MCDM problem using five criteria and five alternatives. The optimized results are: specific energy of 44.19 J/mm³, grinding ratio of 1.04, surface roughness average of 0.41, volume of material removal of 144 mm³ and work-piece surface hardness of 384 HV. Cooling with liquid CO_2 with a micropump is found to be the most effective fluid delivery method for improved grinding of titanium grade 5 alloys.

Advanced MCDM technologies may be complicated due to training, material properties, high costs, data dependence, and regulatory restrictions, potentially deterring smaller producers.

Future research works may focus on applying advanced MCDM techniques, machine learning, taking up sustainability study, etc. for grinding advanced materials effectively through the selection of proper grinding wheel, green grinding fluid and its delivery method, etc.

References

[1] Abhang, L. B., & Hameedullah, M. (2012). Selection of lubricant using combined multiple attribute decision-making method. *Advances in Production Engineering & Management*, 7(1), 39–50.

[2] Azizi, A., Seidi, M., Bahrami, P., & Rabiei, F. (2023). Nickel-based super alloy grinding optimisation using a hybrid multi attribute decision making method based on entropy, AHP and TOPSIS. *Advances in Materials and Processing Technologies*, 11(1), 72.

[3] Banerjee, B., Mondal, K., Adhikary, S., Paul, S. N., Pramanik, S., & Chatterjee, S. (2022). Optimization of process parameters in ultrasonic machining using integrated AHP-TOPSIS method. *Materials Today: Proceedings*, 62, 2857–2864.

[4] Chakraborty, P., Kundu, K., & Mahata, S. (2023). Optimization of grinding environment for surface grinding of low alloy steel using analytic hierarchy process and artificial neural network. *IUP Journal of Mechanical Engineering*, 16(1), 14.

[5] Chatterjee, S., & Chakraborty, S. (2024). A study on the effects of objective weighting methods on TOPSIS-based parametric optimization of non-traditional machining processes. *Decision Analytics Journal*, 11, 100451.

[6] Chattopadhyay, A. B. (2011). Machining and Machine Tools. John Wiley and Sons.

[7] Chaudhury, A., Mandal, B., & Das, S. (2015). Selection of appropriate fluid delivery technique for grinding titanium grade-1 using the analytic hierarchy process. *International Journal of the Analytic Hierarchy Process*, 7(3), 454–469.

[8] Dhanalakshmi, S., & Rameshbabu, T. (2021). An integrated approach for cutting fluid selection using multiple attribute decision making methods. *Journal of Mechanical Engineering and Sciences*, 15(1), 7860–7873.

[9] Dubey, V., Sharma, A. K., Vats, P., Pimenov, D. Y., Giasin, K., & Chuchala, D. (2021). Study of a multicriterion decision-making approach to the MQL turning of AISI 304 steel using hybrid nanocutting fluid. *Materials*, 14(23), 7207.

[10] Dwivedi, P. P., & Sharma, D. K. (2024). A study for an optimization of cutting fluids in machining operations by TOPSIS and shannon entropy methods. *WSEAS Transactions on Fluid Mechanics*, 19, 83–98.

[11] Jayant, A., & Neeru, S. P. (2018). A decision-making framework model of cutting fluid selection for green manufacturing: a synthesis of 3 MCDM approaches. In International Conference on Advanced Engineering Technologies (pp. 1–7).

[12] Li, M., Wang, W., Huang, Y., Yan, S., Zhang, P., & Zou, L. (2024). 3D printed compliance tool incorporated internal-impeller structure for high performance face grinding of titanium alloy. *Journal of Materials Processing Technology*, 329, 118446.

[13] Mahata, S. (2021). Investigation on grindability of different workwheel combinations to evaluate the optimal condition of grinding fluid applications. PhD Dissertation submitted to MAKAUT, WB.

[14] Mahata, S., Mukhopadhyay, M., Kundu, A., Banerjee, A., Mandal, B., & Das, S. (2020). Grinding titanium alloys applying small quantity lubrication. *SN Applied Sciences*, 2(5), 978.

[15] Mandal, B., Majumdar, S., Das, S., & Banerjee, S. (2011). Formation of a significantly less stiff air-layer around a grinding wheel pasted with rexine leather. *International Journal of Precision Technology*, 2(1), 12–20.

[16] Mukhopadhyay, M., & Kundu, P. K. (2018). Development of a simple and efficient delivery technique for grinding Ti-6Al-4V. *International Journal of Machining and Machinability of Materials*, 20(4), 345–357.

[17] Özakin, B. (2023). A comparative study of the selection of cutting fluids used in machining processes by multi criteria decision making (MCDM) methods. *Sādhanā*, 48(4), 204.

[18] Patil, A. S., Sunnapwar, V. K., Bhole, K. S., Oza, A. D., Shinde, S. M., & Ramesh, R. (2022). Effective machining parameter selection through fuzzy AHP-TOPSIS for 3D finish milling of Ti6Al4V. *International Journal on Interactive Design and Manufacturing*, 16, 1–25.

[19] Saaty, T. L. (1980). The Analytic Hierarchy Process. New York: McGraw-Hill.

[20] Shah, M., Modi, Y., Bandhu, D., & Abhishek, K. (2024). Selection of cutting fluids for machining titanium alloys using MCDM methods. In Decision-Making Models and Applications in Manufacturing Environments. Apple Academic Press, (pp. 147–166).

[21] Sinha, M. K., Kishore, K., Archana, & Kumar, R. (2024). Hybrid approach for modelling and optimizing MQL grinding of Inconel 625 with machine learning and MCDM techniques. *International Journal on Interactive Design and Manufacturing*, 18(7), 4697–4713.

[22] Sofuoğlu, M. A. (2021). A new hybrid decision-making strategy of cutting fluid selection for manufacturing environment. *Sādhanā*, 46(2), 94.

[23] Van Thanh, D., Binh, V. D., Duong, V., Tu, N. T., & Van Trang, N. (2024). Using the MAIRCA method for determining the best dressing factors in surface grinding Hardox 500. *Methodology*, 45(2), 50–55.

[24] Wang, Z., Zhang, T., Yu, T., & Zhao, J. (2020). Assessment and optimization of grinding process on AISI 1045 steel in terms of green manufacturing using orthogonal experimental design and grey relational analysis. *Journal of Cleaner Production*, 253, 119896.

[25] Xiao, G., Zhang, T., He, Y., Zheng, Z., & Wang, J. (2024). The robot grinding and polishing of additive aviation titanium alloy blades: a review. *Journal of Intelligent Manufacturing and Special Equipment*, 5(1), 34–54.

34 Study on hole-making in soda-lime glass under varying abrasive jets with SiC abrasive at 4 kg/cm² system pressure

Deb Kumar Adak[1,a], Bidhan Chandra Gayen[2,b], Santanu Das[3,c] and Barun Haldar[4,d]

[1]Assistant Professor, College of Engineering and Management, Kolaghat, Purba Medinipur, West Bengal, India

[2]M.Tech, Kalyani Government Engineering College, Kalyani, Nadia, West Bengal, India

[3]Professor, Kalyani Government Engineering College, Kalyani, Nadia, West Bengal, India

[4]Assistant Professor, College of Engineering, Imam Mohammad Ibn Saud Islamic University, Riyadh, Saudi Arabia

Abstract

Abrasive jet machining finds its wide applications in shaping hard, brittle materials as well as surfacing of different materials to promote spray coating, integrity testing of turbine and propeller blades against erosion, surface hardening by grit blasting etc. In this experimental study following RSM, abrasive jet of silicon carbide grains is used to make hole in soda-lime glass. Effects of size of abrasive grains, stand-off distance (SOD) and discharge of abrasives on material removed per unit time and nozzle wear per unit time are observed at a constant system pressure at 4 kg/cm². ANOVA is done to find out relative significance of each of the parameters, etc. More or less, it is observed that MRR and NWR initially have an increasing trend with a hike in input process variables like grain size, discharge and SOD, attains a maximum value with a certain range of these parameters and then MRR and NWR both decrease with further increase of same input parameters.

Keywords: Abrasive jet drilling, abrasive jet machining, ANOVA, glass, RSM

Introduction

Abrasive jet machining can be a good choice to machine difficult-to-machine hard and brittle materials like soda-lime glass. Finnie [4] reported abrasive jet machining (AJM) to have practiced in Germany as early as in 1931 using dust suspended in smoke. Abrasive particles suspended in a high-velocity stream of air or gas can be used for various applications such as machining, shot peening, cleaning, deburring, micro-machining, and polishing through erosion. With high kinetic energy, abrasive grains strike a surface to remove material through micro-fracturing [9]. In AJM, a localized force is applied, resulting in less heat generation compared to traditional machining processes [3]. Absence of thermal distortion is usually experienced in AJM in contrast to PAM and LBM. On the other hand, Gradeen et al. [6], Srikanth and Rao [15] and Pradhan et al. [12] provided hot abrasive air jet to machine variety of ceramic materials, polymeric materials as well as metallic materials. Alumina and SiC abrasives were widely employed in AJM [2]. Abrasive jet with silicon carbide is used to carry out surface preparation for coating etc., on ductile material also [1]. A literature review on AJM Pawar et al.

[10] highlighted the need for further research on the effect of flow rate on performance, as limited studies have been conducted in this area. Influence of abrasive flow rates with different grain sizes while drilling two types of glasses were studied earlier by Karmakar et al., (2018). Advancements in conventional micro abrasive jet machining showed varying stand-off distance to lead to tapering of holes. Maintaining a constant distance from the start of the process helps reduce taper angles in micro holes, resulting in improved cylindricity [7]. Adopting appropriate process variables is always needed. Pradhan et al. [13] studied the effect of different temperatures on surface generation on K-80 alumina using Al_2O_3 abrasive. Pradhan and Dhupal [11] reported ease of machining K-80 alumina material by erosion using hot abrasives. Prasad et al. [14] presented a multi-objective optimization technique, based on weighted aggregated sum product assessment (WASPAS) method for optimal selection of AJM process parameters to achieve the best AJM performance. An increase in hole diameter was observed with a rise in stand-off distance at a particular work pressure [5]. They reported SiC to have more cutting ability than easily available silica sand. Hassan and

[a]debkumaradak1968@gmail.com, [b]bidhanchandragain@gmail.com, [c]sdas.me@gmail.com, [d]dr.barun.haldar@gmail.com

DOI: 10.1201/9781003663348-34

El-Hofy [8] formulated material removal rate when sharp Al_2O_3 or SiC abrasives strike a fragile surface at a high speed to dislodge tiny particles by brittle fracture.

Schematic representation of abrasive jet drilling along with different terminologies is depicted in Figure 34.1.

The current state-of-the-art literature lacks a low-cost procedure for determining abrasive discharge and its effects on AJM performance.

Development of a cost-effective AJM setup with effective positive abrasive mixing chamber enhancing the efficiency and performance of AJM system is the novelty of the present work. Objectives of this experimental work are to investigate direct and interaction effects of three process variables, i.e., size of abrasive, stand-off distance and discharge of abrasives, on AJM output responses. Box-Behnken method of RSM is followed in the current work. Possibility of making hole in soda-lime glass, a difficult-to-machine material, with AJM using SiC abrasive is explored. Adequacy of experimental model is checked with ANOVA. Additionally, NWR are analyzed to predict nozzle working life. Three levels of SOD, abrasive discharge and size of grain are considered to optimize MRR and

NWR in AJM. It is tried to find out the share of contribution of individual or interactive parameters on MRR and NWR at 4 kg/cm² constant system pressure.

Experimental Details

AJM set up is used to carry out experiments on soda lime glass of 2.37 mm thickness. SiC abrasive with better cutting ability than SiO_2, is used as abrasive particle. Three parameters with its three levels are considered for framing design of experiment with the Box-Behnken strategies in RSM (Table 34.1). Three levels of SOD are 2 mm, 3 mm and 4 mm. Three levels of abrasive flow rate are taken as 250 g/min, 275 g/min and 300 g/min. Similarly, three levels of abrasive grain sizes are 125 µm, 275 µm and 425 µm. Table 34.2 shows MRR and NWR noticed in all of the 15 experimental runs are carried out keeping system pressure constant at 4 kg/cm².

Results and Discussion

Analysis of material removal rate
Fifteen numbers of experiments carried out using the AJM set up and corresponding responses of MRR and

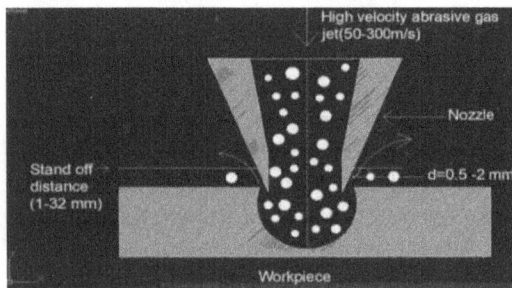

Figure 34.1 Terminology of abrasive jet machining
Source: Author

Table 34.1 Detail of parameters set for experiment set III

Parameter	Low value	Mid-value	High value
Particle size (d) in µm	125	275	425
Discharge (ṁ) in g/min	250	275	300
SOD (δ) in mm	2	3	4

Source: Author

Table 34.2 Design of experiments with MRR and NWR at 4 kg/cm² pressure

Std order	Run order	Particle Size (µm)	SOD (mm)	Discharge (g/min)	MRR (g/min)	NWR (g/min)
1	14	125	2	275	0.29985	0.00388
2	8	425	2	275	0.32532	0.00395
3	3	125	4	275	0.42385	0.00664
4	6	425	4	275	0.31429	0.00398
5	7	125	3	250	0.31739	0.00401
6	2	425	3	250	0.32198	0.00407
7	11	125	3	300	0.31747	0.00428
8	10	425	3	300	0.29882	0.00367
9	15	275	2	250	0.26608	0.00291
10	1	275	4	250	0.32474	0.00419
11	9	275	2	300	0.29934	0.00381
12	4	275	4	300	0.30505	0.00385
13	12	275	3	275	0.49977	0.00861
14	5	275	3	275	0.43946	0.00727
15	13	275	3	275	0.50449	0.00922

Source: Author

NWR are indicated in Table 34.2. Experimented specimen with holes made is depicted in Figure 34.2.

Analysis of variances (ANOVA) is performed after analyzing the responses and tabulated in Table 34.3. For MRR, R-Sq value of 95.48% is obtained from ANOVA. Individual, square or interactive parametric relationships are obtained in regression equation [1].

$$MRR = -17.89 + 0.00345d + 0.3296\delta + 0.0962\dot{m} - 0.000003d^2 - 0.004860\delta^2 - 0.000167\dot{m}^2 - 0.000056d\delta - 0.000002d\dot{m} - 0.000132\delta\dot{m}$$ [1]

A significant parametric contribution towards process responses is judged by their corresponding p-values. Therefore, in Table 34.3, model, linearity, grain size, SOD, square, square term of grain size, square term of SOD, square term of abrasive flow rate, two-way interactive term of grain size and SOD have p value less than 5%. Thus, these parameters are well significant for MRR.

Two-dimensional contour plots and three-dimensional surface plots from Figure 34.3 through Figure 34.8 are analyzed. Surface plots can figure out approximate process parameters where maximum responses occur. In Figure 34.3(a, b), variation of MRR is depicted with respect to size of abrasives and SOD,

Figure 34.2 Soda lime glass samples with drilled holes
Source: Author

Table 34.3 ANOVA table of MRR at 4 kg/cm² pressure

Source	DF	Adj SS	Adj MS	F-value	p-value
Model	9	0.078238	0.008693	11.73	0.007
Linear	3	0.027953	0.009318	12.57	0.009
Grain size, d	1	0.003992	0.003992	5.39	0.068
SOD, δ	1	0.023289	0.023289	31.42	0.002
Discharge, ṁ	1	0.000672	0.000672	0.91	0.385
Square	3	0.067697	0.022566	30.45	0.001
d²	1	0.014492	0.014492	19.55	0.007
δ²	1	0.022327	0.022327	30.13	0.003
ṁ²	1	0.040456	0.040456	54.59	0.001
2-Way Interaction	3	0.005394	0.001798	2.43	0.181
d.δ	1	0.004558	0.004558	6.15	0.056
d.ṁ	1	0.000135	0.000135	0.18	0.687
δ.ṁ	1	0.000701	0.000701	0.95	0.375
Error	5	0.003706	0.000741		
Lack-of-Fit	3	0.001076	0.000359	0.27	0.843
Pure Error	2	0.002629	0.001315		
Total	14	0.081944			

Source: Author

keeping discharge of abrasive constant at 275 g/min. MRR is found to gradually improve with increment of size of grain and reaches the maxima at 138–350μm of grain size. Then it decreases with the further hike in grain size. It may be due to the fact that big size grains give more impact up to a level but further increment in grain size causes less amount of material removal due to dragging effect of large size abrasive particles, at a certain time, which may have reduced MRR as stated by Xiao et al. [16].

MRR slowly improves when SOD increases, reaches the maxima at 2.5–3.8 mm SOD, but after that it again decreases with increasing SOD because initially when SOD increases, abrasive particles are getting enough space to accelerate prior to striking on the job surface. Beyond certain SOD, deceleration of jet takes place and flow becomes diverged, which may cause lower shock on job and hence, low MRR. Surface plot as well as contour plot of Figure 34.4(a,b) indicate change in MRR with size of abrasive and discharge at 3 mm SOD. It clearly shows that MRR is maximum in between 150–350 μm grain size, and MRR decreases at a lower or higher grain size. Initially MRR increases with larger grain size but when grain size is further increased above a certain value, it causes loss of flow velocity at the outlet of nozzle, so total kinetic energy and MRR are reduced.

Within discharge of 262–285 g/min and above 150 μm of grain size, MRR becomes maximum, and it reduces with hike in abrasive discharge. With lower value of discharge, MRR increases with flow rate as increased number grains removes increased amount of material. However, at a high discharge, MRR reduces with increasing discharge. This may be due to more inter-particle collision because of the presence of large number of abrasive grains. Similar observations were also made by Xiao et al. [16].

Surface and contour plots (Figure 34.5a, b) show the variation of MRR with SOD and flow rate keeping grain size constant at 275μm. MRR is maximum in between 2.6-3.7 mm SOD and decreases when it reduces or increases further. It is because initially MRR increases with increasing SOD but when SOD is

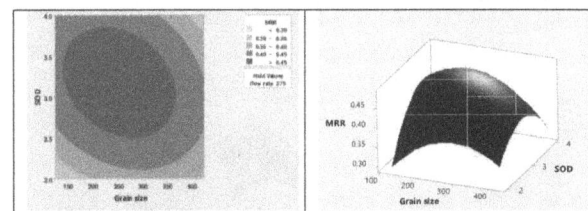

Figure 34.3 (a) Contour, (b) Surface plot of MRR (g/min) relating to grain size (μm) and SOD (mm) at discharge hold value of 275 g/min
Source: Author

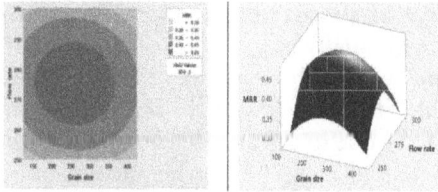

Figure 34.4 (a) Contour, (b) Surface plot of MRR (g/min) relating to size of abrasive (μm) and discharge (g/min) at SOD hold value of 3mm
Source: Author

further increased above a certain value, it may cause loss of flow velocity due to divergence of flow pattern at the outlet of nozzle, so, MRR reduces.

Applying 262-285 g/min discharge with more than 2.6mm SOD, MRR becomes maximum. Beyond this region, MRR gradually decreases. At the lower side of discharge, MRR increases with discharge. Increased number of grains at higher discharge removes increased amount of material, but, at higher side of discharge, MRR gets lowered with a hike in abrasive discharge. More inter-particle collision because of the presence of large number of abrasive particles may be the reason as also opined by Xiao et al. [16].

Analysis of nozzle wear rate
Similar to the case of MRR, ANOVA is done. For NWR, correlation coefficient (R-Sq) of 94.74% is obtained from ANOVA. Individual, square or interactive parametric relationships are obtained in regression equation [2]. Significant parametric contribution towards process output responses are judged by their corresponding least p-values. If p-values for particular parametric contribution happen to be 0.05 or less then that particular parameter will be significant for output responses. Therefore, in Table 34.4, Model, linearity, SOD, square, square term of grain size, square term of SOD, square term of abrasive flow rate, have p value less than 5%. Thus, these parameters are significant for the NWR.

$$NWR = -0.3540 + 0.000065d + 0.01739\delta + 0.002376\dot{m} - 0.000000d^2 - 0.002036\delta^2 - 0.000004\dot{m}^2 - 0.000005d\delta - 0.000000d\dot{m} - 0.000012\delta\dot{m} \quad [2]$$

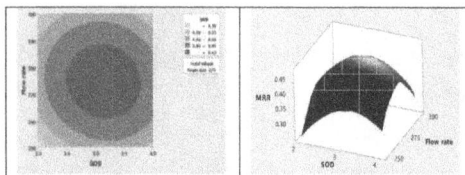

Figure 34.5 (a) Contour, (b) Surface plot of MRR (g/min) relating to SOD (mm) and discharge (g/min) at grain size hold value of 275μm
Source: Author

In Figure 34.6 (a, b), graphical representation about wear rate associated with nozzle with particle size as well as SOD is made at a discharge of 275 g/min. Eq. (2) shows that nozzle wear has a linear and quadratic relation with SOD, size of grain and abrasive discharge. NWR attains maximum value in the region of 180-325μm grain sizes and 2.7–3.5mm SOD.

Observations suggest wearing of nozzle increases with a hike in particle size at comparatively higher velocity due to erosion and abrasion of nozzle wall. But increasing above a certain grain size, due to dragging effect on higher grains, erosion and abrasion reduce wear of nozzle inner wall. Similarly, up to 2.8 mm SOD, accelerated velocity of abrasive jet increases following the principle of usual fluid flow. Thus, nozzle wear is increasing with SOD. Also, with lower value of SOD, nozzle wear increases may be due to rebounding particles from work piece striking back on nozzle. At higher SOD, due to divergence nature of abrasive jet, its velocity decreases as well as striking back on nozzle by rebounding particles decreases. Thus, gross nozzle wearing decreases at higher SOD.

In Figure 34.7 (a, b), variation of NWR with respect to size of abrasive and discharge are observed at a hold value of SOD at 3mm. Contour and surface plots show that NWR is maximum between grain sizes of 180μm to 325μm and 265 to 285 g/min discharge.

NWR becomes high with higher grain size as well as abrasive discharge up to a certain limit, due to increasing particle velocity and particle numbers and sizes. This leads to increasing erosion and abrasion on nozzle surface. For further increment of both discharge and grain size, velocity of abrasive grain reduces due to dragging effect reducing nozzle wear.

Table 34.4 ANOVA for NWR at 4 kg/cm² pressure

Source	DF	Adj SS	Adj MS	F-Value	p-Value
Model	9	0.000051	0.000006	10.00	0.010
Linear	3	0.000018	0.000006	10.45	0.014
Grain size, d	1	0.000002	0.000002	2.67	0.163
SOD, δ	1	0.000016	0.000016	27.96	0.003
Flow rate, ṁ	1	0.000000	0.000000	0.71	0.439
Square	3	0.000045	0.000015	26.63	0.002
d²	1	0.000011	0.000011	19.19	0.007
δ²	1	0.000015	0.000015	26.94	0.003
ṁ²	1	0.000026	0.000026	45.34	0.001
2-Way Interaction	3	0.000002	0.000001	1.38	0.349
d.δ	1	0.000002	0.000002	3.28	0.130
d.ṁ	1	0.000000	0.000000	0.20	0.675
δ.ṁ	1	0.000000	0.000000	0.68	0.448
Error	5	0.000003	0.000001		
Lack-of-Fit	3	0.000001	0.000000	0.28	0.836
Pure Error	2	0.000002	0.000001		
Total	14	0.000054			

Source: Author

Figure 34.8 (a, b) represent change in NWR with flow rate and SOD at a grain size of 275μm. Maximum nozzle wear takes place between 265 to 285 g/min discharge and 2.7 to 3.6mm SOD. Plots show that NWR increases initially with SOD and discharge, attains a maximum nozzle wear rate and further hike in SOD and discharge reduces NWR.

Conclusions

MRR and NWR initially increase with increase of grain size, attains a maximum value with a certain range of grain size and then MRR and NWR both decrease with further increase of size of abrasive. Both MRR and NWR initially increase with a hike of SOD up to an optimum value at a certain range of SOD and then get reduced with further hike in SOD. Both MRR and NWR initially increase with a hike in abrasive discharge, attain a maximum and then reduce with further hike in abrasive flow rate. In future, experimental setup can be modified with hot air abrasive jet drilling on other similar material as well as for surface preparation of different alloy steel. Some of the limitations of this process are little bit of abrasive pollution and not having the facility of measuring jet exit velocity.

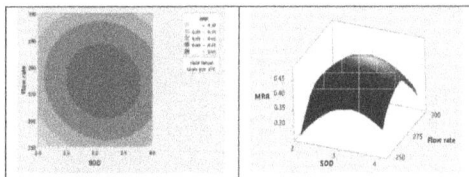

Figure 34.6 a Contour, b surface plot of NWR (g/min) related to grain size (μm) and SOD (mm) at discharge hold value of 275 g/min.
Source: Author

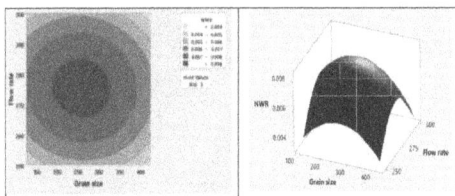

Figure 34.7 a Contour, b Surface plot of NWR (g/min) related to grain size (μm) and discharge (g/min) at SOD hold value of 3mm
Source: Author

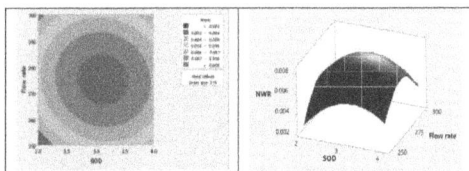

Figure 34.8 a Contour, b Surface plot of NWR (g/min) relating to SOD (mm) and discharge (g/min) at grain size hold value of 275μm
Source: Author

References

[1] Adak, D. K., Pal, V., Das, S., Ghara, T., Joardar, H., al Rasheedi, N., et al. (2023). Surface preparation for coating and erosion MRR of SS 304 using silicon carbide abrasive jet. *Lubricants*, 11(10), 11010010/1-18.

[2] Bharadwaj, S. R., Adithya, V. S., Suryasathvik, S., Rahul, D. B., Chengappa, U. N., Mohan, K. G., et al. (2019). Process parameters of abrasive jet machining. *Journal of Engineering Research and Applications*, 9(11), 46–51.

[3] Chastagner, M. W., Shih, A. J., & Arbor, A. (2007). Abrasive jet machining for edge generation. *Transactions of NAMRI/SME*, 35, 359–366.

[4] Finnie, I. (1960). Erosion of surfaces by solid particles. *Wear*, 3, 87–103.

[5] Ghara, T., Ansari, R., Adak, D. K., Ahmed, M., Das, S., & Haldar, B. (2018). Abrasive jet drilling of porcelain tiles and soda lime glass under different conditions. In Proceedings of the 1st International Conference on Mechanical Engineering, Jadavpur University, Kolkata, India, January 4-6, (pp. 189–208).

[6] Gradeen, A. G., Spelt, J. K., & Papini, M. (2012). Cryogenic abrasive jet machining of poly-dimethyl-siloxane at different temperatures. *Wear*, 274-275, 335–344.

[7] Haldar, B., Ghara, T., Ansari, R., Das, S., & Saha, P. (2018b). Abrasive jet system and its various applications in abrasive jet machining, erosion testing, shot-penning and fast cleaning. *Materials Today: Proceedings*, 5(5), 13061–13068.

[8] Hassan, A., & El-Hofy, G. (2005). Advanced Machining Process, Non-Traditional and Hybrid Machining Processes. McGraw–Hill. Mechanical Engineering Series.

[9] Karmakar, A., Ghosh, D., Adak, D.K., Mandal, B., Das, S., Ahmed, M. and Haldar, B. (2018). Abrasive jet machining of soda lime glass and laminated glass using silica sand. 7th International & 28th All India Manufacturing Technology, Design and Research (AIMTDR) Conference 2018, College of Engineering Guindy, Anna University, Chennai, India, December 13th to 15th, 2018

[10] Mishra, P. K. (2014). Abrasive jet machining: non-conventional machining. The Institution of Engineers (India), Text Book Series, Kolkata, 18th Reprint. 12-13.

[11] Pawar, T., Wagh, S., & Shinde, R. (2005). Literature review on abrasive jet machining. *Engineering Research and General Science*, 3(3), 1047–1051.

[12] Pradhan, S., & Dhupal, D. (2022). Experimental and simulation approach of surface generation on K-80 alumina ceramic by recently developed sustainable FB-HAJM using noval nozzle design. *IMechE Part E:*

Journal of Process Mechanical Engineering, 236(6), 2502–2514.

[13] Pradhan, S., Das, S. R., Nanda, B. K., Jana, P. C., & Dhupal, D. (2020). Experimental investigation on machining of hard stone quartz with modified AJM using hot silicon carbide abrasives. *The Brazilian Society of Mechanical Sciences and Engineering*, 42(11), 1–22.

[14] Pradhan, S., Sahu, S., Das, S. R., & Dhupal, D. (2021). Experimental study and simulation of surface generation during machining of K-80 alumina ceramic in modified abrasive jet machining with different temperature using Al_2O_3 abrasive. *International Journal of Abrasive Technology*, 10(4), 298–329.

[15] Prasad, S. R., Ravindranath, K., & Devakumar, M. L. S. (2018). Experimental investigation and parametric optimization in abrasive jet machining on nickel 233 alloy using WASPAS and MOORA. *Cogent Engineering*, 5(1), 1–12.

[16] Srikanth, D. V., & Rao, M. S. (2014). Application of optimization methods on abrasive jet machining of ceramics. *Industrial Engineering and Technology*, 4(3), 23–32.

[17] Xiao, S., Liping, S., Wei, H., & Xiaolei, W. (2016). A multi-phase micro abrasive jet machining technique for the surface texturing of mechanical seal. *Advanced Manufacturing Technology*, 86, 2047–2054.

35 Experimental investigation on high-speed swine bone drilling for force and temperature minimization

Srija Sarkar[1,a], Soumil Banik[1,b], Kashmira Khatun[1,c], Nilkanta Dhangar[1,d], Biswajit Sing Sardar[2,e], Bikash Banerjee[3,f] and Nripen Mondal[4,g]

[1]Department of Mechanical Engineering, Jalpaiguri Government Engineering College, Jalpaiguri, West Bengal, India

[2]Department of Mechanical Engineering, Ramkrishna Mahato Government Engineering College, Purulia, West Bengal, India

[3]Department of Mechanical Engineering, Abacus Institute of Engineering and Management Hooghly, West Bengal, India

[4]Department of Mechanical Engineering, Kalyani Government Engineering College, Kalyani, Nadia, West Bengal, India

Abstract

This study focused on optimizing drilling parameters for bone surgeries to minimize the risks of thermal osteonecrosis, which can occur when the bone temperature exceeds 47°C. Using swine bones, similar to human bones, and a high-speed mini-CNC machine, the research aimed to identify the best drilling conditions to reduce both temperatures rise and thrust force. The study varied three parameters: feed rate, drill diameter, and spindle speed, based on a Taguchi L27 design. Force and temperature were measured during drilling. The results showed high reliability in the model (97.23% for temperature and 93.13% for thrust force). Through response surface methodology (RSM), the optimal conditions for reducing temperature by 31.84°C were 2.8 mm drill diameter, 17 mm/m feed, and 30,000 rpm spindle speed. For thrust force, the minimum value of 0.0685 N occurred with a 1.5 mm drill, 23,131 rpm speed, and 10.8 mm/m feed rate. Finally, genetic algorithms identified the ideal multi-objective drilling parameters as a 3 mm drill, 29,458 rpm speed, and 8 mm/m feed rate. These findings suggest that higher speeds and feed rates reduce bone temperature during drilling, providing valuable insights for improving surgical outcomes in orthopedic procedures.

Keywords: Bone drilling, experimental study, genetic algorithms, multi-objective, optimization, response surface methodology

Introduction

Contemporary medical science has revolutionized orthopedics, with surgical bone drilling playing a pivotal role in improving patient outcomes in bone grafting, joint replacement, corrective osteotomies, and dental implants. Despite its advancements, however, certain limitations to drilling in human bone must be carefully considered. One of the challenges of bone drilling is the heat generated during the process. Prolonged or excessive heat can lead to thermal necrosis, or cell death, in bone tissue, which impairs its ability to heal. Heat-related complications in nearby soft tissues, such as muscles or nerves, can result in postoperative pain, swelling, or other negative effects. The primary goal of this research was to experimentally study high-speed drilling of swine bone at speeds up to 30,000 rpm, with an emphasis on understanding how drilling parameters influence temperature and force generation. The study then aimed to identify the optimal drilling conditions that minimize temperature and force during the process. Finally, desirability functions and response surface methodology (RSM) were applied to simultaneously minimize temperature and force.

Finally, genetic algorithms (GA) were used to optimize both temperature and force functions in order to determine the optimal bone drilling parameters.

Literature Review

Bone drilling is a vital technique in orthopedic, maxillofacial surgeries, and oral, facilitating precise manipulation, reconstruction of bone, and stabilization. It plays an essential role in a wide range of surgical procedures [3]. The friction between the drill and the bone generates heat [14]. The surgical fixation is delayed by thermal necrosis, which occurs when a temperature rises above 47–50°C and causes tissue to

[a]ss2348@me.jgec.ac.in, [b]soumilbanik@gmail.com, [c]kashmirakhatun6911@gmail.com, [d]niljr800@gmail.com, [e]bishu.sing88@gmail.com, [f]bikashbanerjee25@gmail.com, [g]nripen_mondal@rediffmail.com

DOI: 10.1201/9781003663348-35

lose cells [1]. According to [2], drilling is related to the conversion of mechanical work energy into thermal energy, which results in a brief increase in the temperature of nearby soft tissues and bone above physiologically normal levels. Numerous studies have been done to date, and they all point to the need for the optimization of several parameters to keep heat generation under control and prevent thermal necrosis. Temperature, surface roughness, pullout strength, cutting forces, tool wear, torque, and hole straightness are performance measures that are regarded as essential drilling characteristics. While cutting conditions (such as speed, feed, and hole depth), drill geometric features (such as helix angle, rake angle, clearance angle, tool material, drill diameter, and drill wear), bone-specific parameters (such as bone sex, bone density, and bone material), and irrigation (such as external and internal), are all thought to be crucial factors in conventional bone drilling [5]. Thermocouples can measure temperature more precisely along hole depth (i.e., 1.5 mm) compared to a thermo graphic camera that measures only the surface temperature [6, 7]. A study measured the cutting temperature distribution using thermocouples to evaluate the effect of drill tool material and cutting methods (such as up-cutting and down-cutting) on temperature during vertical milling. The high-speed steel (HSS) tool was found to generate less heat across all cutting types. Additionally, it was noted that the up-cutting temperature of the bone surface was 20% higher than the down-cutting temperature. The issue of temperature variation during bone drilling can also extend to other drill materials that may transfer more heat to the bone [4, 8]. Several studies have employed experimental methods to identify optimal drilling parameters. For instance, drilling experiments on aluminum alloy were conducted; with the best parameters for minimizing burr formation identified using the RSM-GA optimization technique [15]. In another study, drilling of aluminum alloy 6061 was explored using the L27 design, and artificial neural networks (ANN) were used to predict burr formation. The optimal drilling parameters were determined using Taguchi methods and GA [9, 10]. Additionally, a hybrid approach combining RSM, particle swarm optimization (PSO), GA, and the flower pollination algorithm (FPA) has been applied to optimize process parameters across various machining operations [11–13].

Research Gap and Novelty

Many researchers have yet to explore high-speed bone drilling, highlighting a critical research gap that needs to be addressed for a better understanding of temperature increases during the drilling process.

Experimental Setup and Measurement Methodology

The bone drilling experiment was performed using a mini CNC machine operated by a universal G-code. The mechanical properties of the swine bone closely resembled those of human bone. A stepper motor (42SHDC3025-24B) and a drive with a heat sink (1.4 HR-A4988) were used in this setup.

Figure 35.1 illustrates how the CNC machine received instructions from Arduino via a data wire, programmed with general G code. The CNC machine could only move in the Z direction. RPM and feed are controlled by G-code software. Load cell dynamometers were used to measure the load on the specimen held in a vice. An orthogonal drill bit was used to drill bone samples perpendicular to their longitudinal axis. A data acquisition system recorded and displayed force and temperature data for each drilling operation.

As shown in Figure 35.2(a), the dynamometer had a maximum load capacity of 400 N. The drilling bone's temperature was monitored using a thermostat capable of controlling temperatures between (-50 to +110°C) with 0.1°C accuracy. As seen in Figure 35.2(b), this 12 V digital controller has a display and an NTC waterproof temperature sensor (model XH-W1209).

Experimental setup and data collection

In this experimental study, the primary input parameters—feed rate, drill bit diameter, and spindle speed—along with their three levels of variation, were selected

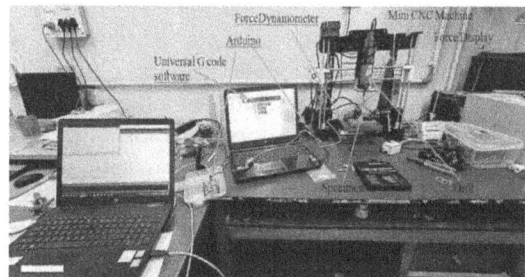

Figure 35.1 Setup for temperature measurement and applied cutting forces in the experiment
Source: Author

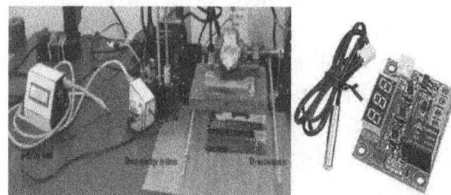

Figure 35.2 (a) Force measurement setup, (b) Digital temperature controller
Source: Author

for the experiment. According to Taguchi's orthogonal array, the L27 experimental run has been performed to minimize the temperature thrust force.

Multi-objective optimization using GA

The primary objective was to determine the ideal bone drilling parameters aiming to minimize both the temperature and force using multi-objective GA. The second-order regression equation of temperature and force has been developed using RSM methods. This two-fitness function was optimized to minimize temperature and force using a GA multi-objective environment in MATLAB 15. This involved fitness evaluation, where an objective function was applied to both original and new chromosomes to minimize maximum temperature and force. This approach ultimately enhanced surgical efficiency and reduced thermal risk, thereby ensuring safer outcomes. Leveraging the capabilities of the GA allowed for a more comprehensive exploration of the parameter space, enabling the minimization of maximum temperature and force in bone drilling.

Results and Discussions

ANOVA analysis for temperature and force

Using experimental data, ANOVA variance analysis was performed using MINITAB 17 at a 95% accuracy level. Model term parameters of greater significance are characterized by their lower P values. If P values below 0.05 exhibit a high degree of significance. The model parameters, including C1 (Drill Bit Diameter), C2 (Spindle Speed), C3 (Feed rate), C1*C2, C1*C3, and C2*C3 are most significant for temperature and the R^2 value (97.23%) aligns well with the Adj R^2 (95.77%). Similarly for the force, model term parameters such as C1, C12, C22, C1*C2, and C1*C3 most significant factors for force and confirm this agreement, displaying an R^2 value of 93.13% in harmony with the Adj R^2 value of 89.50%. RSM-based second-order mathematical equations of temperature and force are given in Equation 1 and Equation 2.

Regression Equation for Temperature...Eq^n 1

$$Temperature = 65.7035 - 18.8697C1 - 0.000174C2 - 0.359467C3 + 3.73667C1^2 - 0.000133C1 * C2 - 1.1e - 8C2^2 - 4.85e - 6C2 * C3 + 0.104C3 * C1 + 0.0022C3^2$$

Regression Equation for Force...Eq^n 2

$$Force = -8.6081 + 14.4117C1 - 0.0002242C2 - 0.8085C3 - 2.11517C1^2 - 0.0001729C1 * C2 + 1.33e - 8C2^2 - 1.233e - 5C2 * C3 + 0.221233C3 * C1 + 0.0353C3^2$$

where C1= drill dia., C2=Spindle speed, C3=feed rate

Figure 35.3(a) and (b) show the comparison between the experimental results and RSM-predicted values for temperature and force, respectively. The experimental and RSM-predicted temperature and force values were found to be closely aligned.

Observation of input parameters influences

Drilling began with the atmospheric temperature being maintained. As drilling commenced, there was a gradual temperature rise. Following the completion of drilling operations, the bone slowly cooled, returning to its original atmospheric temperature. It only took a few seconds for the thermostat to reach its maximum temperature value after finishing the drilling operation. To capture a more accurate and representative pattern of temperature changes during bone drilling, the study selectively analyzed a subset of 15 tests from the total of 27 conducted. By focusing on a smaller, representative sample, the analysis ensured that the observed trends reflect typical performance rather than isolated or atypical instances. Figures 35.4(a), (b), and (c) display temperature fluctuations recorded during sets of five experiments for each drill diameter tested. The random selection of experiments within these sets helps demonstrate the overall temperature response under different conditions, emphasizing the effect of varying drill sizes on heat generation in bone tissue. In the analysis presented in Figure 35.4, the impact of varying spindle speeds and feed rates on temperature during bone drilling is examined using different drill diameters. The findings are presented in three sub-figures, illustrating how these parameters impact the temperature generated during the drilling process.

Figure 35.4(a) displays five force curves for a fixed drill diameter of 1.5 mm. The spindle speeds tested include 10,000, 20,000, and 30,000 rpm, combined with feed rates of 7, 12 and 17 mm/m. The results show that lower feed rates and speeds lead to higher temperatures. The trends observed here align with findings from other studies, emphasizing the need for optimal speed and feed rate combinations to minimize temperature. Figure 35.4(b) shows that experiment

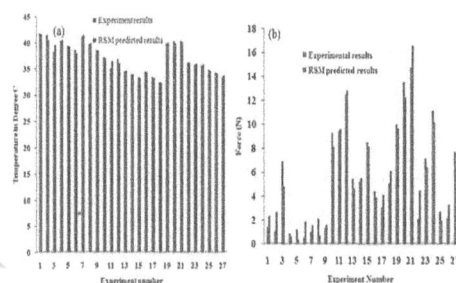

Figure 35.3 Deviation of experimental and RSM predicted results (a) temperature and (b) force

Source: Author

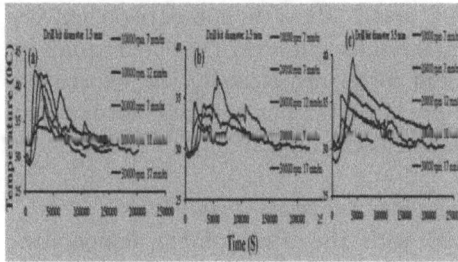

Figure 35.4 Experimental temperature variation for (a) exp. 1, 2, 4, 5, 9 (b) exp. 10, 13, 14, 16, 18 (c) exp. 19, 22, 23, 26, 27
Source: Author

10, conducted at lower feed rate and speed, produced the highest temperature among the tests. In contrast, Experiment 18, with the highest feed rate (17 mm/m) and speed (30,000 rpm) resulted in the lowest temperature during the drilling process. This indicates that increasing both feed rate and speed helps reduce the likelihood of bone damage. The temperature curves shown in Figure 35.4(c) correspond to a drill diameter of 3.5 mm. Experiment 19 recorded the maximum temperature, likely due to a lower spindle speed. In all sub-figures, a consistent trend is observed: higher feed rates and speeds generally lead to lower temperatures.

The downward force exerted during each drilling operation was measured using a dynamometer. For improved representation, 15 tests were randomly chosen from a total of 27, aiming to depict a representative behavior of the force variation.

Figure 35.5 presents an analysis of how varying spindle speeds and feed rates affect thrust force during bone drilling, with different drill diameters considered. The findings are depicted across three sub-figures, showing how these parameters influence the force exerted during the drilling process. Figure 35.5(a) displays five force curves for a fixed drill diameter of 1.5 mm. The spindle speeds tested include 10,000, 20,000, and 30,000 rpm, combined with feed rates of 7, 12, and 17 mm/m. The results show that lower feed rates and speeds lead to higher forces. This is attributed

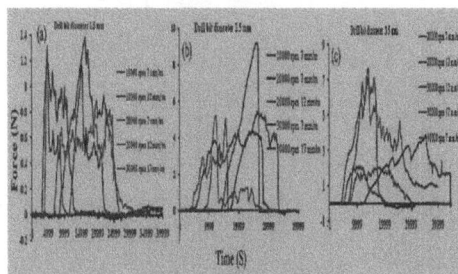

Figure 35.5 Experimental thrust force variation for (a) exp. 1, 2, 4, 5, 9 (b) exp. 10, 13, 14, 16, 18 (c) exp. 19, 22, 23, 26, 27
Source: Author

to the slower cutting action and longer contact time between the bone and the drill, which requires more force for penetration. Lower speeds and feed rates also prolong the drilling duration, as the reduced cutting efficiency means the drill takes longer to complete the hole. The trends observed here are consistent with findings from other studies, highlighting the importance of optimizing feed rate and speed combinations to minimize both drilling time and force. Figure 35.5(b) shows that Experiment 10, conducted at a lower feed rate and speed, resulted in the highest force among the tests, indicating a significant increase in resistance encountered by the drill under suboptimal parameters. In contrast, Experiment 18, with a higher feed rate and speed, completed the drilling in the shortest time.

This demonstrates that increasing both speed and feed rate enhances cutting efficiency, reducing the overall time needed for the procedure. The force curves presented in Figure 35.5(c) are for a drill diameter of 3.5 mm. Experiment 19 recorded the maximum thrust force, likely due to a lower spindle speed, which increases the resistance encountered during drilling. Experiment 27 was conducted at the highest feed rate (17 mm/m) and spindle speed (30,000 rpm), achieved the shortest duration for the drilling process. Across all sub-figures, a consistent pattern emerges: higher feed rates and speeds tend to reduce the force and shorten the drilling time. This is because increased speed enhances the cutting action, while a higher feed rate accelerates material removal. The differences in force behavior between 1.5 mm and 3.5 mm diameters highlight how drill size influences the required thrust force. Larger diameters usually result in higher forces because of the greater contact area with the bone.

RSM & GA based optimization

In this section, we identified the optimal parameters for the bone drilling process by applying the Taguchi principle (smaller is better for temperature and force) alongside RSM. Optimization was achieved using a desirability function to minimize both temperature and force.

Figure 35.6(a) shows that the minimum temperature obtained was around 31.84°C, and the best bone drilling parameters were 2.8 mm drill diameter, 30,000 rpm speed, and 17 mm/m feed rate. Similarly, from Figure 35.6(b), the minimum thrust force of 0.0685 N was reached, and the drill parameters corresponding to this force were 1.5 mm drill diameter, 23,131 rpm, and 10.8 mm/m feed rate. Finally, using GA, a mult objective function combining both the temperature and thrust force regression models (equation 1 and equation 2) was optimized. With a population size of 200, the lower and higher boundary conditions were chosen as [1.5 10000 7] and [3.5 30000 17], respectively. The total number of iterations was set at 5000.

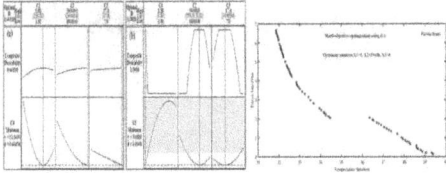

Figure 35.6 RSM based (a) temperature and (b) force Optimization, (c) multi-objective optimization based on GA
Source: Author

Figure 35.6(c) clearly demonstrates that the optimization process identified the ideal drilling parameters for minimizing both force and temperature: a 3 mm drill diameter, a 29,458 rpm speed, and a 8 mm/m feed rate.

Conclusion

The study seeks to preclude thermal necrosis during bone tissue drilling by optimizing drilling parameters. It achieved high model reliability with adjusted R-squared values of 95.77% for temperature and 89.50% for thrust force. The best parameters for reducing temperature (by 31.84°C) were a 1.8 mm drill diameter, 17 mm/m feed rate, and 30,000 rpm speed. To minimize thrust force (0.0685 N), a 1.5 mm diameter, 10.8 mm/m feed rate, and 23,131 rpm speed were optimal. A multi-objective approach identified that a 3 mm diameter, 29,458 rpm speed, and 8 mm/m feed rate effectively minimized both temperature and thrust force.

Future Scope

- Auto feed and speed control with the help of some sensor.
- Investigation on internal or external auto-trigger cooling systems is yet to be explored. This is essential when bone drilling temperatures rise above 47°C.
- To identify best drilling parameters, very limited researchers have implemented hybrid optimization technique and Deep learning methods.
- To implement the findings in real world surgeries.

References

[1] Augustin, G., Davila, S., Mihoci, K., Udiljak, T., Vedrina, D. S., & Antabak, A. (2007). Thermal osteonecrosis and bone drilling parameters revisited. *Archives of Orthopaedic and Trauma Surgery*, 128(1), 71–77.

[2] Bertollo, N., & Robert, W. (2011). Drilling of bone: practicality, limitations and complications associated with surgical drill-bits. *Biomechanics in Applications*, 4, 53–83.

[3] Davim, P. J., Gaitonde, N. V., & Karnik, R. S. (2008). Investigation into the effect of cutting condition on surface roughness in turning of free machining steel by ANN model. *Journal of Materials Processing Technology*, 205, 16–23.

[4] Hein, C., Inceoglu, S., Juma, D., & Zuckerman, L. (2017). Heat generation during bone drilling: a comparison between industrial and orthopedic drill bits. *Journal of Orthopaedic Trauma*, 31, 55–59.

[5] Jamil, M., Rafique, S., Khan, A. M., Hegab, H., Mia, M., Gupta, M. K., et al. (2020). Comprehensive analysis on orthopedic drilling: a state-of-the-art review. *Proceedings of the Institution of Mechanical Engineers, Part H: Journal of Engineering in Medicine*, 234(6), 37–561.

[6] Lee, J., Ozdoganlar, O. B., & Rabin, Y. (2012). An experimental investigation on thermal exposure during bone drilling. *Medical Engineering and Physics*, 34(10), 1510–1520.

[7] Scarano, A., Piattelli, A., Assenza, B., Carinci, F., Donato, L. D., Romani, G. L., et al. (2011). Infrared thermographic evaluation of temperature modifications induced during implant site preparation with cylindrical versus conical drills. *Clinical Implant Dentistry and Related Research*, 13(4), 319–323.

[8] Sugita, N., Osa, T., & Mitsuishi, M. (2009). Analysis and estimation of cutting-temperature distribution during end milling in relation to orthopedic surgery. *Medical Engineering and Physics*, 31(1), 101–107.

[9] Mondal, N., Sardar, B. S., Halder, R. N., & Das S. (2014). Observation of drilling burr and finding out the condition for minimum burr formation. *International Journal of Manufacturing Engineering*, 1, 1–12.

[10] Dey, B., Mondal, N., & Mondal, S. (2018). Experimental study to minimize the burr formation in drilling process with artificial neural network analysis. *IOP Conference Series Materials Science and Engineering*, 377, 1–6.

[11] Leo, K. S. P. (2019). Measurement and uncertainty analysis of surface roughness and material removal rate in micro turning operation and process parameters optimization. *Measurement*, 140, 538–547.

[12] Kalita, K., Shivakoti, I., & Ghadai, K. R. (2017). Optimizing process parameters for laser beam micro-marking using a genetic algorithm and particle swarm optimization. *Materials and Manufacturing Processes*, 32, 1101–1108.

[13] Mondal, N., Mandal, S., & Mandal, C. M. (2020). FPA based optimization of drilling burr using regression analysis and ANN model. *Measurement*, 152, 107327.

[14] Wang, Y., Cao, M., Zhao, X., Zhu, G., Clean, M. C., Zhao, Y., et al. (2014). Experimental investigations and finite element simulation of cutting heat in vibrational and conventional drilling of cortical bone. *Medical Engineering and Physics*, 36(11), 1408–1415.

[15] Mondal, N., Mandal, S., Mandal, C. M., Das, S., & Haldar, B. (2021). ANN-FPA based modelling and optimization of drilling burrs using RSM and GA. GCMM 2021. In Advance in Manufacturing Process, Intelligent Methods and System in Production Engineering, (pp. 180–195).

36 Indigenization for Indian defense forces: opportunities & challenges of smart materials

Ashok Kumar Panda, FIE (India)

Ph D Scholar Poornima University, Jaipur, India

Abstract

The spending of Indian Defense agencies is closely 3% of gross domestic product of the nation and nearly 40% of total allocation is spent on indigenization activity. Both the aspects determine the self-reliance index which is the ratio of expenditure on indigenization to the total expenditure on defense in a financial year. The index for India is currently close to 0.5 against a world average of 3. The rank of India is four amongst 12 Indo Pacific nations in reference to self-reliance in defense sector. The critical overview reflects that opportunity and challenges of smart and modern materials are fundamental factors for success in defense Indigenization since most budgets is involved in procurement, process, research, development and testing of smart and exclusive materials used in customized manner for armed forces. Ongoing global military activities and ambitious program of Atmanirbhar Bharat in defense, India is in a tight spot to avail all opportunities in indigenization sector and mitigate challenges to deliver field requirements in totality by 2030 indigenously. It is established that close to 80% of defense engineering indigenization activities are subject of material innovation either in raw form or semi processed or ready to use off the shelf. However, the most challenge is specification determination as per user requirements, maintaining global standards and to meet qualitative requirements. With an aim to make India a negative import country by 2047 the Indigenization program is focusing on applications of Rare Earth Materials, Meta Materials, Ceramic Matrix Composites and epsilon negative materials as opportunities. The challenges of importing and. exploiting smart materials to support the program are taken care by multiple agencies proactively.

Keywords: Characterization, exploitation, resource mobilization, responsive materials

Introduction

The identification of smart materials for defense applications in a cost-effective manner, satisfying the field users and stringent specifications meeting as specified by standardizing bodies are delicate coordinated activities to meet deliverables. This smart materials aspect is most prominent amongst other factors like provisioning sophisticated machineries, skilled manpower, established methodology, high motivation and availability of money amongst others. As certain raw materials suitable for development as smart materials are concentrated at select countries hence even after having fund, a needy nation may not procure those materials without having a political goodwill with the material resourceful country. Further the smart materials availability should match with a particular type, composition, specifications, grain size and unique character requirements. As a good example, even the least deviation in acoustical property makes a big difference in protection aspect of a tank due to easily traceable by enemy surveillance system with acoustical signature. Product realization, forecast and deployment is only possible after clearance from multiple trial agencies which oversee end result of an indigenized product in own terms of reference negating challenges encountered for smart and modern material provisioning and processing.

The scope of applications of smart materials like smart polymers, ceramics matrix composites (CMCs), Piezoelectric materials, Rare Earth Materials and customized magnetic materials are currently well saturated. The research is on for introducing meta materials, epsilon negative materials (ENM) and mu negative materials (MNM) for applications in defense component level applications to enhance power to weight ratio, mobility, protection, durability of components and efficacy. The replacement scope of conventional materials-based components by reverse engineering with use of smart and intelligent materials are always restricted by limited transfer of technology (ToT) with original equipment manufacturer (OEM). The memorandum of understanding (MoU) hardly covers the ToT coverage on materials characterization front.

The indigenization of defense products is based on the production scheduling in term of firm, planned and forecast modeling. The availability of intelligent and smart materials to meet production schedule

akp.eme@gmail.com

DOI: 10.1201/9781003663348-36

requirements in phased manner through import or indigenously developed needs to be synchronized for undisruptive production. However, the reality is not conducive as envisaged. Consequently, the production units use conventional materials in place of identified smart materials compromising the functional specifications. The requirements of using smart and intelligent materials in defense equipment besides primary functional needs of strength and durability, is to store information for telemetry application, energy efficient service, chemical and nuclear activation at certain operating condition and ecofriendly demands.

Opportunities of Smart Materials Redefined Indian Defense Indigenization Deliverables

As reported in Hyderabad edition of The Times of India dated 18 Jun 2024, the Indian prototype submarine gun Asmita (ASMI) is successfully launched by a Hyderabad based small arms manufacturer Lokesh Machines Limited indigenously. The company supplies the Army, Boarder Security Force (BSF) and National Security Guard (NSG) amongst others. The successes story is based on making available indigenized materials having North Atlantic treaty Organization (NATO) standard adherence. It is reported that the 9x19mm caliber submachine gun ASMI beat competition from internationally renowned Uzi of Israel and German Heckler. The ASMI in Sanskrit means 'Courage' and 'Pride' is suitable for Indian and NATO standards with effective range of 100m, light in weight, rugged, reliable, accurate and 30% cheaper with cost below Rs one lakh. All attributes relate to smart materials selection and integration with the gun for optimization in operation. Finally, what matters in deliverable is the material inducted into the product fulfilling criteria from cost to operational features.

In the Armoured Fighting Vehicle (AFV) sector due to unavailability of specified materials, India imports tanks mostly. Currently project 'Zorawar' which is indigenously made light tank for deployment in High Altitude Area (HAA) deployment. As reported in Hyderabad edition of The Times of India dated 7 July 2024, the light tank developed jointly by Defence Research and Development Organisation (DRDO) with Larsen and Toubro (L&T) are just waiting winter, summer and desert trials in plain and HAA before getting inducted to Indian Army for extensive field use. The 25-ton light tank with high power to weight ratio successes story is credited to availability of customized materials during acquisition, production and induction seamlessly. As per report this agile and versatile light tank is capable of operating at most challenging environment with least logistic support since

embedded with smart and intelligent materials each having multiple functions to perform.

Similarly, the use of rolled homogeneous steel (RHA) in armor of a tank is modified into multi layering with heterogeneous smart materials. Each layer is used specifically to meet predetermined functional requirements like preventing thermal signature, absorbing acoustical signature, storing running hours information and indicative of condition based monitoring (CBM) parameters. The use of epsilon negative materials (ENMs) along with RHA increases thermal conductivity which facilitates faster rate of cooling thus maintain thermal stabilization and never deteriorates the insulation materials used. The mu negative materials (MNM) when subjected to certain optical frequency causes magnetic resonance so that certain optical phenomenon happening surrounding to defense equipment can be identified. The option of using MNM is also contemplated for defense use by R&D.

The Institute of Minerals and Materials Technology (IMMT) and Central Mechanical Research Institute (CMRI) both entities of Council of Scientific and Industrial Research (CSIR) spread heading the customized applications of intelligent and smart materials in component levels for defense applications. As the smart or responsive materials have unique characters like atmospheric hygroscopic monitoring, chemical vapor identification, laser detection, acts as optical sensor and electromagnetic wave monitor and sensor, the uses in defense equipment is diversified and effective in battlefield. The IMMT and CMRI is trying to make smart materials user friendly and reshaping the characteristics to be used effectively in different models of a particular defense equipment.

In the private domain Lam Research India located at Bengaluru with a vision "Nudge an Atom – Move the World" is innovating semiconductor technology indigenously with an aim to support defense indigenization program. The entity believes their work on smart materials will stand with defense equipment for achieving mission accomplishment. The nanotechnology-based software which is the backbone of gun control equipment (GCE) system of AFVs are being delivered shortly to Armed Forces. The potential of Lam Research developing equipment for thin film deposition in optical set ups, plasma etching, photo resist strips using smart materials is cost effective a commendable initiative.

The opportunity of using intelligent and smart materials in complex defense instrumentation system related to navigation, silent mode operation capability, communication and GCE are being introduced in phased manner by R&D organizations. This will place India amongst the top five advanced nations of using smart materials for defense applications. The Metals and Minerals Trading Corporation (MMTC) Limited role is admirable in respect of ensuring specialized material availability by importing minerals and ores in desired form and composition at a competitive price, with desired quality and as per time schedule timeline.

Comprehensively bringing out the current opportunities of smart materials applications through defense indigenization, the following areas are shortlisted as follows: -

a) Autonomous armored vehicles
b) Dedicated and secured communication network
c) Drones, surveillance , and interceptors
d) Night vision and optical devices
e) Test and diagnostic equipment
f) Quantum computing and data analytics
g) Sensors with telemetry
h) Augmented and virtual reality applications
i) Camouflaging and image analysis technology
j) Prescriptive maintenance on IoT platform

Challenges Encountered for Introducing Smart Materials in Defense Indigenization

The application of intelligent and smart materials is extensively researched for space, marine, health care, mobile network, telemetry applications, sensors and actuators for all field, augmented and virtual reality (AR/VR) activities, defense indigenization, cloud computing, machine learning (ML) and artificial intelligent (AI). However, the most challenging application is in defense indigenization uses due to many security and safety regulations in place and restrictions in uses of civil internet network to transfer information from smart materials to assessment centers. The adaptive characteristics of smart materials are both foe and friend for defense indigenization since those materials retain information up to discard stage also calling for security issues. The additional challenging factors are vendors' empowerment, facilitating state-of-art machineries and diagnostic equipment availability, skilling facilities of technicians and adhering to international standards with functional best practices.

The following hurdles are restricting achieving the targeted goals in defense indigenization by introducing smart materials: -

a) Identification, development and introduction of pilot samples requires sufficient fund by R&D sector and time as well due to testing modalities by trial-and-error basis demands time and involvement of experts.

b) Each and every material introduced for large scale use in defense indigenization is required to be standardized for commercial uses and interchangeability. This aspect is challenging as far as defense application are concerned due to such materials may not be commercialized extensively for security reasons.

c) Once R&D sector introduces a particular smart material for use in defense, the scalability of the material is restricted thus causing escalation of price and having limited users' feedback for continual improvement program.

d) The support from financial institutions for import and setting up infrastructure to process smart materials due to cost and profit issue is not encouraging in India.

e) The qualitative checks and user's acceptance of products made for defense using unusual smart materials are very long and not proactive, hence the immediate benefits are not available thus discouraging to progress.

f) The current role of ordnance factories (OFs), Public Sector Undertakings (PSUs) and defense PSUs are not forthcoming to take on projects related to smart materials for defense indigenization due to complicated modalities and non profitable. Further these organizations are made autonomous hence survivability and profit making are prime criteria but not the research.

g) The supply chain difficulties in respect of raw materials to finished products with reference to smart materials requires logistic clearance, restrictions and following standard operating procedures (SOPs) immaculately due to defense indigenization program in India is still considered as a secured and restricted activity.

h) Cyber security issue prevails post adaptability of smart materials on retention of classified information even during storage and discard,

i) The possibility of using smart and conventional materials together may not produce optimized performance due to mismatch of inherent material properties. Other way, it is also difficult to replace all conventional materials with smart ones due to availability and cost issue. Hence integration on material front is a standing complexity.

j) The intrinsic challenge of using smart materials in defense equipment is related to poor mechanical features, delayed response time and restricted environmental stability during multi terrain applications, hence choice of use should be selective.

Emerging Scopes of Smart Material Uses in Defense Indigenization Program

The immediate emerging need and scopes for using smart materials have been finalized by users and R&D entities. The scope relates to different major assemblies, sub-assemblies and component level as well. The application covers armored fighting vehicles (AFVS) up to solider comfort equipment. The comprehensive application areas shortlisted are placed below:

(a) Camouflaging major equipment providing stealth capabilities with information retention features embedded for real time uses.

(b) Inbuilt energy harvesting properties to support silent watch capability of defense equipment without starting engines and avoiding producing thermal signature for identification by adversaries.

(c) Using shape memory alloys (SMA) in a customized manner can activate material, so that variation of surrounding parameters like pressure, temperature and hygroscopicity can change the shape and size of smart materials and correspondingly the physical status can be predicted accordingly with reference to dimensional variation.

(d) Telemetry applications by which the physical onsite parameters can be converted to frequency and pulse mode for transferring to a substantial distance to be used for processing, storing and retrieving as necessary is only feasible by use of smart materials.

(e) Adaptive armor ensures change in smart material characteristics as per situational requirements to meet operational demand. This feature bears a prominent role to play in camouflaging of defense equipment in war fields.

(f) Development of morphing wings using smart materials where the shape of the wings of flying object can be changed as per varied conditions to optimize the desired performance. This feature enhances fuel efficiency, reduce air drag and fetch outstanding aircraft stability.

Financial Implications on Introduction of Sizeable Smart Materials in Defense Equipment

The complete cycle of smart materials procurement to processing for defense indigenization program requires massive financial involvement in particular foreign currency. The support extended by Export – Import Bank of India, Export Credit Guarantee Corporation of India and Small Industries Development Bank of India are noteworthy for facilitating the import of smart materials and related technology. The current decision by Government of India to allow 100% Foreign Direct Investments (FDIs) to augment defense indigenization is a milestone in self-reliance front. The establishment of Technology Development Fund (TDF) is facilitating entrepreneurs to promote defense indigenization at the rate of Rs 50 crores per project. Nearly 200 such technologies are being indigenized under R&D causing substantial saving to Government exchequer and making technologies available for end use.

The outcome of financial support schemes is visible in the following domain:

(a) Joint ownership of private players with DRDO is increasing

(b) The number of Intellectual Property Right (IPR) filing on materials front for defense indigenization is in rise

(c) The sales opportunity of product out of smart materials R&D is feasible

Conclusion

Fostering collaboration over competitiveness is the call of the day in better smart materials solution for successful defense indigenization. This primarily eliminates rigidity of supply chain management and otherwise. Taking stock of field problems holistically with effective data analytics also provides insight for smart materials uses in defense equipment. There is a proposed paradigm shift in the mindset of the researchers for collaborative and co-sharing the infrastructure, facilities and results of various studies so that duplication work is minimized and the research deliverables are accelerated. The outcome of vision 'Make in India' and the reality of 'Made in India' are two distinct concepts of realization. Indian materials science core is poised to exploit the smart materials into chips manufacturing which will strengthen overall indigenization program. Besides semiconductors the smart materials will get into quantum and digital technologies which are in forefront of defense indigenisation. The applications of smart and intelligent in defense sectors also extend to chemical, biological, nuclear and radiological at microscopic level for better results. The IMMT, CMRI & DRDO are leveraging the talents of in house scientist to navigate through the physical properties of smart materials to find optimum applications suitable to all terrain and weather defense uses. The increased need of smart, intelligent, novel and sustainable materials in defense indigenization will soon be a reality with opening of specialized manufacturing corridors, availability of talents and infusion of requisite fund for development.

References

[1] Dhruva Jaishankar; "The Indigenisation of India's Defence Industry"; IMPACT Series [2019]; 1–4.

[2] Ranjit Ghosh; "Indigenisation; Key to Self-Sufficiency and Strategic Capability", Pentagon Press {2016}; 1–12.

[3] Chirisa et al ; "Dynamics of Indian Defence Technology: Indianisation, Indigenisation ,industrialisation, Integration"; The Tribune [2011]; 1–14.

[4] Ashok Jain ;"Strengthening Science and Technology Capacities for Indigenisation of Technology- the Indian experience"; International Journal of Services Technology & Management [2003]; 234–254.

[4] Wan Ramli Wan Daud ;"Indigenisation of Technology and the Challenge of Globalization: The Case of Malaysia"; Research Gate [1999]; 1–23.

[5] Brig Ashok Pathak ;"Indigenisation of Defence Production : India's Journey from Vision to Outcomes"; Vivekananda International Foundation [2022]; 1–7.

[6] Jasjit Singh ;"Indigenisation of Indian Defence Sector"; National Defence Institute (NDI) [2021]; 1–2.

[7] Bikramdeep Singh ;"Defence Indigenisation: Indian Army"; Center of Land Warfare System (CLAWS) Journal [2013]; 1–5.

[8] 8th Edition of Air & Defence India 2022; "Integrating New Technologies and Optimizing Legacy Systems"; Centre of Joint Warfare Studies ; [2022]; 1–7

37 Comparative analysis of genetic algorithms and CDS for optimizing flow shop scheduling under industrial constraints

Nameet Kumar Sethy[1,a], and Dhiren Kumar Behera[2,b]*

[1]Research Scholar, Indira Gandhi Institute of Technology, Sarang, Odisha, India

[2]Associate Professor, Indira Gandhi Institute of Technology, Sarang, Odisha, India

Abstract

In industry, production is frequently organized as a flow shop, methods, and algorithms for optimizing such production processes are highly sought after. While techniques have traditionally concentrated on optimizing a particular goal, it is becoming increasingly vital to address resilient algorithms to give optimal solutions for operational requirements. This study investigates the efficiency of genetic algorithms (GA) and Campbell, Dudek, and Smith algorithm (CDS) in addressing Flow Shop Scheduling Problems (FSSP) under specific constraints, such as maintenance breaks, variable setup times, and job priorities. Implemented in Python, both algorithms were tested on datasets from a steel manufacturing industry, with variations in job and machine configurations. The results demonstrate that while GA consistently yields marginally lower make span values, it does so at a significantly higher computational cost compared to CDS. CDS, on the other hand, provides faster scheduling decisions with minimal increases in computation time, making it more suitable for scenarios requiring quick, efficient solutions. The study concludes that GA is preferable for optimizing complex schedules when computational resources are available, whereas CDS is ideal for simpler or time-sensitive scheduling tasks.

Keywords: CDS, Flow-shop, GA, make span, scheduling

Introduction

A schedule is a visual, physical document that makes time visible and concrete; whether it communicates events or plans for the scheduling of work. The scheduling problem is usually (simplistically) broken down into two stages the first being sequencing or how to determine what job should be picked next. The next stage is the time at which each task will start, and if applicable, end.

The importance of FS scheduling has been highlighted in most manufacturing and service systems besides the information processing environments. To minimize are more objectives of flow shop scheduling. It consists of make-span, tardiness-based objectives, work completion time and flow time. The FSSP scheduling problem is a conventional and existing scheduling problem, which has been broadly studied because of its practical importance.

In this work genetic algorithms (GA) and Campbell, Dudek, and Smith algorithm (CDS) are compared for flow shop scheduling problems with specified limitations. Machine 2's maintenance break variable setup times due dates such critical completion and scheduling priority are constraints. The comparison evaluates how effectively each solution resolves restrictions and optimises scheduling. GA evolves through many job sequences to optimise performance. Instead, CDS creates a realistic timetable using specified criteria and heuristics. We compare GA and CDS findings to determine which method balances restrictions and increases scheduling efficiency, revealing their strengths and weaknesses in practical situations. Section 2 is literature review, methodology is under Section 3. Implementation of research in Section 4. In Section 5, computational experiment statistics on Section 6.

Literature Review

Heuristic approaches for FSS have been extensively studied to optimise makespan and efficiency. Taillard [1] compared heuristic techniques, including the Campbell, Dudek, and Smith (CDS) algorithm, to minimise makespan in PFSSP. Later studies, such as Allahverdi and Soroush [2], stressed the importance of reducing setup times and costs in scheduling and showed the CDS algorithm's efficacy in addressing these issues across different scheduling problems, including PFSSP. Gao et al. [3] added CDS to hybrid algorithms for more complicated flow shop and job shop environments to

[a]nameet158@gmail.com, [b]dkbigit@rediffmail.com

DOI: 10.1201/9781003663348-37

improve makespan optimisation. These studies show that CDS and other heuristic approaches are essential for optimum scheduling in many industrial settings.

Recent permutation flow shop scheduling problem (PFSSP) research has focused on creative ways to reduce makespan and enhance computation time. Another study [4] uses a genetic method to balance makespan and tardiness, improving computational efficiency. Mousighichi and Avci [5] investigates a distributed, no-idle version of PFSSP and shows that genetic algorithms and combinatorial heuristics solve this hard scheduling problem. A genetic algorithm to minimises makespan in hybrid flow shops with sequence-dependent setup durations Ruiz and Maroto [6] introduced an evolutionary method to address this issue, aiming to minimise the make span and processing duration.

In more complex scenarios, Naderi et al. [7] used evolutionary algorithms to PFSSP with machine capacity limitations for makespan minimization. This research demonstrates flow shop scheduling problems are solvable using genetic algorithms, heuristics and current methods.

Flexible flow shop scheduling model is along with the GA and Campbell, Dudek, and Smith (CDS) method and some other heuristics methods to maximize accuracy, minimize makespan, as well as to gain in computing efficiency. PFSSP heuristics including GA and CDS [8]. The pairings were found to optimise makespan across industrial applications. The experiments show that GA and CDS are combining well in flow shop scheduling. Heuristic and algorithmic approaches to the complex FSP have been extensively studied in the literature on job scheduling and sequencing. 4 Campbell, Dudek 5 and Smith [9] presented a heuristic algorithm for the n-job, m-machine sequencing problem on adjacent machines. Reisman et al. [10] provided a review of FSS literature with a statistical review – trends and techniques used. More recently, Rom and Slotnick [11] illustrated the use of genetic algorithms in order acceptance which provides a good example of the applicability of evolution approaches in operational research [12]. Min makespan of PFSSP with a hybrid metaheuristic algorithm, demonstrates that combined techniques are good candidates to produce results. A study by Chaudhry and Drake [13] shows that genetic algorithms can be successfully used to minimize makespan within machine setup times for multi stage flowshops, indicating that the setup consideration is an important dimension of scheduling problems. Additionally, Xiao et al. [14] studied permutation flow shop scheduling with order acceptance and weighted tardiness, thus demonstrating the nature of balancing of two competing criteria in the context of scheduling. Finally, Pan et al. [15] proposed an iterated greedy algorithm for the mixed no-idle permutation flowshop scheduling problem. Together, they reflect the evolution of scheduling paradigms and the tabula rasa efforts to facilitate better efficiency and flexibility in manufacturing and operations management.

Methodology

This paper, addresses the NP hardness of flow shop problem and comparative analysis between two strategies for makespan optimisation along with computation time required.

The problem definition and underlying presumptions are explained below.

Given:

n: Number of jobs.

m: Number of machines.

p_{ij}: Processing time of job j on machine i

Objective:

Makespan:

Minimize $C_{max} = \max_j C_{mj}$

where C_{mj} is the completion time of the last job j on the last machine m.

Assumptions:

1. Deterministic processing times: Processing times for each work on each machine are predetermined and unchanging.
2. Sequential processing: Each job follows the same processing sequence through the machines.
3. No machine breakdown: Machines are assumed to be available at all times, with no breakdowns or maintenance downtime.
4. No pre-emption: Once job starts processing on a machine, it cannot be interrupted.
5. Single job per machine: Each machine can process only one job at a time.
6. Setup times included: Setup times are included in the processing times or considered separately but are fixed and known.
7. Infinite buffer capacity: There is unlimited space for jobs waiting between machines.
8. No job splitting: Jobs cannot be divided into smaller parts to be processed simultaneously on different machines.
9. Job arrival: All jobs are available for processing at time zero (no dynamic arrivals).
10. Flow conservation: Jobs do not leave the system until all processing is completed.

Variables:

- C_{ij}: Completion time of job j on machine i.
- S_{ij}: Start time of job j on machine i.

Constraints:

1. Machine availability: M2 has a maintenance break 12:00 PM to 1:00 PM.
2. Variable setup times: Switching from J3 to J4 on M1 takes an additional 10 minutes.
3. Due dates: J1 must be completed by 5:00 PM.
4. Priority: J2 has a higher priority and should be scheduled first if possible. The steps and explanation of the NEH and CDS algorithm is described below.

CDS algorithm

CDS heuristics constitutes an extension of Johnson's algorithm. The main objective of the heuristics is to reduce the makespan for n tasks and m machines in a deterministic flow shop scheduling problem.

Steps implemented in the algorithm to solve the problem shown in Figure 37.1.

GA algorithm

GA optimization techniques inspired by natural evolution. They work by creating a population of possible solutions, which evolve over time. The best solutions are selected, combined (crossover), and occasionally altered (mutation) to create new solutions. This process repeats until an optimal or satisfactory solution is found. GAs is useful for solving complex problems where traditional methods struggle.

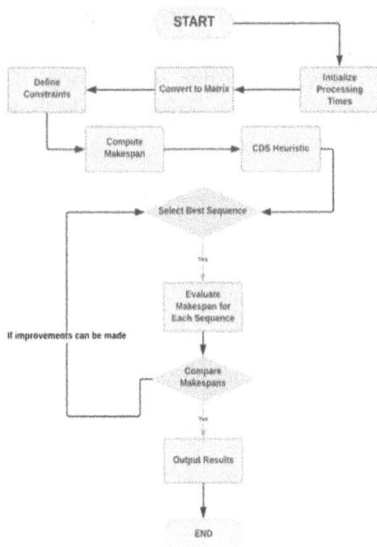

Figure 37.1 Process flow chart of CDS algorithm
Source: Author himself created using lucidchart

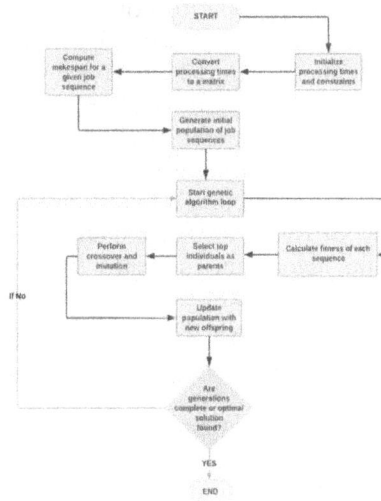

Figure 37.2 Process flow chart of GA algorithm
Source: Author himself created using lucidchart

Assume 2 sets jobs (6,12) each having sets of machines (3,10,20), are to be scheduled.

Implementation

The algorithm was conducted in Python 3.11 on a laptop equipped with a 2.41GHz AMD Ryzen 3 CPU, 4GB of RAM, and operating on Windows 11. Python was chosen for its simplicity, strong libraries, and ability to handle complex computations.

Problem instances were generated using constant variables such as machine speed, setup time, and job release times. Ten iterations were conducted for two job sets across 3, 10, and 20 machines using the critical dispatching strategy (CDS), recording makespan and computation times. The same job sets were then tested with the genetic algorithm (GA) to determine the optimal makespan. Data from an Automatic Filter Press shop in the steel manufacturing industry was utilized to compare the performance of both algorithms regarding makespan and computation time.

Computational Experiments

The overall amount of time needed to finish a series of tasks on a machine is called makespan. Better performance is indicated by lower makespan values.

Two heuristic approaches for scheduling issues are CDS and GA. We will contrast the makespan values of CDS and GA for varying machine and task counts. Where the duration required by the algorithm to calculate the schedule is called computation time. Analysis between the two algorithms (CDS and GA) will provide the effectiveness of the algorithm.

We ran our experiment with objective of minimizing the make-span on the CDS and Genetic Algorithm finding computational time in 6 and12 job having sets of machines (3,10,20) shown in Figure 37.4.

The graph shows a comparison between the makespan values for the CDS algorithm and the GA across different configurations of jobs and machines.

The GA generally provides a better makespan compared to the CDS algorithm, especially as the problem size increases. This makes GA a more efficient choice for more complex scheduling problems, while CDS might be sufficient for smaller or simpler configurations.

The graph comparing the computation time (in seconds) shown in Figure 37.3 for the CDS and GA algorithms across different configurations of jobs and While the GA yields better makespan results than the CDS algorithm, it requires much more computer power. The solution quality-computation time trade-off should choose which algorithm to use.

Statistical Analysis

The mean percentage error shows that GA has a significantly lower error compared to CDS. The standard deviations shown in Table 37.1 indicate variability,

Table 37.1 Statistical analysis of makespan and computational time of CDS & GA

Metric	Value	Value
Mean percentage error	20.0%	19,321.07%
Standard deviation	CDS: 129.10 GA: 140.68	CDS: 0.0034 sec GA: 0.7126 sec
t-statistic	-4.39	-2.79
p-value	0.0071	0.0384

Source: Author himself created using MSExcel tool

with CDS having slightly less variability than GA. In terms of computation time, CDS is much faster than GA. The t-statistics suggest moderate disparities between the two algorithms, with both p-values indicating statistical significance below the 0.05 threshold for CDS and GA.

Statistical significance

This pie chart shown in Figure 37.5 summarizes the headline results of how well GA (green), compared to CDS (orange). This also shows that the two algorithms have advantages in different aspects: GA is superior in optimizing makespan while CDS is superior in minimizing computation time. These factors are to be carefully considered by Stakeholders while selecting an algorithm for a particular scheduling task as it is an integral part of any optimization approach, and may not always align with the optimization goals optimally due to limited computational power or post trivial solutions under stringent time constraints.

Conclusion

Two implementations of a distinct heuristic approach—the combined dispatching system (CDS) and (GA)—were compared in order to explore the makespan optimization efficiency in a flexible job-shop

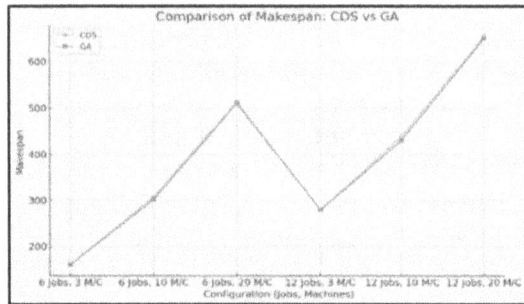

Figure 37.3 Comparison of makespan between CDS and GA

Source: Author himself created using MSExcel tool

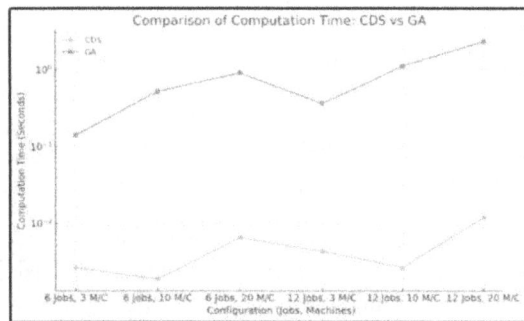

Figure 37.4 Comparison of computation time between GA and CDS

Source: Author himself created using MSExcel tool

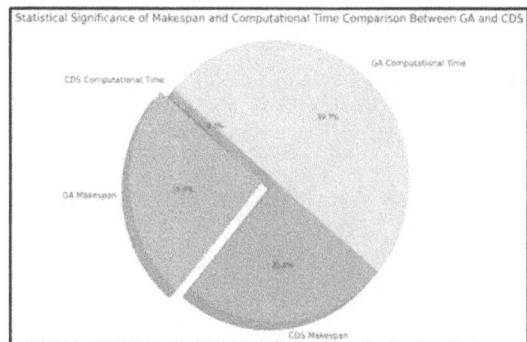

Figure 37.5 Statistical significance of makespan and computational time of CDS and GA

Source: Author himself created using MSExcel tool

scheduling problem (FSSP), which differed in configuration depending on the number and types of jobs and machines involved. Identical sets of benchmark problems were solved using both algorithms and their performance was compared in terms of makespan and computational time. operate more effectively.

Main discoveries
While the CDS algorithm and GA yield the same makespan for six jobs on three machines, their effectiveness varies across scheduling scenarios. For larger, complex problems, GA generally achieves a slightly smaller makespan than CDS, but at the cost of significantly increased computation time. Although GA offers a marginally better mean makespan, it requires around 179 times more processing time than CDS, with high variability in computation time. This makes CDS preferable for shorter scheduling decision periods despite GA's slight advantage in mean makespan.

Novelty of the paper
Comparative examination of GA and Cuckoo search for flow shop scheduling: This work distinctly contrasts GA and CDS within realistic restrictions, emphasising the merits and drawbacks of each method across several scheduling scenarios. Optimisation within industrial limitations: The research enhances the practical significance of flow shop scheduling optimisation by incorporating restrictions like as maintenance breaks, variable setup times, and task priorities

References

[1] Taillard, E. (1990). Some efficient heuristic methods for the flow shop sequencing problem. *European Journal of Operational Research*, 47(1), 65–74.

[2] Allahverdi, A., & Soroush, H. M. (2008). The significance of reducing setup times/setup costs. *European Journal of Operational Research*, 187(3), 978–984.

[3] Gao, J., Sun, L., & Gen, M. (2008). A hybrid genetic and variable neighborhood descent algorithm for flexible job shop scheduling problems. *Computers and Operations Research*, 35(9), 2892–2907.

[4] Tavakkoli-Moghaddam, F. M. R., & Jolai, F. (2023). A bi-objective re-entrant permutation flow shop scheduling problem: minimizing the makespan and maximum tardiness. *Operational Research*, 23(2), 29.

[5] Mousighichi, K., & Avci, M. (2024). The distributed no-idle permutation flowshop scheduling problem with due windows. *Computational and Applied Mathematics*, 43(4), 179.

[6] Ruiz, R., & Maroto, C. (2006). A genetic algorithm for hybrid flowshops with sequence dependent setup times and machine eligibility. *European Journal of Operational Research*, 169(3), 781–800.

[7] Naderi, B., Ruiz, R., & Zandieh, M. (2010). Algorithms for a realistic variant of flowshop scheduling. *Computers and Operations Research*, 37(2), 236–246.

[8] Framinan, J. M., Gupta, J. N., & Leisten, R. (2004). A review and classification of heuristics for permutation flow-shop scheduling with makespan objective. *Journal of the Operational Research Society*, 55(12), 1243–1255.

[9] Campbell, G. H., Dudek, R. A., & Smith, M. L. (1970). A heuristic algorithm for the n job, m machine sequencing problem. *Management Science*, 16(10), B630–B637.

[10] Reisman, A., Kumar, A., & Motwani, J. (1997). Flow-shop scheduling/sequencing research: a statistical review of the literature. *IEEE Transactions on Engineering Management*, 44(3), 316–329.

[11] Rom, W. O., & Slotnick, S. A. (2009). Order acceptance using genetic algorithms. *Computers and Operations Research*, 36 (6), 1758–1767.

[12] Zobolas, G. I., Tarantilis, C. D., & Ioannou, G. (2009). Minimizing makespan in permutation flow shop scheduling problems using a hybrid metaheuristic algorithm. *Computers and Operations Research*, 36(4), 1249–1267.

[13] Chaudhry, I. A., & Drake, P. R. (2009). Minimizing makespan for the multi-stage flow shop with machine setup times using genetic algorithms. *International Journal of Advanced Manufacturing Technology*, 43(3-4), 301–310.

[14] Xiao, Y. Y., Zhang, R. Q., Zhao, Q. H., & Kaku, I. (2012). Permutation flow shop scheduling with order acceptance and weighted tardiness. *Applied Mathematics and Computation*, 218(15), 7911–7926.

[15] Pan, Q. K., Ruiz, R., & Dong, J. Y. (2014). An effective iterated greedy algorithm for the mixed no-idle permutation flowshop scheduling problem. *Omega*, 44, 41–50.

38 Advancing software quality assurance: methods, security, and AI impact

Siddhi Jain[1,a], Sanika Jagiasi[1,b] and Poulami Das[2,c]

[1]Mukesh Patel School of Technology Management and Engineering, SVKM's NMIMS Deemed to be University, Maharashtra, India

[2]Assistant Professor, Mukesh Patel School of Technology Management and Engineering, SVKM's NMIMS Deemed to be University, Maharashtra, India

Abstract

Software quality assurance (SQA) is crucial in ensuring that developed software meets defined requirements and user expectations. This review paper systematically analyzes literature on SQA practices from 2000-2023, drawing insights from 40 research papers sourced from IEEE Xplore, ACM Digital Library, SpringerLink, and ScienceDirect. The review explores the integration of QA practices throughout the software development lifecycle (SDLC), examining aspects such as cost of quality measurement, maturity models, risk assessment, and the role of emerging technologies like AI/ML in security assurance. Key findings emphasize the importance of continuous code review and testing across all phases, collaborative QA approaches, and economic factors influencing QA adoption. Identified gaps include the need for simulation models to predict QA effort and the validation of maturity models across industries. This analysis highlights future research directions to enhance QA frameworks in line with global software industry needs and technological advancements. By synthesizing these insights, the paper provides a reference for both practitioners and researchers aiming to strengthen SQA methodologies, ensuring the development of robust, secure software systems. The paper concludes with an outline of SDLC models, security integration, and the impact of AI in software quality practices.

Keywords: Agile development, artificial intelligence, maturity model, quality metrics, security, software development lifecycle, software quality assurance

Introduction

Over the past decade, there has been significant research on the various kinds of SDLC Models, their applications, implementation of security, and the impact of artificial intelligence (AI) and its applications in the SDLC. However, existing literature lacks a comprehensive review that consolidates the key developments and insights from this growing field of research. The objective of this paper is to fill this gap by providing a systematic and critical analysis of the published work on Software Development Life Cycle models. This review focuses on scholarly articles, conference proceedings, and technical reports published between 2020–2023.

The process of search involved searches on the main databases and electronic libraries with pertinent keywords. Then, the abstracts were filtered for inclusion/exclusion to emphasize the most seminal works. In total, 41 papers were considered appropriate for in-depth analysis, and were processed using a structured and reproducible framework. To put it succinctly, this paper intends to be a reference point

for researchers, practitioners and students who wish to have an overview and depth of work achieved so far in the field of SLDC.

Enhancing SDLC: Security, Autonomy, and Quantum Integration

This research work takes into consideration the integration of security mechanisms, autonomic computing, and quantum technologies in the SDLC models and suggests a quantum software development lifecycle. The focus areas are policy-driven waterfall models for medium-scale secure software, iterative models for large systems with evolving requirements, and hybrid applications based on quantum. It has shown the appropriateness of PDWM and encourages autonomic computing as a self-managing cost-effective means. The research results show that investment in the inclusion of SDLC with security, autonomic computing, and quantum technology can pay off through enhanced robust and self-managed software applications with minimal support and reduced resource consumption

[a]siddhidjain08@gmail.com, siddhi.jain63@nmims.in, [b]sanikaj12@gmail.com, sanika.jagiasi145@nmims.in, [c]dr.poulamidas.cse@gmail.com, poulami.das@nmims.edu

DOI: 10.1201/9781003663348-38

costs, but at an investment cost at the initiation point. Such research opens windows for deeper investigation of agility approaches, as well as holistic testing for high dynamic risk projects and quantum hybrid applications.

Applications of SDLC Methodologies Across Various Domains

This research examines the scope of the applicability of SDLC methodologies across diverse fields such as e-government, enterprise architecture (EA), human-computer interaction (HCI), mental health, educational software, airline management, and agile methodologies for embedded systems. A structured SDLC that is tailor-made according to the special requirements and complexities of a particular project will be crucial to carry out successful software development and deployment.

E-Government and enterprise architecture: The interoperable services from the e-government systems will benefit the most by being built around the EA models, ensuring better cost governance, coordination, and evolutionary support over time in the government contexts.

Human-computer interaction: The use of HCI ideas for the information system design through the SDLC phases means the systems that will emerge will be user-driven, intuitive. Most importantly, with applications sensitive, like in mental health, engaging end-users at any SDLC phase would be best positioned to realize usability coupled with effectiveness in the delivery of interventions.

Agreements on embedded systems agile does support complex, changing requirement problems in embedded systems due to its iterative development and constant feedback mechanisms.

ALM Application lifecycle management ensures a high quality of software using rigorous development, effective management, and structured lifecycle procedures that ensure the SDLC maintains the highest possible level of quality.

Key findings have elicited several emerging themes: E-Government & Enterprise Architecture: The study places much emphasis on the importance of interoperable e-government services, with EA frameworks that would lead to interoperability. EA allows for systems development cost management, coordination among platforms within the government, and sustainable system evolution.

Implementing Security in SDLC Methodologies

Integration of security should always be at every stage of the Software Development Life Cycle to reduce cyber threats and potential vulnerabilities in software. Huge financial losses may occur if significant reputational losses have accompanied these losses, along with accompanying legal losses. If a set of best practices or standards is followed, such as threat modeling, good secure coding, and more importantly, full testing of security, it may enhance and improve the security level on the software at reasonable expenses. Organizations have been given several frameworks, such as SAMM and BSIMM, to help implement security throughout the phases of SDLC. A proper integration for security in the organization will be achieved only through the collaboration of security experts and developers along with clear lines of communication. Good design, coding, testing, and compliance according to research and industry standards demand reliable, safe software systems.

Implementation and Impact of Artificial Intelligence in SDLC

Recent research on how AI is gaining its prominent position attracted the role of AI within software engineering. Many SDLC activities, such as code generation, requirement elicitation, and documentation, have shown impressive efficiency through the application of related domains through tools like ChatGPT and BERT. Big LLMs, like chat GPT and BERT, have state-of-the-art natural language understanding capabilities with domain-specific fine-tuning that may be further extended.

Those are something like Transformers, HuggingFace, and ELMO, though arguably one of the frameworks goes further to optimize LLMs more directly for contextual comprehension and response generation. In one case study, a systematic design where the AI was completely infused in every stage of the SDLC was designed and implemented to develop an e-commerce platform for specific footwear. In that case study, all the phases of the SDLC utilized AI, notably ChatGPT, from start to finish to generate requirements, code, and documentation. Unified modeling language (UML) diagrams played vital roles in transforming those gathered requirements into design specifications. ChatGPT is demonstrated to generate UML diagrams and other relevant artifacts automatically from natural language inputs. Such insights have enabled the creation of useful outputs such as Python code, SQL schemas, and a React-based user interface prototype. This integration of AI into the SDLC opens avenues for further improvements in efficiency, accuracy, and automation in the developmental process of software by creating more intelligent and adaptive engineering processes. Table 38.1 shows various security approaches used in SDLC.

Table 38.1 Security in SDLC approaches

Ref.	Technique Used	Summary in Detail	Merits	Demerits
[1]	Survey Analysis, Comparative Analysis of Agile methodologies.	Recommends pair programming and continuous training for improving security in Agile software development.	Highlights early integration of security to reduce risks in Agile methodologies.	Lacks real-world case studies; narrow focus on specific methodologies.
[2]	Misuse Case Analysis, Penetration Testing, Integrated Security Patterns Guide.	Proposes an integrated secure guide combining Misuse Case Security Patterns and Secure Coding for enhanced security.	Offers comprehensive security guidance for mitigating threats.	Adds complexity to the development process, requiring more resources and expertise.
[3]	Integration of Security, citing Microsoft Security Development Lifecycle (SDL).	Advocates adopting the SSDLC model for SMEs, highlighting the cost-effectiveness of early vulnerability detection.	Promotes the SSDLC for enhancing software security in SMEs.	Lacks real-world case studies and potential bias towards SSDLC without addressing challenges.
[4]	Policy-Driven Waterfall Model (PDWM), Security Policy Framework.	Introduces the Policy-Driven Waterfall Model (PDWM) for integrating security policies into each SDLC phase.	Demonstrates PDWM's effectiveness for addressing vulnerabilities.	PDWM's sequential approach lacks flexibility for projects with changing requirements.
[5]	Software Testing Life Cycle (STLC), modeling languages like Hyperledger Composer's CTO, Smart Contract Development Environments (SCDEs).	Proposes a structured STLC for testing Blockchain-Oriented Software (BOS).	Emphasizes unique testing approaches for BOS development.	Lacks implementation details and empirical validation for the proposed STLC approach.
[6]	Comparison of Waterfall, Spiral, Iterative, and Incremental Models.	Discusses secure software development within various SDLC models like Waterfall, Spiral, Iterative, and Incremental.	Highlights secure software development within Agile methodologies.	Agile's flexibility can impact predictability and timeline accuracy.
[7]	Comparison of SAMM, BSIMM, and CMMI security methods.	Compares security methods like SAMM, BSIMM, and CMMI in the SDLC requirement phase.	Provides a comprehensive comparison of security methods and maturity models for secure software development.	Limited practical implementation and case studies.
[8]	SQL injection mitigation, Authentication management, Access control verification using JSON Web Tokens.	Focuses on secure coding practices like SQL injection mitigation and access control verification in the SDLC.	Offers real-world scenarios and mitigation strategies for improving software security.	Requires additional resources and developer training for effective implementation of secure coding practices.

Source: Author

Conclusion

In conclusion, this review paper has highlighted the critical role of a structured SDLC approach in the successful development of software applications across diverse domains. Tailoring the SDLC to the specific needs of each project ensures that the development process is aligned with its objectives, leading to improved outcomes. Additionally, the incorporation of human-computer interaction (HCI) principles fosters a human-centered design, enhancing user satisfaction and engagement with the system. The involvement of end users throughout the SDLC has been shown to significantly boost the acceptability and productivity of digital solutions. Similarly, maintaining the perspectives of stakeholders during the development and deployment stages is vital for achieving project goals.

Agile methodologies, with their iterative development and continuous optimization, offer considerable advantages, particularly in fields like embedded systems. Furthermore, the role of application lifecycle management (ALM) activities in ensuring software quality and contributing to overall project success cannot be overstated. Together, these practices provide

Table 38.2 Artificial intelligence related approaches in SDLC

Ref.	Technique Used	Summary in Detail	Merits	Demerits
[9]	Swarm Intelligence, Genetic Algorithms, Autonomous Software Code Generation (ASCG), Natural Language Processing, Object-Oriented Programming.	Discusses AI techniques like Swarm Intelligence, Genetic Algorithms, and ASCG to automate processes like code generation and testing in SDLC.	Comprehensive review of AI's role in automating software processes, improving accuracy and efficiency.	Lacks empirical evidence and case studies to support AI's impact; limited discussion on ethical challenges.
[10]	Qualitative Research, Thematic Development, Peer Review.	Proposes a Generative AI-Assisted Software Development (GAASD) model, incorporating AI (ChatGPT, Bard) into SDLC to enhance speed and reliability.	Highlights AI's potential in improving efficiency, speed, and reliability in the SDLC phases.	Small sample size limits generalizability; lacks real-world implementation details.
[11]	Natural Language Processing (NLP), BERT, Domain-specific LLMs (DistilBERT, Code-BERT).	Suggests using AI (BERT, DistilBERT) to automate and improve accuracy in requirement engineering within SDLC.	Offers a modern approach to automating requirement classification, setting a foundation for future advancements in SDLC.	Lacks practical application and detailed real-world testing; further research required for problem resolution.

Source: Author

a holistic framework for driving innovation, enhancing user experiences, and ensuring the quality and effectiveness of digital interventions in various sectors.

Results and Discussion

Increased efficiency with AI in QA: The integration of artificial intelligence (AI) and machine learning (ML) significantly enhances QA processes, automating labor-intensive tasks like code generation, requirements gathering, and documentation. This shift results in improved accuracy and reduced time and costs, with tools like ChatGPT and BERT streamlining these processes across SDLC stages. It has also been experimentally proven that AI considerably reduces labor and capital expenditure in QA with immense future benefits.

Security integration across SDLC phases: Embedding security practices at every SDLC phase—from requirements gathering to maintenance—proactively mitigates vulnerabilities and enhances overall software resilience. Key practices include threat modeling, secure coding, and rigorous testing, supported by frameworks such as SAMM and BSIMM. The results point out the integration of security practices at an early stage in the SDLC to avoid vulnerabilities and build resilience in the system.

Cross-industry application of SDLC models: SDLC models have proven effective across various sectors (e.g., e-government, mental health, education), with domain-specific adaptations that prioritize user-centered design and stakeholder involvement. This cross-industry applicability underscores the flexibility of SDLC methodologies in meeting diverse needs.

Gaps in predictive and validation models: Current gaps in QA include the lack of simulation models to predict QA effort and the need for robust validation of maturity frameworks across industries. Addressing these gaps could improve the precision and adaptability of QA processes, particularly in high-risk and rapidly evolving sectors.

Human-centered design emphasis: Human-Computer Interaction (HCI) principles improve usability and acceptance, especially in sensitive applications like healthcare. Involving end-users throughout development enhances system effectiveness, usability, and satisfaction, creating software solutions that are both functional and user-friendly. Table 38.2 shows various AI related approaches used in SDLC.

References

[1] de Vicente Mohino, J., Higuera, J. B., Higuera, J. R. B., & Montalvo, J. A. S. (2019). The application of a new secure software development life cycle (S-SDLC) with agile methodologies. *Electronics (Switzerland)*, 8(11), 1218. https://doi.org/10.3390/electronics8111218.

[2] Lee, K. H., & Park, Y. B. (2016). Adaption of integrated secure guide for secure software development lifecycle. *International Journal of Security and Its Applications*, 10(6), 145–154. https://doi.org/10.14257/ijsia.2016.10.6.15.

[3] Umeugo, W. (2023). Secure software development lifecycle: a case for adoption in software SMEs. *International Journal of Advanced Research in Computer Science*, 14(01), 5–12. https://doi.org/10.26483/ijarcs.v14i1.6949.

[4] Hussain, S., Anwaar, H., Sultan, K., Mahmud, U., Farooqui, S., Karamat, T., et al. (2024). Mitigating

software vulnerabilities through secure software development with a policy-driven waterfall model. *Journal of Engineering*, 2024(1), 9962691. https://doi.org/10.1155/2024/9962691.

[5] Lilani, S., Modi, J., & Soni, F. (2019). Securing the software development life cycle (SDLC) with a blockchain oriented development approach. In Think India Journal Conference, December 2019.

[6] Hassan, F. M., et al. (2023). Importance of secure software development for the software development at different SDLC phases. https://doi.org/10.31124/advance.23947392.v1.

[7] Shaikh, M., Ali Qureshi, P. H., Shaikh, M., Arain, Q. A., Zubedi, A., & Shaikh, P. (2021). Security paradigms in SDLC requirement phase - A comparative analysis approach. In 7th International Conference on Engineering and Emerging Technologies, ICEET, 2021. https://doi.org/10.1109/ICEET53442.2021.9659614.

[8] Jakimoski, K., Stefanovska, Z., & Stefanovski, V. (2022). Optimization of secure coding practices in SDLC as part of cybersecurity framework. *Journal of Computer Science Research*, 4(2), 31–41. https://doi.org/10.30564/jcsr.v4i2.4048.

[9] Joshi, P., Sorte, B. W., Joshi, P. P., & Jagtap, V. (2015). Use of artificial intelligence in software development life cycle: a state of the art review. *International Journal of Advanced Engineering and Global Technology I*, 3.

[10] Pothukuchi, A. S., Kota, L. V., & Mallikarjunaradhya, V. (2023). Impact of generative AI on the software development lifecycle (SDLC). *International Journal of Creative Research Thoughts*, 11(8): https://ssrn.com/abstract=4536700.

[11] Okonkwo, A., Igah, C., & Onobhayedo, P. (2024). Review of traditional sdlc process, literature and NLP techniques for requirement engineering. https://10.13140/rg.2.2.20705.28002.

39 Applicability of big data in automation industries-intellectual property management

Aranya Nath[1,a], Srishti Roy Barman[1,b], Gautami Chakravarty[2,c] and Anisha Sen[3,d]

[1]Ph.D Scholar, DSNLU Visakhapatnam, Damodaram Sanjivayya National Law University, Andhra Pradesh, India

[2]LLM Student, NLUJAA Assam, National Law University Judicial Academy Assam, India

[3]BA LLB Student, KSOL Bhubaneswar, School of Law KIIT University, Odisha, India

Abstract

The article focuses on big data integration in the automation industries, primarily based on advanced analytics, predictive maintenance, and AI-driven decision-making, which help in business operations in the tech-based arena. The data generated through artificial intelligence applications introduces significant ownership and licensing protection. Further, the big data and automation intersections show how data-driven breakthroughs like machine learning algorithms and industrial Internet of Things (IOT) systems are changing automation's future. It looks at main Intellectual Property issues such as whether data-based inventions can patent how to protect databases, and agreements to share data. Lastly, it stresses the need to update Intellectual Property frameworks to guard new ideas while encouraging teamwork in the global automation field.

Keywords: Big data analytics, data-driven innovation, intellectual property valuation, patent protection

Introduction

Big data has restructured the automation industry by combining analytics, machine learning, and artificial intelligence. It enhances efficiency, predicts equipment problems, and improves product quality. However, intellectual property rights holders provide a monopoly on their inventions for a certain period, subject to certain exceptions. The article discusses the importance of managing and controlling data for commercial and internal purposes and the challenges associated with asserting or claiming data ownership. The data value cycle involving numerous stakeholders exacerbates data ownership difficulties, and cross-border data transfer is crucial for commerce and privacy protection. The absence of privacy and data protection laws and diverse methods across nations threatens fundamental rights and information freedom. Big data applications include predictive maintenance, quality control, supply chain development, demand forecasting robots and automation. IPR security is essential to protect computer algorithms, data analysis techniques, and machine learning models. Patents and trademarks protect ideas, while copyrights protect software and digital content against unlawful copying and piracy. Trade secrets preserve sensitive information, software algorithms and corporate strategies to maintain a competitive advantage. The rest of the article is structured as follows. Section 2 discusses the intersection of Big Data and automation. Section 3 discusses Intellectual Property and its challenges in Big Data. Section 4 discusses case studies. Section 5 discusses the Intellectual Property Frameworks in Big Data Protection. Section 6 concludes the article.

Integration of Big Data in Automation Industries

Sources and types of data

The automation industry relies on data sources, including sensor data, machine-generated data, and human-generated data. Sensor data is crucial for monitoring physical processes, such as temperature, pressure, humidity, and vibration, to ensure peak performance, safety, and problem detection. Electronic data, on the other hand, is generated by machines in automated systems, providing information on equipment health and efficiency. Examining this data enables the identification of patterns and trends, leading to process improvement, preventative maintenance, and improved resource management. Artificial information, on the other hand, is derived from manual inputs, operational choices, and user interactions with automated systems. Operators, quality inspectors, and engineers can contribute to machine-generated data, refining algorithms and improving automation systems. Operator input is essential for incremental improvements in automation processes and ensuring coinciding with real-world realities.

[a]subhamitanath002@gmail.com, [b]srishtisrijaroy@gmail.com, [c]gautamichakravarty21@gmail.com, [d]anishasen005@gmail.com

DOI: 10.1201/9781003663348-39

Data collection and processing

Automation uses Big Data, such as RFID sensors, semi-conductors, and internet-connected devices, to collect and process information associated with industrial automation and control systems. This data is crucial for real-time data collection from distant locations. Data analysis and processing involve cleaning, pre-processing, filtering, coordinating, and transformation procedures. Innovative analytics technologies including statistical analysis, machine learning, and artificial intelligence, assist in identifying trends and relationship identification. Awareness tools help professionals convert complex data into clear graphs, charts, and dashboards, making it easier for practitioners to understand and interpret the content of raw materials.

Key technologies

Numerous fundamental technologies are available in the automation industry to effectively control vast information. The Internet of Things is the technology that connects devices and sensors to collect and share data, making it possible to envisage and manage processes in real-time. This integration enables enhanced efficiency and flexibility of operations and the possibility to control and manage the processes from a distance. Machine learning and artificial intelligence are significant in the automated environment for vast datasets. Such technologies enable the systems to learn from the data, observe trends and make predictions/decisions without writing specific computer codes. For example, artificial intelligence may predict when a fault in a machine may occur, enhance how manufacturing operations execute tasks and improve quality assurance by spotting very minute failures in the systems. Cloud computing and edge computing provide infrastructure and services to store, process and provide analytical services for large volumes of data. Besides, cloud resources enable broad-scope insights and collaboration across various computing application processes and examine data at the edge to reduce

latency and bandwidth requirements. It is significant in automation, where decisions will be immediate. The combination of cloud and advanced computing creates a robust framework for business management.

Integration of big data and automation

Big Data and automation are revolutionizing the industry by incorporating data from various sources. Predictive maintenance is a notable approach, allowing companies to anticipate machinery failures and perform maintenance timely. It shifts from reactive maintenance to proactive and efficient ones and is achieved through algorithms and machine learning models. Big Data analytics also improves quality control and assurance by providing real-time insights into the manufacturing process. It allows businesses to spot flaws early on, reducing the risk of costly recalls and increasing consumer satisfaction. Rashid and Kausik [6] Big Data is also crucial in improving supply chain operations by examining data from suppliers, manufacturers, and distributors. It allows for better demand forecasting, inventory management, and logistical planning. Real-time information on supply chains increases visibility, enabling businesses to monitor shipments, evaluate supplier efficiency, and respond promptly to interruptions. It improves overall supply chain efficiency, lowers costs, and increases customer satisfaction.

IP data assets

Big data integration in automation transforms the recognition of data as a valuable intellectual asset. This data is used to evaluate equipment performance and gain a competitive advantage. It has led to the need for robust protection under intellectual property. Data ownership management is crucial, especially when data is generated and used by numerous stakeholders. Companies must navigate the complex legal landscape of data rights to ensure the security of their information. Companies investing in data analytics are increasingly concerned about the algorithms and methods used to interpret and process this data. Property management is essential for businesses to work with their data while avoiding infringement and competition risks. Standard development strategies for data safety and IP management are crucial for the continuous success of automation products.

Internet of Things (IoT) in the Automation Industry

Definition and characteristics of IoT

The Internet of Things (IoT) is a network of interconnected devices that exchange information over the World Wide Web. These devices, equipped with sensors and advanced technologies, automatically

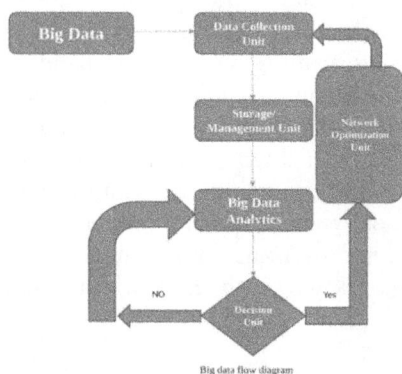

Figure 39.1 Big Data flow diagram
Source: Author

collect and share data without human intervention. IoT is beneficial in the automation sector for providing immediate information, predictive insights, and automated decisions. It enables remote monitoring and control, reducing the need for on-site personnel. Ray [7], IoT's main qualities include connection, intelligence, scalability, and compatibility. Connectivity refers to devices' ability to communicate with each other and central systems, forming a network of interconnected devices. IoT awareness is powered by data analytics and machine learning algorithms, enabling informed decision-making. Scalability is essential as systems can accommodate new devices without compromising performance. Finally, different devices can interoperate within the same IoT ecosystem due to the same protocol manufacturing unit.

Introduction of IoT with big data in automation
Integrating the IoT with big data is a significant innovation that enhances automated operational efficiency and innovation. IoT devices collect vast amounts of data, including sensor readings, machine statuses, environmental variables, and user activities, which can provide patterns, trends, and potential actions for improving automation operations. This data is then sent to unified systems for analysis using Big Data techniques for real-time resource allocation, predictive maintenance, and decision-making. For instance, in manufacturing, IoT devices can monitor machine performance, detect underlying issues, and predict machine failure, reducing time and maintenance costs. The integration also enables the creation of digital twins, virtual counterparts of real assets, processes, or systems, enabling real-time monitoring and modelling of automation systems. It allows proactive improvement and increased production levels [10].

Intellectual Property Challenges in Big Data

Data-related intellectual property rights include copyright, Sui Generis Database Rights, and trade secrets. Trademarks can only be applied to data products, while patents protect applications and business processes that alter and evaluate data. Creative works such as books, music, art, sculpture, films, computer programs, databases, advertising, maps, and technical drawings are protected by copyright legislation. Copyright laws cover databases in most countries but are defined differently and have varied legal interpretations. The TRIPS Agreement WIPO Copyright Treaty offers copyright protection to compilations of data or elements that form a creative production. WIPO Copyright Treaty (WCT) [11], The Berne Convention grants copyright to writers' literary and artistic collections. Berne Convention for the Protection of Literary and Artistic

Works, [2] Data compilation is essential in the context of Big Data. US Copyright Law (17 U.S. Code § 103 - Subject Matter of Copyright) [1] sets a creation produced through collecting and combining previous elements or data, while the European Union's Databases Regulation protects copyright for databases in the EU.

Data ownership
Data ownership in big data and automation is complex, especially when numerous actors are involved in its creation, processing, and use. In IoT industries, data is generated by different companies. At the same time, analytic software is created by third parties. It produces a numerous-layered chain of ownership, posing questions about who owns the data. Data ownership becomes even more problematic when algorithms are used to process it. Co-creation, where numerous participants interact, also raises IP concerns. These unclear distinctions increase uncertainty about legal claims and require precise contractual relationships and legal frameworks. It clearly explains the stakeholder's rights and responsibilities.

Key issues in IPR management for big data
The Big Data presents intellectual property (IPR) challenges that can be legal, technological, and ethical. The management of data ownership is a significant issue, as it involves various aspects of sensors, users, and third-party systems. The authority to handle and profit from this information raises legal questions. Data privacy is crucial as big data contains sensitive information having a competitive advantage. As firms adopt cloud computing and share data with partners, the threat of misuse of trade secrets increases. Owing to that IT laws were enacted to come to the forefront. The challenge is to foster data-driven innovation while protecting personal information. Data quality and integrity are critical for finding patterns and making decisions. Portability and interoperability are essential for data sharing across numerous platforms. Setting industry standards can help resolve data formats and compatibility issues in the intellectual property domain.

Intellectual property management for data license and sharing
Data licensing and sharing add a layer to the intellectual property of big data utilization in automation industries. As businesses depend more on outside information or provide information to another company information protection becomes essential. Such contracts consider who has access to the data and who owns it, as well as what rights the data subject has about its use, processing, modification or transmission. An important issue is that the data might be utilized in a manner not disclosed by the license and can

be stolen by third parties, detracting from the competitive advantage. Further, cross-border data transfer poses legal concerns because the legal requirements on data privacy, security & IPR vary worldwide. For instance, the EU General Data Protection Regulation and international data protection laws consider lawful cross-border data transfers essential. The study contends that businesses must balance measures to oppose data sharing and intellectual property management to avoid exploitation, abuse, or violation.

Patentability of big data technologies

Big Data technology patentability is a rapidly evolving field of intellectual property law requiring businesses to demonstrate their discoveries and innovative solutions to specific challenges and possess control of the physical world. Due to the high speed of invention in this industry, businesses may consider filing provisional patents, which allow a one-year period to further develop the innovation before filing a patent application. Corporations can also seek patents for specific uses of Big Data technology in the automation sector to secure patent protection and gain a competitive advantage. However, patenting Big Data technology may generate issues about patent thickets, which can impede innovation. In mitigating risk, industries could engage in patent pools or licensing arrangements that allow for the pooling of patented innovations while maintaining their intellectual property rights. Evolution of the Patenting of Computer-Related Inventions in India | Managing Intellectual Property, [3].

Case Studies

Bosch's industry 4.0 solutions

Bosch is a leading provider of IoT solutions for Industry 4.0, offering integrated solutions for electronic manufacturing processes. Their proactive upkeep system uses IoT sensors to monitor equipment parameters and predict potential failures, reducing downtime and maintenance costs. It extends the asset's useful life. Smart cities like Chicago, are implementing IoT-based smart street lighting with sensors and cameras. These lights highlighted information on the environment, traffic, and security conditions. Real-time processing of this data can improve traffic control, boost security, and reduce energy consumption, resulting in a more sustainable and pleasant urban experience. Industry 4.0 Solutions from Bosch Rexroth | Bosch Rexroth Hungary [5].

Philips Healthcare's health suite digital platform

Philips Healthcare created the health suite digital platform, which uses IoT technology to collect and examine data from various medical equipment and sensors. This software enables healthcare providers to track patients, detect possible health risks early, and respond proactively, resulting in improved patient outcomes and lower healthcare expenditures. Cleveland Clinic, one of the premier medical institutes in the United States, has integrated IoT devices for remotely tracking patients. Wearable sensors and linked devices enable healthcare practitioners to keep track of symptoms, adherence to medicines, and the activities of patients in real time, resulting in pre-emptive treatments and improved health outcomes.

Industrial internet of things and general electric

General electric (GE) has been a first mover in applying big data to industrial automation with its Predix platform aimed at the industrial IoT. Sensors built into the equipment intake Predix evaluates information leveraged for predictive maintenance, optimal asset performance, and—most importantly—to run a business more effectively across various sectors, including healthcare, energy, and aviation. Therefore Intellectual Property is to patent its software algorithms, data analytics tools, and IoT devices purportedly. The second protection is the protection of GE's computerized databases and machine data actionable insights. Vast data sets that the company keeps technological algorithms used for data analysis projects with the company's hybrid IP strategy and trade secrets. The intellectual property management across its entire data processing and IIoT innovations protect GE's competitive edge in automation [8].

Big data management and Tesla's autopilot

Tesla's Autopilot system utilizes big data from a fleet of vehicles to train AI algorithms and improve autonomous driving capabilities. The company uses sensors, cameras, and radar to collect data and examine its technical system. This data is valuable to Tesla and is tightly controlled as a key intellectual asset. Tesla protects its autonomous driving algorithms through patents and trade secrets, appropriating control of the data generated by its cars while keeping it away from competitors to refine its AI models. This approach allows Tesla to maintain a technological edge in the autonomous vehicle market by managing data and algorithms effectively [12].

Siemens and the role of big data analytics in automation

Mind Sphere, an open IoT operating system, connects machines and devices in industrial settings, enabling Siemens to enhance manufacturing processes through big data analytics. By acquiring big data from operational factory floors, Siemens can determine production rates and energy usage and prevent equipment failures. Siemens' strategic patenting aims to safeguard algorithms and analytical tools, including machine learning models, data integration technologies, and real-time data fusion solutions. Copyright and trade secrets used to protect data analytics algorithms are a company's strengths. Siemens can also use its stringent Intellectual Property framework to license its tools to other manufacturers, cementing its position in the automation industry and ensuring compliance with the law for products benefiting from its technology. Generative Artificial Intelligence Takes Siemens' Predictive Maintenance Solution to the next Level | Press | Company | Siemens [4].

Legal and Ethical Considerations for Big Data

It examines the legal and ethical limitations of applying big data analysis in automation industries, especially in IP management. The electronic nature of data-driven innovations presents challenges because such innovation relates to a universe of ownership and usage rights intricately intertwined with privacy, strong rules on personally identifiable information (PII) and the complexity associated with IP laws. International IP strategies are even more complex as they usually imply compliance with other international legal frameworks (GDPR, sectorial standards) in the cross-border IP and data protection laws. Some ethical issues concerning big data-led automation are related to data ownership, consent and fair access to resources. However, data ownership gets trickier when the owner is unaware of how his/her random slice contributes to an automation system or if he/she does not profit from that minor part. Mainly on the subject of a competitive edge big data and fair use raise significant ethical questions; larger firms can innovate faster than smaller players can keep up. The chapter ends with proposals on how to solve ethical concerns, including documenting the usage of data and avoiding sweeping non-transparent statements that provide room for manoeuvring within the global market emphasizing facilitation in as much access to public datasets while supporting fair competition.

Conclusion and Future Trends

The automation industry relies on intellectual property management (IP) for accurate forecasting and empirical evidence. Conventional IP methods may not accurately capture the intricacies of big data usage patterns, necessitating innovative mechanisms like modular approaches and flexible licensing structures. Policymakers must develop strategies to protect proprietary data without limiting creativity. Suggestions include stronger measures for data protection, data sharing best practices for SMEs, and adherence to international data protection laws. The history of big data in the automation industry is uncertain, but it has the potential for significant improvement. More studies are needed to address big data IP standards and their adaptation in developing economies.

References

[1] 17 U.S. Code § 103 - Subject matter of copyright (n.d.). Compilations and derivative works LII / legal information institute. Retrieved 13 August 2024, Available from: https://www.law.cornell.edu/uscode/text/17/103. 'derivative work', LII / Legal Information Institute. Accessed: Mar. 20, 2025. [Online]. Available: https://www.law.cornell.edu/wex/derivative_work

[2] D. Rai, 'Berne Convention in IPR', iPleaders. Accessed: Mar. 20, 2025. [Online]. Available: https://blog.ipleaders.in/international-conventions-which-shaped-intellectual-property-rights/

[3] 'Patents on Computer-Related Inventions in India', ResearchGate. Accessed: Mar. 20, 2025. [Online]. Available: https://www.researchgate.net/publication/311442534_Patents_on_Computer-Related_Inventions_in_India

[4] 'Siemens brings generative AI to predictive maintenance'. Accessed: Mar. 20, 2025. [Online]. Available: https://www.smart-energy.com/industry-sectors/digitalisation/siemens-brings-generative-ai-to-predictive-maintenance/

[5] 'Digital Transformation in Manufacturing through the lens of Industry 4.0', Bosch Global Software Technologies PVT LTD. Accessed: Mar. 20, 2025. [Online]. Available: https://www.bosch-softwaretechnologies.com/en/explore-and-experience/digital-transformation-in-manufacturing-through-the-lens-of-industry-4-0/

[6] Rashid, A. B., & Kausik, M. A. K. (2024). AI revolutionizing industries worldwide: a comprehensive overview of its diverse applications. *Hybrid Advances*, 7, 100277. https://doi.org/10.1016/j.hybadv.2024.100277.

[7] Ray, P. P. (2018). A survey on Internet of Things architectures. *Journal of King Saud University - Computer and Information Sciences*, 30(3), 291–319. https://doi.org/10.1016/j.jksuci.2016.10.003.

[8] 'Resnick - GE's Industrial Internet of Things Journey.pdf'. Accessed: Oct. 22, 2024. [Online]. Available: https://www.ge.com/digital/sites/default/files/download_assets/GE%20Industrial%20Internet%20of%20Things%20Journey%20%281%29.pdf

[9] Ullah, I., Adhikari, D., Su, X., Palmieri, F., Wu, C., & Choi, C. (2024). 'Integration of data science with the intelligent IoT (IIoT): current challenges and future perspectives - ScienceDirect'. Accessed: Nov. 05, 2024. [Online]. Available: https://www.sciencedirect.com/science/article/pii/S2352864824000269

[10] 'Sheinblatt - THE WIPO COPYRIGHT TREATY.pdf'. Accessed: Mar. 20, 2025. [Online]. Available: https://www.btlj.org/data/articles2015/vol13/13_1_AR/13-berkeley-tech-l-j-0535-0550.pdf.

[11] Wirawanrizkika (2016). 'Competition and Valuation: A Case Study of Tesla Motors', ResearchGate. Accessed: Mar. 20, 2025. [Online]. Available: https://www.researchgate.net/publication/350391969_Competition_and_Valuation_A_Case_Study_of_Tesla_Motors

40 Machine learning based rainfall prediction for weather forecasting: a comparative approach

Ankita Mandal[a], Priti Deb[b] and Manab Kumar Das[c]

Department of Computer Application and Science, Institute of Engineering & Management, Kolkata, University of Engineering & Management, Kolkata, West Bengal, India

Abstract

Out of many challenges related to rainfall prediction, one such challenge with uncertainty, affecting society today is forecasting the amount of rainfall in advance. Predictive rainfall research reduces the maximum financial loss in agriculture and has a significant impact on daily life. Temperature, pressure, wind speed, humidity, and precipitation are the main factors that influence rainfall the most. This research study presents an effective system for prediction of rainfall accurately with the help of various supervised algorithms. In the suggested methodology a large dataset of major cities across India is stored in Microsoft Excel for experimentation, and it is implemented in the Python programming language. This comparative study focuses on three areas: inputs processing, modelling approaches, and utilizing techniques. Finally, the ultimate results enable to compare different machine learning systems' evaluation matrices and analyze meteorological data in order to predict rainfall.

Keywords: Classification algorithms, dataset, machine learning, rainfall prediction, regression

Introduction

Now a days, machine learning (ML) and deep learning techniques play a significant role for predicting the weather and analysis the climate [1–3]. Machine learning techniques have the potential to revolutionize the gradual development of traditional weather prediction [4]. Presently rainfall forecasting becomes essential because doing otherwise could lead to numerous catastrophes. Unpredictable, non-seasonal heavy rainfall can result in flooding, drought, and a major change in crop yield. Eventually, it will have an adverse effect on the ecology and causes harm to human life. Indian economy primarily depends on agriculture, hence accurate prediction of rainfall play a crucial role for successful pre-planning. These days, thanks to advancements in machine learning, we are able to forecast outcomes based on our particular requirements. The field of rainfall prediction has seen significant usage of machine learning as a result of the development of cutting edge technology [5]. The majority cases various algorithms based on machine learning or deep learning are used to forecast the amount of rainfall [6].

The main objective of the present research work is the prediction of rainfall in a particular geographical area, where the users are providing the required data. Parameters that are taken into consideration are-highest and lowest temperature, cloud condition, speed of wind, location, date, humidity etc. These parameters are trained and tested under eight algorithms: decision tree, K-nearest neighbor, random forest, support vector machine, logistic regression, extra trees, XGBoost and LightGBM. Finally, the study predicts the status of rainfall for that particular location.

Related Work

A method for predicting long-term rainfall based on a variety of parameters, including maximum and minimum temperatures, relative humidity during morning and evening, speed of wind, vapor pressure, solar radiation, etc., was proposed by Markuna et al. [7]. The best result was obtained by random forest (RF) out of four different ML models considered in this study: support vector machine (SVM), RF, multiple linear regression (MLR) and multivariate adaptive regression splines (MARS). Pironi et al. [8] established a rapid and accurate method for rainfall prediction. A model implementing ML algorithms for probabilistic rainfall now casting at 10-minute intervals is proposed in this study with short lead durations, ranging from 30 minutes to 6 hours. Patel et al. [9] established a unique technique called gradient boosting, affecting predictability of rainfall with an accuracy of 93%, compared to 90% and 78.5% for random forests and decision trees, respectively. Rahman et al. [10] introduced a new real-time rainfall forecasting system for smart cities that uses a technique called machine learning fusion. This system combines fuzzy logic with

[a]mandalankita92@gmail.com, [b]pritidb@gmail.com, [c]manabdas.aec@gmail.com,

DOI: 10.1201/9781003663348-40

four supervised machine learning algorithms namely DT, NB, KNN and SVM, which considers meteorological data of last 12 years (2005 to 2017) for Lahore city. In order to predict rainfall, Cramer et al. [11] combined the idea of the Markov chain with six other well-known machine learning models. This approach takes into account the data values from 42 distinct cities, each of which had a highly distinctive climate. The effectiveness of daily rainfall prediction using the Ethiopian data was evaluated using extreme gradient boost and RF and MLR. The machine learning model's performance has been evaluated in this study by the Root Mean Squared Error and Mean Absolute Error approaches [12]. Renfei He et al. [13] developed monthly rainfall prediction for long-term planning based on deep learning approach, where rainfall data of Darwin and Perth, Australia was calculated from January 1921 till December 2020. Kumar et al. [14] examined CatBoost, Lasso, ridge, and linear regression for rainfall prediction; CatBoost and XGBoost yielded the best results. The proposed approach considers the weekly rainfall trends in urban metropolitan areas for long-term decisions making. Animas et al. [15] developed a comparable method for rainfall prediction by machine learning and time-series forecasting, taking into consideration climatic data from five major United Kingdom cities in the time period ranging 2000 and 2020. Ridwan et al. analyzes a case study for rainfall forecasting [16]. In order to weigh the station area along with anticipated rainfall, the mean rainfall from 10 stations around the region was calculated using the Thiessen polygon. Sierra et al. established a novel method for long-term rainfall prediction utilizing atmospheric synoptic patterns [17]. This study used 36 years of daily rainfall data to test eight statistical machine learning techniques.

Proposed Methodology

Data exploration and analysis

A diagrammatic representation of the various steps is given in Figure 40.1. Examining the dataset and determining which values are missing is the most useful step in machine learning. Various techniques exist to handle missing values. To complete this task, numerous Python libraries and functions are utilized. In similar ways we also determine whether any missing values exist or not in our data set.

This project had been done using the Indian Weather and Astronomy Data in 2021 (151 cities in all 29 states of India) that contains 29568 data records.

Data pre-processing

In any research context, the process of preparing raw data for a machine learning model is crucial. Here, we

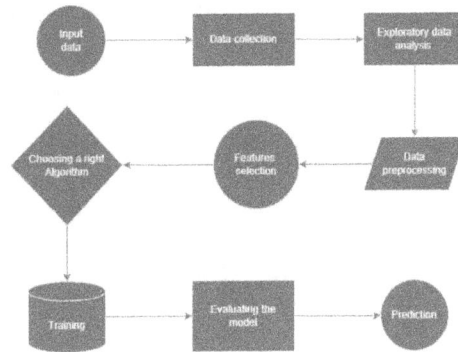

Figure 40.1 Proposed system architecture
Source: Author

eliminate every missing value from our data set and consider 13 features after data pre-processing.

Feature transformation module
Drop irrelevant columns or rows

In this study after removal of few columns, a new set of characteristics are created to predict rainfall. These features included the attributes "time_epoch," "condition," "wind_degree," "precip_mm," "humidity," "cloud," "heatindex_c," "dewpoint_c," and "vis_km."

Correlation analysis

Check the correlation between each feature in the dataset. If any of the features are not found related to the target, then we can revoke that column. In Figure 40.2 shown in the correlation matrix, there are all features related to the result, so we moved forward with these datasets.

Encoding of categorical data into numeric

To convert the category data into a distinct integer value, we used label encoding. This allows us to assign a unique integer to each string or categorical item in order to identify that. The values ['condition',

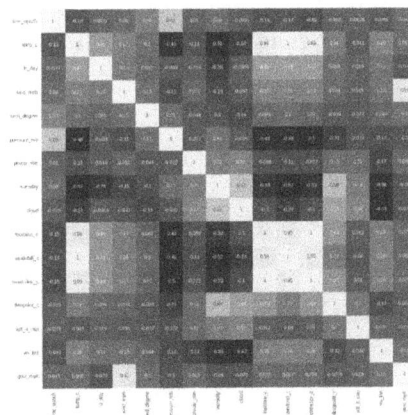

Figure 40.2 Correlation between every feature in dataset
Source: Author

'wind_dir','state', 'city'] are encoded using Label Encoder in this instance.

Scaling the data

A crucial part of our study is scaling the dataset. Due to the wide range of data available, there may be a time complexity issue during the training phase. Therefore, we scale our data according to the range that requires the least amount of processing power to access.

Splitting dataset

The considered dataset is split into 2 sections: training portion and test portion. The use of the training portion is to construct the considered model; thus, a considerable data must be kept for it. The test component is used to validate the model, which means it calculates accuracy based on the test dataset. The data set used in the suggested method is split into two parts: 20% is used for testing and 80% for training.

Feature selection

The final step in this stage is feature selection, during which we identify specific properties that are required to construct the model and exclude or discard any others that are not relevant to the application of algorithms.

Model evaluation

Confusion metrics are used for evaluation of the accomplishment of models. Confusion Matrix is also known as a contingency table. True positive, false positive, true negative and false negative predictions are discriminated against by it. Figure 40.3 shows how the confusion matrix is categorized.

$$\text{Recall} = \frac{\text{TP}}{\text{TP} + \text{FN}}$$

$$\text{Accuracy} = \frac{\text{TN} + \text{TP}}{\text{NP} + \text{FP} + \text{TP} + \text{FN}}$$

First the weather data, which is redeemed, is cleaned, is pre-processed and arranged respectively in the

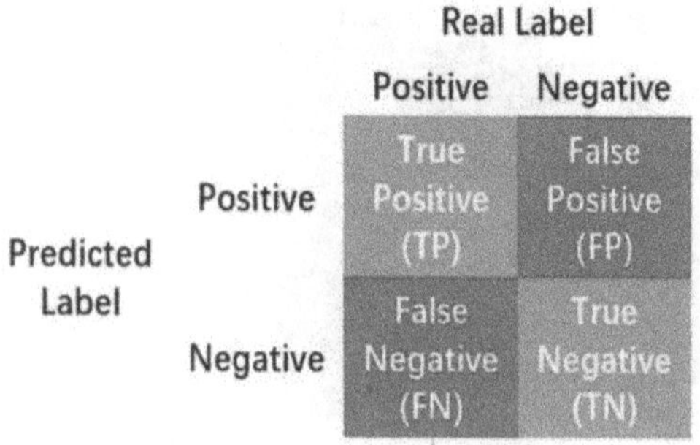

Figure 40.3 Confusion matrix
Source: Author

proposed model. As per Indian Meteorological Department guidelines various categories are formed by designating rainfall data finally. In the present study, machine learning classification and regression methods have been used for building a rainfall forecast method. After processing, the considered data is segregated into training category consisting of 80% of the whole data as well as testing category consisting of the rest 20% data. In this procedure, eight machine learning algorithms are used on the fragmented data. After that, every outcome is examined. Finally, the full, exact data is being shown. The operation of each considered classifiers is explained in the following section.

i)   **Logistic regression:** A method of data analysis known as logistic regression uses the mathematical concepts to detect and establish the relationship between two data factors. Using sigmoid function, this relationship is used to forecast the value of one of those factors dependent on the other. This prediction has a limited number of possible outcomes, either yes or no.

ii) **K-Nearest neighbor:** The k-nearest neighbor algorithm (KNN) is a kind of non-parametric, supervised learning classifier. It uses proximity to make prediction or classification, about individual data point's grouping. The new data points is categorized by this KNN algorithm, which is based on the earlier stored data point's similarity measure.

iii) **Random Forest:** In the form of an ensemble of decision trees, Random forest (RF) algorithm builds a forest while growing the trees. This gives more fickleness. The algorithm run down for the best features from the random subset features while splitting a node. This adds more heterogeneity, therefore resulting in a better model.

iv) **Decision tree:** Decision tree (DT) algorithm is a kind of classification algorithm which works on both unconditional and numerical data. In tree-shaped graph tree-like structures are created by it and that is ease of implementation and of analysis the data. Based on the most important measures, this algorithm helps in separating the data into two or more than two related sets. At first, each attribute's entropy is calculating and then the data is severed, with predictors having minimum entropy or having maximum information gain: The obtained results are much simpler to read and explain. The considered algorithm has accuracy which is much higher than other algorithms as the dataset analyses in the tree as graph.

v)   **Support vector machine:** Support vector machine (SVM) algorithm finds from both the classes the line's closest point. These points are also known to be support vectors. The hyper plane and the distance between the vectors are called margins. Maximize this margin is the goal of SVM. The objective of the SVM algorithm is to discover a hyper plane that, to the best degree possible, This segregates the data points between two classes.

vi) **Extra trees classifier:** Extra trees which is a group supervised machine learning method, short for extremely randomized trees that makes use of decision trees. This is used by the Train Using AutoML tool. Referring to the Decision tree allocation and retrogression algorithm for information about its working. To contain the most important features the tree is then pruned only for making predictions.

vii) **XGBoost classifier:** It is intended for maximum gradient boosting. This approach, which is based on decision trees, is an improvement over gradient boost and random forest techniques. It handles complicated, large-scale datasets with reasonable outcomes by utilizing a variety of optimization techniques. XGBoost is used to fit an initial prediction to a training dataset.

viii) **LightGBM classifier:** It implements a histogram-based method. Using a histogram of the distribution the data is reset into bins. Instead of each data point, the bins are being used to iterate, calculate the gain, and split the data respectively. For a sparse dataset, this method can be optimized as well. Based on decision tree algorithms LightGBM is a fast, distributed, high performance gradient boosting framework.

Results and Discussion

In the framework of this research, Python 3 and Google Colab's Jupyter Notebook are used for all classifier tests and progress. Here a variety of inbuilt libraries are utilized, including Numpy, Seaborn, Matplotlib, Sckit Learn, and Pandas to develop the model

This study approach implemented test strategy using a variety of input data. Here the dataset is partitioned to 80:20 ratio for training testing purpose. A sufficient amount of training as well as testing has been done on the models. Every value can be stated in tabular form, as demonstrated in Table 40.1,which shows the implementation of various classifiers.

LightGBM, XGBoost and SVM are giving generalized results. We consider LightGBM algorithm as our best model. extra trees, decision tree and RF seem too overfit a little.

Hyperparameter tuning

Table 40.1 Accuracy comparison table

Model name	Train_Recall	Test_Recall
XGBoost	0.997626	0.940325
LightGBM	0.994302	0.940325
ExtraTrees	1.000000	0.920434
Random forest	1.000000	0.900542
SVM	0.891738	0.887884
Decision tree	1.000000	0.882459
KNN	0.863248	0.763110
Logistic regression	0.000000	0.000000

Source: Author

Hyperparameter tuning is a fundamental portion of controlling the behavior of a machine learning system. In case we don't accurately tune our hyperparameters, our evaluated demonstrate parameters deliver problematic comes about, as they don't minimize the misfortune work.

Selection of an optimal set of hyperparameters used in learning calculation is called hyperparameter tuning or optimization in machine learning. Parameters whose value is used to regulate the learning handle could be called a hyperparameter.

This study tunes top 2 models i.e. LightGBM and extra trees. Since LGBM gives us the highest accuracy, we have shown the confusion matrix for LGBM for both training and testing dataset. This has been reflected in Table 40.2 and Figure 40.4.

Conclusion

In this paper, investigation and connection of a few pre-processing steps have been done and their effect on the in general execution of the considered classifiers has been discussed. A comparative study of all the relevant classifiers with diverse input information has been done resulting in the influence of input information upon the demonstrate expectations. Based on the work, it can be concluded that Indian climate is dubious and there's no such relationship among precipitation and the particular locale and time

The collected raw information was not arranged for utilizing as input of calculation, hence it had been

Table 40.2 Tune result

Model name	Train_Recall	Test_Recall
Light gradient boosting machine (LGBM)	1.0	0.9584
Extra Trees	0.9838	0.9421

Source: Author

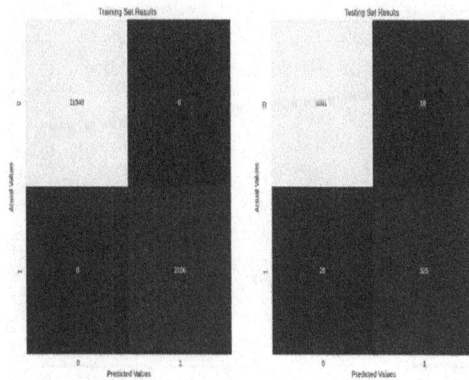

Figure 40.4 Confusion matrix of LGBM
Source: Author

preprocessed physically to suit into the calculation, at that point nourished to the calculation. The assessment comes about of the consideration conducted on the information appears that the anticipated method performs higher than the routine systems in term of precision and prepare running time. The proposed approach yielded the most extreme expectation of 95.84%.

As there is a gigantic portion of information to be processed, profound learning models such as multilayer perceptron, convolutional neural organize etc. can be applied. A comparative think about between the machine learning classifiers and profound learning models is expected to produce some effective outputs with improved precision.

References

[1] Chen, L., Han, B., Wang, X., Zhao, J., Yang, W., & Yang, Z. (2023). Machine learning methods in weather and climate applications: a survey. *Applied Sciences*, 13(21), 12019.

[2] Bauer, P., Dueben, P., Chantry, M., Doblas-Reyes, F., Hoefler, T., McGovern, A., et al. (2023). Deep learning and a changing economy in weather and climate prediction. *Nature Reviews Earth and Environment*, 4(8), 507–509.

[3] Bochenek, B., & Ustrnul, Z. (2022). Machine learning in weather prediction and climate analyses—applications and perspectives. *Atmosphere*, 13(2), 180.

[4] Ben Bouallègue, Z., Clare, M. C., Magnusson, L., Gascon, E., Maier-Gerber, M., Janoušek, M., et al. (2024). The rise of data-driven weather forecasting: a first statistical assessment of machine learning–based weather forecasts in an operational-like context. *Bulletin of the American Meteorological Society*, 105(6), E864–E883.

[5] Hussein, E. A., Ghaziasgar, M., Thron, C., Vaccari, M., & Jafta, Y. (2022). Rainfall prediction using machine learning models: literature survey. *Artificial Intelligence for Data Science in Theory and Practice*, 75–108

[6] Basha, C. Z., Bhavana, N., Bhavya, P., & Sowmya, V. (2020). Rainfall prediction using machine learning & deep learning techniques. In 2020 International Conference on Electronics and Sustainable Communication Systems (ICESC), (pp. 92–97). IEEE.

[7] Markuna, S., Kumar, P., Ali, R., Vishwkarma, D. K., Kushwaha, K. S., Kumar, R., et al. (2023). Application of innovative machine learning techniques for long-term rainfall prediction. *Pure and Applied Geophysics*, 180(1), 335–363.

[8] Pirone, D., Cimorelli, L., Del Giudice, G., & Pianese, D. (2023). Short-term rainfall forecasting using cumulative precipitation fields from station data: a probabilistic machine learning approach. *Journal of Hydrology*, 617, 128949.

[9] Patel, A., Keriwala, N., Soni, N., Goel, U., Bhoj, R., Adhyaru, Y., et al. (2023). Rainfall prediction using machine learning techniques for sabarmati river basin, Gujarat, India. *Journal of Engineering Science and Technology Review*, 16(1), 101.

[10] Rahman, A. U., Abbas, S., Gollapalli, M., Ahmed, R., Aftab, S., Ahmad, M., et al. (2022). Rainfall prediction system using machine learning fusion for smart cities. *Sensors*, 22(9), 3504.

[11] Cramer, S., Kampouridis, M., Freitas, A. A., & Alexandridis, A. K. (2017). An extensive evaluation of seven machine learning methods for rainfall prediction in weather derivatives. *Expert Systems with Applications*, 85, 169–181.

[12] Liyew, C. M., & Melese, H. A. (2021). Machine learning techniques to predict daily rainfall amount. *Journal of Big Data*, 8, 1–11.

[13] He, R., Zhang, L., & Chew, A. W. Z. (2024). Data-driven multi-step prediction and analysis of monthly rainfall using explainable deep learning. *Expert Systems with Applications*, 235, 121160.

[14] Kumar, V., Kedam, N., Sharma, K. V., Khedher, K. M., & Alluqmani, A. E. (2023). A comparison of machine learning models for predicting rainfall in urban metropolitan cities. *Sustainability*, 15(18), 13724.

[15] Barrera-Animas, A. Y., Oyedele, L. O., Bilal, M., Akinosho, T. D., Delgado, J. M. D., & Akanbi, L. A. (2022). Rainfall prediction: A comparative analysis of modern machine learning algorithms for time-series forecasting. *Machine Learning with Applications*, 7, 100204.

[16] Ridwan, W. M., Sapitang, M., Aziz, A., Kushiar, K. F., Ahmed, A. N., & El-Shafie, A. (2021). Rainfall forecasting model using machine learning methods: case study Terengganu, Malaysia. *Ain Shams Engineering Journal*, 12(2), 1651–1663.

[17] Diez-Sierra, J., & Del Jesus, M. (2020). Long-term rainfall prediction using atmospheric synoptic patterns in semi-arid climates with statistical and machine learning methods. *Journal of Hydrology*, 586, 124789.

41 CampusConnect: multi-modal admission assistant

Avishek Gupta[1,a], Anshit Mukherjee[2,b], Sohini Banerjee[1,c] and Sudeshna Das[1,d]

[1]Assistant Professor, CSE Dept., Abacus Institute of Engineering and Management, West Bengal, India

[2]Final year Student, CSE Dept., Abacus Institute of Engineering and Management, West Bengal, India

Abstract

There are many prospective college students who often have various queries about the process of admission, life in campus and academic schedules. It is very difficult to answer these questions manually, especially during peak application periods and is also a time-consuming procedure for admission staffs. So, it is the need of time to develop an intelligent chatbot system that has the potential to supply precise and timely responses to various enquiries from students which will surely upgrade the overall admissions experience. There are many existing voids and there is also lack of research in the development of chatbots that offer personalized responses and not generic responses. Other voids like multimodal input, real time adaptation in tackling complex queries to name a few are also big voids that exists in this domain. This study put forward the novel algorithm of multimodal chatbot for college admissions experience that overcomes all the discovered voids. The performance of the chatbot is tested through simulating tools. The chatbot utilizes natural language processing (NLP) and machine learning algorithms to understand queries and generate suitable responses. The chatbot allows users to interact either via text or voice inputs. The chatbot's knowledge base is completely trained with dataset from Kaggle site and some manually gathered dataset of frequently asked questions and answers during admission process.

Keywords: Artificial intelligence (AI), chatbot system, multimodal input, natural language processing (NLP) for predictive words

Introduction

The admission procedure in colleges is a crucial stage for prospective students, often fraught with vagueness and complexity. A student has various questions like admission requirements, campus life, and academic schedules, and the need for timely and precise information is the need of the hour. Traditional strategies of addressing these inquiries, primarily through manual responses by admission staff, can be inefficient and overwhelming, especially during peak application time. This situation highlights the urgent demand for innovative solutions that can ease the procedure of admission. Despite progress in technology, existing chatbot solutions in the domain of higher education frequently provide generic responses that fail to address the individual needs of students. Most chatbots focus on answering queries without the feature of personalized interactions based on user profiles, preferences or past interactions. This lack of personalization often creates disengagement and frustration among users as they are not provided with tailored support they require. In addition to this, many chatbots only operate through text, limiting accessibility for users who often prefer voice interactions. This absence of multimodal input capabilities restricts user experience and fails to cater to needs of diverse group of users. Also, the current literature in this domain shows a significant gap in the ability of chatbots to learn from interactions and also adapt it in real-time. There are many existing frameworks that do not incorporate machine learning algorithms that can help the chatbots to upgrade their responses based on user feedback and query patterns. This limitation stops the capability of the chatbot to provide accurate and contextually suitable answers over time, particularly when faced with complex inquiries which require deeper understanding of context. Also, the user experience (UX) Daswani et al. [3] design principles are also overlooked while building educational chatbots. If we focus on UX design, it can surely increase user satisfaction among prospective students. The motivation behind this study is to address these identified gaps by developing a multimodal chatbot in particularly designed for college admissions. The main objective of this research is to create a chatbot that supports both text and voice interactions implementing advanced natural language processing (NLP) techniques for understanding complex queries and incorporating real-time learning

[a]avishekgupta.aiem@jisgroup.org, [b]anshitmukherjee@gmail.com, [c]sohinibanerjee.aiem@jisgroup.org, [d]sudeshnadas.aiem@jisgroup.org

DOI: 10.1201/9781003663348-41

capabilities to generate responses based on interactions with users. This paper contributes to the domain by presenting a novel approach to developing chatbot that fills the voids stated previously. By utilizing multimodal input capabilities, personalized interactions and advanced machine learning algorithms, the proposed chatbot enhances user engagement and improves the overall efficacy of the procedure of admission. Section 2 gives the previous literature work associated in this field of chatbot, highlighting the drawbacks of existing research. Section 3 is the Methodology section where we presented our novel algorithm with proper explanation of how our algorithm overcome previous mentioned voids. Section 4 is going to discuss the experimental results concluded by utilizing simulating tools. Section 5 gives the conclusion and future scope section. Figure 41.1 provides a schematic diagram of our proposed framework on CampusConnect.

Literature Review

Smith and Brown [10] in the year 2024 opted a design-based research approach to build a multimodal chatbot capable of processing text inputs. The chatbot was tested with a group of prospective students during admission cycle but efficiency achieved is 72%. The advantage of this work is the system made more accessible to users with various preferences and disadvantages is its efficiency and since it is limited to single institution this lack's generalizability.

Lee and Patel [7] in the year 2024 proposed a design of the chatbot using machine learning algorithms that adapted responses based on user profiles. The advantage of this work is that it is the first paper in this domain in the direction of using machine learning algorithm and the disadvantage is it lacks a personalized approach followed by multimodal input strategy, and overlooked the UX design principles.

Johnson and Wang [5] in the year 2024 implemented a real-time learning model for chatbot that utilized user feedback to improve its accuracy over time but still the efficacy of the model is 81%. The advantage

of this work is that the system is capable enough to learn from interactions, but the disadvantage is that the system lacks multimodal input strategy with personalized approach, less efficiency, didn't focus on UX design principles and also the model consumes too much computational resources which creates barrier for many institutions to purchase.

Garcia and Kim [4] in the year 2024 focused on the UX design principles for the development of a chatbot for college admission procedure and the model achieved efficacy of 82%. The advantage of this work is that the emphasis on UX led more intuitive interface led to more user satisfaction and enhancement but failed to incorporate machine learning approach along with multimodal input strategy for further ease of the process and also the study didn't focus on long-term effectiveness of the chatbot instead on initial user impressions.

Thompson and Zhao [11] in the year 2024 utilized a mixed-methods approach to evaluate the effectiveness of multimodal chatbot in understanding student inquiries during the process of admission and quantitative data indicated a 20% reduction in email inquiries to the admission office, while qualitative feedback mentioned the chatbot's ability to handle complex questions and the efficacy achieved is 78%. The advantage is that multimodal capabilities improved accessibility and user satisfaction while the limitation is that the study rely on self-supported data which introduced biasness and as well it lacks UX design principles and personalized approach.

While the existing studies have made remarkable progress in this domain but still today there is no chatbot that has personalized approach for multimodal input strategy obeying UX design principles and using machine learning algorithms. So, to overcome this void we put forward our innovative algorithm on chatbot designed specifically admission assistant purpose.

Methodology

The proposed CampusConnect chatbot system responds to queries in two ways. They are domain specific responses (include responses regarding the specific institutes being inquired for) and apologetic responses (include responses to queries which seem tough to retrieve, being quietly answered with an apologetic response).

Our novel algorithm is as follows:

1. START
2. Define the database of query-response pairs as a set: D = {(q_i, r_i) | q_i is a query r_i is the corresponding response }.

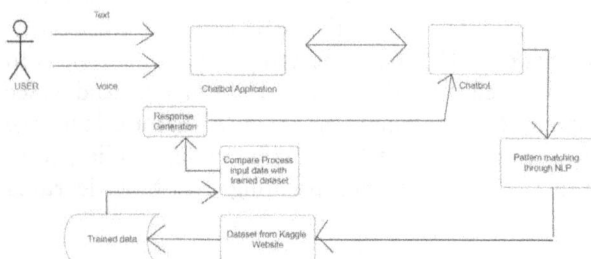

Figure 41.1 Proposed model
Source: Author

3. Define the NLP model as a function: N: X→Y where X is the input space of user queries and Y is the output space of predicted responses.

4. Define the user input processing function: P: U→V where P(u) = {preprocess_voice(u) if is_voice(u), preprocess_text(u) otherwise}. where U is the set of user inputs and V is the set of processed inputs.

5. Define the preprocessing function: Q: V→W *where* Q(v) = join({stem(w) | w ∈ filter(tokenize(v), stopwords)}) where W is the set of cleaned and stemmed inputs.

6. Define the response generation function: R: W→Y where R(w) = {$\mu(w, D)$ if $\mu(w, D) \neq \varnothing$, N(w) if $\mu(w, D) = \varnothing$, "I'm sorry, I cannot assist", otherwise}.

7. Define the pattern matching function: : W × D→Y where μ (w, D) = {r_i if ∃ (q_i, r_i) ∈ D such that $q_i \subseteq W$, $\varnothing$ otherwise}.

8. Define the chat function: C: U→Y where $C(u)$ = R(Q(P(u)))

9. Define the feedback loop function: *F*: U × Y→R where F(u,r) = {1 if user feedback is positive, 0 if user feedback is negative}.

10. Update the database based on user feedback: D ← D U {(u, r) | F(u,r) = 1}.

11. Incorporate user experience design principles:
    a. Define the user interface improvement function: *UI*: V→V'
       UI(v) = v' such that v' is the improved interface based on user feedback and design principles.
    b. Iteratively update the user interface: V ← {UI(v) | v ∈ V}.

12. Implement Q-learning for real-time adaptation: Q:S×A→R *where* $Q(s, a) \leftarrow Q(s,a) + \alpha(r + \gamma \max_{a'} Q(s', a') - Q(s, a))$ where s is the current state (processed user input), a is the action taken (response given), r is the reward received (user feedback), s' is the next state, α is the learning rate, γ is the discount factor.

13. END

Before moving to how we implemented the algorithm first let's look into how our novel algorithm overcomes the void stated previously. The algorithm begins by actually stating necessary components in Line 2 and 3 which includes the database of query-response pairs and the NLP Model which will be utilized to tackle user queries. Line 4 and 5 recovers the void of multimodal input strategy. Line 4 states that the function P(u) actually processes user inputs recognizing whether the input is voice-based or text-based and the important part is despite the type of input it applies preprocessing to guarantee that the data is cleaned and organized. Line 5 tokenizes the input, removes stop words and stems the remaining words and this is a crucial step for transformation of raw user input into a suitable format for further analysis. Line 6 fills the void of tackling complex queries. Line 6 and 7 implements personalization in responses. Line 6 talks about function $R(w)$ which generates responses based on the processed input. Actually, it first checks for existing patterns in the database and if no match is found it uses the NLP model to predict for an apt response. Line 7 uses the $\mu(w,D)$ function to check it the processed input matches any known queries in the database while returning corresponding responses if detected. Line 8 implements chat function $C(u)$ which integrates all the previous functions and then process the user input to produce suitable responses. Line 9 and 10 uses feedback loop to continuously upgrade the chatbot automatically. Line 9 states the feedback loop function $F(u,r)$ which captures user feedback indicating the response was satisfactory and this feedback is essential for improving the performance of the chatbot. Line 10 fills the void of utilization of information analytics. It uses an update strategy for the database based on feedback from user which enables the chatbot to leverage data analytics to improve its responses. By continuously updating its knowledge base, the chatbot can better understand the behaviour of the user and their preferences, and this addresses the void of underutilization of data analytics in existing systems. Line 11 states that the algorithm incorporates user experience design principles by defining an interface improvement function. This function iteratively upgrades the client interface based on client feedback, guaranteeing that interface based on client feedback, ensuring that the chatbot remains user-friendly and engaging. The incorporation of Q-learning in line 12 provides a structured strategy for the chatbot to learn structured approach for the chatbot to learn from user interactions and the update function of Q-learning Adamopoulou and Moussiades [1] adjusts the Q-values based on user feedback, and thus allowing the chatbot to optimize over time.

For our algorithm we collected the dataset from **Kaggle** website (https://www.kaggle.com/datasets/pandanup/college-admission-data-set) which contains comprehensive details about college admission. We have pre-processed and normalized the dataset values utilizing Z score normalization technique. Among hardware we used HP G8 Z5 Workstation, SQLite3 database and software like **Python** (3.8.10), **TensorFlow** (2.6.0), **Keras** (2.6.0), **NumPy** (1.19.2), **Pandas** (1.1.3), **NLTK** (3.6.3), **spaCy** (3.1.0), **Scikit-learn** (0.23.2), **imgaug** (0.4.0) as major recognizable tools for our project purpose. We have used Pandas to load the dataset from SQLite3 database. The data was then pre-processed using NLTK and spaCy to clean the text followed by tokenization and then removal of stopwords. The Keras functional API was then used to state our NLP framework. Then we trained our model with 70% dataset followed by testing with 30% dataset. Then we receive feedback as stated in line 9 and 10. And in line 12 we use Q learning algorithm for real time adaptation and improvement based on feedback received from line 10.

Empirical Results

Table 41.1 shows the results we obtained after training and testing our algorithm CampusConnect through simulation tools against other advanced algorithms like Khanmigo Amin et al. [2], Extraaedge Shanmugam et al. [9], Salesforce CRM [6] in this domain. Our algorithm from Figure 41.2 achieves an accuracy of 92.571% which indicates that our chatbot is powerful at correctly identifying user intents and supplying appropriate responses. With the precision of 93.132% and recall of 91.807% our algorithm describes the strong capability to reduce false positives while correctly recognizing true positives which is important in this domain. The F1-Score of 0.9241 and ROC AUC score of 0.9452 shows that CampusConnect makes a good balance between precision and recall showing that our chatbot maintains high accuracy while retrieving relevant information. Our algorithm has a training time of 2.825 hours which is significantly less than other systems and optimal choice for quicker updates and adaptations to the model. The inference time per response of 28 ms indicates faster timely responses. CampusConnect uses GPU memory of 2.1 GB making our algorithm more resource-efficient and

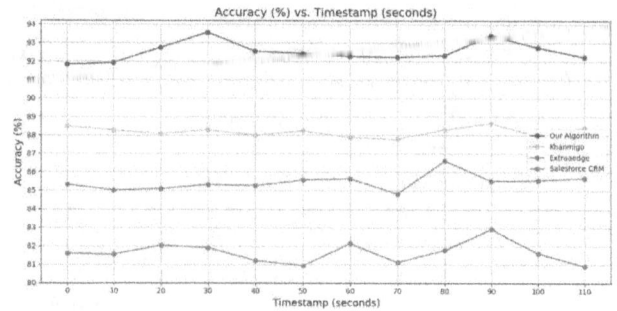

Figure 41.2 Graph for accuracy (%) versus timestamps (seconds) for all algorithms mentioned in Table 1
Source: Author

suitable for deployment in scenarios with limited computational resources. Latency of 45 and throughput of 112 responses/sec further indicates the responsiveness of our chatbot making it optimal choice in handling multiple user queries simultaneously without significant delay. The computational complexity of 2.8×10^9 FLOPS shows that CampusConnect balances performance with efficiency permitting to operate effectively without excessive computational demands. Our algorithm has an energy-efficiency of 0.85 responses per Joule which indicates that our chatbot is designed to be environmentally friendly and cost-effective in terms of energy consumption.

Conclusion and Future Scope

In this paper we introduced our algorithm CampusConnect. The purpose of this proposed chatbot is to manage the questionnaire by providing a user-friendly system to address queries from parents and college students. With this interface, parents and students may ask questions and receive responses in

Table 41.1 Empirical results obtained after comparing various algorithms with our algorithm in this domain

Metric	Our algorithm	Khanmigo	Extraaedge	Salesforce CRM
Accuracy (%)	92.571	88.245	85.472	81.762
Precision (%)	93.132	89.562	87.217	83.423
Recall (%)	91.807	86.957	83.723	79.524
F1-s	0.9241	0.8817	0.8542	0.8135
ROC AUC Score	0.9452	0.9126	0.8973	0.8621
Training time (hours)	2.825	4.126	5.227	6.343
Inference time per response (ms)	28	35	42	51
GPU Memory usage (GB)	2.1	2.7	3.4	4.2
Latency (ms)	45	58	71	89
Throughput (responses/sec)	112	93	78	61
Computational complexity (FLOPS)	2.8×10^9	3.5×10^9	4.1×10^9	5.2×10^9
Energy efficiency (responses/Joule)	0.85	0.71	0.59	0.47

Source: Author

plain English. Their inquiries are promptly and effectively answered by the chatbot, which also makes pertinent connections for them. We can use voice search functionality. Voice input and written output from the system will be provided by users. Chatbots on the internet save time and effort. By 2020, more than 85% of client interactions will be managed remotely, predicts Gartner [8]. Time and money savings are two of the reasons this machine was created. The system primarily uses AI and ML concepts to answer user queries because users can write in a variety of ways. The system's job is to arrange the words and match them with the database; only if a sentence that is fully matched with the database does it provide the user with the relevant answer; otherwise, it automatically saves the user's query and, upon viewing it, the admin can respond appropriately through the Chatbot system. Following the successful completion of the college chatbot, we can conduct further study and use it in many industries, such as banking and medicine etc.

References

[1] Adamopoulou, E., & Moussiades, L. (2020). An overview of chatbot technology. In IFIP International Conference on Artificial Intelligence Applications and Innovations, (pp. 373–383). Springer. https://doi.org/10.1007/978-3-030-49161-1_31.

[2] Amin, K. M., Alturki, N., Alramlawi, S., & Alhejori, K. (2022). Interacting with educational chatbots: a systematic review. *Education and Information Technologies*, 27(6), 8153–8183. https://doi.org/10.1007/s10639-022-11177-3.

[3] Daswani, M., Desai, K., Patel, M., Vani, R., & Eirinaki, M. (2020). CollegeBot: a conversational AI approach to help students navigate college. In Artificial Intelligence and Machine Learning for Multi-Domain Applications, (pp. 49–60). Springer. https://doi.org/10.1007/978-3-030-60117-1_4.

[4] Garcia, T., & Kim, S. (2024). Focusing on UX design principles in the development of a college admissions chatbot. *Journal of Human-Computer Interaction*, 40(3), 301–315.

[5] Johnson, M., & Wang, X. (2024). Implementing a real-time learning model for a college admissions chatbot. *Computers and Education*, 178, 104–120.

[6] Koundinya, H., Palakurthi, A. K., Putnala, V., & Kumar, A. (2020). Smart college chatbot using ML and Python. In 2020 International Conference on Smart Electronics and Communication (ICOSEC), (pp. 1–5). IEEE. https://doi.org/10.1109/ICSCAN49426.2020.9262426.

[7] Lee, C., & Patel, R. (2024). Designing a machine learning-based chatbot for personalized college admissions inquiries. *International Journal of Artificial Intelligence in Education*, 34(2), 123–140.

[8] Ram Mohan, C. B., Divi, A. B., Venkatesh, A., Teja, B. S., & Kotha, M. K. (2019). Chatbot for university resource booking. *International Journal of Scientific Research in Computer Science, Engineering and Information Technology*, 5(2), 113–116.

[9] Shanmugam, R., Jena, S., & Gaur, V. (2020). College information chat-bot system based on natural language processing. *Journal of Xidian University*, 14(5), 825–831. https://doi.org/10.37896/jxu14.5/086.

[10] Smith, J., & Brown, A. (2024). A design-based research approach to building a multimodal chatbot for college admissions. *Journal of Educational Technology*, 14(1), 45–60.

[11] Thompson, E., & Zhao, L. (2024). Evaluating the effectiveness of a multimodal chatbot for college admissions inquiries: a mixed-methods approach. *Educational Technology Research and Development*, 72(1), 55–72.

42 Study on engine performance using various blends of jatropha and lemon peel oil biodiesel with diesel

Kalyan Mukherjee[1,a], Pritam Bhattacharjee[1,b], Arindam Mukherjee[1,c] and Manik Chandra Das[2,d]

[1]Assistant Professor, Department of Automobile Engineering, Dr. Sudhir Chandra Sur Institute of Technology and Sports Complex, Dum Dum, West Bengal, India

[2]Associate Professor, Department of Industrial Engineering and Management, Maulana Abul Kalam Azad University of Technology, Haringhata, West Bengal, India

Abstract

In search of an alternate fuel for diesel, it has been found that biodiesel derived from edible, non-edible vegetable oil and animal fat is the best alternative renewable energy source, but due to having some operational issues that cause a decrease in the performance of the engine, and to mitigate these operational challenges, biofuel derived from plant leaves, wood, seeds, and grass may be used. In the present study, diesel, lemon peel oil (LPO) biofuel, and jatropha biodiesel (JB) blended in varied proportions to generate nine diverse fuel combinations, such as D60J10L30, D60J20L20, D60J30L10, D70J10L20, D70J15L15, D70J20L10, D80J5L15, D80J10L10, and D80J15L5. The Diesel RK program was used to simulate a single-cylinder 4-stroke diesel engine virtually run by all the different nine varied ternary fuel combinations combined with the diesel at maximum load to predict the engine performance characteristics. The simulated result shows that the D60J10L30 fuel combination produced the maximum brake thermal efficiency (BTE) among all the fuels, which is 6.24% greater than the regular diesel, and again, the D60J10L30 fuel blend exhibits the lowest BSFC among all the fuel combinations used in this simulation, which is 3% lower than the diesel.

Keywords: Diesel engine, diesel RK, engine performance, jatropha biodiesel, lemon peel oil

Introduction

The global economy is heavily reliant on fossil fuels, and their consumption is increasing. Due to the need for highly efficient equipment, the transportation and heavy sectors rely heavily on fossil fuel usage. The researchers made an important observation in their study, Milano et al. [7], regarding the finite supply of fossil fuels, which will run out in a few decades and contribute to today's fuel price volatility. Aside from this, fossil fuels are based on hydrocarbon chemicals, which release more hazardous pollutants, resulting in climate change, acid rain, and other environmental consequences [11]. Given the aforementioned considerations, researchers are seeking a renewable, cost-effective, and environmentally benign alternative fuel to diesel, the fuel of choice for the past two decades.

Biodiesels are a significant renewable energy source because of their high oxygen content, lack of sulphur and aromatics, and lack of need for engine modifications [16, 17]. A reduced knocking tendency and shorter ignition delay are the outcomes of an increase in oxygen concentration, which, in turn, minimizes the uncontrolled combustion propensity.

An additional benefit related to increasing the oxygen concentration of the combustion chamber is that fuel burns more completely, releasing more thermal energy during combustion thus enhancing the combustion process. An oxygen-enriched state makes fuel molecules more sensitive and interacts better with oxygen molecules in the combustion chamber, enabling faster combustion and more burning of fuel at the same stoichiometric air-fuel ratio. Biodiesel derived from many sources, including edible and non-edible vegetable oil and animal fats, is a viable option for diesel engines and has already been used in advanced countries [3]. However, employing biodiesel in diesel engines produces various operational issues, such as increased viscosity, decreased heat capacity, gum deposition, and auto oxidation, which reduce engine endurance [2].

To mitigate the adverse effect, biofuel derived from various plant woods, leaves, grass, and vegetable seeds has properties similar to diesel. These biofuels have low viscosity and boiling point temperatures, allowing for more efficient atomisation and evaporation. In addition, the thermal properties of all these

[a]kalyan.aue08@gmail.com, kalyan.mukherjee@dsec.ac.in, [b]pritam.bhattacharjee@dsec.ac.in, [c]arindam.mukherjee@dsec.ac.in, [d]manikdas1@gmail.com

DOI: 10.1201/9781003663348-42

low-viscosity biofuels are substantially equal or comparable with traditional diesel fuels [9].

In a study by Kumari et al. [1], the emission parameters of diesel engines were examined with diesel fuel containing 10%, 20%, and 30% LPO by volume. The findings displayed that diesel fuel containing 20% LPO reduced CO, HC, and smoke emissions by 25%, 25.6%, and 15.44%, respectively, in comparison with diesel fuel alone. The addition of LPO nano emulsions to CI engines was examined by Ramar et al. [10]. The LPO was obtained by steam distillation and transformed into biodiesel through transesterification. In order to examine the emission properties of CI engines, a mixture consisting of 80% diesel fuel and either 50 or 100 ppm of Al_2O_3 nanoparticles was introduced into the biodiesel by a nano emulsion technique. The findings showed that the engine's HC, CO, and NO_x emissions decreased. In another investigation carried out by Vellaiyan and Amirthagadeswaran [15], the authors investigated the acceptability of LPO and water emulsion as ecologically friendly fuels for diesel engines. The LPO was extracted by steam distillation and melded with diesel at various volumetric proportions. Furthermore, the addition of a surfactant emulsified the water. The experimental findings demonstrated that the use of emulsified LPO led to a 9.8% improvement in BSFC and an 11.6% increase in BTE. Additionally, it resulted in reductions of 18.7% in HC emissions, 33.3% in CO emissions, and 26% in NO_x emissions. Sivalingam et al. [13] noticed that while a diesel engine was powered by emulsified diesel merged with multi-walled carbon nanotubes and nanoparticles as a fuel, CO, HC, NO_x, and smoke emissions were diminished by 7.83%, 20.68%, 27.7%, and 37.3%, correspondingly, at the maximum load, in comparison with diesel as a fuel.

Consequently, from the entire literature, it appears that the introduction of biodiesel in CI engines establishes specific functional difficulties, and to counter these potential problems, biofuel derived from plant woods, leaves, and vegetable seeds may be applied. LPO comes under this sort of biofuel, which was employed in diesel engines by various researchers in conjunction with nanoparticle additives, liquid additives, and sometimes emulsions of nanoparticles. In the present research, LPO mixed with JB and diesel to forecast the functioning features of a diesel engine at peak load with the aid of a virtual simulation program, Diesel RK, which was not done previously.

Materials and Methods

In the current study, nine distinct ternary fuel combinations were developed by blending diesel, JB, and LPO in 80% to 60%, 5% to 30%, and 5% to 30%, respectively and their properties presented in Table 42.2. A diesel RK simulation program is employed to anticipate the optimum fuel mixture, which enhances the functioning attributes of the engine. All the simulations were carried out by entering the details of the Kirloskar TV1 single-cylinder diesel engine and the description of the engine summarised in Table 42.1.

Simulation method
The diesel RK program is meant to create an engine by providing its core data, producing a novel endeavour, and storing it. It is vital to keep operating systems revised, regulate the engine's temperature and RPM, and test the spray nozzle's functionality. The component configuration and combustion chamber geometry are developed, and the engine's piston bowl arrangement is an input. Accuracy and study into injection quality are crucial, and the simulation results should be carried out after assessing the configuration settings Mukherjee et al. [8]

Simulation model
The Diesel RK program is an engine simulation tool that uses the first law of thermodynamics to evaluate the engine operation, emanations, and combustion features. In this present inquiry, factors like pressure, temperature, etc. are stated for different crank orientations. Partially empirical correlations dependent on the experimental data were employed to calculate the transfer of heat as well as friction for the engine. A multi-zone model has been introduced to improve accuracy. The conservation equations used in this simulation are presented in this section after Fiveland and Assanis [5].

Conservation of mass
The mathematical expression for the conservation of mass in an open system is as follows:

$$\frac{dm}{dt} = \sum_j \dot{m}_j \tag{1}$$

Conservation of species
During combustion, several species formed on a mass basis inside the chamber. These species may be quantified using the following equation:

$$\dot{Y}_j = \sum_j \left(\frac{\dot{m}_j}{m}\right)\left(Y_i^j - Y_i^{cyl}\right) + \frac{\Omega_i W_{mw}}{\rho} \tag{2}$$

Conservation of energy
The generalised conservation of energy expression for an ensemble that is open could potentially be represented mathematically as follows:

$$\frac{d(mu)}{dt} = -p\frac{dv}{dt} + \frac{dQ_{ht}}{dt} + \sum_j \dot{m}_j h_j \tag{3}$$

Equivalence ratio

The proportion of the empirical air fuel proportion to the balanced air fuel quantity is one way to characterise it, and it can be expressed mathematically via the equation that follows:

$$\varepsilon = \frac{\left(\dot{m}_a / \dot{m}_f\right)}{\left(\dot{m}_a / \dot{m}_f\right)_S} \qquad (4)$$

Engine friction modelling

To assess the engine brake power and brake mean effective pressure (BMEP), it is necessary to accurately model the engine's friction and analyze the different losses associated with engine friction. There are a variety of engine friction theories accessible in Diesel-RK, including the McAulay et al. [6] model and the Chen and Flynn [4]. The engine friction idea that Flynn and Chen devised has been used. In a simple calculation, all of the associated losses are considered FMEP.

$$FMEP = \alpha + \beta P_{max} + \gamma S_p \qquad (5)$$

Brake power

Brake Power is characterised by the amount of power which obtained at engine crankshaft and mathematically can be written as:

$$P_b = T\omega \qquad (6)$$

Brake mean effective pressure

It is described as a ratio of quantity of work delivered to the displacement volume of the cylinder and numerically given as:

$$BMEP = \frac{(Q_S - Q_R)}{V_S} \qquad (7)$$

Specific fuel consumption

It refers to the quantity of fuel employed to generate every unit of power and mathematically can be written as:

$$SFC = \frac{m_f}{P_b} \qquad (8)$$

Result and Discussions

Multiple alternative fuels are evaluated by examining the engine's performance and emission attributes. Several academics have looked at BTE and BSFC as the most crucial performance indicators because they quantify how effectively alternative fuels function in engines. The Diesel RK program was used to model engines with nine potential ternary fuel combinations and diesel, and the results are presented and analysed in this part. Diesel RK software fuel gallery feeds LPO and JB in varied quantities and with appropriate parameters for the simulation. The Diesel RK software contains basic features of data from single-cylinder, 4-stroke diesel engines.

BTE

BTE is a measure of quantity of fuel consumption and the calorific value of the fuel. Figure 42.1 shows the fluctuation in BTE for all nine different fuel combinations, including diesel fuel at maximum load. It has been perceived from the figure that when the LPO concentration rises in the fuel combination, BTE also increases. D60J10L30 fuel blend in which 30% by volume LPO mixed with 60% diesel and 10% JB by volume, and it generated maximum BTE among all the fuels used in this simulation and which is 6.24% greater than the BTE produced by regular diesel. This may be due to increased combustion characteristics, an improved atomization process, and reduced LPO viscosity. The low viscosity and low boiling point temperature accelerate the process of air fuel mixing, which may lead to improved evaporation [14]. From Figure 42.1, it has also been found that all the ternary fuel combinations generated more BTE in comparison to conventional diesel because they had a higher amount of oxygen content, resulting in the availability of more oxygen in the combustion chamber during combustion, leading to complete combustion and improving BTE.

BSFC

Figure 42.2 illustrates the BSFC variation for various fuel combinations, including diesel. BSFC is referred to as the proportion of the quantity of fuel needed to generate every unit of brake power. The figure shows that as the amount of LPO by volume increased in

Table 42.1 Engine specification details

Model	Kirloskar TV1
Description	4 Stroke diesel engine
Type of injection	Direct injection
No. of cylinders	1
Rated power	5.2 kW
Bore	87.5 mm
Stroke	110 mm
Compression ratio	17.5:1
Speed	1500 rpm
Type of cooling	Liquid cooled
Fuel injection	23^0 before TDC
Length of connecting rod	110 mm

Source: Aruna Kumari et al. [1]

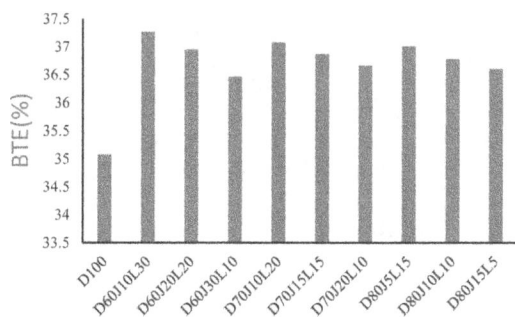

Figure 42.1 Variation of BTE with different fuel blends
Source: Author

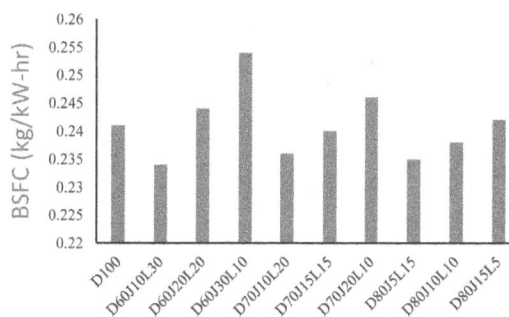

Figure 42.2 Variation of BSFC with different fuel blends
Source: Author

the ternary fuel combination, the BSFC decreased. Although diesel has the highest calorific value among the fuels used in this simulation, the D60J10L30 fuel exhibits the lowest amount of BSFC, which is 3% lower than regular diesel. The low viscosity and boiling point temperature of LPO may explain the observed phenomenon, which enhances the atomization process and accelerates the air fuel mixing process [12].

Conclusion

This study forecasts the performance of a diesel engine, virtually driven by diesel, and nine distinct

ternary fuel blends by blending diesel, LPO, and JB in varying proportions with the aid of a Diesel RK program. Performance of an engine may be determined by finding the two most essential performance characteristics, such as BTE and BSFC, depending on the researcher's point of view. The main conclusions were reached from the current evaluation as follows:

➢ Biodiesel is a sustainable energy source that may be employed in diesel engines as an alternative to diesel because of its oxygen content, which enhances its combustion process and leading to a complete combustion.
➢ The use of biodiesel in a diesel engine introduces various operational challenges such as high vis-

Table 42.2 Comparative properties of tested fuel blends

Properties	D60J20L20	D60J30L10	D60J10L30	D70J15L15	D70J20L10	D70J10L20	D80J10L10	D80J15L5
Carbon composition	0.8532	0.8408	0.8656	0.8574	0.8512	0.8636	0.8616	0.8554
hydrogen Composition	0.1183	0.12115	0.11545	0.120225	0.12165	0.1188	0.12215	0.12357
oxygen Composition	0.025	0.0363	0.0137	0.01975	0.0254	0.0141	0.0145	0.02015
sulfur Fraction	0.00848	0.00772	0.00924	0.00886	0.00848	0.00924	0.00924	0.00886
low heating value (MJ/Kg)	39.902	38.801	41.003	40.5515	40.001	41.102	41.201	40.6505
Cetane Number	42.4	46.2	38.6	43.8	45.7	41.9	45.2	47.1
Density of fuel @ 323K (Kg/m3)	840.4	841.6	839.2	837.8	838.4	837.2	835.2	835.8
Molecular mass of fuel	197.648	212.224	183.072	195.736	203.024	188.448	193.824	201.112
dynamic Viscosity co-efficient (Pa-sec)	0.00464	0.00436	0.00492	0.00423	0.00409	0.00437	0.00382	0.00368
Specific Vaporization heat (KJ/Kg)	240	239	241	242.5	242	243	245	244.5

Source: Mukherjee et al. [18]

cosity, limited thermal capacity, gum deposition, and autooxidation, all of which affect the engine's performance.

➢ Biofuel generated from plant leaves, seeds, wood, and grass has a low viscosity and low boiling point temperature, which increase the atomisation process.

➢ BTE rose with the growth in concentration of LPO in the ternary fuel combination, and it has been observed that the D60J10L30 fuel blend generated the greatest quantity of BTE among all the fuels utilized in this simulation, which is 6.24% greater than diesel.

➢ BSFC reduced with the increase in concentration of LPO in the ternary fuel blend, and it also has been discovered that D60J10L30 shows the lowest level of BSFC among all the fuels utilized in this simulation, which is 3% lower than diesel.

This simulation study predicted that a ternary fuel combination of diesel, LPO, and JB may be utilised in a diesel engine to increase the performance of the engine, and the D60J10L30 fuel combination is an ideal fuel combination among the different fuels employed in this simulation. The engine performance parameters, including BTE and BSFC measured in this study using Diesel RK simulation software in a virtual environment, will be experimentally validated with all ternary fuel blends on the same engine, whose data was input into the Diesel RK simulation software.

References

[1] Aruna Kumari, A., Sivaji, G., Arifa, S., Sai Mahesh, O., Raja Rao, T., Venkata Kalyan, S., et al. (2022). Experimental assessment of performance, combustion and emission characteristics of diesel engine fuelled with lemon peel oil. *International Journal of Ambient Energy*, 43(1), 3857–3867.

[2] Ashok, B., Raj, R. T. K., Nanthagopal, K., Krishnan, R., & Subbarao, R. (2017). Lemon peel oil–A novel renewable alternative energy source for diesel engine. *Energy Conversion and Management*, 139, 110–121.

[3] Atadashi, I. M., Aroua, M. K., & Aziz, A. A. (2010). High quality biodiesel and its diesel engine application: a review. *Renewable and Sustainable Energy Reviews*, 14(7), 1999–2008.

[4] Chen, S. K., & Flynn, P. F. (1965). Development of a single cylinder compression ignition research engine. SAE Technical Paper.

[5] Fiveland, S. B., & Assanis, D. N. (2000). A four-stroke homogeneous charge compression ignition engine simulation for combustion and performance studies. *SAE Transactions*, 109(3), 452–468.

[6] McAulay, K. J., Wu, T., Chen, S. K., Borman, G. L., Myers, P. S., & Uyehara, O. A. (1966). Development and evaluation of the simulation of the compression-ignition engine. *SAE Transactions*, 74, 560–593.

[7] Milano, J., Ong, H. C., Masjuki, H. H., Silitonga, A. S., Chen, W. H., Kusumo, F., et al. (2018). Optimization of biodiesel production by microwave irradiation-assisted transesterification for waste cooking oil-calophyllum inophyllum oil via response surface methodology. *Energy Conversion and Management*, 158, 400–415.

[8] Mukherjee, K., Bhattacharjee, P., Roychowdhury, J., Das, B., Roy, S., & Das, M. C. (2023). Numerical investigation for performance and emission characteristics of a diesel engine fueled with soybean methyl ester biodiesel-diesel blend. *Journal of Decision Analytics and Intelligent Computing*, 3(1), 257–269.

[9] Naik, S. N., Goud, V. V., Rout, P. K., & Dalai, A. K. (2010). Production of first and second generation biofuels: a comprehensive review. *Renewable and Sustainable Energy Reviews*, 14(2), 578–597.

[10] Ramar, K., Subramani, Y., Paramasivam, K., Jayaraman, J., Krishnakanth, P., & Yadav, K. A. (2020). Performance, emission and combustion characteristics of the diesel engine powered by the nano emulsion of the lemon peel oil biodiesel. In AIP Conference Proceedings, (Vol. 2311, no. 1, pp. 1–7).

[11] Silitonga, A. S., Mahlia, T. M. I., Kusumo, F., Dharma, S., Sebayang, A. H., Sembiring, R. W., et al. (2019). Intensification of Reutealis trisperma biodiesel production using infrared radiation: Simulation, optimisation and validation. *Renewable Energy*, 133, 520–527.

[12] Siva, R., Munuswamy, D. B., & Devarajan, Y. (2019). Emission and performance study emulsified orange peel oil biodiesel in an aspirated research engine. *Petroleum Science*, 16, 180–186.

[13] Sivalingam, A., Perumal Venkatesan, E., Roberts, K. L., and Asif, M. (2023). Potential effect of lemon peel oil with novel eco-friendly and biodegradable emulsion in un-modified diesel engine. *ACS Omega*, 8(21), 18566–18581.

[14] Velavan, A., Saravanan, C. G., Vikneswaran, M., Gunasekaran, E. J., & Sasikala, J. (2020). Visualization of in-cylinder combustion flame and evaluation of engine characteristics of MPFI engine fueled by lemon peel oil blended gasoline. *Fuel*, 263, 116728.

[15] Vellaiyan, S., & Amirthagadeswaran, K. S. (2020). Compatibility test in a CI engine using lemon peel oil and water emulsion as fuel. *Fuel*, 279, 118520.

[16] Vellaiyan, S. (2019). Enhancement in combustion, performance, and emission characteristics of a diesel engine fueled with diesel, biodiesel, and its blends by using nanoadditive. *Environmental Science and Pollution Research*, 26(10), 9561–9573.

[17] Yuan, X., Ding, X., Leng, L., Li, H., Shao, J., Qian, Y., et al. (2018). Applications of bio-oil-based emulsions in a DI diesel engine: the effects of bio-oil compositions on engine performance and emissions. *Energy*, 154, 110–118.

[18] Mukherjee, K., Bhattacharjee, P., Mukherjee, A., Ghosh, B., Das, M.C. (2024). Comparative performance of various fuel blend for diesel engine in emission perspective. *International Journal of Scientific Research in Science and Technology*, 11(18), 145-152.

43 The impact of crude oil prices on exchange rates: an econometric analysis

Tran Trong Huynh[1], Bui Thanh Khoa[2,a]

[1]Department of Mathematics, FPT University, Hanoi, Vietnam

[2]Faculty of Commerce and Tourism, Industrial University of Ho Chi Minh City, Ho Chi Minh city, Vietnam

Abstract

This study developed and compared econometric models to understand the dynamics of exchange rate returns, concentrating on the Ho Chi Minh City Stock Exchange using monthly data. Using ordinary least squares regression, the research investigated the impact of oil market returns on foreign exchange returns. Results indicated a positive correlation, with increases in oil market returns significantly influencing foreign exchange gains in the same direction. This highlights the key role of commodity prices, particularly oil, in determining currency values, critical for investors and policymakers in emerging markets like Vietnam. Further, the study contrasted the predictive ability of the OLS model with a support vector regression model. While both models performed well, the OLS model was superior in forecasting accuracy, ascribed to the linear relationships among the examined variables, well-captured by OLS.

Keywords: Crude oil prices, econometric analysis, exchange rate returns, volatility

Introduction

Forecasting exchange rate returns is an important topic in finance, due to its significance for investors, policymakers and international business activities [12]. Exchange rates play a crucial role in the global economy, affecting international trade balances, capital flows and macroeconomic stability [6]. Although basic models like uncovered interest rate parity and interest rate parity have quantified the linear relationship between exchange rate differentials and interest rate differentials across countries, their empirical validity remains debated. The forecasting performance of these models has not been consistently superior to a Random Walk model. Some studies suggest incorporating macroeconomic variables may enhance the predictability of exchange rate returns [7].

The topic of building exchange rate forecasting models remains debated to this day, especially since the seminal work of Meese and Rogoff [10] showed that models based on economic fundamentals did not outperform a random walk model for exchange rate prediction. Since then, the random walk model has been considered a basic benchmark for comparison. Nevertheless, subsequent studies continued exploring the impact of different economic and financial indicators to build exchange rate forecasting models, often with conflicting results. Research findings highlight the complexity of foreign exchange markets, which are influenced by many factors ranging from macroeconomic conditions to market psychology and geopolitical risks [5].

Recent advances in machine learning have opened new avenues for exploring exchange rate predictability as an alternative to traditional quantitative economic models. Our study focuses on building a model to forecast exchange rate returns (rex) using inputs like volatility (vex), skewness (skex), WTI crude oil returns (rwti) Khoa et al. [9], and geopolitical risk (lgpr) [4]. These inputs were chosen based on fundamental theories and prior research in international finance.

Volatility (vex) is an important measure in financial markets, reflecting the variation in trading prices over time [2]. For currencies, higher volatility typically signals greater uncertainty or risk, which can significantly impact returns. Skewness (skex) measures the asymmetry of the profit distribution around its mean value. In traditional quantitative models, the common assumption for the output variable is a normal distribution, implying skewness is zero. However, in reality skewness can vary. Positive skewness indicates a distribution with many small losses and a few extremely large gains, while negative skewness indicates the opposite. Investors face risks from currency returns deviating from a normal distribution, impacting risk management strategies [1].

WTI crude oil returns (rwti) serve as an exogenous price shock. In a closed economy, a negative oil price shock can shift the supply curve left, resulting in lower economic output. Moreover, in an open economy, the

[a]buithanhkhoa@iuh.edu.vn

DOI: 10.1201/9781003663348-43

output decline causes the domestic currency to depreciate relative to foreign currencies. Oil price volatility can significantly impact the global economic landscape, influencing inflation, growth and exchange rates [11]. Finally, geopolitical risk (lgpr) is incorporated by taking the logarithm to capture its non-linearity. Geopolitical instabilities can prompt strong and sudden market reactions. Investors tend to delay investment projects and wait for opportunities as geopolitical risks rise, which can in turn affect exchange rates.

This study aims to expand on previous research by incorporating several exogenous shocks in exchange rate return forecasts, such as geopolitical risk and oil prices. Firstly, the study employs a linear regression model to evaluate the impact of exogenous shocks. Subsequently, it utilizes support vector regression (SVR) with different parameters to capture nonlinear relationships between variables in the model. SVR is chosen for its ability to handle nonlinear relationships well and operate efficiently in multidimensional space. This model is especially suitable for financial time series data, which is often noisy and unstable [8]. Finally, the study compares the performance of SVR and OLS using the RMSE metric. This study chose OLS and SVR regression because they are traditional regression methods that are favored for their simplicity and good interpretability in linear models, suitable for basic financial analysis. SVR was chosen because of its ability to handle nonlinear relationships effectively, especially when financial data is highly volatile. Compared to ARIMA and GARCH, which are commonly used for time series and volatility, OLS and SVR are more flexible in applying to both linear and nonlinear relationships without being too complicated. Advanced machine learning models such as neural networks require large data sets, which are not suitable for the current scope of this study.

Although there has been extensive research on exchange rate return predictability, many limitations remain to be addressed. This study aims to improve those limitations by developing and testing the effectiveness of forecasting models. The regression analysis shows oil market returns tend to be inversely related to foreign exchange market returns. Meanwhile, geopolitical risk does not have a significant impact on foreign exchange markets. Despite testing various parameters with the SVR model, the results still indicate OLS has higher forecasting accuracy.

Data and Method

Data description
The data set utilized in this study comprises monthly observations of exchange rates and crude oil prices spanning from January 1, 2001, to January 1, 2024.

The exchange rate data specifically focuses on the Vietnamese Dong to US Dollar (VND/USD) rate, and the crude oil price data pertains to West Texas Intermediate (WTI) crude oil. Both data sets were sourced from www.investing.com, a comprehensive financial platform providing real-time data, quotes, charts, financial tools, and news across global markets.

In addition to these financial metrics, the geopolitical risk index (GPR) was incorporated to evaluate the impact of geopolitical uncertainty on currency and commodity markets. This index was retrieved from the geopolitical risk website, available at https://www.policyuncertainty.com/gpr.html, which compiles the index based on the frequency of newspaper articles that cover geopolitical tensions.

From these primary data sources, several derived variables were computed for inclusion in our econometric models:

- Rate of return of exchange rate (rex): This variable denotes the monthly rate of return on the VND/USD exchange rate, computed to reflect the percentage change from one month to the subsequent month.
- Volatility of exchange rate (vex) is defined as the mean of the squared daily exchange rate returns for each month. This metric indicates exchange rate volatility and acts as a barometer of market uncertainty and risk.
- Skewness of exchange rate returns (Skex): This statistic quantifies the asymmetry in the distribution of monthly exchange rate returns relative to their mean, offering insights on the propensity of returns to diverge from a normal distribution.
- Rate of return of crude oil West Texas Intermediate (WTI): This variable measures the monthly rate of return for WTI crude oil, emphasizing the investment performance of this commodity over time.
- Log of geopolitical risk (lgpr): The natural logarithm of the GPR index was employed to alter this variable, leveling its distribution and maybe improving the linear relationships with other variables in the regression models.

Research model
This study extends the conclusions of previous research that examined multiple facets of financial markets, including the predictability of exchange rates and the impact of external economic factors such as oil prices and geopolitical threats. This paper proposes a regression model that aims to encapsulate the complexities of market dynamics by integrating many variables recognized as major

predictors in existing literature. These encompass exchange rate volatility, the skewness of exchange rate returns, crude oil return rates, and the influence of geopolitical risks. The model is articulated as follows:

$$(OLS): rex_t = \beta_0 + \beta_1 vex_t + \beta_2 skex_t + \beta_3 rwti_t + \beta_4 lgpr_t + \varepsilon_t$$

Where: $\beta_{0,1,2,3,4}$ are regression coefficients and ε_t is the error term.

This paper conducts a comprehensive analysis to evaluate the effectiveness of SVR in predicting exchange rate returns, utilizing diverse parameter configurations to enhance forecasting performance. SVR is distinguished for its adaptability and strength in managing non-linear data patterns, rendering it appropriate for intricate financial datasets where variable correlations may not be linearly apparent.

Three types of kernels are utilized in the SVR models to evaluate their prediction performance based on varying assumptions regarding the data structure: Linear kernel (Presumes a linear correlation among the variables.) It is straightforward and effective for datasets where the interrelationship among the variables is anticipated to be linear. Radial basis function (RBF) kernel (effective for modeling non-linear interactions). It can project the input features into a higher-dimensional space where they become more distinguishable; Polynomial kernel transforms the original features into a polynomial feature space, enabling the detection of more intricate patterns in the data. Each kernel type is assessed across a hyperparameter grid to identify the most optimal configuration. This study altered the hyperparameters, which include:

- Cost (C): Controls the trade-off between achieving a low error on the training data and minimizing the model complexity for better generalization. The values tested are 1, 0.1, 0.01, and 0.001.
- Epsilon (ε): Defines the breadth of the epsilon-insensitive tube, delineating the margin wherein no penalties are imposed for errors. The epsilon values examined are 0.5, 0.1, 0.01, and 0.001.

A total of 48 different SVR models are tested, combining each kernel with the various settings of cost and epsilon. This comprehensive approach allows the study to meticulously assess which combinations yield the best predictions for exchange rate returns, thus providing a robust analysis of the predictive power of SVR under varying conditions.

Specifically, the model is formulated as follows:

$$(SVR): E(rex_t) = f(vex_t, skex_t, rwti_t, lgpr_t)$$

Model evaluation

In this study, we partitioned the dataset into a 70:30 training and testing split to assess the effectiveness of our predictive models. We employed Root Mean Square Error (RMSE) as the primary metric for evaluation, defined by the equation:

$$RMSE = \sqrt{\frac{1}{n}\sum_{i=1}^{n}(y_i - \hat{y}_i)^2}$$

where y_i and $\hat{y}_i$ represent the actual and predicted values, respectively.

Result

The dataset comprises five variables: skex, vex, rwti, lgpr, and rex, each showcasing varied statistical properties. The descriptive statistics of variables are shown in Table 43.1. The mean values indicate that most variables hover close to zero, particularly vex and rex, suggesting modest average returns and volatility. However, lgpr, the log of geopolitical risk, averages 4.6163, indicating its comparatively larger scale. The standard deviations reveal that skex and rwti exhibit the most variability among the variables, with values of 1.0943 and 0.1144, respectively. Notably, the skewness metrics present significant asymmetry in the data distribution, especially for vex and rex, with values of 9.1931 and 2.6222, suggesting a strong rightward tail. This is further supported by the ranges seen in minimum and maximum values, where skex and rex range widely between negative and positive extremes. Such statistics hint at the potential need for data transformation or specialized modeling approaches to address non-normality and leverage effects in further analyses.

The regression analysis in Table 43.2 provides insights into the relationships between several predictors and the dependent variable, with an R-square value of 0.402, indicating that the model explains approximately 40.2% of the variance. The variable skex shows a strong positive impact on the dependent variable, with a coefficient of 0.003 and a statistically significant p-value of 0.000. Similarly, vex significantly affects the dependent variable with a coefficient of 138.521, underscoring a substantial positive

Table 43.1 Descriptive statistics of variables

	skex	vex	rwti	lgpr	rex
Mean	0.1629	0.0000	0.0059	4.6163	0.0019
St. Dev.	1.0943	0.0000	0.1144	0.3299	0.0082
Skewness	0.7276	9.1931	-0.5601	1.6110	2.6222
Minimum	-3.1026	0.0000	-0.7819	3.9233	-0.0423
Maximum	4.0708	0.0002	0.6333	6.2394	0.0684

Source: Calculation result from data

Table 43.2 OLS regression results

	Coefficients	Standard error	t Stat	P-value
Intercept	-0.002	0.005	-0.326	0.744
skex	0.003	0.000	9.676	0.000
vex	138.521	19.786	7.001	0.000
rwti	-0.006	0.003	-1.722	0.086
lgpr	0.001	0.001	0.476	0.635
R Square	0.402			

Source: Calculation result from data

influence. In contrast, rwti has a slightly negative effect with a coefficient of -0.006, though its p-value of 0.086 suggests only marginal significance, indicating that increases in rwti might slightly decrease the dependent variable. lgpr contributes a statistically insignificant positive effect with a p-value of 0.635. These findings highlight the importance of skex and vex as influential predictors while suggesting that further model refinement and inclusion of additional variables could enhance predictive accuracy.

Following the division of the dataset into training and testing sets, we calculated parameters on the training set and used these parameters to predict outcomes on the testing set. The result shows that the OLS model achieved an RMSE of 0.007450 on the testing set. This indicates a high level of accuracy in predicting outcomes, showing the OLS model as a reliable baseline. For the SVR model:

- Linear kernel: Showed the best performance with a cost (C) of 0.001 and epsilon of 0.01, resulting in an RMSE of 0.006653 on the training set and 0.007476 on the testing set, suggesting excellent model accuracy and generalization from training to testing.
- Polynomial (Poly) kernel: With a cost of 0.01 and epsilon of 0.01, this kernel achieved an RMSE of 0.006644 on the training set and 0.007565 on the testing set, demonstrating robust predictions close to those of the linear kernel.
- The radial (Rbf) kernel, configured with a cost of 0.01 and an epsilon of 0.01, achieved an RMSE of 0.006589 on the training set and 0.007485 on the testing set, demonstrating marginally superior performance in training compared to the polynomial kernel, while remaining equivalent in testing.

These findings demonstrate that although the OLS model serves as a robust baseline, the SVR models, especially those utilizing linear and radial kernels, deliver comparable accuracy with meticulous parameter adjustment. This indicates

significant potential for employing SVR in contexts where intricate management of data linkages is essential. Research indicates that gains from the oil market align with those from the foreign exchange market, attributable to the phenomenon where elevated oil prices induce import inflation due to rising energy and transportation expenses. This may necessitate the State Bank of Vietnam to modify monetary policy to regulate inflation, impacting the VND/USD exchange rate [3].

In contrast to prior studies, including Iyke et al. [5], which identified numerous findings regarding the influence of political risk on foreign exchange markets. The effects in South Asia are beneficial, although some are not statistically significant. Our research aligns with the findings of Iyke et al. [5]. Geopolitical risks may not significantly affect Vietnam's exchange rate due to its stable geographical location, multilateral trade relations, and flexible exchange rate policies from the State Bank of Vietnam. Foreign direct investment (FDI) flows into Vietnam are mainly focused on long-term projects, which are less sensitive to short-term fluctuations. In addition, Vietnam is less directly affected by major geopolitical conflicts, helping the exchange rate remain more stable than other markets that are susceptible to geopolitical risks. The GPR index can be replaced by indicators measuring market sentiment, extracted from specialized newspapers and social networking sites. These news are quickly reflected in the forex market faster than geopolitical risks.

Nonetheless, our study accounted for other factors, including the oil market, volatility, and skewness risk.

Conclusion

This study has notably enhanced the discussion on exchange rate forecasting by utilizing a comprehensive analytical framework that combines both conventional and sophisticated econometric models. Our research has elucidated the complexities of the relationship between the Vietnamese dong (VND/USD) exchange rate and many macroeconomic variables, including crude oil prices, market volatility, and geopolitical threats.

Our findings confirm a significant link between oil prices and the exchange rate, supporting prior research indicating that elevated oil prices might generate inflationary pressures that influence monetary policy and exchange rate stability. The comparative analysis of the OLS and SVR models revealed that, although the OLS model serves as a solid baseline, SVR models demonstrated competitive efficacy, especially with linear and radial kernels, indicating their capability in managing complex and non-linear relationships present in financial data.

Nonetheless, the study possesses certain drawbacks. The statistical insignificance of geopolitical threats indicates that other unassessed factors may influence exchange rates. The data's scope, limited to monthly observations, may neglect shorter-term oscillations that could provide more profound insights. Using high-frequency data helps the model more accurately reflect the market's immediate reactions to oil price fluctuations, thereby improving forecast accuracy and increasing its usability for financial and policy-making applications.

Further research may investigate the integration of high-frequency data to more effectively capture these patterns and broaden the investigation to more developing market currencies to confirm the generalizability of the findings. Broadening the array of predictors, including global economic data or sentiment analysis from news outlets, may improve the predicted accuracy of the models created in this study.

References

[1] Ayadi, M. A., Cao, X., Lazrak, S., & Wang, Y. (2019). Do idiosyncratic skewness and kurtosis really matter? *The North American Journal of Economics and Finance, 50*, 101008.

[2] Bali, T. G., Brown, S. J., & Tang, Y. (2017). Is economic uncertainty priced in the cross-section of stock returns? *Journal of financial economics*, 126(3), 471–489.

[3] Beckmann, J., Czudaj, R. L., & Arora, V. (2020). The relationship between oil prices and exchange rates: Revisiting theory and evidence. *Energy economics*, 88, 104772.

[4] Hao, X., Ma, Y., & Pan, D. (2024). Geopolitical risk and the predictability of spillovers between exchange, commodity and stock markets. *Journal of Multinational Financial Management*, 73, 100843.

[5] Iyke, B. N., Phan, D. H. B., & Narayan, P. K. (2022). Exchange rate return predictability in times of geopolitical risk. *International Review of Financial Analysis*, 81, 102099.

[6] Jamil, M. N., Rasheed, A., Maqbool, A., & Mukhtar, Z. (2023). Cross-cultural study the macro variables and its impact on exchange rate regimes. *Future Business Journal*, 9(1), 9.

[7] Khoa, B. T., & Huynh, T. T. (2022). Predicting exchange rate under UIRP framework with support vector regression. *Emerging Science Journal*, 6(3), 619–630. https://doi.org/10.28991/esj-2022-06-03-014.

[8] Khoa, B. T., & Huynh, T. T. (2023). The value premium and uncertainty: an approach by support vector regression algorithm. *Cogent Economics and Finance*, 11(1), 2191459. https://doi.org/10.1080/23322039.2023.2191459.

[9] Khoa, B. T., Huynh, T. T., & Huong, N. T. D. (2023). Predicting returns of exchange rate from oil prices: machine learning approach. In Paper Presented at the 2023 IEEE 8th International Conference for Convergence in Technology (I2CT).

[10] Meese, R. A., & Rogoff, K. (1983). Empirical exchange rate models of the seventies. *Journal of International Economics*, 14(1-2), 3–24. https://doi.org/10.1016/0022-1996(83)90017-x.

[11] Mollick, A. V., & Sakaki, H. (2019). Exchange rates, oil prices and world stock returns. *Resources Policy*, 61, 585–602.

[12] Narayan, P. K., Sharma, S. S., Phan, D. H. B., & Liu, G. (2020). Predicting exchange rate returns. *Emerging Markets Review*, 42, 100668.

44 The impact of regularization on financial forecasting models

Tran Trong Huynh[1], Bui Thanh Khoa[2,a]

[1]Department of Mathematics, FPT University, Hanoi, Vietnam

[2]Faculty of Commerce and Tourism, Industrial University of Ho Chi Minh City, Ho Chi Minh city, Vietnam

Abstract

This study analyzed data from August 2000 to April 2024 to investigate how local and global economic factors, especially geopolitical risks and gold prices, influence Vietnam's VN-Index (VNI) stock market. The data exhibited significant autocorrelation and heteroskedasticity. We utilized Lasso regression to tackle these complexities and diminish model intricacy, serving as an alternative to conventional ordinary least squares (OLS) regression. The optimization of the lambda regularization parameters within a rolling window framework improved forecasting precision, assessed through mean absolute error (MAE) and root mean square error (RMSE). This methodological rigor yielded a more precise and resilient model to encapsulate the intricate dynamics of the VNI. The results illustrated the efficacy of advanced econometric techniques such as Lasso regression in forecasting financial markets, especially in emerging economies.

Keywords: Geopolitical risks, lasso regression, market prediction, regularization parameters, VN-index

Introduction

Stock markets are essential to the global economy, acting as indices of economic vitality and enablers of capital distribution. These marketplaces enable enterprises to get cash and offer investors avenues for wealth accumulation, rendering the prediction of market indexes both crucial and exceedingly difficult [4]. The VN-Index (VNI), the principal stock index for the Vietnamese market, embodies these dynamics and serves as a distinctive case study owing to its responsiveness to both local and global economic influences [9].

The comprehension and forecasting of the VNI's behavior is intricate owing to the interaction of several components. The index is significantly affected by domestic transitions, policy modifications, and economic indicators inside the Vietnamese economy. It responds to fluctuations in worldwide markets, particularly attuned to the U.S. market, which serves as an indicator of global economic trends. Furthermore, Vietnam's expanding gold market frequently exhibits an inverse correlation with the VNI, signifying its function as a refuge amid economic instability [2]. This inverse link underscores the duality of investor behavior in pursuing development via shares and safeguarding with gold.

Moreover, geopolitical risks pose a substantial issue, since regional tensions and global wars may destabilize markets [1]. The VNI is vulnerable to market-specific risks, including volatility and skewness, which denote the frequency and size of market fluctuations and the asymmetry of investment returns. These factors enhance the index's volatility, rendering forecasting a challenging endeavor.

Notwithstanding comprehensive study, a completely predictive model of the VNI continues to be unattainable. Conventional econometric models frequently inadequately account for the comprehensive range of influencing variables due to their complexity and the non-linear interrelationships among them [10]. This problem is exacerbated by statistical concerns like serial correlation and heteroskedasticity, which result in uneven variability of the index across time, contravening the normal assumptions of homoscedasticity in regression models.

This work utilizes Lasso regression to address these issues, a contemporary statistical method suitable for high-dimensional situations and multicollinearity among variables. Lasso, or least absolute shrinkage and selection operator, improves model accuracy by selecting reducing less significant predictor coefficients to zero, so simplifying the model while preserving crucial variables [6]. This methodology is especially advantageous in financial modeling, where several variables may possess predictive capability, however only a select number are essential.

Our methodology employs a rolling window technique for the training and evaluation of the Lasso model. This approach perpetually modifies the temporal frame utilized by the model, therefore capturing dynamic fluctuations and any structural disruptions within the data series [7]. This method offers a more

[a]buithanhkhoa@iuh.edu.vn

DOI: 10.1201/9781003663348-44

resilient and flexible model compared to conventional static models.

This study reveals an inverse association between gold prices and the VNI. In times of significant market volatility or economic decline, investors typically shift from equities to gold, therefore solidifying their reputation as a safe-haven asset. This behavior highlights the intricate interdependencies in financial markets, where many asset classes may function both as investment possibilities and as risk management instruments. The document is organized into four primary sections to thoroughly address these issues. The initial section, the Introduction, delineates the significance of the stock market and the difficulties inherent in predicting the VNI. The Methodology section delineates the data utilized, the particulars of the Lasso regression technique, and the justification for the rolling window approach. The Results and Discussion section analyzes the findings, highlighting their practical significance for investors and policymakers. Ultimately, the Conclusion encapsulates the principal findings and proposes directions for further inquiry to enhance the prediction models of the VNI.

Method

The dataset included in this study comprises monthly data from August 2000 to April 2024. The dataset includes characteristics pertinent to financial market dynamics, including the S&P 500 index, gold prices, VNI, and the geopolitical risk index (GPR). Financial data was obtained from www.investing.com, providing a dependable foundation for the investigation. The GPR index data, which quantifies geopolitical risk by the frequency of newspaper articles, was sourced from https://www.policyuncertainty.com/gpr.html.

Definitions of Variables:

- S&P 500: Monthly return rate of the S&P 500 index.
- rgold: Monthly return rate of gold prices.
- logGPR: The natural logarithm of the geopolitical risk index is utilized to standardize the data and mitigate skewness.
- lagvni: The lagged monthly rate of return of the VNI, indicative of the performance from the preceding month.
- vvni: The monthly volatility of the VNI, determined as the mean of squared daily return rates.
- skvni: The skewness of the VNI monthly returns, signifying asymmetry in the distribution of returns.

The fundamental econometric model utilized to assess the influence of these factors on the VNI is articulated by the subsequent regression equation.

$$vni_t = \beta_0 + \beta_1 sp500_t + \beta_2 rgold_t + \beta_3 logGPR_t + \beta_4 lagvni_t + \beta_5 vvni_t + \beta_6 skvni_t + \varepsilon_t$$

where ε_t represents the model's error term, and β_i coefficients indicate the impact of each predictor.

The study initially employs an ordinary least squares (OLS) regression to estimate the relationships between the VNI and the included predictors. Lasso regression is also applied to ensure robustness and address potential issues of multicollinearity and overfitting. Lasso regression is especially beneficial in models with several variables, since it conducts variable selection by reducing less significant coefficients to zero. Lasso regression reduces multicollinearity by zeroing out less significant variable coefficients, leaving just the most important predictors. This feature is useful for high-dimensional data like financial time series, because OLS may yield incorrect estimates due to linked factors. Lasso reduces MAE and RMSE better than OLS through regularization. Your study shows that the optimized Lasso model (lambda = 1) decreases these mistakes, improving forecast dependability. Regularization prevents Lasso from overfitting and improves generalization on new data. Overfitting datasets with complicated relationships can cause OLS forecasts to be highly variable. The rolling window technique with Lasso improves adaptation for volatile financial settings.

Rolling window analysis: A fixed-length rolling window methodology is employed for model validation, utilizing a 60-month window. This approach assesses the model's prediction capability over time by persistently modifying the training and testing datasets.

Regularization and model selection: The Lasso regularization parameter, lambda, is adjusted across several levels to examine its impact on model performance. The values tested include 0, 0.1, 0.25, 0.5, 0.75, 1, 1.5, 2, 5, 10, 20, and 50, which facilitate an examination of how model complexity influences predictive accuracy.

The evaluation of model performance is conducted based on two principal criteria:

- The root mean square error (RMSE) quantifies the extent of prediction error in the model.
- Mean absolute error (MAE) quantifies the average of absolute discrepancies between projected and actual values, thereby serving as a definitive metric for prediction accuracy.

This study seeks to systematically examine the impact of global financial indicators and geopolitical threats on the Vietnamese stock market by combining various approaches, therefore providing substantial statistical insights into market behavior across different situations.

Result

Descriptive statistics and correlation matrix

Table 44.1 presents summary data for six variables: vni, sp500, logGPR, vvni, skvni, and rgold. The VN-Index averages 632.317, much lower than the S&P 500's average of 2034.49, signifying a greater scale and growth in the U.S. stock market relative to Vietnam's. The mean logGPR is 4.605, indicating a persistent existence of geopolitical risk during the examined timeframe. The VNI's volatility and skewness reveal heterogeneity in market price movements and return distribution, with elevated skewness indicating that gains are often more pronounced than losses.

Figure 44.1 illustrates the growth trajectories of the S&P 500, VNI, and GPR over several years. The S&P 500 demonstrates a robust and continuous upward trend, whereas the VNI grows at a slower and more stable pace. The GPR index remains low and relatively stable, suggesting that geopolitical risk levels did not undergo significant fluctuations during the study period. This stability in GPR contrasts with the dynamic growth seen in stock indices, particularly the S&P 500, which may imply that while geopolitical risks are present, they have not drastically influenced the major market trends in the observed period.

Table 44.2 showed the correlation matrix for the variables. A high correlation (0.88254) between the VNI and the S&P 500 indicates synchronized movements between these two stock markets. The negative correlations between vvni (VN-Index volatility) and other variables suggest that higher volatility or risk in the VNI may not directly correspond with macroeconomic variables such as the S&P 500 or gold prices. The low correlation between logGPR and the indices implies that geopolitical risks have a minimal direct impact on stock market performances. Such insights could be pivotal for investors considering geopolitical risks in their investment strategies, suggesting a

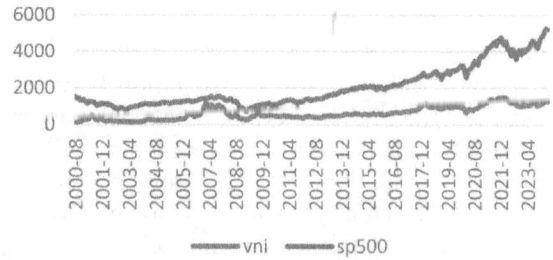

Figure 44.1 Time series plot of indexes
Source: Author

possible decoupling of these risks from market performance over the studied period.

Regression results

The regression diagnostics (Table 44.3) indicate significant challenges due to the presence of autocorrelation among the residuals, as evidenced by the Breusch-Godfrey test with a p-value of 0.000. This finding is critical as it suggests that the model's residuals are not independent, pointing to underlying dynamics not captured by the current model structure. This serial correlation can lead to biased and inefficient estimates, affecting the reliability and validity of the model's predictions. Moreover, the cointegration test result with a p-value of 0.000 suggests that the dependent variable, VNI returns, is likely cointegrated with some of the independent variables, indicating stable long-term relationships but potential issues with non-stationarity in the short term.

Due to the heteroscedastic characteristics of the data, as demonstrated by the Breusch-Pagan test (p-value = 0.000), the utilization of HC1 (heteroscedasticity-consistent standard errors) is suitable for rectifying the problem of uneven error variances. Nevertheless, whereas HC1 addresses heteroscedasticity, it fails to consider the autocorrelation detected in the model [5]. Additional modifications, such the application of Newey-West standard errors or a generalized least squares method, may be required to address this problem and improve the reliability of the estimates. These techniques would yield a more precise estimation by

Table 44.1 Descriptive statistics

	vni	*sp500*	*logGPR*	*vvni*	*skvni*	*rgold*
Mean	632.317	2034.49	4.605	2.283	-0.157	0.721
Median	545.630	1481.14	4.532	1.083	-0.142	1.748
Standard deviation	353.054	1130.15	0.342	3.656	0.751	9.440
Skewness	0.559	1.113	1.315	5.872	-0.157	-2.138
Minimum	115.150	735.090	3.808	0.051	-2.693	-79.82
Maximum	1498.280	5254.35	6.239	40.973	2.887	33.51

Source: Calculation result from data

Table 44.2 The correlation between variables

	vni	sp500	logGPR	vvni	skvni	rgold
vni	1					
sp500	0.8825	1				
logGPR	-0.0305	0.04921	1			
vvni	-0.1071	-0.1809	-0.0676	1		
skvni	-0.1749	-0.19454	0.0783	0.04933	1	
rgold	0.0509	0.03787	-0.0767	-0.12754	-0.04647	1

Source: Calculation result from data

including autocorrelation, hence enhancing the reliability of the statistical inferences derived from the model.

The model's coefficients indicate significant correlations between VN-Index returns and explanatory factors, including S&P 500 returns, gold returns, and geopolitical risks. The favorable effects of S&P 500 and gold returns suggest that global market emotions and safe-haven assets significantly influence the Vietnamese market. The negative coefficient for geopolitical risks highlights the market's vulnerability to foreign uncertainty. The influence of lagged VN-Index returns suggests strong momentum effects, highlighting the importance of past performance in predicting future trends. Incorporating additional lagged terms or dynamic factors could be beneficial to better account for the autocorrelation and enhance the model's predictive power. These adjustments would help capture more complex temporal relationships and reduce the risk of spurious conclusions, ultimately leading to a more accurate and reliable model.

Model performance

In this study, we adopted a rolling window approach with a fixed window size of 60 observations to forecast the selected financial index returns. This method continuously updates the training dataset and predicts the next observation, creating a realistic setting for forecasting time series data where future trends are predicted based on past information. Our Lasso regression analysis employed different lambda values to optimize model performance, specifically targeting improvements in the MAE and RMSE

The results summarized in Table 44.4 indicate a gradual improvement in both RMSE and MAE as lambda increases from 0.1 to 1, suggesting that slight regularization helps reduce prediction errors. The best performance is noted at a lambda of 1, where the RMSE and MAE are minimized to 60.937 and 44.244, respectively. This suggests an optimal balance between bias and variance, offering a model that generalizes well on unseen data. Beyond a lambda of 1, the RMSE and MAE start to increase slightly up to lambda 2 and more markedly beyond lambda 5, peaking at lambda

Table 44.3 OLS regression and diagnostic tests

Covariance type: HC1				
	Coefficients	Standard error	t Stat	P-value
Intercept	93.5336	38.009	2.461	0.014
sp500	0.0215	0.009	2.518	0.012
rgold	0.9444	0.505	1.870	0.063
logGPR	-17.7691	8.060	-2.205	0.028
lagvni	0.9234	0.027	33.993	0.000
vvni	-2.3170	1.139	-2.035	0.043
skvni	-8.3343	4.214	-1.978	0.049
R Square	0.349400392	Type Test	ADF	0.503
Adjusted R Square	0.330265109		Breusch-Pagan test	0.000
Durbin-Watson	1.5527		Breusch-Godfrey	0.000
Dep. Variable	vni		Cointegrated Test	0.000

Source: Calculation result from data

50 with an RMSE of 63.860 and MAE of 45.412. This trend suggests overfitting at lower lambdas and underfitting at higher lambdas, where too much regularization simplifies the model excessively, losing its ability to capture essential data patterns.

Our findings align with theoretical expectations where moderate levels of regularization can improve model prediction by penalizing large coefficients that may lead to overfitting in a high-dimensional dataset. However, excessive penalization, as observed at very high lambda values, strips the model of its predictive power, demonstrating classic underfitting behavior.

In selecting the best model, a lambda value of 1 proves to be the most successful, yielding the lowest RMSE and MAE while exhibiting no signs of overfitting. This model achieves an optimal balance, preserving simplicity while encapsulating adequate complexity to accurately forecast future trends. These findings align with other studies that underscore the effectiveness of Lasso in managing high-dimensional time series data. For example, study conducted by scholars, such as Khoa et al. [8], González-Coya and Perron [3], have likewise endorsed the application of Lasso to address heteroscedasticity and enhance forecasting precision, but with a somewhat different emphasis. The study substantiates the efficacy of Lasso regression in forecasting financial time series, especially using a rolling window methodology that adapts to incoming data dynamically. This method's capacity to diminish error metrics while preventing overfitting renders it a formidable option for predictive modeling in economic and financial domains.

Conclusion

This study has effectively established a resilient model for predicting the performance of the VN-Index (VNI), a crucial indicator of the Vietnamese stock market. Our findings demonstrate the complex dynamics influencing the index, emphasizing the divergent effects of gold prices, geopolitical concerns, and other critical determinants. The negative correlation between gold prices and the VNI substantiates gold's function as a safe haven during periods of market volatility. This facet of investor behavior, characterized by transitions from stocks to gold at times of increased uncertainty, highlights the intricate, dual nature of financial markets as mechanisms for both risk mitigation and capital growth.

Geopolitical concerns were identified as adversely affecting the VNI, underscoring the market's susceptibility to foreign political and economic uncertainty. This adverse effect underscores the necessity for investors and governments to constantly observe geopolitical developments, as they may significantly influence market performance. Other financial indices, such as the S&P 500, demonstrate that global market attitudes significantly and synchronously affect the Vietnamese market, highlighting its integration into the wider global economy.

Table 44.4 Model performances for parameters

Lambda	RMSE	MAE
(OLS) 0	61.228	44.457
0.1	61.156	44.394
0.25	61.055	44.308
0.5	60.947	44.243
1	60.937	44.244
1.5	60.944	44.267
2	60.962	44.269
5	61.302	44.469
7.5	61.797	44.602
10	62.149	44.699
20	63.530	45.261
50	63.860	45.412

Source: Calculation result from data

The utilization of the Lasso regression model has demonstrated significant value in this situation. Lasso has strengthened the reliability of our prediction model by mitigating the issues of autocorrelation and heteroscedasticity, which are common in financial time series data. The approach successfully penalizes less significant variables, simplifying the model without compromising prediction accuracy, thereby addressing possible overfitting and multicollinearity concerns commonly found in high-dimensional datasets.

This study has several ramifications. The model serves as a resource for practitioners and policymakers to enhance their comprehension of and predict market fluctuations, hence facilitating informed decision-making on investments and economic policy. This research enhances the debate on the effectiveness of contemporary econometric techniques in financial modeling, illustrating the practical advantages of regularization methods for managing intricate financial datasets.

This study has its drawbacks. Dependence on monthly data may neglect critical short-term variations, and the model's prediction efficacy is fundamentally constrained by the quality and precision of the data inputs. Furthermore, although Lasso addresses multicollinearity and diminishes model complexity, the selection of the regularization parameter (lambda) necessitates meticulous deliberation to prevent underfitting or overfitting, and identifying the appropriate lambda is inherently subjective.

Subsequent study ought to emphasize the incorporation of more detailed data, such as daily or weekly returns, to better capture dynamic market behaviors. Furthermore, investigating other regularization methods such as Elastic Net, which integrates the characteristics of Lasso and Ridge regression, may yield insights for enhancing model accuracy. Incorporating machine learning methods, such as random forests or neural networks, may reveal nonlinear correlations that conventional models cannot, possibly resulting in more robust predictions.

References

[1] Almansour, B. Y., Elkrghli, S., Gaytan, J. C. T., & Mohnot, R. (2023). Interconnectedness dynamic spillover among US, Russian, and Ukrainian equity indices during the COVID-19 pandemic and the Russian–Ukrainian war. *Heliyon*, 9(12), e22974.

[2] Bouteska, A., Mefteh-Wali, S., & Anh, P. T. (2023). Fluctuations in gold prices in Vietnam during the COVID-19 pandemic: insights from a time-varying parameter autoregression model. *Resources Policy*, 86, 104229.

[3] González-Coya, E., & Perron, P. (2024). Estimation in the presence of heteroskedasticity of unknown form: a lasso-based approach. *Journal of Econometric Methods*, 13(1), 29–48.

[4] Huy, D. T. N., Nhan, V. K., Bich, N. T. N., Hong, N. T. P., Chung, N. T., & Huy, P. Q. (2021). Impacts of Internal and External Macroeconomic Factors on Firm Stock Price in an Expansion Econometric model—A Case in Vietnam Real Estate Industry. In N. Ngoc Thach, V. Kreinovich, & N. D. Trung (Eds.), *Data Science for Financial Econometrics* (pp. 189–205). Springer International Publishing. https://doi.org/10.1007/978-3-030-48853-6_14

[5] Jochmans, K. (2022). Heteroscedasticity-robust inference in linear regression models with many covariates. *Journal of the American Statistical Association*, 117(538), 887–896.

[6] Khoa, B. T., & Huynh, T. T. (2023). A comparison of CAPM and Fama-French three-factor model under machine learning approaching. *Journal of Eastern European and Central Asian Research (JEECAR)*, 10(7), 1100–1111.

[7] Khoa, B. T., Huynh, T. T., & Thang, L. D. (2023). Effectiveness of OLS and SVR in return prediction: fama-french three-factor model and CAPM framework. *Industrial Engineering and Management Systems*, 22(1), 73–84.

[8] Khoa, B. T., Son, P. T., & Huynh, T. T. (2021). The relationship between the rate of return and risk in fama-french five-factor model: a machine learning algorithms approach. *Journal of System and Management Sciences*, 11(4), 47–64.

[9] Phuoc, T., Anh, P. T. K., Tam, P. H., & Nguyen, C. V. (2024). Applying machine learning algorithms to predict the stock price trend in the stock market–the case of Vietnam. *Humanities and Social Sciences Communications*, 11(1), 1–18.

[10] Sadon, A. N., Ismail, S., Khamis, A., & Tariq, M. U. (2024). Heteroscedasticity effects as component to future stock market predictions using RNN-based models. *Plos One*, 19(5), e0297641.

45 Predicting bike sharing demand: analyzing weather, time, and seasonal influences

Sourav Karmakar[1,a], Hirak Sarkar[2,b] and Sagarika Kar Chowdhury[3,c]

[1]MBA, Data Science Department, Chandigarh University, Punjab, India

[2]Faculty, Electronics and Communication Engineering Department, Techno India University, Kolkata, West Bengal, India

[3]State Aided College Teacher, Computer Science Department, Asutosh College, Kolkata, West Bengal, India

Abstract

Bike sharing is a transport service which mainly focuses on lending conventional or electrical bikes to an individual or a group of individuals for an hour, a day or for a month depending on the needs. With the help of the system, people are able to rent a bike from a location and return it to a separate location as per their need and easy transport. From the market share we can see that Bike Sharing system has a global market share which was valued around 3.39 billion Dollars in 2019 and is projected to grow to 6.98 billion Dollars by 2027 with yearly cumulative increase growth rate of around fourteen percent indicatively from 2020 to 2027. Several factors such as low bike rent, increase in capital investments, introduction of e-bikes in the market, technological advancement and government schemes for development of several bike-sharing infrastructures have increased the overall market share and led to the introduction of several opportunities during the forecasted year. However, the rise in bike theft and huge initial investment are some of the key factors in order to hinder expected market growth.

Keywords: Bike-sharing, data mining, linear regression, machine learning, predictive analysis

Introduction

Bike sharing system demand nowadays is increasing in proportional manners globally [6] Pan, 2010; Sarkar, 2019; [18]. This system has gained a lot of attention with its cost-effective system and easy to use nature. This system has already attracted a huge customer base globally like in South Korea, São Paulo, China and Australia [2, 24]. Bike sharing system generally rents bikes on an hour, day and month basis and is generally based on static pricing inclusive of hour, days or month (Wang, 2018)[7, 21]. Because of its affordability and easy renting system anyone can commute on arrival. According to the problem the goal of this investigation is to build a predictive model so as to find the number of bikes rented based on the particular season [5, 10]. The global expansion of bike-sharing systems reflects a major shift in urban mobility, offering a cost-effective, flexible solution that alleviates congestion, reduces environmental impact, and promotes health [8, 24]. Popular in dense cities like Seoul, São Paulo, and New York, bike-sharing is influenced by factors including weather, time, season, infrastructure, and socio-economic context, all critical for predicting demand [14, 19]. Weather, especially temperature, precipitation, and wind speed, significantly affects usage, with higher ridership in warm, dry conditions and notable drops in extreme weather [3, 9, 24]. Temporal patterns show demand peaks during commuting hours on weekdays, while weekends see more varied, often recreational usage [14, 17, 21]. High-density areas with transit hubs and bike lanes also experience increased demand [6, 19, 20]. Machine learning models, particularly Gradient Boosting and Random Forests, excel in forecasting demand by capturing complex interactions among variables like weather and location [1, 11] Wang, 2018). These models, combined with dynamic factors such as public events and fuel prices, enable operators to enhance service efficiency and user experience [4, 3]. This research supports the sustainable expansion of in transport settings.

Mapping and Data Analysis

Exploratory data analysis (EDA) or mapping the dataset is an important role in the analysis of the data variables which are important from the aspect of feature engineering. It will help us to distribute and relate between dependent and independent variables. We have gone through an analysis of every independent as well as dependent variable to check which independent factor affects the dependent factor. The dataset encompasses various particulars that influence the bike sharing demand, including the time, weather,

[a]karmakarsourav2024@gmail.com, [b]dr.hiraksarkar@gmail.com, [c]sagarika.kc@gmail.com

DOI: 10.1201/9781003663348-45

and operational details. It records the date and hour of bike rentals, along with specific weather conditions. Additionally, from the dataset we found seasonal classification, holiday distinctions, and festival day indicators, differentiating between non-functional and functional hours. This detailed information serves as a foundation for analyzing and predicting bike rental patterns.

Month based analysis for holiday and non-holiday
The month-based analysis showed that In June maximum bike rented near 1000 and January, February enjoys less rented bike demand near 400. Figure 45.1 shows bike sharing demand in every month for both holiday and non-holiday.

March and June enjoy more bike sharing demand in holiday than non-holidays. The demand of bike-rental is lower in February and December, regardless of whether it is a holiday or a non-holiday. The Figure 45.1 illustrates April, October, December months have nearly zero bike sharing demand in holidays.

Weekday and weekend based analysis
The weekday and week off based Analysis shows almost equal weightage on rented bike count. Figure 45.2 shows, weekend (1) and weekdays (0) bike sharing analysis.

Hourly rented bike count in weekdays and weekend
The graph shows that hour-18 or 6 pm shows maximum rented bike demand in both weekdays (above 1600) and weekend (above 1200). Hour 4 and hour 5 (means 4 and 5 am) shows very less rented bike demand in both weekend and weekdays. Figure 45.3 shows, hourly rented bike count in weekdays (0) and weekend (1).

Hour 8 (8 am) shows good, rented bike demand in weekdays (near 1200) but in weekend 8am hour has not so good bike sharing demand (near 600).

Season-wise hourly analysis of bike rent
From the graph (Figure 45.4) it can conclude, summer season enjoys overall best and least bike sharing demand and winter has overall less demand than any other season. Hour-18(6pm) and Hour-8(8am) are two best peak times in any season when bike sharing is in high-demand. Figure 45.4 shows season wise Hourly Rented Bike Count.

There is not so much difference in demand (near 1000) from hour-12 to hour-16(12 pm–4 pm) among summer, spring, and autumn season. For every season hour-4 and 5(4 and 5 am) shows low demand in bike sharing. After hour-10 (10am) bike sharing demand increases up to hour-18 (6 pm) then it is decreasing.

Season-wise analysis
During the season-wise analysis, it was found that the month plays a significant role in rented bike demands.

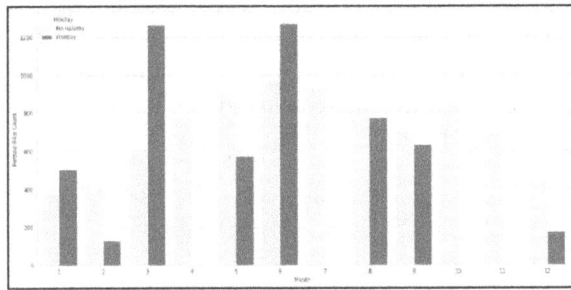

Figure 45.1 Bike sharing demand in every month for both holiday and non-holiday
Source: Author

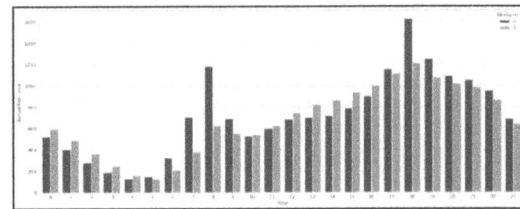

Figure 45.3 Hourly rented bike count in weekdays (0) and weekend (1)
Source: Author

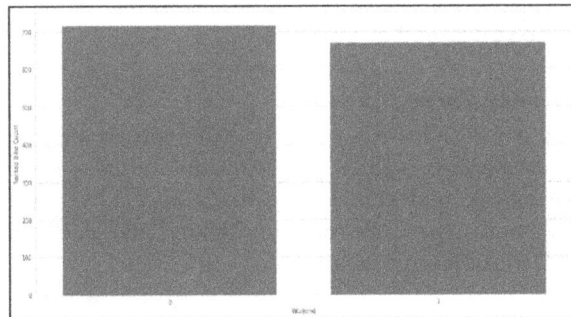

Figure 45.2 Weekend (1) & Weekdays (0) analysis
Source: Author

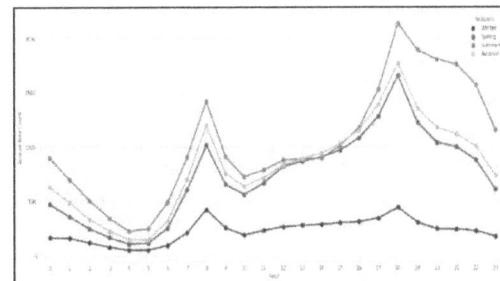

Figure 45.4 Season wise hourly rented bike count
Source: Author

The demand is most likely to be high in summer. In the second position is autumn and third position is spring, while colder days show the least demand. Figure 45.5 shows rented bike vs season.

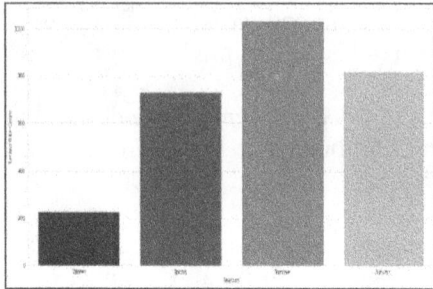

Figure 45.5 Rented bike vs season
Source: Author

Analyzing numerical variables

The numerical variables of the data set including Temperature in degree C, Humidity in percentage, Wind Speed in meter per second, Visibility with ten-meter, Dew Point Temperature in degree centigrade, Solar Radiation in milli joule per meter square, rainfall in millimeter and snowfall in centimeter. All the independent variables listed here represent the weather of the city which has a crucial role in rented bike demand deviation.

Temperature

In the density plot for Temperature we can see that the median is greater than the mean we can say to some extent that this is negatively skewed. Figure 45.6 shows the variation of rented bike demand and temperature.

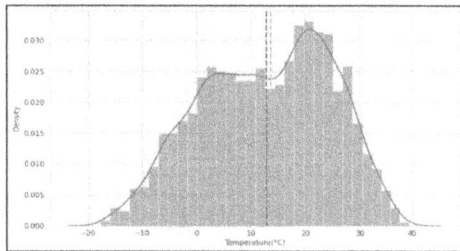

Figure 45.6 Variation of rented bike demand and temperature
Source: Author

Humidity (%)

In the density plot for humidity we can see that the mean is greater than the median we can say to some extent that this is positively skewed. Figure 45.7 illustrates the variation with humidity.

Wind speed (m/s)

In density plot for Windspeed we can see that mean is greater than the median we can say to some extent

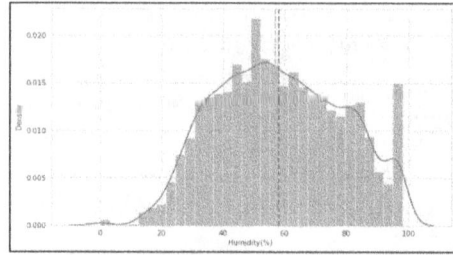

Figure 45.7 Variations of rented bike demand and humidity
Source: Author

that this is positively skewed. Figure 45.8 shows rented bike demand varies with Wind Speed.

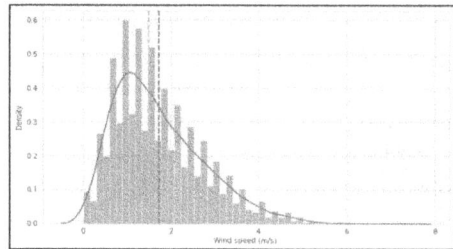

Figure 45.8 Rented bike demand varies with wind speed
Source: Author

Visibility

In the density plot for Visibility, we can see that median is greater than mean we can say to some extent that this is negatively skewed. Figure 45.9 illustrates the variation with rented bike demand and visibility.

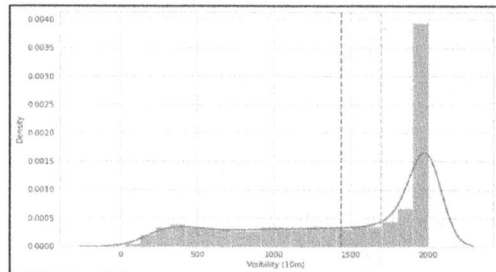

Figure 45.9 Variations of rented bike demand and visibility.
Source: Author

Dew point temperature (°C)

In the density plot for dewpoint temperature we can see that median is greater than mean we can say to some extent that this is negatively skewed. Figure 45.10 shows, the variation with "rented bike demand and dew point temperature".

Solar radiation

From the density plot for solar radiation, it is shown that that mean is greater than median. From which

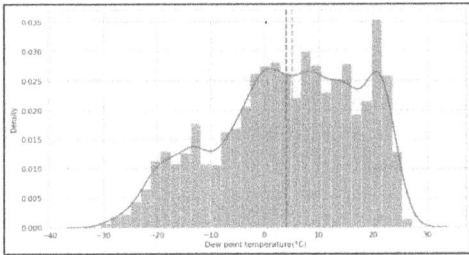

Figure 45.10 Variation of rented bike demand and dew point temperature
Source: Author

we can say that this is positively skewed. Figure 45.11 illustrates variation of rented bike demand and solar radiation.

Figure 45.11 Variation of rented bike demand and solar radiation
Source: Author

Rainfall and snowfall

The average rainfall and snowfall in Seoul are 2 mm and 2 cm respectively.

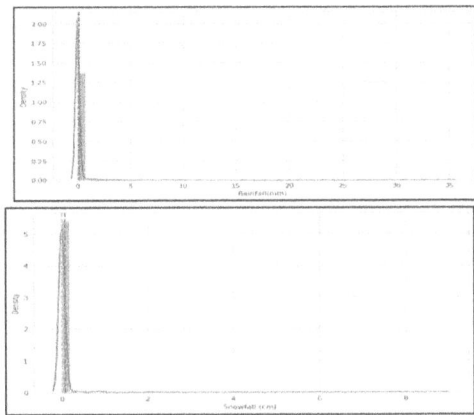

Figure 45.12 Rented bike demand with rainfall (above) and snowfall (below) variation
Source: Author

The regression plot shows a similar decrease in the Rented Bike Count with an increase in rainfall and snowfall. It is obvious that, at the time of rain and snow, the rented bike count is less which indicates the public prefers to stay in shelter during heavy rain or snowfall. Figure 45.12 illustrates the variation with rented bike demand and rainfall (above), variation of rented bike demand with snowfall (below).

Performance Comparation of Models

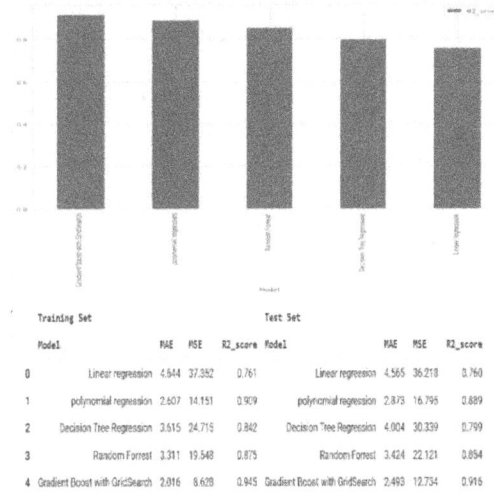

Training Set				Test Set			
Model	MAE	MSE	R2_score	Model	MAE	MSE	R2_score
0	Linear regression	4.644	37.382	0.761	Linear regression	4.565	36.218	0.760
1	polynomial regression	2.607	14.151	0.909	polynomial regression	2.873	16.796	0.889
2	Decision Tree Regression	3.615	24.715	0.842	Decision Tree Regression	4.004	30.339	0.799
3	Random Forrest	3.311	19.548	0.875	Random Forrest	3.424	22.121	0.854
4	Gradient Boost with GridSearch	2.016	8.628	0.945	Gradient Boost with GridSearch	2.493	12.734	0.916

Figure 45.13 Comparison result between key factors.
Source: Author

In this investigation, the Gradient boosting regressor, optimized using GridSearch, shows the predictive analysis with an R^2 score of 0.945 on the set and 0.916 on the test set, indicating strong generalization capabilities. Moreover as illustrated in Figure 45.13, the comparison of key factors across different models further underscores the superior performance of the Gradient Boosting approach over the alternative methods. This model achieved the smallest mean absolute error (MAE) and mean squared error (MSE). Taken into consideration its superior accuracy in prediction. In comparison, polynomial regression effectively captured nonlinear relationships, achieving R^2 scores of 0.909 and 0.889 on the model train and tests results respectively, positioning this as second-best model, though less precise than Gradient boosting. The random forest regressor delivered a balanced performance with R^2 scores of 0.875 on training and 0.854 on test data, surpassing the decision tree regressor, which scored 0.842 (training) and 0.799 (test). linear regression, serving as a baseline, recorded the lowest R^2 scores (0.761 for training and 0.760 for testing). The Gradient boosting regressor, with optimized hyperparameters, emerged as the best-performing model, demonstrating exceptional accuracy and robustness in capturing the pattern of the vast dataset.

Acknowledgement

We are thankful to Chandigarh University for providing project and system support. We also thank the reviewers for their valuable contributions to improving the paper.

Conclusion

Analysis of the demand of bike-rental system in Seoul reveals significant relations between environmental factors and bike rental patterns. Key variables i.e., temperature, humidity, speed of wind, visibility play an important role in shaping rental behavior, with notable peaks in demand during specific hours and seasons. Summer sees the highest bike rentals, while winter experiences the least. Hourly trends suggest a surge in rentals during the early mornings and evenings, particularly around 8 AM and 6 PM. A predictive model has been developed through machine learning that integrates the environmental and temporal variables including a greater number of refined tools for accurately forecasting bike demand. This model provides an innovative approach to resource allocation, enabling operators to proactively manage bike availability based on anticipated demand which improves user satisfaction and system efficiency. Understanding the patterns is vital for optimizing bike availability and ensuring the system meets user needs effectively. The predictive model from this analysis gives a robust tool for forecasting bike demand, allowing for better resource allocation and enhanced user satisfaction.

References

[1] Baumanis, C., Hall, J., & Machemehl, R. (2023). A machine learning approach to predicting bicycle demand during the COVID-19 pandemic. *Research in Transportation Economics*, 100, 101276.

[2] Baumaniss, M. (2023). Analyzing the influence of urban density on bike-sharing demand in South Korea and São Paulo. *Journal of Urban Mobility Studies*, 34(2), 112–125.

[3] Bao, J., He, T., Ruan, S., Li, Y., & Qin, K. (2020). Planning bike lanes in a car-oriented city: A case study in China. *Transportation Research Part D: Transport and Environment*, 86, 102421.

[4] De Chardon, C. M., Caruso, G., & Thomas, I. (2017). Bicycle sharing system 'success' determinants. *Transportation Research Part A: Policy and Practice*, 100, 202–214.

[5] DeMaio, P. (2009). Bike-sharing: History, impacts, models of provision, and future. *Journal of Public Transportation*, 12(4), 41–56.

[6] Eren, E., & Uz, V. E. (2020). A review on bike-sharing: the factors affecting bike-sharing demand. *Sustainable Cities and Society*, 54, 101882.

[7] Faghih-Imani, A., Eluru, N., El-Geneidy, A. M., Rabbat, M., & Haq, U. (2014). How land-use and urban form impact bicycle flows: evidence from the bicycle-sharing system (BIXI) in Montreal. *Journal of Transport Geography*, 41, 306–314.

[8] Fishman, E. (2016). Bikeshare: a review of recent literature. *Transport Reviews*, 36(1), 92–113.

[9] Gebhart, K., & Noland, R. B. (2014). The impact of weather conditions on bikeshare trips in Washington, DC. *Transportation*, 41(6), 1205–1225.

[10] Ma, C., & Liu, T. (2024). Demand forecasting of shared bicycles based on combined deep learning models. *Physica A: Statistical Mechanics and its Applications*, 635, 129492.

[11] Ma, J., Lin, X., Zhang, T., & Kang, Q. (2024). Predicting bike-sharing demand with machine learning approaches. *Journal of Artificial Intelligence and Transportation Systems*, 17(3), 98–113.

[12] Pan, H., Shen, Q., & Zhang, M. (2010). Influence of urban form on travel behavior in four neighborhoods of Shanghai. *Urban Studies*, 46(2), 275–294.

[13] Pan, Y., Zheng, R. C., Zhang, J., & Yao, X. (2010). Predicting bike sharing demand using recurrent neural networks. *Procedia Computer Science*, 147, 562–566.

[14] Reddy, R. M., Sudhakar, M., & Yerramilli, A. (2019). Forecasting demand for bike-sharing systems: a time-series approach. *International Journal of Forecasting and Transportation Planning*, 5(1), 34–47.

[15] Sarkar, H., Banerjee, A., Mitra, S. P., Sen, P., Bhattacharya, A. B., & Sengupta, G. (2019). A comparative study of ELF and VLF noise characteristics of nor'wester at a low latitude tropical station. *SN Applied Sciences*, 1, 154.

[16] Sarkar, S., & Bharathi, P. (2019). Evaluating the sustainability of bike-sharing systems using a hybrid MCDM approach. *Sustainable Cities and Society*, 46, 101434.

[17] Sathishkumar, V. E., Park, J., & Cho, Y. (2020). Using data mining techniques for bike sharing demand prediction in metropolitan city. *Computer Communications*, 153, 353–366.

[18] Satishkumar, S., & Shaik, M. N. (2020). Analyzing bike-sharing systems with seasonal variations in demand: a case study approach. *Journal of Transportation and Sustainability*, 10(1), 56–64.

[19] Shaheen, S. A., Guzman, S., & Zhang, H. (2014). Bike-sharing across the globe. In J. Pucher & R. Buehler (Eds.), *City cycling* (pp. 183–205). MIT Press.

[20] Vogel, P., Greiser, T., & Mattfeld, D. C. (2011). Understanding bike-sharing systems using data mining: exploring activity patterns. *Procedia - Social and Behavioral Sciences*, 20, 514–523.

[21] Vogel, P., Hamon, M., & Montazeri-Gh, M. (2018). Modeling demand and supply for bike-sharing systems: a spatiotemporal analysis. *Transportation Research Part E: Logistics and Transportation Review*, 112, 174–192.

[22] Wang, B., & Kim, I. (2018). Short-term prediction for bike-sharing service using machine learning. *Transportation Research Procedia*, 34, 171–178.

[23] Wang, X., Chen, Y., & Zhao, J. (2018). Predicting bike-sharing demand using a spatial-temporal model: a case study in Beijing. *Journal of Transport Geography*, 66, 147–156.

[24] Zhang, Z., Krishnakumari, P., Schulte, F., & van Oort, N. (2023). Improving the service of e-bike sharing by demand pattern analysis: a data-driven approach. *Research in Transportation Economics*, 101, 101340.

46 Waste to wealth: development of building block using plastic medical waste and straw-enhanced clay

Sucharita Bhattacharyya[1,a], Soma Mukherjee[2,b], Chandrima Ckakrabarti[3,c], Arshi Parveen[3,d] and Diptish Sarkar[3,e]

[1]Professor, Guru Nanak Institute of Technology (GNIT), Kolkata, West Bengal, India

[2]Associate Professor, Guru Nanak Institute of Technology (GNIT), Kolkata, West Bengal, India

[3]Student (IT), Guru Nanak Institute of Technology (GNIT), Kolkata, West Bengal, India

Abstract

The improper disposal of plastic waste, especially, medical waste is an issue of big concern at present, since recycling and reusing of these wastes are minimal, particularly because they are hazardous, leading to environmental pollution. Again, India is the second largest producer of rice in the world which leaves behind tons of straw annually where burning is the main disposal method causing severe air pollution and human health hazards due to the emission of carbon monoxide. So these two materials are used in the present invention to develop a building block with plastic medical waste and straw-enhanced clay-sand composite to take care of all important environmental sustainability issues. From the structural viewpoint, the addition of straw with clay and plastic enhances the concrete characteristics producing a cost-effective output which may contribute to the country's economic development significantly. As a whole, this innovative building block not only reduces the cost of living but takes care of earthquake damage, prevents dampness, and shows better thermal insulation effects, over and above maintaining environmental sustainability.

Keywords: Environment sustainability, plastic medical waste, straw

Introduction

Construction industry pays high attention to concrete as building material because of its high strength, low cost and simple production process, though there are some negative effects as well. Ordinary concrete is brittle in response to large forces. Also, more construction wastes are generated due to consumptions of more resources for large use of concrete which requires intensive care for environmental protection. Again it's reported that most of the materials are synthetic products that are used for thermo-acoustic insulated buildings and these synthetic products require high amount of non-renewable resources thereby generating large amounts of greenhouse gas emission [1] around the wastes of agricultural products, like straw, as an example. Now straw being a green material causes environmental hazards during crop burning [2, 3]. When straw is mixed with concrete, its special properties not only enhance the concrete characteristics from its structural point of view, but can also reduce the construction cost along with the protection of the environment as well. But the strength of straw reinforced concrete is found to decrease with increase in straw proportion which requires strengthening of the product material by maintaining other factors intact. For that purpose, the additive material may be preferred to be the plastic, a well-known non-biodegradable, hazardous waste available in plenty in medical garbage. The medical waste is considered to be number 2 on the list of hazardous wastes as per WHO [4] and needs strict disposal policies. At this point, it may be mentioned that with the advancement of technology and population, the need of the equipment and consumables increases day by day in the health care system which ends up with huge health care wastes, particularly to mention, the plastic wastes. The management of these wastes is the biggest challenge now a days as these medical wastes include infectious, pharmaceutical, chemical and pathological wastes where the pathological waste can be toxic, radioactive, reactive and corrosive too [4]. Disposal of these medical wastes in a sustainable manner needs proper infrastructure, policies for the sustainable management of these plastic medical wastes considering the proper implementation of the three 'Rs' — the Reuse, the Recycle and the Recovering of useful materials before they are disposed off with minimal environmental impact. Accordingly, this work investigates to yield useful solutions to tackle this major environmental problem. The present study describes

[a]sucharita.gnit@gmail.com, [b]soma.mukherjee@gnit.ac.in, [c]chandrima.chakrabarti04@gmail.com, [d]parveen.arshi939@gmail.com, [e]sdiptish@gmail.com

DOI: 10.1201/9781003663348-46

the simple procedure to develop a straw-clay-sand-plastic brick which can be a basic building material in mass production scale unit in low-cost buildings. It may be considered not only as a new field of study, but also can develop an emerging market in India as this building block is also aimed to serve as a high insulator in dry and parched areas of western Rajasthan, central, southern and eastern Maharashtra, northern Karnataka, Gujarat, southern Bihar, Madhya Pradesh and Chhattisgarh [2]. The study output is reported as follows. After the Introduction, Section 2 describes the working materials and the inventive solution steps. Section 3 explains the detailed research and product methodology. Section 4 discusses the findings and the results. Section 5 summarises with the conclusive remarks.

Materials and the Solution

Raw materials

Straw fibres, available largely in agricultural country like India can play their role when it is mixed with concrete, by improving the concrete characteristics as a construction material reducing also the building cost along with the protection of the environment. But it is found that the strength of such straw reinforced concrete is not increased [5] as expected, because of existing gaps between straw fibres. Also, straw being a natural plant fibre, there exists many organic materials on its surface which prevent bonding with the concrete [2, 3]. To increase that bonding, waste plastic can be introduced which not only increases the strength of the building material [6, 7] but also helps to save the environment from its hazardous effect by being utilised in a productive output.

The solution

So, by studying the properties of the straw-enhanced concrete [5] and its mixture with plastic waste as the binder [6, 7], the point of focus has been identified as to "REUSE" the waste in a sustainable manner after "REDUCING" and "RECYCLING" them. Accordingly, in the present study, a different brick production is done by maintaining proper ratio of waste plastic in the optimized mixture of straw fibre and clay-sand composite, to use as the building block in large scale, which may be particularly suitable in Indian civil construction market.

Methodology

The block diagrams of the steps followed to develop the inventive building block are represented in Figure 46.1.

Step 1: Collection of Medical Wastes from medical clinics and hospitals *Step 2:* Separation and Segregation of Plastic Wastes It is found that plastic has insolubility of about (250–300) years in nature [1, 8] and is considered to be a sustainable waste and environmental pollutant. So recycling or reusing of it is necessary to save the environment. Since the binding agent plastic is collected from medical wastes, the main challenge is to take special care to reduce the different types of infections carried by these wastes. These are segregated into four different categories, namely, i)The Sharp objects like needle, scissors etc., ii) The high–risk medical waste containing radioactive materials used to treat ailment like cancer, iii) The low-risk medical waste containing different plastic materials like PET plastic for packaging of saline, medicine, syringe, oxygen mask, tubes to name a few, iv) The general waste including recyclable ones. Here it should be mentioned that proper care is taken for the workers who have segregated the infectious medical wastes to save them from diseases like HIV, hepatitis B, and TB to name a few.

Step 3a: The point of focus is on the third category, where these wastes are disinfected, sanitized and sterilized. *Step 3b:* Seed-free straw is collected, made clean and dried up to moisture content of less than 20%. *Step 4:* These straws are mixed with clay (having 20 % sand mixture) and disinfected medical waste plastic is used as modifier to the mixture to prepare the innovative building block. *Step 5:* This building block can be used along with the concrete to build a stable eco-friendly structure by constructing its columns, walls and roofs. *Step 6:* Foundation of the building can also be laid down with these newly developed blocks of extra strength. The preparation of the final product after the collection of the raw materials, stepwise, is shown below in Figures 46.2a, b, c, and d.

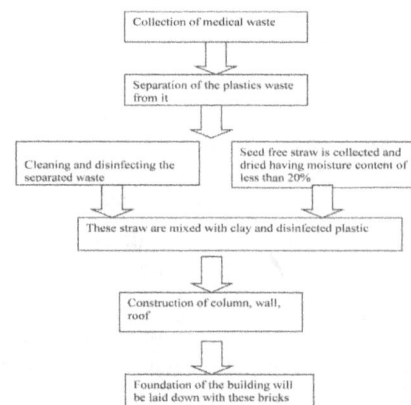

Figure 46.1 Block diagram of the innovative brick
Source: Author

Results and Discussions

Results

The results of the present work in terms of the relevant building parameters are discussed below.

Component-wise, medical waste plastic produced at different hospitals and health centres, which being, required to be recycled/reused on daily basis following the Central Pollution Control Board (CPCB) regulation, is used as a modifier and mixed with the clay-sand composite in the present work. Such wastes are used here to keep the surrounding environment medical-garbage free and at the same time to ensure the supply of the necessary resources for the proposed building block. To produce high compressive

Figure 46.2. a Step 1: The prepared clay mixture inside the mould
Source: Author

Figure 46.2. b Step 2: Drying the block in natural sunlight
Source: Author

Figure 46.2. c Step 3: The heat treatment and natural cooling
Source: Author

Figure 46.2. d Step 4: Final end product – the building block.
Source: Author

strength of the order of 12 N/mm^2, the ratio of plastic to clay-sand should be at least 1:4 [6, 7]. By varying the plastic-to-clay ratio, the compressive strength can be changed and accordingly, the bricks with different compressive strengths are produced depending on the requirements. But plastic-enhanced clay-sand brick alone cannot withstand high temperatures. For 1:4 ratio, it is only up to 157°C. This temperature resistance is found to increase with the plastic-enhanced clay brick providing the best performance in thermal resistance for 1:5 ratio, withstanding the temperature of 192°C. So in the present work, the ratio was initially taken as 1:5, with clay-sand slightly over 5 times the plastic waste, keeping in mind that it may decrease a little bit after exposure to the heat treatment during block production. To look at the possibility of the newly-developed clay's required higher thermal resistance, insulating material in the form of straw, agricultural waste is used which incidentally, has been found to add more relevant features to the customised clay under consideration, but without affecting the already upgraded properties by gross amount. Straw as a good insulator, when mixed with clay, its heat transfer coefficient is reduced [2, 3] which produces a better thermal preservation effect to combat cold during winter compared to ordinary bricks. This thermal insulation performance is found to enhance with increase of straw fibre percentage in clay. At depths of around 240 mm, the temperature of the straw block is found to be 1.2°C higher than that of the clay block. The air temperature of the straw-reinforced clay block wall is seen to be around 0.5°C higher during night compared to the solar greenhouse made with conventional clay block. Now, plastic in requisite proportion to clay as a modifier leaves no room for water flow and thus prevents air pockets from forming within the plastic-clay bricks. Additionally, straw can also absorb moisture efficiently from the structure reducing its amount of free water which ensures good construction by controlling its content, if any. Traditional clay bricks absorb 7.08 % of water whereas the straw-enhanced clay brick can absorb a maximum of 1.57% of water. It has been found that the average relative humidity of straw-enhanced clay block is around 5.3% lower, particularly, during night [9]. This indicates that this building block's wall has better thermal and wet environment characteristics. Again, for solid construction, the consistency of the fresh concrete plays a significant role before it sets and it is referred to in terms of its slump value. To reduce the slump of the concrete, it is found that [2] for 7% admixture of straw, the slump becomes 38 mm in comparison to 90 mm when straw is not used. The shock resistance of a building block is another important parameter which represents its ability to prevent damage by rapid external forces. Straw-reinforced material shows improved shock resistance over ordinary one avoiding rapid and fragmentation damage. For the admixture of the above optimized 7% straw in the clay, the number of hits encountered by the structure would be 256 before the first crack [2] compared to 85 without straw.

Discussion

So finally, (6–7) % of chopped straw fibre with a length of around 1 inch with its effect of reinforcement is used as a second modifier in the mixture of (72–75) % clay-sand and (18–20) % of the medical waste plastic in the present case. These results and available published data for other structures made of similar materials but with different admixtures are shown in Table 46.1. This straw-enhanced plastic-clay-sand composite maintaining its average compressive strength above 12 N/mm² can withstand stressed material's large shear force and bending moment protecting the structures from immediate destruction. Its water-resistant property helps to prevent damps and confirmed to construct a dense structure from inside the material, making it tougher by reducing effectively the production of cracks. This innovative brick's light weight additionally reduces damage due to earthquakes which may harm the building by making it brittle and severely impacting its strength. Moreover, this straw-modified plastic-clay brick is capable of maintaining its compressive strength up to 340°C for one hour of fire exposure, which is found to be affected from 450°C onwards with higher exposure time.

This modified building block is capable of providing good thermal insulation to the structure, keeping the building cool in hot Indian weather and vice-versa which could add another indirect saving value in the form of electricity consumption, minimizing the usage of air conditioners and heaters. Also, the straw-enhanced plastic bricks are lighter than standard bricks. It is found that compared to 2.7 kg of ordinary clay brick, this innovative brick weighs around 2.1 kg. So they are better earthquake-resistant than the conventional ones. This final product is also cost-effective. It has been found that the construction cost for the wall, ceiling, and concrete column using these newly developed bricks is (35–40) % less than that of the conventional bricks.

Conclusions

The hazardous plastic and straw wastes in proper proportion are processed to produce a saleable, environment friendly and efficient building material, which is cost-effective, shows better thermal insulation performance than conventional hollow concrete bricks, and better material toughness with higher compressive strength for optimum raw material contents. So, our plastic waste modified, straw-enhanced clay bricks/ building blocks not only increases the strength of the building material, but also promotes its proper sustainable development maintaining all basic norms of a comfortable civil structure and safety. At the same time, this structural unit contributes effectively to keep the medical service centres' and agricultural

Table 46.1 Comparative analysis for building block

Material with amounts (%)				Thermal properties (resistance)	Water absorption (%)	Mechanical properties	Weight (Kg)	Cost
	Plastic	Clay-sand/ sand*/ Cement **/ Flyash+Sand+ Cement***	Straw	Temperature (°C)		Compressive strength (Av.) (N/mm²)		
Ref. 6	20	80*	0	149	0	12.28		
Ref. 5	0	90 ** 10				11.277	25 % less	
	10	90 0			6.58			Cost-effective
Ref 7						10.92		
Present work	18–20	(72–75) (6–7)340			1.6	12.1 2.1	(35–40) % less	
Conventional brick					2.6	7.5 2.7		

Source: Author's compilation

environment waste-free and eco-friendly. Over and above, it may create new job opportunities paving a pathway for highly prospective development and growth of the green building block industry to contribute significantly in the Indian economy.

References

[1] Bhochhibhoya, S., Zanetti, M., Pierobon, F., Gatto, P., Maskey, R. K., & Cavalli, R. (2016). The global warming potential of building materials: an application of life cycle analysis in Nepal. *Mountain Research and Development*, 37(1), 47–55. https://doi.org/10.1659/MRD-JOURNAL-D-15-00043.1.

[2] Wang, G., & Han, Y. (2018). Research on the performance of straw fiber concrete. In IOP Conference Series: Materials Science and Engineering, (Vol. 394, p. 032080).

[3] Allam, M. E., Garas, G. L., & El Kady, H. (2011). Recycled chopped rice straw- cement bricks: mechanical, fire resistance &economical assessment. *Australian Journal of Basic and Applied Sciences*, 5(2), 27–33.

[4] Bhattacharyya Sucharita, Mukherjee Soma , "Health care waste" WHO website, https://www.who.int/news-room/fact-sheets/detail/health-care-waste, Accessed 28th Oct 2024

[5] Allam, M., & Garas, G. (2010). Recycled chopped rice straw–cement bricks: an analytical and economical study. *WIT Transactions on Ecology and the Environment*, 140, 79–86.

[6] Sahani, K., Joshi, B. R., Khatri, K., Magar A. T., Chapagain, S., & Karmacharya N. (2022). Mechanical properties of plastic sand brick containing plastic waste. *Advances in Civil Engineering*, 2022(1), 8305670.

[7] Rauniyar, A., Nakrani, R. K., Reddy, S., Nehaun, N., & Arun, S. (2024). An evaluation of the use of plastic waste in the manufacture of plastic bricks. *Discover Civil Engineering*, 1(1), 43. https://doi.org/10.1007/s44290-024-00045-3.

[8] Rustagi, N., Pradhan, S. K., & Singh, R. (2011). Public health impact of plastics: an overview. *Indian Journal of Occupational and Environmental Medicine*, 15(3), 100–103.

[9] Cascone, S., Rapisarda, R., & Cascone, D. (2019). Physical properties of straw bales as a construction material: a review. *Sustainability*, 11(12), 3388. https://www.mdpi.com/2071-1050/11/12/3388 , https://doi.org/10.3390/su11123388.

47 Geospatially augmented flood risk assessment incorporating machine learning approaches: a case study in the Teesta river basin

Deepanjan Sen[1,2,a] and Swarup Das[2,b]

[1]Department of Computer Applications, Dr. B. C. Roy Academy of Professional Courses, Durgapur, West Bengal, India

[2]Department of Computer Science and Technology, University of North Bengal, Siliguri, West Bengal, India

Abstract

Climate change is exacerbating the risks of floods a more significant threat to global stability and ecosystems, particularly in vulnerable regions like the Teesta River Basin in India with an estimated flood risk rate of 85%. The study generates comprehensive flood susceptibility maps by combining other important geospatial elements with digital elevation model (DEM) derived from Cartosat-1 satellite data. This study employed a novel approach implementing strategic data augmentation and Boruta analysis to optimize the selection of flood-conditioning factors, with advanced machine learning techniques- K-nearest neighbors (KNN), Naïve Bayes (NB), and decision trees (DT), ensuring the robustness of the predictive models for the study area. To determine flood-prone area prediction, these models were assessed using accuracy, precision, recall, F1-score, and receiver operating characteristics area under the curve (ROC-AUC). The findings show that the KNN model outperforms the NB and DT models in terms of accuracy of 91.62% and demonstrating robustness with ROC-AUC of 97.80%, 94.20%, 95.00% in training, testing and validation respectively, effectively delineating areas with varying levels of flood risk. This research provides flood risk management insights and a methodological framework for flood preparedness and resilience in flood-prone regions.

Keywords: Boruta analysis, digital elevation model, flood susceptibility, spatial data augmentation

Introduction

Climate change is exacerbating floods, which threaten global stability and ecosystems. Ecosystem degradation and human growth in flood-prone areas increase flooding risk. Floods killed 7 million people between 1900 and 2023, out of 38.5 million killed by natural catastrophes [1]. Floods have destroyed farms, homes, public infrastructure, and livestock. With an estimated flood risk rate of 85%, the Indian subcontinent faces challenges particularly in the Himalayan Teesta River Basin [2]. Digital elevation models (DEMs) are vital for precise identification of flood-prone regions as they offer critical data on the topography and altitude of the land. Cartosat-1 satellite data (Carto DEM) has been applied to build and analyze flood risk assessment models for the Teesta River Basin [3]. The objectives of this study include the following:

- Creating a flood inventory with spatial augmentation.
- A detailed analysis of geospatial and hydrological characteristics to identify crucial factors that affect flood risk.
- Developing flood risk assessment models based on machine learning, evaluating the performance of these models.
- Applying the developed model to assess flood risk, and providing recommendations for flood risk management.

The devastating floods in India have wiped out vast swaths of farmland, houses, and public buildings, killing millions of people and destroying infrastructure. According to a 1927 report by P.C. Mahalanobis, the Himalayan foreland of North Bengal had been battling floods for nearly 150 years, demonstrating the magnitude of this natural calamity [4]Click or tap here to enter text.. West Bengal, the region with the most affected land area and population, is particularly affected. Riverine floods account for about 60% of India's disaster, with rain and cyclone impacts accounting for the rest. Advances in remote sensing data, geographic information systems (GIS) techniques, and other technological platforms have resulted in a diverse set of tools for urban flood risk modelling. The aim is to improve flood risk assessment methods in the Teesta River Basin for efficient preparation for

[a]sen.deepanjan92@gmail.com, [b]sd.csa@nbu.ac.in

DOI: 10.1201/9781003663348-47

disasters and the encouragement of sustainable development in vulnerable areas and safeguarding of the welfare of the people living in the basin.

Study Region

Crucially important in South Asia's socio-hydrological system is the Brahmaputra's tributary- Teesta River Basin. Forming Asia's biggest river system from the Tso Lhamo Lake in the Sikkim Himalayas, it originates Notable channel alterations have occurred along the Teesta's geomorphological history, including a devastating flood in 1787 that caused it to split off from the Ganga and join the Brahmaputra [5]. The West Bengal Teesta Basin is the focus of this research. Originating from Sikkim's Lachen Chu and Lachung Chu, the Teesta joins the Rangit River near Sevoke in West Bengal [6]. The Teesta wanders across the picturesque areas of Jalpaiguri and Cooch Behar. Near Dahagram, the river forms the border between India and Bangladesh [7]. The 414-kilometer Basin drains 12,370 square kilometers; the West Bengal section is especially of interest, depicted in Figure 47.1. Focusing on a four-administrative-block region comprising 2239.39934 square kilometers, 123 kms in length and with a latitudinal range of 26°58'35.502" N to 26°13'50.039" N and a longitudinal range of 88°23'54.964" E to 89°3'47.511" E., this paper investigates flooding risk in the Teesta River Basin of West Bengal. Particularly in this important sector, the aim is to offer significant insights to enhance flood susceptibility maps and develop flood risk reduction strategies thereby strengthening the resilience of this sensitive area.

Resources and Strategies

Methodologies adopted

The present study was conducted in six stages to assess the flood susceptibility of the proposed area, depicted in Figure 47.2.

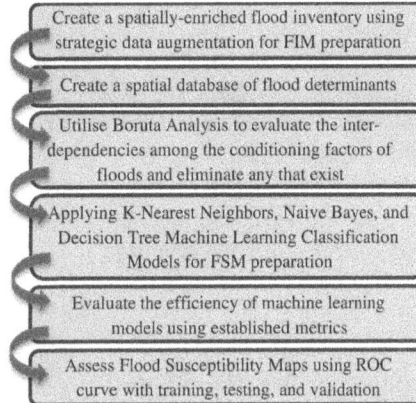

Figure 47.2 Methodological flowchart
Source: Author

Flood inventory map preparation

Maps of flood susceptibility help mitigate risk and prepare for disasters. Accurate models require a comprehensive flood inventory map. A suitable FIM for investigating potential flooding in a target area is described in this paper [8]. Hi-level fact-finding included site surveys, satellite imagery, and flood records. Detailed cross-verification identified 317 flood regions. While preserving event details, strategically and manually augmented datasets with 1:3 ratios of original data sources increased dataset size. This method simulated

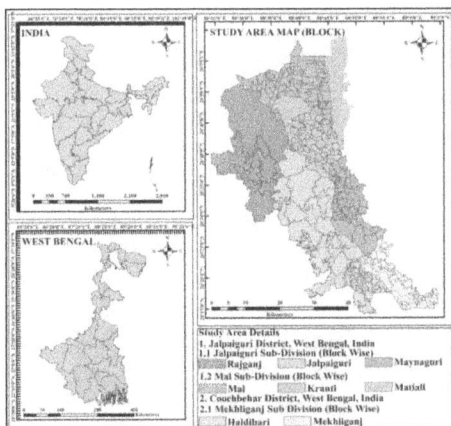

Figure 47.1 Location:(a) India Map, (b) West Bengal Map, and (c) Study Area Map
Source: Author

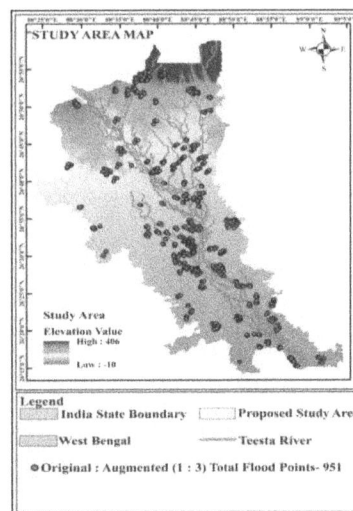

Figure 47.3 Study area map with data augmentation
Source: Author

flood covers around every identified flood point, including hypothetical flooded areas outside boundaries, depicted in Figure 47.3. This augmentation method improved model generalizability and flood limit uncertainty in areas with limited data. Modelling techniques were applied to the generated dataset. Researchers and disaster management authorities can use this comprehensive approach to create flood inventory maps to create accurate flood hazard maps, improving preparedness and mitigating strategies for flood-prone areas. On Arc-GIS, a similar number of non-flood data (0s) was selected for binary analysis. Missing values have been handled through Universal Kriging method. Following stratified random partitioning, 15% of the dataset has utilized for testing, 15% has used for validation, while 70% has used for training.

Spatial dataset formation

Each flood susceptibility analysis parameter affects mapping differently. Choosing the right parameters requires extensive literature research, field specialist consultation [9]. This research created a geospatial database of 12 flood conditioning factors that have been provided in Table 47.1 to build thematic layers.

FDF: Flood determinant factor; El: Elevation; Sl: Slope; PrCurv: Profile curvature; MRN: Melton ruggedness number; MrVBF: Multi-Resolution Valley Bottom Flatness; AR: Annual Rainfall; LULC: Land use land cover; Geol: Geology; CND: Channel network distance; TST: Terrain surface texture; Geom: Geomorphons; NRSC: National remote sensing center; ISRO: Indian Space Research Organization; IMD: India Meteorological Department; ESRI: Environmental Systems Research Institute; GSI: Geological Survey of India

Table 47.1 Spatial dataset details

FDF	Spatial details	Data source
El	30 × 30 m	NRSC, ISRO
Sl	30 × 30 m	NRSC, ISRO
PrCurv	30 × 30 m	NRSC, ISRO
MRN	30 × 30 m	NRSC, ISRO
MrVBF	30 × 30 m	NRSC, ISRO
AR	High Resolution Gridded data	IMD
LULC	ESA Sentinel-2 imagery	ESRI
Soil	Polygon data (vector data)	FAO world soil data
Geol	Reference map (1:50000)	BHUKOSH, GSI
CND	30 × 30 m	NRSC, ISRO
TST	30 × 30 m	NRSC, ISRO
Geom	30 × 30 m	NRSC, ISRO

Source: Author

Flood causative factors

Elevation: River depth and excess water release depend on river elevation. Lower elevations are more prone to flooding due to groundwater leakage. *Slope:* Flood susceptibility mapping includes slope, which indicates territory steepness and horizontal plane tilt and sharpness. Slopes improve water flow over the surface and decrease groundwater seepage. *Profile curvature:* Depending on vertical slope steepness, profile curvature affects stability and erosion. Negative profile curvature indicates a slow drop, while positive indicates an abrupt one. Straight slopes are zero-curvature profiles [10]. *Melton ruggedness number (MRN):* Scientific investigation uses the MRN to assess Teesta River basin terrain roughness. Measurement of watershed height discrepancies [11]. *Multi-resolution valley bottom flatness (MrVBF):* This study uses the MrVBF index to assess flatness and low elevation on a multiscale. Flat, lower valley bottoms are found using this strategy. The sigmoid/logistic transformation normalizes elevation percentiles and terrain slope angles. To capture lowest point features and expose floodwater accumulation sensitivity, this approach aggregates updated values across DEM smoothing layers [12]. *Annual rainfall:* There is a direct correlation between rainfall and floods, as the likelihood of flooding rises in tandem with the amount of rainfall. *Land use land cover (LULC):* Land cover affects runoff, evaporation, transpiration, and infiltration. The study area has seven land cover types: Water, trees, flooded vegetation, crops, developed areas, bare ground, rangeland. *Soil:* Soil types significantly affect floodwater amounts due to their effects on infiltration, percolation, and runoff [13]. *Geology:* Geological changes affect flood possibilities in a location; hence geology stimulates flood risk. The current investigation's geological map was acquired from BHUKOSH, a Geological Survey of India open-source site with Quaternary, Undivided Precambrian, and Neogene sedimentary rocks. *Channel network distance (CND):* CND is the distance between river or stream locations. Hydrological research is more exact since it considers natural water flow routes. Distance calculation is useful in ecological studies, flood risk assessment, and hydrological modelling. *Terrain surface texture:* In a given area, terrain surface texture specifies topographic element spatial arrangement (peaks, pits, ridges, valleys). It describes the hard, complex terrain. Calculates the frequency and distribution of height deviations in a neighborhood [14]. *Geomorphons:* It classifies landforms using pattern recognition to analyze elevation and visibility changes and create landform feature maps. There are 498 geomorphon patterns in 10 landforms: flat, peak, ridge, shoulder, spur, slope, hollow, foot slope, valley, and pit. In low-resolution DEMs, this is crucial [15].

Boruta analysis

A critical preprocessing step in the development of robust predictive models is feature selection. Identifying the most important predictors is crucial in complex hydrological systems like the Teesta River Basin, where numerous topographical, morphometrical, and hydrological variables are often considered. This study used Boruta, a wrapper-based feature selection algorithm, to address this challenge.

Boruta operates by assessing the significance of primary features in relation to randomly shuffled (shadow) attributes within a random forest framework. Features that have a significantly higher level of importance than their corresponding shadow attributes are considered relevant and kept for further modelling [16]. This study endeavors to compile a subset of the most influential flood risk predictors by utilizing Boruta on the dataset, thereby enabling the creation of machine learning models that are both precise and frugal, displayed in Figure 47.3 and 47.4.

Machine learning classifiers

K-Nearest neighbors (KNN): Cover and Hart's 1968 KNN algorithm is a popular non-parametric flood susceptibility method. It is known for its simplicity and spatial classification tasks. K-NN is employed in flood risk assessment to forecast areas susceptible to flooding by examining patterns within the data, the flood determinant variables. KNN is a supervised classification method that stores all data and classifies new data points using distance functions. The Hamming distance is used for discrete variables, but the Euclidean distance is recommended for continuous variables. K-NN is flexible, simple, and beneficial in flood susceptibility studies, especially when used with ensemble machine learning models. This method is computationally expensive, noise-sensitive, and requires careful 'k' selection [17].

Naïve Bayes (NB): NB is a probabilistic classifier that relies on Bayes' theorem. It operates on the assumption that the features utilized in flood susceptibility analysis are mutually independent. This assumption is frequently employed in flood susceptibility analysis to estimate the likelihood of floods in a certain region. It takes into account several flood determinant factors, that may classify places as high, moderate, or low risk for flooding by assessing the probability of each feature's contribution to a flood occurrence and is highly efficient for integrating several spatial datasets, where the correlation between covariates and flood incidence is conditionally autonomous. When combining spatial information with conditionally independent variables and flood occurrence, it works well. The equation is as follow:

$$P\left(c|x\right) = \frac{P\left(x|c\right).P(c)}{P(x)} \qquad (1)$$

The NB method assumes that predictor cost (x) neutrally affects a category (c) of predictor values. The assumption of 'conditional independence' is as follows: P(c) represents the class prior probability, P(x) represents the predictor prior probability, P(x|c) is the likelihood of observing the feature vector x given class c, and P(c|x) represents the target posterior probability [18].

Decision Tree (DT): Decision Tree analysis is commonly used in flood susceptibility study because to its simplicity, interpretability, and usefulness with categorical and numerical data. Previous flood data combined with flood inducing features can train a Decision Tree in flood susceptibility analysis. All nodes in the tree represent decisions based on these factors, predicting the chance of flooding in a certain area. A flowchart-like decision tree has internal nodes reflecting dataset attributes. The branches indicate feature-based decision rules. Leaf nodes forecast outcomes or classes (e.g., flood-prone or not). The tree is formed by recursively partitioning the dataset by the feature that delivers the maximum information gain or impurity decrease, such as Gini impurity or entropy. Tracing decision processes from root to leaf is the goal to create a model that can predict flood risk for new data points [19].

Results and Discussion

Spatial distribution of flood susceptibility maps

The flood susceptibility map is essential for locating flood-prone areas in a region. This comprehensive evaluation of the Teesta River Basin's flood risk using KNN, NB, and DT, is enabled by the distinct perspectives that each of these algorithms offers on flood susceptibility, displayed in Figure 47.5.

The KNN model rated 32.55% of the area very low for flood susceptibility, 25.13% low, 11.38% moderate, 11.61% high, and 19.32% very high. This distribution is balanced, with a large portion of the area in the very low and low susceptibility categories. NB model

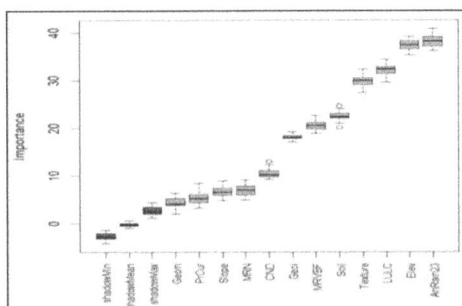

Figure 47.4 Boruta analysis

Source: Author

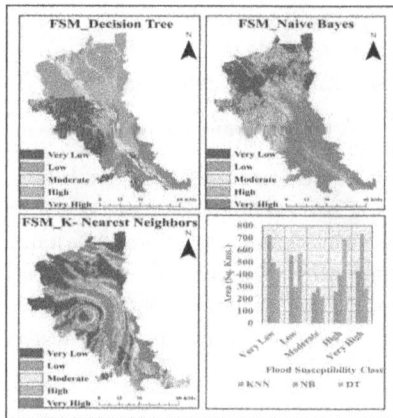

Figure 47.5 Flood susceptibility maps and its spatial distribution
Source: Author

displayed a clear distribution with 22.58% of the area classified as very low, 13.33% as low, 13.32% as moderate, 17.73% as high, and 33.03% as very high. This model indicated that a larger portion of the area was highly vulnerable to flooding, implying that NB may be more sensitive to elements causing such high flood risks. The DT model showed that 20.23% of the area had very low susceptibility, 25.69% low, 10.00% moderate, 31.03% high, and 13.05% extremely high. Unlike NB, this distribution has a larger low and high susceptibility area and a lower very high percentage. The DT model may prioritize moderate to high susceptibility over extreme susceptibility.

Performance evaluation and accuracy assessment
Accuracy, Kappa, pPrecision, recall, F1-score, and ROC-AUC measure flood susceptibility model performance [20]. Tables 47.2 and 47.3 compare metrics for KNN, NB, and DT models. The accuracy metric assesses model classification accuracy. The KNN model predicts flood susceptibility with the highest accuracy 91.62%. The Naïve Bayes model has the lowest accuracy 74.60% and the DT follows with 82.63%. KNN exhibits the highest Kappa value 0.8325 in this case, suggesting a substantial degree of agreement between the predicted and actual values. The Naïve Bayes model has the lowest Kappa 0.4921, indicating moderate agreement. The DT model is middle with 0.6526 Kappa.

Precision and recall determine the model's flood-prone location detection success. With a recall of 0.9312 and a precision of 0.9041, KNN can accurately and comprehensively identify flood-prone areas. Naïve Bayes has higher recall (0.8959) but lower precision (0.6893), indicating greater sensitivity. The Decision Tree model exhibits satisfactory performance, with a precision of 0.8234 and a recall of 0.8307. KNN

Table 47.2 Performance evaluation metrics

Metrics	KNN	NB	DT
Accuracy	0.9162	0.7460	0.8263
Kappa	0.8325	0.4921	0.6526
Precision	0.9041	0.6893	0.8234
Recall	0.9312	0.8959	0.8307
F1 Score	0.9175	0.7791	0.8270

Source: Author

has the highest F1-score of 0.9175, which confirms its superiority in precision and recall. While Naïve Bayes and DT have lower F1-scores (0.7791 and 0.8270, respectively), they still offer useful insights.

In Table 47.3 and Figures 47.6–47.8, ROC-AUC analysis shows the KNN model's excellent performance across all datasets: 97.80% in training, 94.20% in testing, and 95.00% in validation high AUC values show this model's reliability in separating flood-prone from non-flood-prone regions. The AUC values of Naïve Bayes and DT models are lower, with Naïve Bayes having 85.10% in training, 83.70% in testing, and 88.70% in validation, while DT has 85.40%, 82.20%, and 83.20%.

Conclusion and Future Directives

This study illustrates the critical efficacy of geospatially enhanced flood risk assessment models in the Teesta River Basin, with the K-Nearest Neighbours (KNN) algorithm as the most advanced predictive methodology. A multi-model approach was employed

Table 47.3 ROC-AUC analysis

Utilized models	Training	Testing	Validation
KNN	97.80	94.20	95.00
Naïve Bayes	85.10	83.70	88.70
Decision Tree	85.40	82.20	83.20

Source: Author

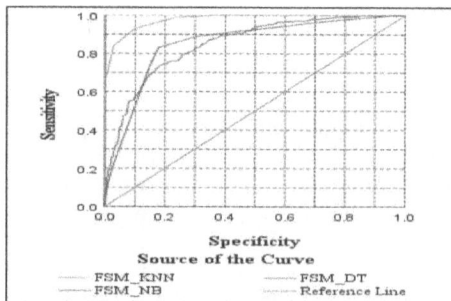

Figure 47.6 AUC curve on training
Source: Author

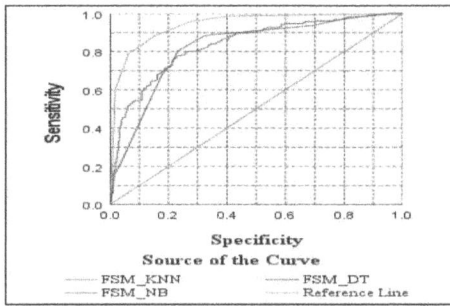

Figure 47.7 AUC curve on testing
Source: Author

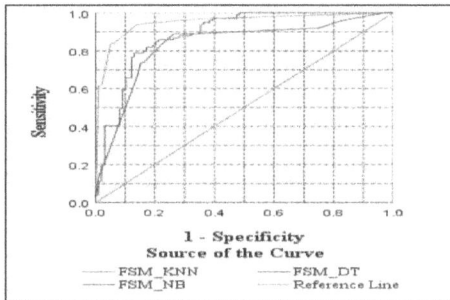

Figure 47.8 AUC curve on validation
Source: Author

to reveal nuanced insights into flood susceptibility in the comprehensive analysis.

Principal Research Conclusions:

- The KNN model attained an accuracy of 91.62% in predicting flood risk.
- Spatial risk mapping designated 35% of the basin as high-vulnerability areas.
- Accurate geospatial integration with digital elevation models (DEMs) and data driven techniques improved predictive capabilities.

Methodological importance: The comparative analysis of KNN, NB, and DT models underscored the significance of multi-model evaluation. The K-NN has provided a thorough risk assessment, whereas the NB has identified the extreme susceptibility zones and the DT has pointed out the moderate to high-risk areas. The amalgamation of sophisticated machine learning methodologies with geospatial analysis establishes a comprehensive framework for comprehending flood risk dynamics, yielding essential insights for disaster management and mitigation strategies.

Future Research Directions: Improving upon the existing approach necessitates-

- Ensuring accuracy in various environmental settings.
- Application of advanced machine learning methods.

- Using deep learning and EL methods into investigation.
- Creation of more sophisticated forecasting algorithms.

These long-term paths for research will play a crucial role in managing flood risks on a global scale, assisting with plans for adjusting to climate change, encouraging long-term growth in delicate ecosystems. This research is a giant leap forward in the quest for evidence-based methods of flood risk assessment and prevention.

References

[1] CRED: EM-DAT (2024). http://www.emdat.be (accessed August 20, 2024).

[2] Central Water Commission (2024). Ministry of Jal Shakti, Department of Water Resources, River Development and Ganga Rejuvenation, GoI.. https://cwc.gov.in/ (accessed August 20, 2024).

[3] Datla, R., & Mohan, C. K. (2021). A novel framework for seamless mosaic of Cartosat-1 DEM scenes." *Computers & Geosciences* 146 (October 10, 2020): 104619. https://doi.org/10.1016/j.cageo.2020.104619.

[4] Mahalanobis, P. C. (1927). Report on rainfall and flood in North Bengal during the period 1870-1922. In Proceedings of the Indian Science Congress (Bombay).

[5] Majumdar, R. C. (1971). History of Ancient Bengal. G. Bharadwaj, Calcutta.

[6] Mukhopadhyay, S. C. (1984). The Tista Basin: A Study in Fluvial Geomorphology. K.P. Bagchi.

[7] Rudra, K. (2018). Rivers of the ganga–brahmaputra–meghna delta: an overview. In Rivers of the Ganga-Brahmaputra-Meghna Delta: A Fluvial Account of Bengal. Geography of the Physical Environment, (pp. 1–14). Springer, Cham. https://doi.org/10.1007/978-3-319-76544-0_1.

[8] Antzoulatos, G., Kouloglou, I. O., Bakratsas, M., Moumtzidou, A., Gialampoukidis, I., Karakostas, A., et al. (2022). Flood hazard and risk mapping by applying an explainable machine learning framework using satellite imagery and GIS data. *Sustainability*, 14(6), 3251. https://doi.org/10.3390/su14063251.

[9] Arabameri, A., Rezaei, K., Cerdà, A., Conoscenti, C., & Kalantari, Z. (2019). A comparison of statistical methods and multi-criteria decision making to map flood hazard susceptibility in Northern Iran. *Science of The Total Environment*, 660, 443–458. https://doi.org/10.1016/j.scitotenv.2019.01.021.

[10] Dohnal, F., Hubacek, M., & Simkova, K. (2019). Detection of microrelief objects to impede the movement of vehicles in terrain. *ISPRS International Journal of Geo-Information*, 8(3), 101. https://doi.org/10.3390/ijgi8030101.

[11] Melton, M. A. (1965). The geomorphic and paleoclimatic significance of alluvial deposits in Southern Ari-

zona. *The Journal of Geology,* 73(1), 1–38. https://doi.org/10.1086/627044.

[12] Gallant, J. C., & Dowling, T. I. (2003). A multiresolution index of valley bottom flatness for mapping depositional areas. *Water Resources Research,* 39(12). https://doi.org/10.1029/2002WR001426.

[13] Böhner, J., & Selige, T. (2006). Spatial prediction of soil attributes using terrain analysis and climate regionalisation. In SAGA-Analyses and Modelling Applications. Goltze.

[14] Hu, G., Dai, W., Xiong, L., & Tang, G. (2020). Classification of terrain concave and convex landform units by using TIN. In Proceedings of the Geomorphometry 2020 Conference.

[15] Jasiewicz, J., & Stepinski, T. F. (2013). Geomorphons — a pattern recognition approach to classification and mapping of landforms. *Geomorphology,* 182, 147–156. https://doi.org/10.1016/j.geomorph.2012.11.005.

[16] Kursa, B. M., & Rudnicki, R. W. (2010). Feature selection with the Boruta package. *Journal of Statistical Software,* 36(11), 1–13.

[17] Guo, G., Wang, H., Bell, D., Bi, Y., & Greer, K. (2003). KNN model-based approach in classification. In On the Move to Meaningful Internet Systems 2003: CoopIS, DOA, and ODBASE, (pp. 986–996). https://doi.org/10.1007/978-3-540-39964-3_62.

[18] Zhang, H. (2004). The optimality of naive Bayes. *American Association for Artificial Intelligence,* 1(2), 3.

[19] Kundapura, S., Aditya, B., & Apoorva, K. V. (2023). Feature elimination and comparative assessment of machine learning algorithms for flood susceptibility mapping in Kerala, India. In 2023 IEEE 2nd International Conference on Data, Decision and Systems (ICDDS), (pp. 1–5). https://doi.org/10.1109/ICDDS59137.2023.10434786.

[20] Dey, S., & Das, S. (2023). Performance Assessment of multivariate statistical and bagging ensembles in landslide susceptibility mapping: case study of national highway-10. In 2023 International Conference on Device Intelligence, Computing and Communication Technologies, (DICCT), (pp. 284–289). https://doi.org/10.1109/DICCT56244.2023.10110089.

48 Structural behavior and ductility variations in multi-story RC frames under seismic loading

Suvendu Kar

Assistant Professor, Abacus Institute of Engineering and Management, MAKAUT, West Bengal, India

Abstract

This study investigates the variations in the ductility factor of reinforced concrete (RC) framed structures with different story heights using nonlinear static analysis (NSA) in ETABS. The empirical formula developed by Eduardo Miranda provides a basis for calculating the ductility factor, primarily influenced by the natural time period of the structure and the prevailing soil conditions. However, discrepancies were observed between results obtained from NSA, suggesting that additional factors beyond the structure's height influence the ductility factor. The study further explores how ductility factors vary with changes in story height and time periods through modal analysis. Findings indicate significant variations in the ductility factor with an increase in stories, underscoring the importance of incorporating comprehensive factors in predictive models.

Keywords: Ductility factors, modal analysis, nonlinear static analysis, soil conditions, time periods

Introduction

The seismic performance of buildings is a critical aspect of structural engineering, especially in earthquake-prone regions. Among the key factors influencing the design of earthquake-resistant structures is the ductility factor [5], which quantifies a structure's ability to undergo inelastic deformations while maintaining its load-carrying capacity. The ductility factor is essential for assessing the seismic resilience of reinforced concrete (RC) framed structures, allowing engineers to design buildings that can dissipate energy and avoid catastrophic failure during seismic events. Research in this field has led to the development of various empirical formulas for estimating the ductility factor, one of the most notable being the formula proposed by Eduardo Miranda. Miranda's approach considers the natural time period, ductility ratio, and soil conditions as primary determinants of the ductility factor. His work established a foundational understanding of how these parameters affect the structural performance of RC framed buildings under seismic loads. However, subsequent studies have indicated that additional factors, such as building height, number of stories, and variations in structural stiffness, also play significant roles in influencing the ductility factor. For instance, Mahmoudi and Abdi [6] found that increasing the number of stories results in a decrease in both overstrength and ductility factors in steel frames with TADAS devices. Similarly, Asgarian and Shakrgozar [3] observed that in Buckling Restrained Braced Frames (BRBFs), ductility factors decrease as the number of stories increases, highlighting the complexity of accurately predicting ductility. Further investigations by Lin and Chang [4] into damping reduction factors and Miranda [8] into strength reduction factors emphasized the nuanced effects of structural and material properties on seismic performance. These studies collectively underscore the necessity for a more refined approach that incorporates a broader range of variables affecting ductility beyond the standard parameters. Despite these advancements, the empirical models often fail to account for the combined impact of multiple factors, such as building height and story configuration, which are crucial in realistic design scenarios.

This study aims to address these gaps by exploring the variations in the ductility factor of regular RC framed structures with different story heights on flat ground. Utilizing Nonlinear Static Analysis Pushover Analysis (NSAPA) in ETABS, the research compares simulation results with manual calculations based on Miranda's formula. The objective is to identify discrepancies and refine the understanding of how various structural attributes, including story height, influence ductility. By integrating findings from previous literature and conducting detailed analyses, this study seeks to contribute to the development of more accurate predictive models for seismic design, ultimately enhancing the resilience and safety of RC framed structures in earthquake-prone areas.

suvendukar.aiem@jisgroup.org

DOI: 10.1201/9781003663348-48

Response Reduction Factor (R)

Most modern design codes incorporate the nonlinear response of a structure indirectly through a 'Response Reduction/Modification Factor' (R) [7], enabling designers to use a linear elastic force-based approach while considering nonlinear behavior and deformation constraints. This concept was first introduced in ATC-3-06 to decrease the elastic base shear force (V_e) derived from elastic analysis, leading to the determination of the design base shear (V_d). In the Indian standard IS 1893, it is referred to as the "response reduction factor," whereas in ASCE7, it is known as the "response modification coefficient." Eurocode 8 describes this factor as the "behavior factor."

$$R = R_S \times R_\mu \times R_r$$

where, represents the overstrength factor, R_μ denotes the ductility reduction factor, and R_r is the redundancy factor. The degree of inelastic deformation that a structural system endures under a particular ground motion or lateral loading is characterized by the displacement ductility ratio, often denoted as "μ" (ductility demand). This ratio is defined as the quotient of the maximum absolute relative displacement Δ_u experienced by the structure compared to its yield displacement Δ_y, providing an indication of the extent to which the structure can deform plastically before failure. The ductility factor (R_μ) is an essential measure that reflects the overall nonlinear response of a structural system with respect to its capacity for plastic deformation. It is calculated by dividing the base shear corresponding to an elastic response (V_e) by the maximum or ultimate base shear experienced during an inelastic response (V_u). This factor is crucial for understanding how well the structure can accommodate inelastic deformations while maintaining stability under extreme loading conditions. The calculation of the ductility reduction factor is performed using Miranda's approach. This method involves specific procedures and formulas designed to accurately estimate the ductility reduction based on the system's response characteristics. By applying Miranda's method, the study aims to provide a precise assessment of the ductility reduction factor, which is vital for evaluating the structural system's performance and safety under seismic events or other significant loading scenarios.

$$R_\mu = \frac{\mu - 1}{\varphi} + 1 \geq 1$$

Here, $\emptyset = 1 + \frac{1}{12T - \mu T} - \frac{2}{5T} exp\left[-2\left(lnT - \frac{1}{5}\right)^2\right]$

Where, φ is a function of μ, T and the soil conditions at the site. The details of which may be seen in

Miranda and Bertero. The overstrength factor (R_S) is introduced as a measure of the built in overstrength in the structural system and is obtained by dividing the ultimate base shear, V_u by the design base shear (V_d) and is obtained as,

$$R_S = \frac{V_u}{V_d}$$

Numerical Study

For this study, seven distinct structural frames: two, three, four, five, six, seven, and eight story buildings all of 4 bays (Spacing of each Bay- 4m) were designed by hand to accurately satisfy all 'dimension and loads' provisions of the respective codes [1]. These frames were then modeled in ETABS software where the structures are simulated and analyzed under various conditions. The first procedure of the analysis was to do a modal analysis in order to establish the Fundamental Natural Period (T_a) of each of the building. This period is a key factor that determines the behavior of a structure under dynamic loads, which perhaps arise due to earthquake hence a key parameter. The Fundamental Natural Period was then estimated through ETABS and in a manual format using the normative formula obtaining from the Indian Standard IS 1893 PART: 2016 [2] clause no. 7. 6. 1. Thus, comparison with manual calculations assists in cross checking the analysis and result. Following the modal analysis, a Nonlinear Static Analysis, referred to as Pushover Analysis, was conducted. This type of analysis is used to evaluate the potential seismic performance of the structures by incrementally applying lateral forces until the buildings reach their ultimate capacity. From this analysis, key performance indicators such as the yield displacement (Δ_y) and the ultimate displacement (Δ_u) were derived. These parameters are crucial for understanding the ductility and deformation characteristics of the buildings under seismic loading conditions. In regard to this research, the height of the base storey was standardized at 3.2 meters high and consecutive storey was 3 meters high with up to 8th storey. These dimensions were selected in order to accurately represent real multi storey building constructions and to make sure that the models used were as realistic as possible.

Nonlinear static analysis [10]

Performance-based seismic design NSPA often referred as pushover analysis is an essential tool in evaluating the capacity and inelastic response of structures to seismic excitations. This analysis involves gradually applying horizontal forces to a building model up to a predetermined displacement with an aim of

emulating an earthquake. In the particular provision of Pushover Analysis, which is an analysis [11] of the response of structures beyond their elastic limits, possible energy dissipation, avoidance of collapse, and the protection of the occupants when a structure experiences an earthquake, may be ascertained. The other important aspect of Pushover Analysis is the tracking and extension of the process of formation of plastic hinges. plastic hinges are areas of localized yielding in members such as beams and/or columns wherein the material deforms plastically when it has reached its yield point. These hinges separate the elastic and inelastic behavior of structures which re-distribute internal forces and seismic energy through acceptable deformations. The manner in which these hinges are formed and the sequence in which they occur is very important in determining the likely failure modes in the structure and possible areas of weakness which may be of immense help in ascertaining the ductility factor or lack of fit in the structural system of the building. Pushover Analysis evaluates structural performance across distinct levels, each corresponding to a different extent of damage and post-earthquake usability of the building:

Immediate occupancy (IO): Ideally at this performance level the structure is not significantly damaged and remains safe to occupy right after an earthquake. Indeed, there may be non-structural damage, but the primary load-bearing capacity is fairly well compromised so that the building may be used as intended without the need for extensive follow-up work. **Life safety (LS):** The prescriptive provision, as known as the life safety level is dedicated to the protection of occupants' lives during an earthquake. At this level substantial inelastic deformations and plastic hinge formations at critical sections are permitted, but the structure should have adequate stash strength to carry dead loads and to provide safe egress. Extensive structural deterioration is evident while structural conditions do not pose a risk to the integrity of a building. **Collapse prevention (CP):** The last one is CP level that indicates specifically that the structure can barely withstand more loads. At this stage, all the plastic hinges are formed and the building possesses comparatively small portions of its original ability to carry the lateral load. The building may be structurally unsound and may need to be demolished, nevertheless, it remains steady and intact long enough to enable occupants to get out unharmed. To take into consideration the impact of the higher modes of vibration, particularly on the tall and complicated structures a method known as the square root of the sum of the squares (SRSS) is used. This approach integrates the response of many vibrational modes as it avails

the total response of the building structure to seismic excitation by taking the square root of the sum of the square of individual modal responses. This method in turn improves the order of the analysis and thus avoids losing higher mode effects. The procedure of Nonlinear Static Pushover Analysis typically involves modeling the structure in software ETABS, incorporating non-linear material properties and defining potential plastic hinge locations. Lateral load patterns are applied to simulate seismic forces, which are incrementally increased until the structure reaches its ultimate capacity or target displacement. The structural response is tracked through the formation of plastic hinges and plotted on a capacity curve (pushover curve), which illustrates the relationship between base shear and roof displacement. By evaluating this curve, engineers can determine if the structure meets the desired performance levels of IO, LS, and CP.

Result and Analysis

The study involved conducting a nonlinear static (pushover) analysis on reinforced concrete (RC) framed structures of varying heights (2 to 8 storeys) using ETABS software. The ductility factor, a key parameter influencing the seismic performance [9] of structures, was calculated using both IS code recommendations and Miranda's empirical formula, and the results were compared with those obtained from the pushover analysis. The primary aim was to observe how changes in storey height influence the ductility factor. Figure 48.1 provides a graphical representation of the base share in relation to the observed displacement across all the frames. This illustration compares how the base share correlates with the displacement measurements recorded for each frame, offering a comprehensive view of the structural behavior and performance under various conditions.

1 provides a detailed presentation of the data collected through the use of ETABS software and codal formulas. A few recommendations of IS 1893-2016 on ductility calculation bear focus related to basic time period, ductility factor and building heights; any structural plan generally requires additional ductility at increased heights and hence needs modification. This table is designed to support a thorough examination of the overall dataset by showcasing all relevant calculations. By compiling the results in this manner, it allows for an in-depth analysis and comparison of the data, aiding in the assessment of the structural performance and the accuracy of the applied methods. From the graph and table, the following observations are made:1. The seismic performance of RC building frames can be effectively assessed from the parameter

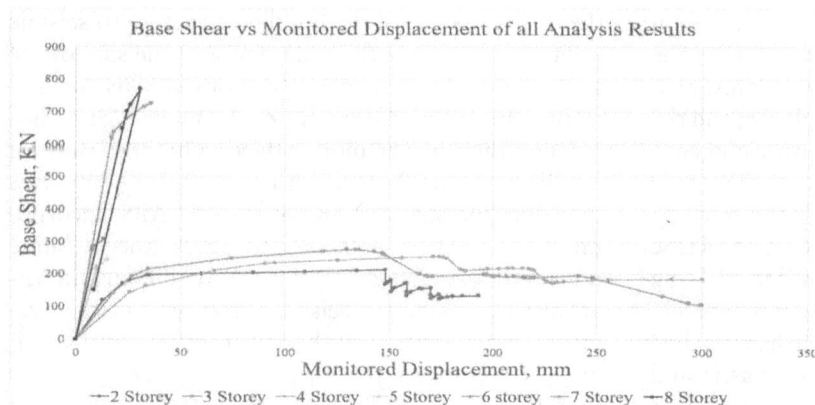

Figure 48.1 Base share vs monitored displacement
Source: Author

Table 48.1 Detailed presentation of the data collected through the use of ETABS and codal formula

Sl. No	Storey	Fundamental Time Period (T) (IS CODE)	Fundamental Time Period (T) (Modal Analysis)	Max Disp. (Δ_u) (mm)	Yield Disp. (Δ_y) (mm)	Ductility Ratio (μ)	Ductility Factor R_μ (IS CODE)	Ductility Factor R_μ (Pushover Analysis)
1	2	0.294	1.473	121.17	26.539	4.566	3.488	5.26
2	3	0.396	2.988	133.84	26.442	5.061	4.165	4.38
3	4	0.489	4.977	171.52	25.491	6.728	5.64	6.528
4	5	0.577	3.723	14.966	10.013	1.494	1.526	1.486
5	6	0.660	3.441	13.253	7.816	1.695	1.8	1.685
6	7	0.740	3.041	35.809	17.168	2.085	2.34	2.076
7	8	0.818	3.875	30.627	25.255	1.212	1.28	1.2

Source: Author

known as the ductility factor. Its calculation is computed as per the empirical formula by Eduro Miranda which has parameters of Natural time period, ductility ratio and site soil conditions. However, the time period given from ETABS modal analysis is slightly more than per IS code results. With the pushover analysis, it shows that for 2-4 storey buildings, the ductility factor higher than with Miranda formula and then from 5 until 8 story building. These seven RC framed structures have been studied; it can be observed that this behaviour is common.

2. Analysis of the ductility factor using modal analysis is not consistent, but it shows overall that taller buildings seem to have a lower ductility factor suggesting less seismic energy dissipating capacity as per IS Code 1893 (Part 1) 2016 [2] That could mean taller structures are less bendy and more brittle. 3-5 and 6-8 story buildings behave more ductile; however, these are generic structures that could be designed to higher or lower levels of isolation. Additional study is necessary

to understand the relationship between building height, ductility level and seismic performance.

3. Different frames design, materials and construction methods result also in differences in top displacement and yield displacement in buildings. These factors also influence ductility, that is, leeward deformation, and seismic energy management, which complicates standardization.

4. On average, the yield over maximum displacement ratio does not exhibit a strong hallmark in any of this data. However, the 4-story and 7-story frames experience ductility ratios that are high compared to percent angles except for one likely due to design/structural/material differences. The severity of this concept can be assessed more effectively if the ductility ratios are considered as a function of building height and frame systems, which calls for further research.

5. Modal analysis often calculates the fundamental time period that differs from IS Code 1893-2016 values, which is dependent on structure's height,

frame design, material properties and configuration. Dynamic properties are considered in modal analysis, whereas IS Code uses standardized equations, resulting in discrepancies because of different frame specifications.

Summary and Conclusion

In this study, nonlinear static analysis (NSA) in ETABS is used to examine how the ductility factor of reinforced concrete (RC) framed structures changes with building height. Results are compared across seven frames (2 to 8 stories) against Eduardo Miranda's empirical formula. Results indicate that taller buildings tend to have lower ductility and thus lower seismic energy absorption, while mid-rise buildings exhibit irregular increases resulting from unique design features [12]. Empirical and analytical results differ, indicating that factors such as natural time period, stiffness, and material properties affect ductility, and that better predictive models are needed.

References

[1] IS 456:2000 (2000). Plain and Reinforced Concrete- Code of Practice. (fourth revision). New Delhi: Bureau of Indian Standards.

[2] IS 1893(PART I):2016 (2016). Criteria for earthquake resistant design of structures. (Sixth Revision), New Delhi: Bureau of Indian Standards.

[3] Asgarian, B., & Shokrgozar, H. R. (2003). BRBF Response Modification Factor. Tehran, Iran: K.N. Toosi University of Technology.

[4] Lin, Y. Y., & Chang, K. C. (2003). Study on damping reduction factor for buildings under earthquake ground motions. *Journal of Structural Engineering*, 129(2), 206–214. 10.1061/(ASCE)0733-9445129:2(206).

[5] Mahmoudi, M., & Zaree, M. (2010). Evaluating response modification factors of concentrically braced steel frames. *Journal of Constructional Steel Research*, 66(10), Pages 1196–1204.

[6] Mahmoudi, M., & Abdi, M. G. (2012). Evaluating response modification factors of TADAS frames. *Journal of Constructional Steel Research* 71 162–170. doi:10.1016/j.jcsr.2011.10.015

[7] Mondal, A., Ghosh, S., & Reddy, G. R. (2013). Performance-based evaluation of the response reduction factor for ductile RC frames. *Engineering Structures*, 56, 1808–1819.

[8] Miranda, E. (1993). Site-dependent strength-reduction factors. *Journal of Structural Engineering*, 119(12), 3503–3519.

[9] Barakar, S. A., Husein Malkawi, A. I., & Al-Shatnawi, A. S. (1997). A step towards evaluation of the seismic response reduction factor in multistorey reinforced concrete frames. Natural Hazards: *Journal of the International Society for the Prevention and Mitigation of Natural Hazards*, Springer; International Society for the Prevention and Mitigation of Natural Hazards, vol. 16(1), 65–80.

[10] Federal Emergency Management Agency (FEMA). (2000). Prestandard and Commentary for the Seismic Rehabilitation of Buildings (FEMA 356 / November 2000). Washington, D.C.: Federal Emergency Management Agency.

[11] Jalayer, J., & Cornell, C. A. (2009). Alternative nonlinear demand estimation methods for probability-based seismic assessments. Earthquake Engineering & Structural Dynamics 38(8), 951–972, DOI:10.1002/eqe.876.

[12] Karavasilis, T. L., Bazeos, N., & Beskos, D. E. (2008). Estimation of seismic inelastic deformation demands in plane steel MRF with vertical mass irregularities. *Engineering Structure*, 30(11), 3265–3275.

49 Possible challenges in assigning IPR to AI driven inventions

Anuska Pramanick[1,a], Asmita Rajput[1,b], Mahuya Hom Choudhury[2,c], Arpita Chakraborty[3,d], Jyoti Sekhar Banerjee[3,e] and Siddhartha Bhattacharyya[4,f]

[1]Department of IT, Bengal Institute of Technology, Kolkata, West Bengal, India

[2]Patent Information Centre, WBSCST, Kolkata, West Bengal, India

[3]Department of CSE (AI & ML), Bengal Institute of Technology, Kolkata, India

[4]Faculty of Electrical Engineering and Computer Science, VSB Technical University of Ostrava, Czech Republic

Abstract

In this study, the topic of intellectual property rights for AI-based innovations is examined. Using its "WIPO Conversation on IP and AI," the WIPO agency has started to interact with the stakeholders. In terms of safeguarding their AI innovations, South Korea, US, Japan, and the China are at the forefront. The policies of their respective patent offices will also be significant. Although the IP rules haven't exactly kept up with the speed of these AI-based innovations. Companies like Google have developed their own guidelines to safeguard their intellectual property in the meantime. The rules regulating AI innovations must, in the grand scheme of things, be able to safeguard and honoring the creator for the sake of society. Additionally, it needs to ensure that the advantages of the innovation are distributed fairly to all facets of society. This work makes a unique contribution by discussing potential difficulties in awarding AI-driven creations intellectual property rights.

Keywords: Infringement and intellectual property rights, invention, world intellectual property organization

Introduction

Rapid artificial intelligence (AI) developments have many unexpected and complicated effects on society and the economy. These developments may have far-reaching effects on productivity, employment, and competitiveness by altering the very nature of production for a broad variety of goods and services. However, although these results are likely to have an impact, artificial intelligence may also alter the nature of innovation itself, with far-reaching ramifications that may eventually overshadow the direct benefit [1–3].

Deep learning and robotics are the AI fields that are expanding the quickest, whereas the transportation sector is the one where AI patents are expanding the quickest. Telecommunications are next in line. Next are personal gadgets, medical and life science. Top five applicants for AI patents are Toshiba (Japan), Samsung (South Korea), Microsoft (US), NEC (Japan) and IBM (US). The Chinese Academy of Sciences (CAS) is the leading organization for filling patents based on artificial intelligence. Since 2013, the number of applications for AI patents has increased significantly.

In regard to IP protection for AI-based innovations, the issue of ownership arises. When a person comes up with a creative action, they are creating an innovation. That person or his or her employer is the rightful owner of the innovation [4–6]. If AI is being utilized to create an innovation, then the creator and proprietor of that creation is the person employing AI. It is not enough to just possess the equipment to claim ownership. The topic of whether or not an AI creation that requires no human involvement may be patented has interesting implications. Does it become public domain if "a machine" cannot apply for a patent? It has been stated clearly by the United States Patent and Trademark Office (USPTO), the European Patent Office (EPO), and the United Kingdom Intellectual Property Office (UKIPO) that artificial intelligence cannot be listed as an inventor on patent applications [7]. The patent application mandates that the inventor identify himself as the creator. A patent application has to contain the applicant's name and address as well as the name and description of the invention. The law thus assumes that only naturally occurring persons could be inventors. Still, artificial intelligence could one day be acknowledged as a separate legal

[a]mailofanuska@gmail.com, [b]asmitarajput0510@gmail.com, [c]mhc123ster@gmail.com, [d]chakraborty_arpita2006@yahoo.com, [e]tojyoti2001@yahoo.co.in, [f]dr.siddhartha.bhattacharyya@gmail.com

DOI: 10.1201/9781003663348-49

entity. Acknowledgement of artificial intelligence as an innovator has no obvious legal or other restrictions. Patents guarantee that those engaged in artificial intelligence can profit from their work. An inventor in artificial intelligence might want patent protection for their work. Applying for a patent, it has happened that artificial intelligence was listed as the inventor.

Literature Review

The corpus of research on the difficulties of granting intellectual property rights (IPR) to artificial intelligence-powered ideas is investigated in this work using the literature review. Emphasizing the need of building frameworks to solve ethical, legal, and practical problems that develop as artificial intelligence technologies inspire more creativity, it looks at these concerns.

The impact of national and international intellectual property restrictions on AI-generated solutions during the COVID-19 epidemic has been examined, with an eye on a temporary strategy to improve IP accessibility for essential health resources [6]. It aims to maximize advantages by means of a utilitarian approach, therefore balancing public health needs with intellectual rights. The paper [2] summarizes the intellectual, legal, and financial concerns brought about by artificial intelligence as well as emphasizes the need of new laws concerning the ownership and protection of works created by AI. Emphasizing human authorship and inventiveness in IP law, the author [3] investigates artificial intelligence and intellectual property within the framework of the fourth industrial revolution. Expanding copyrights might, it argues, restrict fundamental liberties, inventiveness, and cultural diversity. It advises filling legal gaps using contracts and fair usage instead of complicating intellectual property legislation. The concept of "public property from the machine" is offered as a legal framework for AI-generated works that fulfil particular criteria so benefiting several stakeholders. The paper looks at present legal systems on artificial intelligence authorship and offers legislative recommendations to encourage innovation. Ultimately, it seeks to improve discussions on artificial intelligence and intellectual property by providing viewpoints to diverse experts and thereby developing universal concepts inside the global IP system. Furthermore, the paper [4] investigates the challenges artificial intelligence-generated content provides for copyright law and advocates modifying legal systems to manage the interaction among creativity, intellectual property, and machine learning. The authors [7] evaluate the status of knowledge about intellectual property rights for artificial intelligence and investigate possible approaches for intellectual property control inside a social setting.

Conflict Between Traditional Systems and Artificial Intelligence Based Solutions

A critical inquiry about the safeguarding of AI and its outputs is the adequacy of existing intellectual property (IP) protections [11–15].

One of the main points of dispute in this field is the widespread perception that the welfare of individuals is the primary aim of intellectual property protection. This compassionate viewpoint was represented in Dentons' AI poll, wherein 58% of participants believed that the intellectual property rights should belong to the system's end user and 20% to the system's developer. Just 4% of respondents said that ownership should belong to the AI system (see Figure 49.1). Is this method compliant with legal standards? Mixing and rearranging the fundamental principles of distinct IP rights as previously known might be one way to discover the proper tools and strike the correct balance.

Protecting AI Systems

Distinct features from regular software programs and copyright protection
Unlike conventional computer programming, in which the programmer has defined the process, artificial intelligence algorithms are made to be able to discern the optimum course of action from the data they are trained on.

In general, copyright only shields elements that best display the creative genius of the author. Since the source code embodies the author's creative expression, it is frequently protected by copyright laws. But the algorithm as a whole is frequently not protected since it lacks proper authorship because it makes 'autonomous' decisions.

In other words, just like any other computer program, an AI's core programming may be covered by

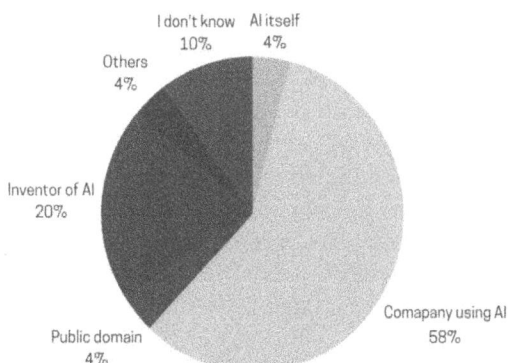

Figure 49.1 Survey results depict who won the intellectual property rights of AI systems
Source: Author

copyright rules. However, copyright protection alone does not stop an individual from using the same algorithm with distinct source code to develop an AI system. This may also pose challenges to intellectual property protection for peer-reviewed research or regulatory submissions, which often need public access and assessment of the foundational test data that informed the development of the product or study.

Database protection

Because AI systems rely on data for creating value, the data itself—including its selection and organization—may be considered an intellectual innovation that might be protected by IP laws. The investment in building a database is a unique asset, and certain governments have the right to safeguard it. However, even while compilations are considered intellectual works *protected* by copyright, the material that is incorporated in them is frequently not covered by copyright protection.

Protection of the patent

Some countries consider some software program-related innovations patentable, whereas others do not. Although mathematical equations and computer programs are not usually patentable, "computer-implemented innovations" with technical consequences are eligible for EU patent protection [13].

Trade secrets

The AI algorithm could be subject to trade secret enforcement as AI systems are often not sufficiently protected by copyright or patent. On the other hand, each nation has its own set of rules regarding trade secrets. Trade secrets often fall under the laws on torts, privacy, confidentiality, or unfair competition. The use of fair and honest methods of discovery, such as independent invention or reverse engineering, is not prohibited by this protection. Reverse engineering of software in AI is both technically challenging and legally perilous since contractual prohibitions to do so could be seen as unlawful and unenforceable in certain countries.

Trade secret protection is an impediment to information and technology exchange because of the inherent secrecy of trade secrets. Therefore, from the perspective of rights holders and policymakers, it may be better to combine patent protection and copyright with trade secret protection, as is often done with conventional software and related inventions.

Securing AI Interventions

Depending on the kind of AI technology used, either computer-generated or with human assistance can produce the outcomes. The current IP framework may still be applicable as human creativity may result in 'AI-assisted' solutions, and the AI system can be seen of as a tool similar to an artist's brush.

An 'AI-generate' outcome is one produced by artificial intelligence free of human intervention. In this situation, artificial intelligence may change its behavior midway through operation in reaction to fresh information or events. However, the issue of how much human contribution is meaningful remains unanswered.

Can works created by AI be eligible for copyright protection?

AI is now able to produce results that are regarded as creative works as it grows more advanced. The Google's Magenta—NSynth's sounds, Robot Poet Deniz Ylmaz's poetry, and Next Rembrandt's artwork are a few instances from recent times. At present, the legislation does not provide a single law, particularly to protect the creative works generated by AI legally.

The European Parliament's 2020 declaration that 'the degree of human interaction' and 'autonomy of AI' should be taken into account when deciding who is entitled to what intellectual property is not shocking. 'AI-assisted human inventions' might be protected by copyright in some circumstances, but 'AI-generated works' might not. Up to this point, only the author's original works have been considered worthy of legal protection. Sui generis database rights, on the other hand, may serve to safeguard databases made possible by AI, rewarding the developers of such systems rather than the humans who came up with the idea.

Can a patent be issued for an idea generated by AI?

Concerning AI-created innovations, similar difficulties and questions occur. Indeed, as compared to copyright the patent system is less concentrated on human innovation. But the peculiarities of self-driving computational works also put moral rights of the creator, innovation, and the skilled artist under strain.

Who is the creator or author of content produced by AI?

The checklist that would serve as a legal definition of the person most deserving the title of 'creator' or 'inventor' of AI-generated works is still a matter of conflict. Is it, for example, the AI programme user or the AI programme developer?

Legislators must prioritize changing IP protection in light of these changes in order to encourage a climate of encouraging innovators, whoever they may be in the end, in order to foster creativity and innovation.

frame design, material properties and configuration. Dynamic properties are considered in modal analysis, whereas IS Code uses standardized equations, resulting in discrepancies because of different frame specifications.

Summary and Conclusion

In this study, nonlinear static analysis (NSA) in ETABS is used to examine how the ductility factor of reinforced concrete (RC) framed structures changes with building height. Results are compared across seven frames (2 to 8 stories) against Eduardo Miranda's empirical formula. Results indicate that taller buildings tend to have lower ductility and thus lower seismic energy absorption, while mid-rise buildings exhibit irregular increases resulting from unique design features [12]. Empirical and analytical results differ, indicating that factors such as natural time period, stiffness, and material properties affect ductility, and that better predictive models are needed.

References

[1] IS 456:2000 (2000). Plain and Reinforced Concrete-Code of Practice. (fourth revision). New Delhi: Bureau of Indian Standards.

[2] IS 1893(PART I):2016 (2016). Criteria for earthquake resistant design of structures. (Sixth Revision), New Delhi: Bureau of Indian Standards.

[3] Asgarian, B., & Shokrgozar, H. R. (2003). BRBF Response Modification Factor. Tehran, Iran: K.N. Toosi University of Technology.

[4] Lin, Y. Y., & Chang, K. C. (2003). Study on damping reduction factor for buildings under earthquake ground motions. *Journal of Structural Engineering*, 129(2), 206–214. 10.1061/(ASCE)0733-9445129:2(206).

[5] Mahmoudi, M., & Zaree, M. (2010). Evaluating response modification factors of concentrically braced steel frames. *Journal of Constructional Steel Research*, 66(10), Pages 1196–1204.

[6] Mahmoudi, M., & Abdi, M. G. (2012). Evaluating response modification factors of TADAS frames. *Journal of Constructional Steel Research* 71 162–170. doi:10.1016/j.jcsr.2011.10.015

[7] Mondal, A., Ghosh, S., & Reddy, G. R. (2013). Performance-based evaluation of the response reduction factor for ductile RC frames. *Engineering Structures*, 56, 1808–1819.

[8] Miranda, E. (1993). Site-dependent strength-reduction factors. *Journal of Structural Engineering*, 119(12), 3503–3519.

[9] Barakar, S. A., Husein Malkawi, A. I., & Al-Shatnawi, A. S. (1997). A step towards evaluation of the seismic response reduction factor in multistorey reinforced concrete frames. Natural Hazards: *Journal of the International Society for the Prevention and Mitigation of Natural Hazards*, Springer; International Society for the Prevention and Mitigation of Natural Hazards, vol. 16(1), 65–80.

[10] Federal Emergency Management Agency (FEMA). (2000). Prestandard and Commentary for the Seismic Rehabilitation of Buildings (FEMA 356 / November 2000). Washington, D.C.: Federal Emergency Management Agency.

[11] Jalayer, J., & Cornell, C. A. (2009). Alternative nonlinear demand estimation methods for probability-based seismic assessments. Earthquake Engineering & Structural Dynamics 38(8), 951–972, DOI:10.1002/eqe.876.

[12] Karavasilis, T. L., Bazeos, N., & Beskos, D. E. (2008). Estimation of seismic inelastic deformation demands in plane steel MRF with vertical mass irregularities. *Engineering Structure*, 30(11), 3265–3275.

49 Possible challenges in assigning IPR to AI driven inventions

Anuska Pramanick[1,a], Asmita Rajput[1,b], Mahuya Hom Choudhury[2,c], Arpita Chakraborty[3,d], Jyoti Sekhar Banerjee[3,e] and Siddhartha Bhattacharyya[4,f]

[1]Department of IT, Bengal Institute of Technology, Kolkata, West Bengal, India

[2]Patent Information Centre, WBSCST, Kolkata, West Bengal, India

[3]Department of CSE (AI & ML), Bengal Institute of Technology, Kolkata, India

[4]Faculty of Electrical Engineering and Computer Science, VSB Technical University of Ostrava, Czech Republic

Abstract

In this study, the topic of intellectual property rights for AI-based innovations is examined. Using its "WIPO Conversation on IP and AI," the WIPO agency has started to interact with the stakeholders. In terms of safeguarding their AI innovations, South Korea, US, Japan, and the China are at the forefront. The policies of their respective patent offices will also be significant. Although the IP rules haven't exactly kept up with the speed of these AI-based innovations. Companies like Google have developed their own guidelines to safeguard their intellectual property in the meantime. The rules regulating AI innovations must, in the grand scheme of things, be able to safeguard and honoring the creator for the sake of society. Additionally, it needs to ensure that the advantages of the innovation are distributed fairly to all facets of society. This work makes a unique contribution by discussing potential difficulties in awarding AI-driven creations intellectual property rights.

Keywords: Infringement and intellectual property rights, invention, world intellectual property organization

Introduction

Rapid artificial intelligence (AI) developments have many unexpected and complicated effects on society and the economy. These developments may have far-reaching effects on productivity, employment, and competitiveness by altering the very nature of production for a broad variety of goods and services. However, although these results are likely to have an impact, artificial intelligence may also alter the nature of innovation itself, with far-reaching ramifications that may eventually overshadow the direct benefit [1–3].

Deep learning and robotics are the AI fields that are expanding the quickest, whereas the transportation sector is the one where AI patents are expanding the quickest. Telecommunications are next in line. Next are personal gadgets, medical and life science. Top five applicants for AI patents are Toshiba (Japan), Samsung (South Korea), Microsoft (US), NEC (Japan) and IBM (US). The Chinese Academy of Sciences (CAS) is the leading organization for filling patents based on artificial intelligence. Since 2013, the number of applications for AI patents has increased significantly.

In regard to IP protection for AI-based innovations, the issue of ownership arises. When a person comes up with a creative action, they are creating an innovation. That person or his or her employer is the rightful owner of the innovation [4–6]. If AI is being utilized to create an innovation, then the creator and proprietor of that creation is the person employing AI. It is not enough to just possess the equipment to claim ownership. The topic of whether or not an AI creation that requires no human involvement may be patented has interesting implications. Does it become public domain if "a machine" cannot apply for a patent? It has been stated clearly by the United States Patent and Trademark Office (USPTO), the European Patent Office (EPO), and the United Kingdom Intellectual Property Office (UKIPO) that artificial intelligence cannot be listed as an inventor on patent applications [7]. The patent application mandates that the inventor identify himself as the creator. A patent application has to contain the applicant's name and address as well as the name and description of the invention. The law thus assumes that only naturally occurring persons could be inventors. Still, artificial intelligence could one day be acknowledged as a separate legal

[a]mailofanuska@gmail.com, [b]asmitarajput0510@gmail.com, [c]mhc123ster@gmail.com, [d]chakraborty_arpita2006@yahoo.com, [e]tojyoti2001@yahoo.co.in, [f]dr.siddhartha.bhattacharyya@gmail.com

DOI: 10.1201/9781003663348-49

Is it not essential to consider human contribution?
When discussing assigning Intellectual Property Rights (IPR) to AI inventions, it's important to find a balance between protecting both human contribution and AI-generated interventions. While AI can come up with great solutions on its own, we must also recognize the critical role of human designers and developers in creating these technologies. Protecting human contributions ensures inventors are rewarded for their creativity and hard work. Acknowledging human contributions is the key to ensuring those people stay driven, feel a sense of recognition, and remain dedicated to their tasks. This approach supports both human innovation and ethical responsibility in AI. By creating systems that value both AI outputs and human involvement, we can make a collaborative environment that appreciates both human and technological progress.

AI as a Potential Danger Itself in Safeguarding IPRs

Text, music, images, and movies are just a few of the copyrighted materials AI can rapidly and effortlessly create and imitate. Stated differently, without author permission, AI systems are practically able to copy any form of creation, including literature and art. As so, it becomes quite challenging to tell the original from the illegal replica versions of a work. This raises major ownership and rights issues since it blurs the lines between the original creator's work and the AI duplicates, so affecting intellectual property rights and maybe leading to ethical and legal issues.

One aspect of the challenges AI presents in safeguarding intellectual property rights (IPRs) is automated content generation, which begs questions about ownership and authoring. Since artificial intelligence systems often require big datasets, which may include private, confidential, or sensitive data, privacy and data security issues are more pressing. Unauthorized access to or use of such data could jeopardize trade secrets and other forms of IPR. Moreover, deepfakes and content manipulation represent a major threat since they enable the creation of quite realistic fakes ranging from stolen artwork to fake trademarks and patent applications. Such tampering damages confidence in original works and makes IPR more difficult to enforce. Moreover, the use of artificial intelligence begs a lot of ethical issues about duty, justice, and the danger of abuse that might boost prejudice or damage. These problems necessitate a balanced approach that makes use of AI's benefits while protecting intellectual property by building and enforcing solid legal and ethical standards.

Responding to the reckless data scraping, the dubious legal climate surrounding artificial intelligence training datasets, and the illegal use of protected assets, authors and content creators have filed legal action. Together with comedian Sarah Silverman, authors Christopher Golden and Richard Kadrey have sued OpenAI and Meta alleging that their works were used to teach AI models without permission [16]. The accusations state that datasets derived from websites known as "shadow libraries", which included their copyrighted works, were utilized to teach OpenAI's ChatGPT and Meta's Llama.

Popular generative AI models, such as Meta's open-source Llama, were largely trained using unauthorized copies of well-known author' research, according to an investigation by the online magazine The Atlantic. This contains models from the charity EleutherAI, such as BloombergGPT and GPT-J. A broader dataset known as the Pile, which was publicly accessible online until recently, included the pirated books, which amounted to almost 170,000 volumes produced in the previous 20 years [17].

Legislative Approaches

The topic of IP law pertaining to AI has been brought up by legislators several times across the globe [8]. Many nations consider the development, protection, and use of IP in AI crucial to national competitiveness, although no AI-specific legislation has been enacted as of the book's publication.

The IP issues around AI were not addressed in the draught AI Act.

Many nations have begun to consider and debate how to deal with IP in AI:

- The United Kingdom has finished soliciting feedback on artificial intelligence and intellectual property for the year 2020. Trade secrets, designs, trademarks, patents, copyright and related rights, and cross-IP issues were all discussed.
- On June 23, 2021, the Italian government issued its strategic guidelines on industrial property for the period 2021-2023. The guidelines stipulate that "all opportunities shall be taken into account to conduct a comprehensive analysis of the relationship between emerging technologies, such as AI, and industrial property. This will enable the identification of the most effective protection systems and the appropriate forms of protection for products that are directly or indirectly derived from AI technologies" [9, 10].
- To better accommodate artificial intelligence and the other technology, Japan is debating how to update its patent system. It is described how artificial intelligence (AI) is used in law enforcement.

- Ireland's National AI Strategy, which was released in 2021, claims that efforts are currently being made at the federal level to assess the legal gaps for AI, particularly the system of intellectual property laws.
- China's 'New Generation Artificial Intelligence Development Plan', which was released in 2017, advocates industry standardizing of AI technologies and enhancement of IP rights protection. It also implies building an artificial intelligence patent pool to support the application and dissemination of fresh AI technologies.
- The 2020 National AI Initiative Act in the United States establishes an organized government initiative that will promote AI research and applications for economic and national security objectives. It will also examine intellectual property protection in relation to innovation promotion.

Conclusion

The kinds of IP protection designed for software are insufficient to secure AI systems because of the complexity that sets AI apart from traditional software. The issue of who owns ideas developed or supported by AI is controversial since conventional forms of IP protection are human-centered. According to the Dentons AI study, 86% of respondents thought that legislation was necessary to clarify IP protection in the context of AI, and 45% of participants indicated that legislation was urgently needed. Theoretical debate on AI's legal ramifications from both a local and international legal standpoint is still going strong. The development of the legal norms and concepts that will serve as the foundation for the regulations is currently ongoing. The terms and conditions specified in the partnership agreement may not match the court's decision in the case of a dispute, and there may be problems and discrepancies relating to AI systems and/or their advances.

Businesses should assess how each state should provide rights to AI and works produced by or using AI. They should next devise a strategy that combines various IP-rights tactics with contractual safeguards to efficiently secure the intellectual property (and data) produced by their AI systems.

Declarations

According to the corresponding author, there is no funding for this study and no conflicts of interest.

References

[1] Cockburn, I. M., Henderson, R., & Stern, S. (2018). The Impact of Artificial Intelligence on Innovation, (Vol. 24449). Cambridge, MA, USA: National Bureau of Economic Research.

[2] Zekos, G. I., & Zekos, G. I. (2021). AI and Legal Issues. Economics and Law of Artificial Intelligence: Finance, Economic Impacts, Risk Management and Governance, (pp. 401–460).

[3] Kop, M. (2019). AI & intellectual property: Towards an articulated public domain. *Texas Intellectual Property Law Journal*, 28, 297.

[4] VT, V. N. (2024). THE NEXUS OF MACHINE LEARNING, CREATIVE PROCESS, AND IPR LAW: IMPLICATIONS AND LEGAL CONSIDERATIONS. *Journal of Higher Education Theory and Practice*, 24(1) (pp. 279-289).

[5] Nyaboke, Y. (2024). Intellectual property rights in the era of artificial intelligence. *Journal of Modern Law AMD Policy*, 4(2), 58–72.

[6] Kokane, S. (2021). The intellectual property rights of artificial intelligence-based inventions. *Journal of Scientific Research*, 65(2).

[7] Chhavi, S., & Reeta, S. (2020). AI-generated inventions and IPR policy during the COVID-19 pandemic. *Legal Issues in the Digital Age*, (2), 63–91.

[8] Somaya, D., & Varshney, L. R. (2018). Embodiment, anthropomorphism, and intellectual property rights for AI creations. In Proceedings of the 2018 AAAI/ACM Conference on AI, Ethics, and Society (pp. 278–283).

[9] Davies, C. R. (2011). An evolutionary step in intellectual property rights–Artificial intelligence and intellectual property. *Computer Law and Security Review*, 27(6), 601–619.

[10] Tull, S. Y., & Miller, P. E. (2018). Patenting artificial intelligence: issues of obviousness, inventorship, and patent eligibility. *RAIL*, 1, 313.

[11] Lim, D. (2018). AI & IP innovation & creativity in an age of accelerated change. *The Akron Law Review*, 52, 813.

[12] Flynn, S. (2020). Wipo conversation on intellectual property (IP) and artificial intelligence (AI) (pp.2–10).

[13] da Cruz Fernandes, D. (2020). Mining digital treasures: text and data mining interfering with copyrights in the EU and in the US: which legal system is more favourable to TDM users? (Master's thesis, Universidade Catolica Portuguesa (Portugal)).

[14] https://www.dentons.com/en/insights/articles/2022/june/8/legal-regulation-of-artificial-intelligence-in-kazakhstan-and-abroad.

[15] Gupta, R., Choudhury, M. H., Mahmud, M., & Banerjee, J. S. (2023). Patent analysis on artificial intelligence in food industry: worldwide present scenario. In Doctoral Symposium on Human Centered Computing (pp. 347–361). Singapore: Springer Nature Singapore.

[16] https://www.livemint.com/news/world/comedian-sarah-silverman-2-authors-file-copyright-infringement-lawsuits-against-meta-openai-sue-11688949711521.html.

[17] https://www.theatlantic.com/technology/archive/2023/08/books3-ai-meta-llama-pirated-books/675063/.

For Product Safety Concerns and Information please contact our EU
representative GPSR@taylorandfrancis.com
Taylor & Francis Verlag GmbH, Kaufingerstraße 24, 80331 München, Germany

www.ingramcontent.com/pod-product-compliance
Lightning Source LLC
Chambersburg PA
CBHW081057220326
41598CB00038B/7126